# Physiological Plant Ecology

# Symposia of
# The British Ecological Society

# Physiological Plant Ecology

*The 39th Symposium of the British Ecological Society*
*held at the University of York*
*7–9 September 1998*

EDITED BY

## MALCOLM C. PRESS
Department of Animal and Plant Sciences
University of Sheffield

## JULIE D. SCHOLES
Department of Animal and Plant Sciences
University of Sheffield

and

## MARTIN G. BARKER
Department of Plant and Soil Science
University of Aberdeen

**Blackwell
Science**

© 1999 the British Ecological Society and Published for them by Blackwell Science Ltd
Editorial Offices:
Osney Mead, Oxford OX2 0EL
25 John Street, London WC1N 2BL
23 Ainslie Place, Edinburgh EH3 6AJ
350 Main Street, Malden
    MA 02148 5018, USA
54 University Street, Carlton
    Victoria 3053, Australia
10, rue Casimir Delavigne
    75006 Paris, France

Other Editorial Offices:
Blackwell Wissenschafts-Verlag GmbH
Kurfürstendamm 57
10707 Berlin, Germany

Blackwell Science KK
MG Kodenmacho Building
7–10 Kodenmacho Nihombashi
Chuo-ku, Tokyo 104, Japan

First published 1999

Set by Excel Typesetters Co., Hong Kong
Printed and bound in Great Britain by
MPG Books Ltd, Bodmin, Cornwall

The Blackwell Science logo is a
trade mark of Blackwell Science Ltd,
registered at the United Kingdom
Trade Marks Registry

A catalogue record for this title is available from the British Library

ISBN 0-632-05493-X (hardback)
      0-632-05491-3 (pbk)

Library of Congress Cataloging-in-publication Data

British Ecological Society. Symposium (39th: 1998: University of York)
    Physiological plant ecology: the 39th Symposium of the British Ecological Society, held at the University of York, 7–9 September 1998/edited by Malcolm C. Press, Julie D. Scholes, and Martin G. Barker.
      p.   cm.
      Includes bibliographical references.
      ISBN 0-632-05493-X (hardback)
      ISBN 0-632-05491-3 (pbk.)
      1.  Plant ecophysiology Congresses.
      I.  Press, Malcolm C.
      II.  Scholes, Julie D.
      III.  Barker, Martin G.   IV.  Title.
      QK717 .B75 1998
      581.7 — dc21                    99-29393
                                                    CIP

DISTRIBUTORS
    Marston Book Services Ltd
    PO Box 269
    Abingdon, Oxon OX14 4YN
    (Orders:  Tel:  01235 465500
                    Fax:  01235 465555)

USA
    Blackwell Science, Inc.
    Commerce Place
    350 Main Street
    Malden, MA 02148 5018
    (Orders:  Tel:  800 759 6102
                            781 388 8250
              Fax:  781 388 8255)

Canada
    Login Brothers Book Company
    324 Saulteaux Crescent
    Winnipeg, Manitoba R3J 3T2
    (Orders:  Tel:  204 837 2987)

Australia
    Blackwell Science Pty Ltd
    54 University Street
    Carlton, Victoria 3053
    (Orders:  Tel:  3 9347 0300
              Fax:  3 9347 5001)

For further information on
Blackwell Science, visit our website:
www.blackwell-science.com

# Contents

## Responses to global environment change

## Ecosystems

## Integration and scaling

# List of Contributors

*D.D. Ackerly*
Department of Biological Sciences, Stanford University, Stanford, CA 94305, USA

*J.M. Anderson*
Photobioenergetics Group, Research School of Biological Sciences, Institute of Advanced Studies, The Australian National University, Box 475, Canberra, ACT 2601, Australia

*M.C. Ball*
Ecosystem Dynamics Group, Research School of Biological Sciences, Institute of Advanced Studies, The Australian National University, Box 475, Canberra, ACT 2601, Australia

*P.W. Barnes*
Department of Biology, Southwest Texas State University, San Marcos, TX 78666-4616, USA

*F.A. Bazzaz*
Department of Organismic and Evolutionary Biology, Harvard University, Cambridge, MA 02138, USA

*A. Borland*
Department of Agricultural and Environmental Science, University of Newcastle upon Tyne, NE1 7RU, UK

*M.M. Caldwell*
The Ecology Center and Department of Rangeland Resources, Utah State University, Logan, UT 84322-5205, USA

*M.J. Crawley*
Department of Biology, Imperial College of Science, Technology and Medicine, Silwood Park, Ascot, SL5 7PY, UK

*W.J. Davies*
Department of Biological Sciences, Lancaster University, Bailrigg, Lancaster, LA1 4YQ, UK

*J.J.G. Egerton*
Ecosystem Dynamics Group, Research School of Biological Sciences, Institute of Advanced Studies, The Australian National University, Box 475, Canberra, ACT 2601, Australia

*J.R. Ehleringer*
Department of Biology, University of Utah, Salt Lake City, UT 84112-0840, USA

*J.F. Farrar*
School of Biological Sciences, University of Wales Bangor, Bangor, Gwynedd, LL57 2UW, UK

*A.H. Fitter*
Department of Biology, University of York, York, YO10 5DD, UK

*S.D. Flint*
The Ecology Center and Department of Rangeland Resources, Utah State University, Logan, UT 84322-5205, USA

**R. Gebauer**
Department of Biology, Keene State College, Keene, NH 03469, USA

**J. Gillon**
Department of Environmental Research and Energy Research, Weizmann Institute of Science, 76100 Rehovot, Israel

**D.J.G. Gowing**
Institute of Water and Environment, Cranfield University, Silsoe, Bedfordshire, MK45 4DT, UK

**J. Grace**
Institute of Ecology & Resource Management, The University of Edinburgh, Kings Buildings, Mayfield Road, Edinburgh, EH9 3JU, UK

**H. Griffiths**
Department of Agricultural and Environmental Science, University of Newcastle upon Tyne, NE1 7RU, UK

**K. Harwood**
Department of Agricultural and Environmental Science, University of Newcastle upon Tyne, NE1 7RU, UK

**Ch. Körner**
Institute of Botany, University of Basel, Schönbeinstrasse 6, CH-4056 Basel, Switzerland

**J.A. Lee**
Sheffield Centre for Arctic Ecology, Department of Animal and Plant Sciences, University of Sheffield, Sheffield, S10 2TN, UK

**S.P. Long**
Departments of Crop Sciences and Plant Biology, University of Illinois, 190 Edward R. Madigan Laboratory, 1201 West Gregory Street, Urbana, IL 61801, USA

**T.A. Mansfield**
Department of Biological Sciences, Institute of Environmental and Natural Sciences, Lancaster University, Lancaster, LA1 4YQ, UK

**K. Maxwell**
Department of Agricultural and Environmental Science, University of Newcastle upon Tyne, NE1 7RU, UK

**C.B. Osmond**
Photobioenergetics Group, Research School of Biological Sciences, Institute of Advanced Studies, The Australian National University, Box 475, Canberra, ACT 2601, Australia

**J.S. Pate**
Botany Department and Centre for Legumes in Mediterranean Agriculture, The University of Western Australia, Nedlands, WA 6907, Australia

**R.W. Pearcy**
Section of Evolution and Ecology, Division of Biological Sciences, University of California, Davis, CA 95616, USA

**M.C. Press**
Department of Animal and Plant Sciences, University of Sheffield, Sheffield, S10 2TN, UK

**D.J. Read**
Department of Animal and Plant Sciences, University of Sheffield, Sheffield, S10 2TN, UK

*S. Schmidt*
Department of Botany, University of Queensland, Brisbane, QLD 4072, Australia

*J.D. Scholes*
Department of Animal and Plant Sciences, University of Sheffield, Sheffield, S10 2TN, UK

*S. Schwinning*
Department of Biology, University of Utah, Salt Lake City, UT 84112-0840, USA

*P.S. Searles*
The Ecology Center and Department of Rangeland Resources, Utah State University, Logan, UT 84322-5205, USA

*M.A. Sobrado*
Ecosystem Dynamics Group, Research School of Biological Sciences, Institute of Advanced Studies, The Australian National University, Box 475, Canberra, ACT 0200 Australia and Laboratorio de Biologia Ambiental de Plantas, Departamento Biologia de Organismos, Universidad Simon Bolivar, Apartado 89.000, Caracas 1080 A, Venezuela

*G.R. Stewart*
Faculty of Science, University of Western Australia, Nedlands, WA 6907, Australia

*K.A. Stinson*
Department of Organismic and Evolutionary Biology, Harvard University, Cambridge, MA 02138, USA

*M.J. Unkovich*
Botany Department and Centre for Legumes in Mediterranean Agriculture, The University of Western Australia, Nedlands, WA 6907, Australia

*F. Valladares*
Centro de Ciencias Medioambientales, CSIC, Serrano 115 dpdo., 28006 Madrid, Spain

*J.R. Watling*
Department of Animal and Plant Sciences, University of Sheffield, Sheffield, S10 2TN, UK

*J. Wilson*
Department of Agricultural and Environmental Science, University of Newcastle upon Tyne, NE1 7RU, UK

*F.I. Woodward*
Department of Animal and Plant Sciences, University of Sheffield, Sheffield, S10 2TN, UK

# History of the British Ecological Society

The British Ecological Society is a learned society, a registered charity and a company limited by guarantee. Established in 1913 by academics to promote and foster the study of ecology in its widest sense, the Society currently has around 5000 members spread around the world. Members include research scientists, environmental consultants, teachers, local authority ecologists, conservationists and many others with an active interest in natural history and the environment. The core activities are the publication of the results of research in ecology, the development of scientific meetings and the promotion of ecological awareness through education. The Society's mission is:

> To advance and support the science of ecology and publicize the outcome of research, in order to advance knowledge, education and its application.

The Society publishes four internationally renowned journals and organizes at least two major conferences each year plus a large number of smaller meetings. It also initiates a diverse range of activities to promote awareness of ecology at the public and policy maker level in addition to developing ecology in the education system, and it provides financial support for approved ecological projects. The Society is an independent organization that receives little outside funding.

British Ecological Society
26 Blades Court
Deodar Road, Putney
London SW15 2NU
United Kingdom
Tel.: +44 (0)20 8871 9797 Fax: +44 (0)20 8871 9779
E-mail: general@ecology.demon.co.uk
ULR : http://www.demon.co.uk/bes

The British Ecological Society is a limited company, registered in England No. 15228997 and a Registered Charity No. 281213.

# Preface

The last decade has seen rapid and major advances in our understanding of the physiological ecology of plants. This volume contains 22 chapters that review some of these advances. Authors were asked to review briefly major recent developments in their subject area and to show how their studies have contributed towards the advancement of our knowledge. The papers were presented at the 39th symposium meeting of the British Ecological Society held at the University of York in September 1998, which was attended by more than 250 delegates from 23 countries.

The chapters are arranged under five headings: resource acquisition and utilization; interactions between organisms; responses to global environmental change; ecosystems; and integration and scaling. It was never our intention to cover all areas of physiological plant ecology. Rapid advances in both the plant and ecological sciences, together with the availability of new technology from the molecular to the ecosystem level, mean that the scope of the discipline is now broader than ever. Rather, our aim was to allow key practitioners to illustrate some of the major advances and to provide accounts that would be accessible to those interested in physiological plant ecology, from senior level undergraduates upwards.

Individuals will no doubt identify several recurrent themes in the volume, but we highlight six here. First, early studies of the physiological ecology of plants were skewed towards measurements of photosynthesis. Although this volume certainly testifies to the importance of this process, it also demonstrates that an understanding of post-photosynthetic processes can be at least as critical, if not more so, than rates of gas exchange in attempting to explain performance and survivorship. Second, the volume demonstrates the importance of examining below- as well as above-ground processes and shows how our understanding of plant responses to changes in the edaphic environment have been modified in recent years. Third, it is now clear that a complete understanding of the physiological ecology of plants cannot be gained without understanding their interactions with other species and trophic levels. Fourth, progress in many areas has been underpinned by advances in technology. This ranges from the use of molecular techniques to understand nutrient uptake by roots and mycorrhizas to the use of micrometerological techniques to determine ecosystem $CO_2$ fluxes. Fifth, the importance of scaling up has been recognized and addressed in various ways, including the use of stable isotopes as a means of integrating short term processes. Finally, concerns about the impact of global climate change have been an important stimulus for much ecophysiological research. Such studies not only provide information that can be used by modellers for climate impact studies but they also further our basic science understanding and demonstrate the importance of linkages between leaf-level and ecosystem-level processes.

We would like to express our gratitude to the contributors and all who participated in the symposium meeting for generating a lively and stimulating atmosphere. We are also grateful to the referees who provided valuable comments on the chapters and to the following who provided advice during the planning of the meeting: Alastair Fitter, John Grace, Jonathan Graves, John Lee, Steve Long and Ian Woodward. Finally, we would like to thank the British Ecological Society for generously supporting the symposium and Hazel Norman, the Executive Secretary, and her staff for all their hard work.

*M.C. Press*
*J.D. Scholes*
*M.G. Barker*
June 1999

# Chapter 1

# Compromising efficiency: the molecular ecology of light-resource utilization in plants

*C.B. Osmond,[1] J.M. Anderson,[1] M.C. Ball[2] and J.J.G. Egerton[2]*

## Introduction

Light, unlike most other resources for autotrophy, displays daily, and often momentary, excursions through several orders of magnitude. Although some plants can modulate these excursions in photon flux by chloroplast and leaf movements (Koller 1990; Brugnoli & Björkman 1992), response to limiting or excessive light in the short and long term lies principally in biophysical and biochemical molecular mechanisms within chloroplast thylakoid membranes. Light-use efficiency is thus a primary process at the smallest scale in the space–time continuum of autotrophy (Osmond 1989), one that is currently best understood in terms of molecular ecology in the ecosystem of pigment–protein complexes that comprise photosystem II (PSII).

This chapter cannot do justice to more than a fraction of the large body of remarkable research in photosynthesis published since our last attempt at a comprehensive review (Anderson & Osmond 1987). Instead the chapter focuses on a paradigm shift in understanding photosynthetic activity in natural environments; the causes and consequences of inefficiencies in light-resource utilization that afford photoprotection and flow from photoinactivation. It concentrates on the relatively rapidly relaxing components of the molecular ecosystem and their effects on light-resource utilization in leaves and plants within minutes to hours, but it also briefly considers processes that facilitate acclimation over hours to days.

Although the proportion of global photosynthesis that occurs under light limitation in the shade is not clear, much past ecophysiological research into light-resource use focused on the maximum intrinsic efficiency of photosynthetic $CO_2$ fixation (processes defining the initial slope of Blackman's light–response curve). It has been proposed, for example, that differences in efficiency of light use (the quantum yield of $CO_2$ fixation in air) account for plant distribution according

[1] *Photobioenergetics Group, Research School of Biological Sciences, Institute of Advanced Studies, The Australian National University, Box 475, Canberra, ACT 2601, Australia. E-mail: osmond@rsbs.anu.edu.au*
[2] *Ecosystem Dynamics Group, Research School of Biological Sciences, Institute of Advanced Studies, The Australian National University, Box 475, Canberra, ACT 2601, Australia*

to photosynthetic pathways, both today and through evolutionary history (Ehleringer & Monson 1993). However, the abundance of high-quantum-requiring Crassulacean-acid-metabolism (CAM) plants in shaded tropical under-storey and epiphytic habitats (Skillman & Winter 1997) is just one example that questions whether we really understand the role of quantum yield in plant success. With access to adequate instrumentation for field evaluation of photosynthesis in full sunlight, there has been a growth of ecophysiological interest in the regulation of light use efficiency in the most productive leaves; those most often exposed to high irradiance. Thus, much present research is focused on intrinsically inefficient processes in photosynthesis; on disposal of the photons harvested but not used in bright light. The extent of this disposal problem is most readily judged by compar-ing the area between an extrapolation of the initial slope of the light–response curve of photosynthesis and the light-saturated curve with the area below the same curve (Fig. 1.1; cf. Björkman & Demmig-Adams 1994).

To a large extent, these rapid advances and new insights can be ascribed to the pervasive and persuasive field exploitation of pulse-modulated chlorophyll fluor-escence measurement systems, conceived by Bradbury and Baker (1981) but real-ized by Schreiber *et al.* (1986, 1994). At room temperature, these devices monitor fluorescence arising principally from PSII, and thus our emphasis on the structure of this water-splitting, charge-separating complex has as much to do with it being

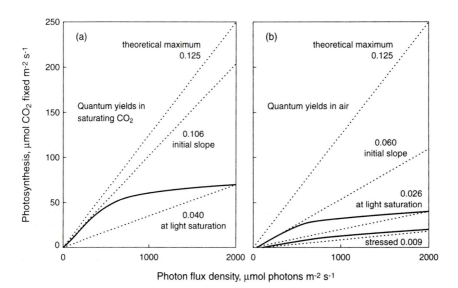

**Figure 1.1** Light–response curves of photosynthetic $CO_2$ fixation in leaves of a hypothetical $C_3$ plant (heavy filled lines) in (a) saturating $CO_2$ and (b) air. The dashed lines show the attenuation of quantum yield from the theoretical maximum owing to limitations of electron transport at light saturation (in both a and b). The effect of Rubisco oxygenation in air on quantum yield is shown by comparing the initial slopes in (a) and (b), and the effect of stress by comparing the upper and lower curves in (b).

measurable as it has to do with its role in the primary events of light-resource utilization. Photosystem I (PSI) is less easily studied in leaves, is less likely to be rate limiting in light use, but is particularly important in photosynthetic responses to chilling stress (Sonoike 1996). PSI also lies at the heart of oxidative metabolism in photosynthesis (Osmond & Grace 1995), the understanding of which will become more important in the future.

Reductionist approaches inevitably lead us to engage sometimes hazy concepts of population biology that feature prominently in current thinking about the molecular ecology of PSII. Thus we envisage the binary possibilities among excited pigment–protein complexes (e.g. functional or non-functional states; quenching or non-quenching processes), and search for significant stoichiometries among key molecular components in thylakoid membranes. Sometimes these relationships may be more difficult to define than living or dead, or short-lived or long-lived individuals in a plant community, but the approaches are similar when attempting to understand photosynthetic efficiency or ecosystem function. Although the approach in this chapter will often be a reductionist one, recent research of a few close colleagues, with the energy (and insight) to explore these phenomena in the field, will also be drawn on. It is these colleagues who, in remarkable ways, have allowed ecophysiology to chart new paths in photosynthesis research.

Physiological plant ecologists like to promote their discipline with accounts of success in the face of natural selection in the abiotic environment with less attention to the implications of competition and other biotic factors. However, success often involves compromise as much as fine tuning and, in principle, survivors in a particular habitat should be measured against the often more numerous unsuccessful genotypes. Only then can we fully understand acclimation, the changes in functions needed to facilitate adaptation, and which, if they are inadequate, lead to exclusion. Thus, in the context of light-resource utilization, we hope to inject some balance into a discussion in which there is currently much emphasis on the efficacy of photoprotective processes at light saturation in every survivor in every habitat. There is also merit in understanding the role of photoinactivation, which may explain in part why other plants do not survive in the habitats in question. Both processes compromise the efficiency of light-resource utilization in photosynthesis, and their interactions at the molecular level are well characterized by the ceaseless contest implicit in the symbolism of Yin and Yang (Osmond 1994). It is convenient to think of photoprotection as the bright side of the contest, and photoinactivation as the dark side, anchored in photochemical and biochemical processes, respectively. Effective photoprotection minimizes photoinactivation; failure of photoprotection produces greater photoinactivation, and in the minds of some, photoinactivation is the ultimate form of photoprotection.

## Semantics and criteria

This review is a good occasion to abandon earlier attempts to plot paths through the semantic minefield of photoinhibition from a historical perspective (Osmond 1994; Osmond & Grace 1995). The changes in our vocabulary have been so complex that we will limit ourselves to the terms photoprotection and photoinactivation and summarize the chlorophyll fluorescence criteria that differentiate them *in vivo*. Photoprotection principally involves events in the antenna complexes of PSII, whereas photoinactivation involves PSII reaction centre structure and function.

Photoprotection refers to a light-dependent, often rapidly reversible decline in the efficiency of primary photochemistry indicated by a decline in the ratio of variable to maximal fluorescence ($F_V/F_M$), accompanied by a decline in intrinsic fluorescence in weak light ($F_O$). The latter indicates reduced fluorescence lifetimes in the antennae as a result of increased thermal dissipation of excitation there, rather than transfer to the reaction centre in which the excitation is dissipated through photochemistry. Also signalled by an increase in non-photochemical fluorescence quenching ($q_N$ or NPQ) measured by pulse-modulated fluorometers during illumination, photoprotection mitigates photoinactivation by relieving 'excitation pressure' in the reaction centre of PSII.

Photoinactivation refers to a light-dependent, slowly reversible decline in primary photochemistry ($F_V/F_M$), often accompanied by an increase in $F_O$, well correlated with a decline in the population of functional reaction centres (Park *et al.* 1995). The rise in $F_O$ indicates an increase in fluorescence lifetimes in the antennae where excitation builds up when reaction centre function is impaired during prolonged exposure to excess light. Excess light, indicated by a low value of photochemical fluorescence quenching ($q_P$) measured with pulse-modulated fluorometers during illumination, signals excitation pressure in the reaction centre of PSII and highly reduced electron carriers between the photosystems.

Photoprotection is thought to minimize photoinactivation, but because both processes may be engaged at the same time, to differing extents in different organisms and habitats, fluorescence criteria may not always be very discriminating. That is, extensive photoprotection may hide significant photoinactivation because both lead to a decline in $F_V/F_M$, and the decline in $F_O$ as a result of photoprotection may mask the increase in $F_O$ caused by photoinactivation. These problems persist in interpretations of both pulse-modulated and dark-adapted fluorescence data. Relaxation experiments (Demmig & Winter 1988; Franklin *et al.* 1992) can help distinguish the two events (relaxation times tend to increase from photoprotection, through photoinactivation to photoacclimation), and are most useful if combined with inhibitor treatments (Thiele *et al.* 1996, 1998). The least equivocal criterion is the saturating flash yield of $O_2$ evolution (Chow *et al.* 1991), in which photoprotection should not register, but in which photoinactivation should produce a decline. Unfortunately, the method is not yet available for field use.

## Efficiency of light utilization in leaves: responses to metabolism and stress

The paradigm shift in studies of light-resource use is best illustrated by commencing with traditional light–response curves of $CO_2$ fixation in a hypothetical $C_3$ plant (Fig. 1.1), and moving onto light–response curves for $O_2$ evolution and fluorescence parameters in *Arabidopsis thaliana* (Fig. 1.2). The full efficiency of photon capture in light-limited conditions is a robust but rarely exercised property of the photosynthetic apparatus of $C_3$ plants. The intrinsic quantum yield only approaches the theoretical maximum of 8 photons per mol $CO_2$ assimilated during light-limited photosynthesis in atmospheres of saturating $CO_2$ (Fig. 1.1a,b; upper broken lines). Plants from diverse habitats show intrinsic quantum yields some 85% of theoretical maximum ($0.106 \pm 0.001$; *vs.* 0.125 $CO_2$ photon absorbed⁻¹; Fig. 1.1a, initial slope of heavy line) during $CO_2$-saturated photosynthesis (Björkman & Demmig 1987). This suggests that about 15% of electron flow may be drained to alternative electron acceptors, such as $O_2$, and to other metabolic events such as nitrate and sulphate reduction under these conditions. The actual quantum yield of $CO_2$ fixation at $CO_2$ saturation in full sunlight (2000 µmol photons m⁻² s⁻¹; Fig. 1.1a, lower broken line) is only about 35% of the maximum, because of the limitations of electron transport capacity, and the photoprotective processes discussed below.

In air, the intrinsic high efficiency of light-energy utilization in leaves is inevitably compromised by Rubisco oxygenation and the subsequent costs of photorespiratory carbon cycling in $C_3$ plants (Fig. 1.1b, initial slope of upper heavy line). At light saturation in air, the actual quantum yield of $CO_2$ fixation is reduced by the same limitations of electron transport and photoprotection. The quantum yield of $CO_2$ fixation in air is further reduced by water stress and other conditions that lower intercellular $CO_2$ concentrations through stomatal closure (thereby increasing photorespiratory carbon cycling), or lead to photoinactivation of PSII as discussed below (Fig. 1.1b, lower heavy line). Metabolic $CO_2$-concentrating mechanisms that mitigate the Rubisco-imposed inefficiency in light-resource utilization, such as those in $C_4$ and CAM plants (Leegood *et al.* 1997), come at a further cost.

It is now commonplace to deduce light utilization in photosynthesis from fluorescence rather than from measurements of gas exchange, especially in the field. This is appropriate, given that photon accounting made possible by these methods is a better approach to light-resource utilization than the highly variable bioenergetic basis of carbon economy in air in $C_3$ plants. The insights of Genty *et al.* (1989) and colleagues, have played pivotal roles in developing these approaches, which have the advantage that they can reveal the full extent of photon utilization in the futile cycles of photorespiratory carbon cycling. However, they too rely on assumptions about photon distribution between photosystems, do not yet account effectively for alternative electron flows, and need further validation by direct measurements of gross $CO_2$ and $O_2$ exchanges (Maxwell *et al.* 1998).

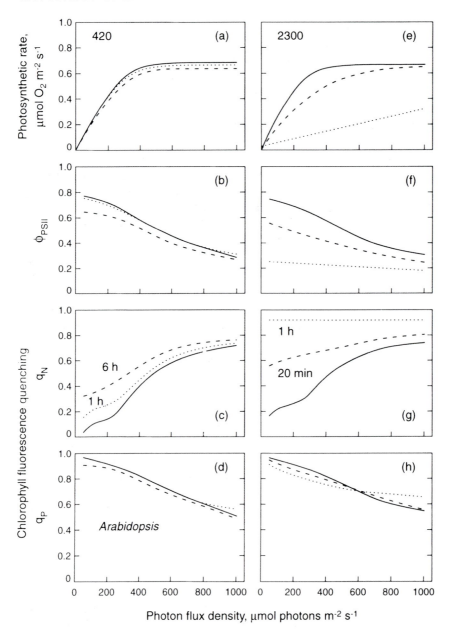

**Figure 1.2** Photosynthetic and chlorophyll fluorescence parameter profiles in *Arabidopsis thaliana* leaves grown at 230 μmol photons m$^{-2}$ s$^{-1}$, then treated at light saturation (420 μmol photons m$^{-2}$ s$^{-1}$) for 1 hour and 6 hours (a–d), or in excess light (2300 μmol photons m$^{-2}$ s$^{-1}$) for 20 minutes or 1 hour (e–h). Redrawn from Russell *et al.* (1995).

In spite of its preferred role in molecular plant science research, the environmental physiology of autotrophy in *A. thaliana* is almost unknown. Fortunately it proves to be unexceptional when examined under controlled environmental conditions (Russell *et al.* 1995). Light–response curves of photosynthetic $O_2$ evolution at $CO_2$ saturation in *A. thaliana* grown in moderate light (230 µmol photons) show that up to 6 hours exposure at light saturation (420 µmol photons $m^{-2} s^{-1}$) had very little effect on the efficiency of photosynthesis (Fig. 1.2a). This conclusion was confirmed by fluorescence-deduced analysis of PSII efficiency ($\Phi$PSII, Fig. 1.2b) which closely matches a curve that could be constructed from the slopes of lines drawn from the dark value to points on the curves in Fig. 1.2(a). Profiles of the coefficients of chlorophyll fluorescence quenching show small increases in $q_N$ at low light (Fig. 1.2c), but $q_P$ did not respond to the treatment (Fig. 1.2d).

When these plants were exposed to excess light (equal to full sunlight) for 20 minutes or 1 hour, marked reductions in the initial slopes of the light–response curve of photosynthetic $O_2$ evolution were observed (Fig. 1.2e), corresponding to decreased efficiency of PSII as deduced from fluorescence measurements (Fig. 1.2f). Low $\Phi$PSII at low light was correlated with high values of $q_N$ (Fig. 1.2g), but again the $q_P$ profiles were unaltered (Fig. 1.2h). Evidently, light stress had produced large changes in thermal dissipation in PSII antennae, but remaining photochemically functional reaction centres behaved normally. The molecular ecology behind these responses will now be examined, before considering plants of greater ecophysiological interest.

## Molecular ecology of photosystem II structure and function

In higher plants, most PSII complexes assemble as a relatively well ordered molecular ecosystem of PSII dimers in thylakoid membranes. The composition of this ecosystem of more than 25 individual polypeptides (encoded by both chloroplast and nuclear genomes) is such that thylakoids stack to varying extents as grana, in which the spatial arrangement in both membrane planes is crucial to function. Fewer PSII complexes with smaller light-harvesting antennae (probably PSII monomers) are located with PSI complexes in the stroma-exposed grana margins and in the thylakoids that link granal stacks. This lateral heterogeneity in the distribution of PSII and PSI complexes in most terrestrial plants ensures that most PSII complexes are not in direct contact with PSI complexes, an arrangement that limits energy spillover from slower PSII reactions to the faster reactions of PSI. Planar views of functional PSII complexes now being reconstructed using cryomicroscopy and other techniques provide evidence that the LHCII–PSII supercore complexes (comprising the major proteins of PSII, minor polypeptides, cytochrome $b_{559}$ and about 100 chlorophylls) exist as dimers *in vivo* (Hankamer *et al.* 1997). The major outer light-harvesting antennae (LHCII trimers; light-harvesting chlorophyll protein complex II.) deliver exciton energy to the photosynthetic reaction centre via linking, minor monomeric LHCs (CP29, CP26 and CP24). Light-dependent, $\Delta$pH-driven changes in composition of xanthophyll pigments associated with these minor chlorophyll–protein complexes are now known to lower the efficiency of excitation transfer to PSII and to confer photoprotection.

Overexcitation of the whole system in strong light hastens the loss of reaction centre function, principally ascribed to the failure of one short-lived and crucial component, the D1 protein (the least stable gene product in this ecosystem under light stress). In the fluid membrane environment of the PSII ecosystem, the non-functional D1 protein (possibly in one of the two centres in each dimer) is recognized and repaired after monomers migrate to the non-appressed domains of thylakoids. Here D1 protein is degraded, new D1 protein is synthesized on chloroplast ribosomes, processed and assembled, first with its heterodimeric partner, the D2 protein, and then with chlorophyll–protein complexes (van Wijk et al. 1997), to restore PSII function.

The molecular ecology of chlorophyll–protein complexes in a PSII monomer in thylakoids of shade and sun plants is shown schematically in Fig. 1.3. Shade plants have larger populations of LHCII trimers per PSII reaction centre and hence lower chlorophyll $a/b$ ratios (Anderson 1986) and lower concentrations of xanthophyll pigments on a chlorophyll basis (Demmig-Adams et al. 1995). Bassi et al. (1993) showed that, in maize, 80% of the xanthophyll pigments associated with PSII are 'strategically' located in the minor chlorophyll–protein complexes, between the major peripheral antenna and the PSII reaction centre core. In A. thaliana, some 50% of total xanthophylls are located here (Hurry et al. 1997).

## Fates of photons absorbed but not used

Excitation within LHCII–PSII complexes is partitioned between utilization in photochemistry after delivery to the reaction centre P680, and dissipation as heat in the antennae before or after it reaches the reaction centre. Thermodynamically, the majority of photons absorbed but not used in photochemistry are dissipated as heat (a very small proportion is re-emitted as fluorescence), irrespective of whether excitation is 'grounded' (quenched) in the antennae, or in non-functional PSII centres. However, 'excitation pressure' builds up in PSII whenever photon absorption and delivery exceeds the capacity of metabolism to consume ATP generated by electron transport and the [H+] concentration in the thylakoid lumen increases (Schönknecht et al. 1995). Circumstances that lead to high $\Delta$pH (acid thylakoid lumen) promote conversion of violaxanthin (V) to zeaxanthin (Z), via antheraxanthin (A), and produce a strong correlation between Z+A and NPQ (Gilmore & Yamamoto 1993). The acid lumen also facilitates protonation of a binding site in one or more of the inner antenna chlorophyll–protein complexes (Gilmore 1997) that binds the less polar Z and A. Gilmore et al. (1995) showed that these complexes lose their capacity to transfer excitation to the reaction centre because, after binding Z or A, they become traps with about the same half-time for fluorescence decay (inner antenna +Z, $t_{1/2} \approx 10\,\text{ps}$) as the PSII reaction centre itself (PSII reaction centre, $t_{1/2}$, 8–20 ps). The mechanisms that permit this competitive situation to function remain conjectural, but Chow (1994) predicted, and Frank et al. (1994)

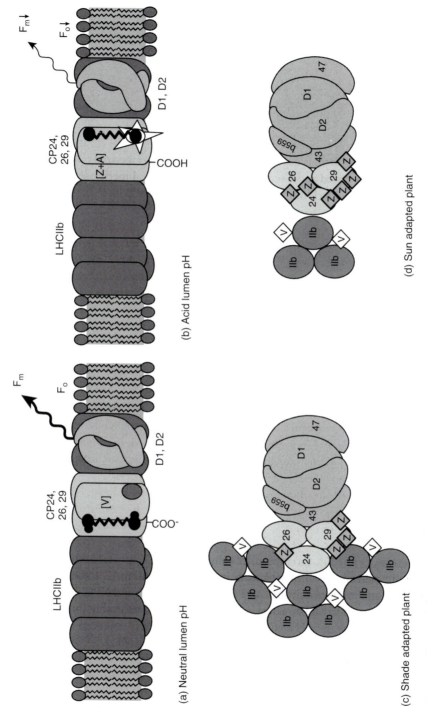

**Figure 1.3** Schematic location of xanthophyll pigments under neutral and acidic conditions in the thylakoid lumen, in cross-section (a,b) and stoichiometric relationships with respect to LHCs in shade and sun plants (c,d) in planar view.

9

observed, that molecular orbital overlap was possible between the more conjugated zeaxanthin and chlorophyll *a*, permitting direct energy transfer, from the latter to the former, and dissipation of excitation as heat. These changes in pigment composition and conformational state of the antennae pigment–protein complexes enable them to compete with the reaction centre as sinks for excitation, thus relieving the excitation pressure on PSII.

Genetic analysis has confirmed an essential role for zeaxanthin in thermal dissipation, showing that high ΔpH is not itself adequate (Niyogi *et al.* 1997). These data also contest the view that thylakoid ΔpH-dependent excitation dissipation is principally due to LHCII aggregation (Horton *et al.* 1996) and that zeaxanthin, or the absence of violaxanthin, simply amplifies this process. The molecular ecology of the pigments and proteins in the inner antenna is unclear and stoichiometry of the xanthophyll/chlorophyll pool is variable. Gilmore (personal communication) suggests as little as 1 molecule of zeaxanthin or antheraxanthin per chlorophyll protein complex converts the inner antenna to excitation traps that are competitive with the reaction centre.

This frontline, still controversial area of photosynthesis research continues to evolve from a decade of convincing ecophysiology, following the observation of Demmig *et al.* (1987) that conversion of violaxanthin to zeaxanthin was correlated with the development of photon excess as photosynthesis became light saturated. This insight provided a context for the previously identified xanthophyll cycle, now one of the most exhaustively established hypotheses with correlative ecophysiological evidence from most life forms and environments (Demmig-Adams & Adams 1992, 1996; Björkman & Demmig-Adams 1994).

### Photoinactivation and replacement of non-functional PSII reaction centres

Like many other biological machines, PSII reaction centres have a finite functional life. Some non-functional centres will always be present, and in some plants a large subpopulation (some 20%, possibly located at the margins of grana) seem more readily inactivated than others (Park *et al.* 1996a). The rate of decline in PSII function in leaves shows reciprocity between irradiance and the duration of illumination (Park *et al.* 1995), i.e. the loss of PSII function, is dependent on photon dose, whether delivered at different fluxes for a fixed time or at a particular flux for different times. Photoinactivation of individual PSII reaction centres is inevitable after $10^6$–$10^7$ photons have been absorbed (Park *et al.* 1995, 1996a). This means that even in weak light ($100 \, \mu mol \, photons \, m^{-2} \, s^{-1}$), about $10^6$ PSII complexes per $mm^2$ of leaf surface are inactivated every second (Lee *et al.* 1999).

The decline in the population of functional PSII centres is greatly accelerated by lincomycin, an inhibitor of chloroplast protein synthesis and hence an inhibitor of repair processes, showing that the rate of photoinactivation reflects a balance between damage and repair (Park *et al.* 1995). The extent of photoinactivation is mirrored by a rise in $F_O$, possibly due to the accumulation of excitation in the antenna, in spite of photoprotection. The weakest link in the PSII reaction centre is the D1 protein, as first recognized by Kyle *et al.* (1984), and this 32-kDa polypeptide remains the focus in mechanisms of photoinactivation. The D1 protein, together with the D2 protein in the heterodimeric reaction centre of PSII, binds all the redox components involved in photosynthetic charge separation leading to plastoquinone reduction. It also has a key role as plastoquinone receptor on the stromal side of the thylakoid, as well as in stabilizing the water-splitting complex on the lumen side. Ironically, this most rapidly turned over chloroplast gene product, with a critical role in the most energetically charged environment of living systems, has been dubbed 'the suicide polypeptide'. Because PSII centres are inactivated after processing about $10^7$ photons, it is not surprising that each PSII centre turns over at least once a day (Anderson *et al.* 1997).

The molecular mechanism of the initial steps leading to inactivation *in vivo* is not yet known, but it seems to depend on the generation and maintenance of increased concentrations of the primary radical pair, P680$^+$ Pheo$^-$, and on the different ways charge recombination is regulated under varying environmental conditions (Anderson *et al.* 1998). Although drainage of excitation in the antennae is probably the major photoprotective mechanism, avoiding excitation pressure and the generation of extremely high oxidizing potential P680$^+$, thus prolonging the life of functional PSII centres, other dissipative processes in the reaction centre itself may also be important. No matter how effective the photoprotection, some excitation will reach non-functional centres, and cause unstable charge separations that will also be dissipated as heat, either after transfer back to the antenna or by recombination in the centre itself.

Turnover of D1 in the PSII reaction centre increases linearly with irradiance up to light saturation. However, repair processes in leaves saturate at low irradiance regardless of the growth irradiance, maximum photosynthetic rate or antenna size (Park *et al.* 1996a; Anderson *et al.* 1997), and are usually able to prevent net loss of PSII under non-stress conditions (Anderson & Aro 1994). It is thought that as excitation pressure builds up beyond light saturation, non-functional PSII complexes accumulate in the appressed grana membrane regions because replacement can no longer keep up with inactivation. The formation of huge granal stacks in chloroplasts of deeply shaded leaves may be a device for storing PSII centres rendered non-functional by excess light, awaiting the slower repair processes in these leaves (Anderson & Aro 1994). Such non-functional centres continue to dissipate excitation as heat and may serve a photoprotective role with respect to neighbouring functional PSII complexes in the grana (Öquist *et al.* 1992). This photoprotective mechanism may be particularly important at low temperature when the enzymic reactions of electron transport, xanthophyll cycle interconversion, and D1 protein turnover are impaired (Öttander *et al.* 1993).

## Photoprotection and photoinactivation of PSII in vivo

Photoprotection does not always fully mitigate photoinactivation, as is well illustrated in the *A. thaliana* experiments of Fig. 1.2. Transfer from growth irradiance ($230 \, \mu$mol photons m$^{-2}$ s$^{-1}$) to light saturation and to full sunlight produced approximately twofold and fourfold increases in A+Z, respectively (Fig. 1.4a,b).

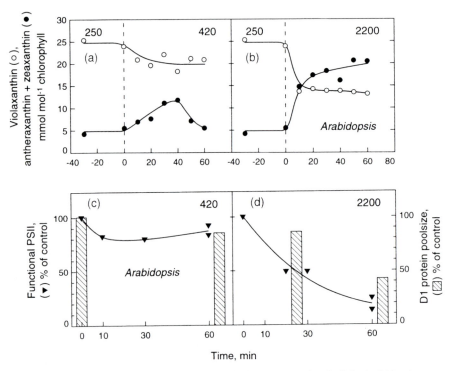

**Figure 1.4** Dynamics of photoprotection (a,b) and photoinactivation (c,d) in *Arabidopsis thaliana* leaves following transfer from growth irradiance to: (a) saturating (420 μmol photons m$^{-2}$s$^{-1}$); (b) excess (2200 μmol photons m$^{-2}$s$^{-1}$) irradiance. Upper panels (a,b) show the time-courses of xanthophyll pigment interconversion and lower panels (c,d) show the loss of PSII function by two different methods. Redrawn from Russell *et al.* (1995).

The twofold increase in A+Z over 1 hour made little difference to ΦPSII or q$_N$ at light saturation (Fig. 1.2b,c), but nevertheless there was a small decrease (~20%) in functional PSII centres and D1 protein content (Fig. 1.4c). However, a fourfold increase in A+Z after 20 minutes at full sunlight decreased ΦPSII by about 25% in weak light, and raised q$_N$ to about the same value as that in controls at light saturation (Fig. 1.2f,g). In spite of the high A+Z pool, functional PSII centres declined by 50% over the period of change in xanthophylls, and declined further to only 20% after 1 hour (Fig. 1.4d). Other experiments that show that Z-enriched ABA mutants of *A. thaliana* are no better photoprotected than wildtype (Hurry *et al.* 1997) indicate that fine stoichiometries, rather than large pools, may be critical to photoprotection. Changes in the xanthophyll stoichiometries of the antenna population within LHCII–PSII dimers are not yet understood. There would seem to be little advantage in retaining Z-rich LHCs within a non-functional PSII monomer, for example, and comprehensive models of the molecular ecosystem are needed.

Early studies of photoinhibition in leaves made it clear that photosynthetic and photorespiratory metabolism effectively mitigate photoinactivation (Powles

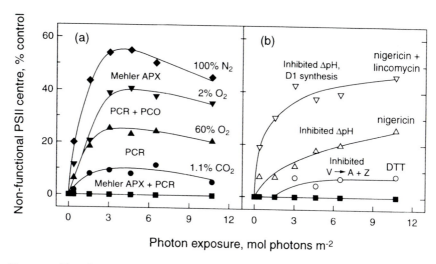

**Figure 1.5** Photoinactivation of PSII in pea leaves. The accumulation of non-functional PSII centres: (a) under different metabolic conditions that sustain different rates of electron transport; (b) in the presence of inhibitors of photoprotection or of chloroplast protein synthesis. Redrawn from Park *et al.* (1996b,c).

1984). Park *et al.* (1996c) demonstrated the cumulative protective effects of electron transport and carbon metabolism by controlling gas composition around pea leaves (Fig. 1.5a). The fastest and most extensive photoinactivation is found in $N_2$. As little as 2% $O_2$ provides significant relief of excitation pressure through photosynthetic electron transport to $O_2$ via the Mehler ascorbate peroxidase pathway. Photorespiratory carbon metabolism at 60% $O_2$ halves the extent of non-functional PSII centre formation in pea leaves in strong light, and saturating $CO_2$ alone reduces photoinactivation by 80%. The requirement for some electron transport to $O_2$, in addition to carbon metabolism alone, is evident in the control line (1.1% $CO_2$ in 21% $O_2$ in air). These seemingly futile photorespiratory pathways actually sustain a high rate of 'non-assimilatory' electron transport and high $\Delta pH$, thereby contributing to photoprotection (Schreiber *et al.* 1994) and preserving a functional photosynthetic apparatus. In much the same way, metabolic acclimation during cold hardening enables leaves to sustain higher photosynthetic rates, $\Delta pH$ and photoprotection at low temperatures (Öquist & Huner 1993), whereas unhardened leaves are more susceptible to photoinactivation (Thiele *et al.* 1996).

However it is now clear that metabolic sinks rarely account for more than 20–30% of photoprotection at light saturation. Xanthophyll and $\Delta pH$-dependent photoprotection predominates under most natural conditions, and Park *et al.* (1996a) showed that inhibition of component photoprotective mechanisms also stimulates photoinactivation (Fig. 1.5b). Thus, dithiothreitol (DTT) (an inhibitor of $V \rightarrow A+Z$) and nigericin (an uncoupler that prevents build up of high $\Delta pH$) increased the rate and extent of formation of non-functional PSII centres with increasing light dose in peas. When the repair cycle was further inhibited with

13

lincomycin, the rate and extent of non-functional PSII formation matched that observed during illumination in $N_2$ (Fig. 1.5a). Similar conclusions were reached earlier in the seaweed *Ulva rotundata*, using different criteria (Franklin *et al.* 1992; Osmond *et al.* 1993).

## Dynamics of photoprotection and photoinactivation in natural habitats

One can imagine that, in the short term, rapid photoprotection afforded by zeaxanthin-dependent reduction of photosynthetic efficiency could be especially important to rain forest understorey plants exposed to 10–100-fold increases in photon fluxes during sunflecks. Equally, one can imagine that rapid restoration of maximum efficiency would be important after the sunfleck has passed. Adams *et al.* (1998) found that *Stephania japonica* exposed to repeated sunflecks (from about 50 to 1500 µmol photons $m^{-2} s^{-1}$) in the understorey of an open eucalypt forest converted some 65% of xanthophylls to Z+A after the first large sunflecks and retained these levels throughout the day. In this condition PSII efficiency remained negatively correlated with photon flux density (PFD) during subsequent brief sunflecks, consistent with the notion that $\Delta pH$ might fine tune the dissipation of excess excitation in the antennae as a result of the persistent high levels of zeaxanthin. It seems likely that significant improvements in our understanding of these short-term responses to sunflecks can be obtained by rapid response gas exchange methods (Valladares *et al.* 1997), especially if $CO_2$ and $O_2$ exchange are combined with optical techniques (Laisk & Oja 1998).

The dynamics of xanthophyll interconversion during longer sunflecks varies between species. Observations of rain forest understorey sunflecks of 15–20 minutes duration, involving an increase from 10 to 2000 µmol photons $m^{-2} s^{-1}$, revealed at least three different responses (Watling *et al.* 1997). *Castanospora alphandii* converted 50% of xanthophylls to Z+A during the first 15-minute sunfleck of the day, but showed relaxation to control levels within 30 minutes after the sunfleck (Fig. 1.6). *Alocasia macrorrhiza* was just as responsive to the first sunfleck, but retained 70% Z+A over the next hour, reminiscent of *S. japonica* above. The gap pioneer *Omalanthus novo-guineensis* also converted xanthophylls to 50% Z+A after 30 minutes of exposure to high light in a small gap, but leaf movement reduced photon interception drastically and the xanthophyll pool relaxed to 80% V over the next 20 minutes. Turgor-based leaf movements seem to be an important feature of light stress avoidance in rain forest understorey plants (Lovelock *et al.* 1994; Krause & Winter 1996; Thiele *et al.* 1998).

Adams *et al.* (1999) also examined *S. japonica* and *Smilax australis* exposed to full sunlight for half a day at the edge of a eucalypt forest. These plants retained some 30–40% A+Z overnight and rapidly converted most of the remaining V to A+Z as PFD increased during the morning. Fluorescence-based calculations of light use showed that as much as 50–75% of absorbed photons were wasted as heat during the morning, and that high-efficiency photosynthesis was restored only slowly during the afternoon shading of these plants (Fig. 1.7). These, and

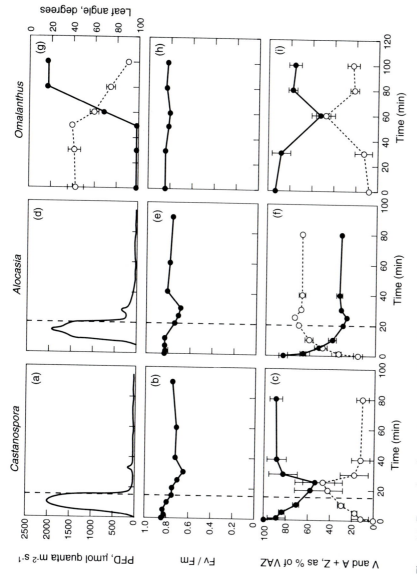

**Figure 1.6** Dynamics of xanthophyll pigment interconversion in leaves of rain forest tree seedlings during sunflecks *in situ*. The data for *Castanospora alphandii* (a–c) and *Alocasia macrorrhiza* (d–f) show the time-course of change in incident PFD due to the sunfleck, corresponding changes in $F_V/F_M$ and in concentrations of the xanthophylls, V (closed circles) and A + Z (open circles, $n=3$). The broken line shows the end of each sunfleck. The data for *Omalanthus novo-guineensis* (g–i) show increase in leaf angle from horizontal (g, ○) and PFD on a horizontal surface (g, ●) during exposure to full sunlight in a small gap, as well as corresponding changes in other parameters. Redrawn from Watling *et al.* (1997).

15

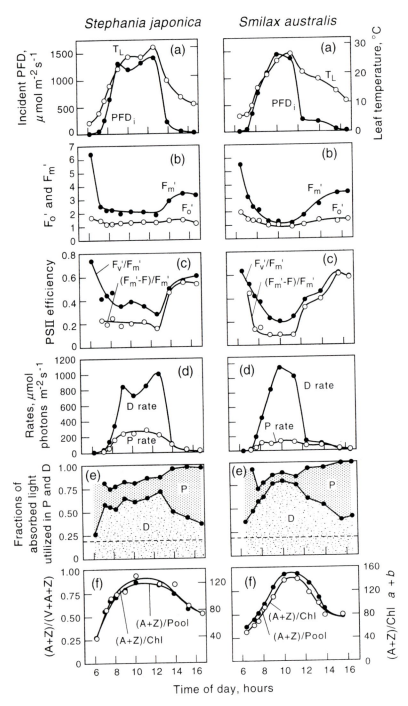

**Figure 1.7** Dynamics of light-resource utilization in leaves of *Stephania japonica* and *Smilax australis*, in response to diurnal changes in irradiance and leaf temperature at the margins of a eucalypt forest (a). (b) and (c) show chlorophyll fluorescence data used to calculate the proportions of light used in photochemistry (P) and dissipated as heat (D) in (d) and (e). Xanthophyll pigment pool size dynamics are shown in (f) Redrawn from Adams *et al.* (1999).

numerous other studies, show that plants exposed to bright light, or even weak light at low temperature (Adams *et al.* 1995), in their natural habitats exercise a remarkable capacity to waste excess excitation through xanthophyll and $\Delta$pH-related photoprotective mechanisms. The near universality of this phenomenon has provided a significant stimulus in ecophysiological research. It also seems adequate to explain many multiple-stress-related responses, such as nitrogen limitation at high light (Demmig & Winter 1988; Verhoeven *et al.* 1996), and winter stress in evergreens (Adams *et al.* 1995; Öttander *et al.* 1995).

Slow relaxation of zeaxanthin-associated photoprotection seems to come with a cost, but it must be remembered that the rate of carbon acquisition at low efficiency at light saturation may still be much greater than high-efficiency $CO_2$ fixation under light limitation. Persistent low efficiency in low light may not matter very much in understorey plants if most of the light is received in long, strong sunflecks; much depends on their amplitude and duration. However, sustained photoprotection or chronic photoinactivation can influence growth in plant canopies where most photosynthetic carbon assimilation occurs under limiting light. Here, assimilation is depressed by lower quantum yield, even in the absence of changes in light-saturated photosynthetic capacity (Ort & Baker 1988). There are consistent observations of strong correlations between plant growth and cold-induced reduction in photosynthetic efficiency (measured as $F_V/F_M$) in plants intercepting the same amount of light under field (Farage & Long 1991; Ball *et al.* 1997) and laboratory conditions (Königer & Winter 1991; Laing *et al.* 1995).

It is now clear that in many field studies persistently low values of $F_V/F_M$ in the morning signal sustained photoprotection, and that few plants persist in habitats in which chronic photoinactivation occurs. The extent to which photoinactivation is a factor in excluding plants from particular habitats is largely unknown, and is probably best evaluated by resorting to reciprocal transplant techniques of a former era in plant ecophysiology. We foresee important opportunities in ecophysiological research as the two sides of this coin are examined. Almost certainly, there will be subtle seasonal, habitat and genetic differences that expose the limits to photoprotection, some of which are suggested in the *Eucalyptus pauciflora* study in Fig. 1.8.

Egerton and Ball (Fig. 1.8) followed seasonal changes in $F_V/F_M$ in leaves of young snow gums in a subalpine woodland (1000 m) in mountainous terrain in SE Australia, and for part of the study shaded some plants with vertical panels of 50% shadecloth, positioned so as not to influence minimum temperatures. As expected, midday $F_V/F_M$ declined further in the high-light-exposed leaves in autumn and winter, and this decline was usually accompanied by a decline in $F_O$, suggesting effective photoprotection. Recovery through spring and summer was interrupted by episodic drought, but $F_V/F_M$ returned to values of about 0.8 in early autumn, i.e. to high efficiency of PSII. However, when unhardened plants experienced severe frosts in autumn, or high temperatures during droughts, marked depression of $F_V/F_M$ was accompanied by elevated $F_O$, indicative of photoinactivation (Fig. 1.8). A slowly relaxing decline in $F_V/F_M$ sometimes observed in plants growing in the

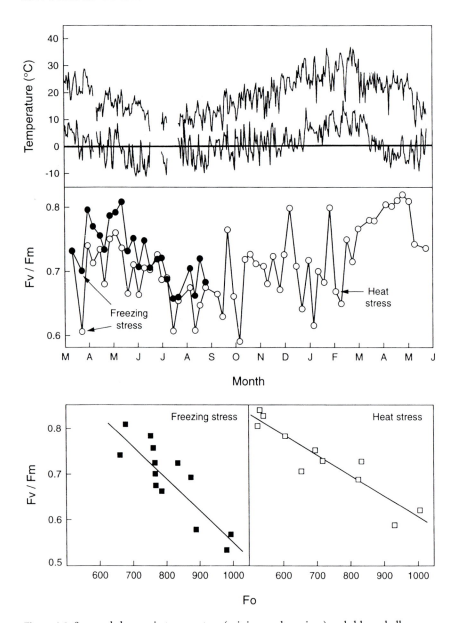

**Figure 1.8** Seasonal changes in temperature (minima and maxima) and chlorophyll fluorescence parameters in leaves of *Eucalyptus pauciflora* at 1000 m elevation in SE Australia (35°45′ S, 148°59′ E). Some seedlings were protected from full sunlight in the morning by panels of 50% shade-cloth (closed circles, centre panel). During episodic freezing and drought events in autumn and spring, respectively, decline in $F_V/F_M$ was correlated with an increase in $F_O$, indicative of temporary photoinactivation in the course of the seasonal progression of photoprotection in the long-lived leaves of this tree. J.J.G. Egerton (unpublished data).

rain forest understorey, insensitive to DTT, but abolished by inhibitors of protein synthesis, probably also reflects photoinactivation (e.g. *Dryobalanops lanceolata*, Scholes *et al.* 1997; *Brosimium alicastrum*, Thiele *et al.* 1998). The interactive effects of high light and high temperature stress that lead to necrosis in *Alocasia* evidently proceed through the whole catena of photoprotection and photoinactivation, and may be most important shortly after gaps form (Mulkey & Pearcy 1992; Königer *et al.* 1998). Clearly, the full panoply of Yin-Yang in photoprotection and photoin-activation is most likely to be uncovered at the margins of plant distributions, and during extreme aseasonal events.

## Photoacclimation

Light-dependent alteration of the molecular ecosystem of PSII, and the stoichio-metry of the two photosystems, remains one of the most challenging areas for research in light-resource utilization. Ecophysiological studies of acclimation during leaf ontogeny and canopy development continue to indicate the limits to the process in intact, natural systems (Ögren & Sundin 1996; Kitajima *et al.* 1997; Scholes *et al.* 1997). Although the changes in composition of the photosynthetic apparatus in response to light quantity and quality are now well known (Anderson *et al.* 1995), underlying molecular mechanisms responsible for these changes are not. Durnford and Falkowski (1997) listed three main hypotheses to account for the increase in antenna LHCs in response to low light that involved either a pho-toreceptor, a feedback effect in chlorophyll synthesis, or a feedback from photosyn-thetic activity. In terrestrial plants light-quantity-sensitive chromophores seem to sense day length, not the more variable aspects of daylight (Anderson *et al.* 1995). Of the feedback hypotheses, most evidence currently favours the second, the redox regulation of gene expression. Most of this evidence comes from experiments with unicellular algae, and it is by no means clear that the same priorities prevail in long-lived cells in leaves of higher plants.

Maxwell *et al.* (1994) proposed that *Chlorella* cells sensed excitation pressure in PSII whenever the rate of photon delivery exceeded electron transport to PSI. They found that cells (with low chlorophyll per cell) that developed under low light and temperature resembled those that developed under high light. Using selective inhibitors of photosynthetic electron transport, Escoubas *et al.* (1995) concluded that *Lhc* gene expression responded to the redox-state of the inter-photosystem plastoquinone pool. When low-light-acclimated cells of *Chlamydomonas rein-hardtii* with little endogenous NPQ were transferred to high light, measurements of $(1-q_p)$ indicated that the plastoquinone pool became fully reduced (Shapira *et al.* 1997). Under this potentially photoinactivating stress the *psbA* gene encoding D1 protein synthesis was up-regulated 5–10-fold, Rubisco synthesis was put on hold, cell division ceased and chlorophyll concentration per cell declined several-fold. In longer-lived leaves of higher plants most earlier studies focused on the importance of increased Rubisco synthesis during acclimation to high light (Anderson & Osmond 1987). Perhaps such studies need to be revisited, with particular attention being paid to the processes engaged beyond early photoprotection and photoinactivation, which permit reconstruction of an acclimated photosynthetic apparatus in existing leaves, or lead to accelerated senescence of these leaves and their replacement with leaves better acclimated to the changed light environment.

Nevertheless, a picture is emerging in which light-resource utilization depends on 'a nested series of photoacclimatory responses' (Durnford & Falkowski 1997), perhaps most responsive to electron transport, but in which causes and effects are still unclear. Under 'excess light', the $\Delta pH$ across the thylakoid membrane that drives the proton pump of the ATPase builds up when ATP synthesis exceeds consumption. This high $\Delta pH$ transduces photoprotective mechanisms that deflect excitation from the PSII reaction centres and, in the short term, slow the rate at which the particularly energetic species, which arise continuously during primary charge separation in PSII, photoinactivate the D1 protein. In the longer term, the redox state of PSII acceptors ($q_p$) or the electron transport capacitor (plastoquinone) signals adjustments in gene expression; stimulating synthesis of the most vulnerable protein of the LHCII–PSII dimer (and presumably some other proteins) while repressing synthesis of LHCII and Rubisco. Precisely how populations of pigment–protein components in the light reactions of the antennae and reaction centres, and the enzyme proteins in dark reactions of metabolism are coordinated in response to such 'generic' signals, is not clear.

## Acknowledgements

We are grateful to A. Gilmore and F. Chow for advice on simplifications that they find at times unsettling, to J. R. Watling and W. Adams III for permission to redraw recently published data and to A. Gilmore for providing Fig. 1.3.

## References

Adams, W.W. III, Demmig-Adams, B., Verhoeven, A.S. & Barker, D.H. (1995). 'Photoinhibition' during winter stress: involvement of sustained xanthophyll cycle-dependent energy dissipation. *Australian Journal of Plant Physiology*, **22**, 261–276.

Adams, W.W. III, Demmig-Adams, B., Logan, B.A., Barker, D.H. & Osmond, C.B. (1999). Rapid changes in xanthophyll cycle-dependent energy dissipation and photosystem II efficiency in two vines, *Stephania japonica* and *Smilax australis*, growing in the understorey of an open *Eucalyptus* forest. *Plant, Cell and Environment*, **22**, 125–126.

Anderson, J.M. (1986). Photoregulation of the composition, function, and structure of thylakoid membranes. *Annual Review of Plant Physiology*, **37**, 93–136.

Anderson, J.M. & Aro, E.-M. (1994). Grana stacking and protection of photosystem II in thylakoid membranes of higher plants under sustained high irradiance: an hypothesis. *Photosynthesis Research*, **41**, 315–326.

Anderson, J.M. & Osmond, C.B. (1987). Shade–sun responses: compromises between acclimation and photoinhibition. In *Photoinhibition. Topics in Photosynthesis 9* (Ed. by D.J. Kyle, C.B. Osmond & C.J. Arntzen), pp. 1–38. Elsevier, Amsterdam.

Anderson, J.M., Chow, W.S. & Park, Y.-I. (1995). The grand design of photosynthesis: acclimation of the photosynthetic apparatus to environmental cues. *Photosynthesis Research*, **46**, 129–139.

Anderson, J.M., Park, Y.-I. & Chow, W.S. (1997). Photoinactivation and photoprotection of photosystem II in nature. *Physiologia Plantarum*, **100**, 214–223.

Anderson, J.M., Park, Y.-I. & Chow, W.S. (1998). Unifying model for the photoinactivation of Photosystem II *in vivo* under steady-state photosynthesis. *Photosynthesis Research*, **56**, 1–13.

Ball, M.C., Egerton, J.J.G., Leuning, R., Cunningham, R.B. & Dunne, P.J. (1997). Microclimate above grass adversely affects spring growth of seedling snow gum (*Eucalyptus pauciflora*). *Plant, Cell and Environment*, **20**, 155–166.

Bassi, R., Pineau, B., Dainese, P. & Marquardt, J. (1993). Carotenoid-binding proteins of photosystem II. *European Journal of Biochemistry*, **212**, 297–303.

Björkman, O. & Demmig, B. (1987). Photon yield of $O_2$ evolution and chlorophyll fluorescence characteristics at 77 K among vascular plants of diverse origins. *Planta*, **170**, 489–504.

Björkman, O. & Demmig-Adams, B. (1994). Regulation of photosynthetic light energy capture, conversion, and dissipation in leaves of higher plants. In *Ecophysiology of Photosynthesis* (Ed. by E.-D. Schulze & M.M. Caldwell), pp. 17–47. Springer-Verlag, Berlin.

Bradbury, M. & Baker, N.R. (1981). Analysis of the slow phases of the *in vivo* chlorophyll fluorescence induction curve. Changes in the redox state of photosystem II electron acceptors and fluorescence emission from photosystem I and II. *Biochimica et Biophysica Acta*, **63**, 542–551.

Brugnoli, E. & Björkman, O. (1992). Chloroplast movements in leaves: influence on chlorophyll fluorescence and measurements of light-induced absorbance changes related to ΔpH and zeaxanthin formation. *Photosynthesis Research*, **32**, 23–35.

Chow, W.S. (1994). Photoprotection and photoinhibitory damage. *Advances in Molecular Cell Biology*, **10**, 151–196.

Chow, W.S., Hope, A.B. & Anderson, J.M. (1991). Further studies on quantifying photosystem II *in vivo* by flash-induced oxygen yield in leaf discs. *Australian Journal of Plant Physiology*, **18**, 397–410.

Demmig, B. & Winter, K. (1988). Light response of $CO_2$ assimilation, reduction state of Q and radiationless energy dissipation in intact leaves. *Australian Journal of Plant Physiology*, **15**, 151–162.

Demmig, B., Winter, K., Krüger, A. & Czygan, F.-C. (1987). Photoinhibition and zeaxanthin formation in intact leaves. A possible role for the xanthophyll cycle in the dissipation of excess light energy. *Plant Physiology*, **84**, 218–224.

Demmig-Adams, B. & Adams, W.W. III (1992). Photoprotection and other responses of plants to light stress. *Annual Review of Plant Physiology and Plant Molecular Biology*, **43**, 599–626.

Demmig-Adams, B. & Adams, W.W. III (1996). Xanthophyll cycle and light stress in nature: uniform response to excess direct sunlight among higher plant species. *Planta*, **198**, 460–470.

Demmig-Adams, B., Adams, W.W. III, Logan, B.A. & Verhoeven, A.S. (1995). Xanthophyll cycle-dependent energy dissipation and flexible photosystem II efficiency in plants acclimated to light stress. *Australian Journal of Plant Physiology*, **22**, 249–260.

Durnford, D.G. & Falkowski, P.G. (1997). Chloroplast redox regulation of nuclear gene transcription during photoacclimation. *Photosynthesis Research*, **53**, 229–241.

Ehleringer, J.R. & Monson, R.K. (1993). Evolutionary and ecological aspects of photosynthetic pathway variation. *Annual Review of Ecology and Systematics*, **24**, 411–439.

Escoubas, J.-M., Lomas, M., La Roche, J. & Falkowski, P.G. (1995). Light intensity regulation of *cab* gene transcription is signalled by the redox state of the plastoquinone pool. *Proceedings of the National Academy of Sciences USA*, **92**, 10237–10241.

Farage, P.K. & Long, S.P. (1991). The occurrence of photoinhibition in an over-wintering crop of oilseed rape (*Brassica napus* L.) and its correlation with changes in crop growth. *Planta*, **185**, 279–286.

Frank, H.A., Cua, A., Chynwat, V., Young, A., Gosztola, D. & Wasielewski, M.R. (1994). Photophysics of the carotenoids associated with the xanthophyll cycle in photosynthesis. *Photosynthesis Research*, **41**, 389–395.

Franklin, L.A., Levavasseur, G., Osmond, C.B., Henley, W.J. & Ramus, J. (1992). Two components of onset and recovery during photoinhibition of *Ulva rotundata*. *Planta*, **187**, 399–408.

Genty, B., Briantis, J.M. & Baker, N.R. (1989). The relationship between the quantum yield of photosynthetic electron transport and quenching of chlorophyll fluorescence. *Biochimica et Biophysica Acta*, **180**, 302–319.

Gilmore, A.M. (1997). Mechanistic aspects of xanthophyll cycle-dependent photoprotection in higher plant chloroplasts and leaves. *Physiologia Plantarum*, **99**, 197–209.

Gilmore, A.M. & Yamamoto, H.Y. (1993). Linear models relating xanthophylls and lumen acidity to non-photochemical fluorescence quenching. Evidence that antheraxanthin explains zeaxanthin-independent quenching. *Photosynthesis Research*, **35**, 67–78.

Gilmore, A.M., Hazlett, T.L. & Govindjee (1995). Xanthophyll cycle dependent quenching of photosystem II chlorophyll *a* fluorescence. Formation of a quenching complex with a short fluorescence lifetime. *Proceedings of the National Academy of Sciences USA*, **92**, 2273–2277.

Horton, P., Ruban, A.V. & Walters, R.G. (1996). Regulation of light harvesting in green plants. *Annual Review of Plant Physiology and Plant Molecular Biology*, **47**, 655–684.

Hurry, V., Anderson, J.M., Chow, W.S. & Osmond, C.B. (1997). Replacement of epoxy-xanthophylls by zeaxanthin in the ABA-deficient mutants of *Arabidopsis thaliana* L (Heynh.) does not affect chlorophyll fluorescence quenching, or sensitivity to photoinhibition *in vivo*. *Plant Physiology*, **113**, 1001–1011.

Kitajima, K., Mulkey, S.S. & Wright, S.J. (1997). Seasonal leaf phenotypes in the canopy of a tropical forest: photosynthetic characteristics and associated traits. *Oecologia*, **109**, 490–498.

Koller, D. (1990). Light-driven leaf movements. *Plant, Cell and Environment*, **13**, 615–632.

Königer, M. & Winter, K. (1991). Carotenoid composition and photon-use efficiency of photosynthesis in *Gossypium hirsutum* L. grown under conditions of slightly suboptimum leaf temperatures and high levels of irradiance. *Oecologia*, **87**, 349–356.

Königer, M., Harris, G.C. & Pearcy, R.W. (1998). Interaction between photon flux density and elevated temperatures on photoinhibition in *Alocasia macrorrhiza*. *Planta*, **205**, 214–222.

Krause, G.H. & Winter, K. (1996). Photoinhibition of photosynthesis in plants growing in natural forest gaps. *Botanica Acta*, **109**, 456–462.

Kyle, D.J., Ohad, I. & Arntzen, C.J. (1984). Membrane protein damage and repair: selective loss of a quinone-protein function in chloroplast membranes. *Proceedings of the National Academy of Sciences USA*, **81**, 4070–4074.

Laing, W.A., Greer, D.H. & Schnell, T.A. (1995). Photoinhibition of photosynthesis causes a reduction in vegetative growth rates of dwarf bean (*Phaseolus vulgaris*) plants. *Australian Journal of Plant Physiology*, **22**, 511–520.

Laisk, A. & Oja, V. (1998). *Dynamics of Leaf Photosynthesis. Rapid-response Measurements and their Interpretations.* CSIRO Publishing, Melbourne.

Lee, H-Y, Chow, W.S. & Hong, Y-N (1999) Photoinactivation of photosystem II in leaves of *Capsicum annuum*. *Physiologia Plantarum*, **105**, 377–384.

Leegood, R.C., von Caemmerer, S. & Osmond, C.B. (1997). Metabolite transport and photosynthetic regulation in C$_4$ and CAM plants. In: *Plant Metabolism* (Ed. by D.T. Dennis, D.H. Turpin, D.D. Lefebvre & D.B. Layzell), pp. 341–369. Addison Wesley Longman, Harlow.

Lovelock, C.E., Jebb, M. & Osmond, C.B. (1994). Photoinhibition and recovery in tropical plant species: response to disturbance. *Oecologia*, **97**, 297–307.

Maxwell, D.P., Falk, S., Trick, G.C. & Huner, N.P.A. (1994). Growth at low temperatures mimics high light acclimation in *Chlorella vulgaris*. *Plant Physiology*, **105**, 535–543.

Maxwell, K., Badger, M.R. & Osmond, C.B. (1998). A comparison of CO$_2$ and O$_2$ exchange patterns and the relationship with chlorophyll fluorescence during photosynthesis in C$_3$ and CAM plants. *Australian Journal of Plant Physiology*, **25**, 45–52.

Mulkey, S.S. & Pearcy, R.W. (1992). Interactions between acclimation and photoinhibition of photosynthesis of a tropical forest understorey herb, *Alocasia macrorrhiza*, during simulated canopy gap formation. *Functional Ecology*, **6**, 719–729.

Niyogi, K.K., Björkman, O. & Grossman, A.R. (1997). The roles of specific xanthophylls in photoprotection. *Proceedings of the National Academy of Sciences USA*, **94**, 14162–14167.

Ögren, E. & Sundin, U. (1996). Photosynthetic responses to variable light: a comparison of species from contrasting habitats. *Oecologia*, **106**, 18–27.

Öquist, G. & Huner, N.P.A. (1993). Cold-hardening-induced resistance to photoinhibition of photosynthesis in winter rye is dependent upon an increased capacity for photosynthesis. *Planta*, **189**, 150–156.

Öquist, G., Chow, W.S. & Anderson, J.M. (1992). Photoinhibition of photosynthesis represents a mechanism for the long-term regulation of photosystem II. *Planta*, **187**, 450–460.

Ort, D.R. & Baker, N.R. (1988). Consideration of photosynthetic efficiency at low light as a major determinant of crop photosynthetic performance. *Plant Physiology and Biochemistry*, **26**, 555–565.

Osmond, C.B. (1989). Photosynthesis: from the molecule to the biosphere. In: *Photosynthesis* (Ed. by W.R. Briggs), pp. 5–18. W.R. Liss, New York.

Osmond, C.B. (1994). What is photoinhibition? Some insights from comparisons of shade and sun plants. In *Photoinhibition: Molecular Mechanisms to the Field* (Ed. by N.R. Baker & J.R. Bowyer), pp. 1–24. Bios Scientific, Oxford.

Osmond, C.B. & Grace, S.C. (1995). Perspectives on photoinhibition and photorespiration in the field: quintessential inefficiencies of the light and dark reactions of photosynthesis? *Journal of Experimental Botany*, **46**, 1351–1362.

Osmond, C.B., Ramus, J., Levavasseur, G., Franklin, L.A. & Henley, W.J. (1993). Fluorescence quenching during photosynthesis and photoinhibition of *Ulva rotundata* Blid. *Planta*, **190**, 91–106.

Öttander, C., Hundal. T., Andersson, B., Huner, N.P.A. & Öquist, G. (1993). Photosystem II reaction centres stay intact during low temperature photoinhibition. *Photosynthesis Research*, **35**, 191–200.

Öttander, C., Campbell, D. & Öquist, G. (1995). Seasonal changes in photosystem II organisation and pigment composition in *Pinus sylvestris*. *Planta*, **197**, 176–183.

Park, Y.-I., Chow, W.S. & Anderson, J.M. (1995). Light inactivation of functional photosystem II in leaves of peas grown in moderate light depends on photon exposure. *Planta*, **196**, 401–411.

Park, Y.-I., Anderson, J.M. & Chow, W.S. (1996a). Photoinactivation of functional Photosystem II and D1-protein synthesis *in vivo* are independ-

ent of the modulation of the photosynthetic apparatus by growth irradiance. *Planta*, **196**, 300–309.

Park, Y.-I., Chow, W.S., Anderson, J.M. & Hurry, V.M. (1996b). Differential susceptibility of Photosystem II to light stress in light-acclimated peas leaves depends on the capacity for photochemical and non-radiative dissipation of light. *Plant Science*, **115**, 137–149.

Park, Y.-I., Chow, W.S., Osmond, C.B. & Anderson, J.M. (1996c). Electron transport to oxygen mitigates against the photoinactivation of Photosystem II *in vivo*. *Photosynthesis Research*, **50**, 23–32.

Powles, S.B. (1984). Photoinhibition of photosynthesis induced by visible light. *Annual Review of Plant Physiology*, **35**, 15–34.

Russell, W.A., Critchley, C., Robinson, S.A. *et al.* (1995). Photosystem II regulation and dynamics of the chloroplast D1 protein in *Arabidopsis* leaves during photosynthesis and photoinhibition. *Plant Physiology*, **107**, 943–952.

Scholes, J.D., Press, M.C. & Zipperlen, S.W. (1997). Differences in light energy utilisation and dissipation between dipterocarp rain forest tree seedlings. *Oecologia*, **109**, 41–48.

Schönknecht, G., Neimanis, S., Katona, E., Gerst, U. & Heber, U. (1995). Relationship between photosynthetic electron transport and pH gradient across the thylakoid membrane in intact leaves. *Proceedings of the National Academy of Sciences USA*, **92**, 12185–12189.

Schreiber, U., Schliwa, U. & Bilger, W. (1986). Continuous recording of photochemical and non-photochemical chlorophyll fluorescence quenching with a new type of modulation fluorometer. *Photosynthesis Research*, **10**, 51–62.

Schreiber, U., Bilger, W. & Neubauer, C. (1994). Chlorophyll fluorescence as a non-invasive indicator for rapid assessment of *in vivo* photosynthesis. In *Ecophysiology of Photosynthesis* (Ed. by E.-D. Schulze & M.M. Caldwell), pp. 49–70. Springer-Verlag, Berlin.

Shapira, M., Lers, A., Heifetz, P.B. *et al.* (1997). Differential regulation of chloroplast gene expression in *Chlamydomonas reinhardtii* during photoacclimation: light stress transiently suppresses synthesis of the Rubisco LSU protein while enhancing synthesis of the PSII D1 protein. *Plant Molecular Biology*, **33**, 1001–1011.

Skillman, J.B. & Winter, K. (1997). High photosynthetic capacity in a shade-tolerant Crassulacean acid metabolism plant. *Plant Physiology*, **113**, 441–450.

Sonoike, K. (1996). Photoinhibition of photosystem I; its physiological significance in the chilling sensitivity of plants. *Plant Cell Physiology*, **37**, 239–247.

Thiele, A., Schirwitz, K., Winter, K. & Krause, G.H. (1996). Increased xanthophyll cycle activity and reduced D1 protein inactivation related to photoinhibition in two plant systems acclimated to excess light. *Plant Science*, **115**, 237–250.

Thiele, A., Krause, G.H. & Winter, K. (1998). *In situ* study of photoinhibition of photosynthesis and xanthophyll cycle activity in plants growing in natural gaps of the tropical forest. *Australian Journal of Plant Physiology*, **25**, 189–195.

Valladares, F., Allen, M.T. & Pearcy, R.W. (1997). Photosynthetic responses to dynamic light under field conditions in six tropical rainforest shrubs occurring along a light gradient. *Oecologia*, **111**, 505–514.

Verhoeven, A.S., Demmig-Adams, B. & Adams, W.W. III (1996). Enhanced employment of the xanthophyll cycle and thermal energy dissipation in spinach exposed to high light and N stress. *Plant Physiology*, **113**, 817–824.

Watling, J.R., Robinson, S.A., Woodrow, I.E. & Osmond, C.B. (1997). Responses of rainforest understorey plants to excess light during sunflecks. *Australian Journal of Plant Physiology*, **24**, 17–25.

van Wijk, K.J., Roobol-Boza, M., Kettunen, R., Andersson, B. & Aro, E.-M. (1997). Synthesis and assembly of the D1 protein into photosystem II: processing of the C-terminus and identification of the initial assembly partners and complexes during photosystem II repair. *Biochemistry*, **36**, 6178–6186.

# Chapter 2

# Acquisition, partitioning and loss of carbon

*J.F. Farrar*

## Introduction

Higher plants acquire their carbon by photosynthesis, and quite properly photosynthesis has occupied a prominent place in the minds of plant scientists. Because the net acquisition of carbon approximates to growth, understanding the carbon fluxes which result in growth is inseparable from understanding growth itself. Indeed, there is an unspoken assumption in large areas of plant science, 'understand photosynthesis, and you understand both plant growth and how it responds to the environment'. This chapter will contend that understanding photosynthesis is only part of understanding plant growth.

There are three reasons for thinking that post-photosynthetic processes are important too. First, whole-plant photosynthesis is partly determined by partitioning of carbon and dry weight into new leaf area, and partly by regulation of photosynthetic rate per unit leaf area by internal factors; second, respiration is of major quantitative importance; and third, recent evidence suggests that growth of the whole plant is partly regulated by sinks. The story of the net acquisition of carbon by plants must of necessity take account of the whole plant and a wide variety of processes. Selected processes will be dealt with in turn. The message of this chapter can be stated easily: plant growth is always co-limited, internally by sources, transport system and sinks, and externally by more than one environmental resource.

## Photosynthetic acquisition of carbon

The capture of $CO_2$ in whole-plant photosynthesis is a function of the total leaf area, the photosynthetic capacity of each unit of that leaf area, and the extent to which the prevailing environmental conditions allow that capacity for photosynthesis to be expressed. The direct effects of environment are well known (Lawlor 1993; Baker 1996) and will not be dwelt on here, but the other two determinants will be considered.

## Regulation of photosynthesis per unit leaf area

The regulation of the capacity for photosynthesis of a leaf is a topic of current and increasing interest. The capacity of photosynthesis per unit leaf area ($A_{max}$) can

*School of Biological Sciences, University of Wales Bangor, Bangor, Gwynedd, LL57 2UW, UK. E-mail: j.f.farrar@bangor.ac.uk*

vary in response to demand from sinks. Although previous explanations for altered rates of photosynthesis have proposed stomatal aperture, phosphate status of the cytosol and feedback regulation by metabolites such as fructose 2,6-bisphosphate to be critical in relating sink demand and source function, these mechanisms are likely to be of more importance in the short term (minutes to hours) than in the long term (days to seasons) (Stitt 1991, 1996). They do not allow for adjustment of the amount of photosynthetic machinery, so that potentially scarce resources such as nitrogen can be re-allocated if not needed for photosynthetic proteins, and the amount of machinery which has to be maintained with a respiratory cost does not exceed that needed. Selection would have been expected to favour a mechanism of photosynthetic regulation which minimized unnecessary storage of nitrogen and respiratory loss of carbon (but not necessarily an 'optimal' use of nitrogen, whatever that may be).

Further, there is clear evidence that photosynthesis is indeed regulated at the level of gene expression and the amount of machinery present. Always implicit in the different amounts of machinery in sun and shade leaves (Evans 1996), expression of genes such as the small (rbcS) and large (rbcL) subunits of Rubisco, and the chlorophyll a/b binding protein (cab) has now been shown to be variable (Koch 1996; Pollock & Farrar 1996; Jang & Sheen 1997). The broad argument is that when carbohydrates accumulate in source leaves because their photosynthetic production exceeds export, they reduce the expression of key genes encoding photosynthetically important proteins. Whilst the broad hypothesis is clear, details of the mechanisms are not yet certain. One possibility is centred on hexokinase as a sugar sensor, partly due to analogy with yeast (Jang & Sheen 1997; Smeekens & Rook 1997; Smeekens 1998). There are still uncertainties with this hypothesis. The actual signal molecule may be a hexose phosphate, but alternatively it may be that the act of phosphorylation of the enzyme itself initiates the signal transduction pathway. In addition, it is not clear how the signal is integrated over time. The sugar content of a leaf can change 5- to 10-fold over a day, so does the plant sense the sugar content at one time in this cycle, or measure some kind of average content? Is it the flux through hexokinase rather than a sugar concentration that is important for signal transduction? Is integration achieved by continuously varying the activity of the signal transduction pathway in response to the current and changing sugar content, with the half-lives of the mRNA and the protein itself providing the integration? A further problem is rooted in the heterogeneity of source leaves: presumably regulation of genes encoding photosynthetic proteins is largely confined to mesophyll cells, yet about half the cells in a leaf are not mesophyll. Because the sugar content varies hugely between cell types (Koroleva et al. 1997, 1998) and phloem in particular can have high sugar contents, there is little reason to hope for success in attempting to correlate whole-leaf sugars with expression of photosynthetic genes.

One circumstance in which photosynthetic production exceeds export is when sinks are incapable of using the sugar at the rate at which it is being produced. They may be cold; they may be nitrogen deficient; they may be growing at the maximum rate of which they are genetically capable; for whatever reason, in sum they cannot metabolize or store sugar as fast as it is produced. If they cannot continue to import it indefinitely, it will accumulate in source leaves and downregulate photosynthesis. If this mechanism works *in vivo*, plants have a simple and elegant method for linking sink demand to the supply of carbon from photosynthesis. An argument will be given below for a feedforward mechanism operating to produce more sophisticated control.

One consequence of coarse control of the photosynthetic machinery is that nitrogen may be committed to the machinery only when rates of $CO_2$ fixation are high. Some workers have suggested that the allocation of nitrogen can be optimized such that when sinks limit photosynthesis, nitrogen can be allocated to sinks to relieve constraints on their growth. This suggestion would be supported if expression of genes such as *rbcS* and *cab*, whose proteins are abundant and so lock up a great deal of nitrogen, were under the control of a nitrogen-containing resource compound. Some genes encoding proteins of photosynthesis and of nitrogen metabolism are indeed regulated by nitrate or specific amino acids (Koch 1997; Paul & Driscoll 1997; Stitt 1999).

### Allocation to leaf area
The extent to which newly acquired carbon is allocated to the production of new leaf area has a key role in determining the plant's future photosynthetic capacity, a fact long enshrined in the growth–analysis relationship RGR = NAR × SLA × LWR (where RGR = relative growth rate, NAR = net assimilation rate, SLA = specific leaf area, LWR = leaf weight ratio), or more recently the fuller equation of Poorter *et al.* (1992) and Poorter and Garnier (1996).

The regulation of the allocation of carbon to new leaf growth is not understood. The functional equilibrium model (Thornley 1976; Brouwer 1983) is successful at describing how leaf area might be preferentially expanded under low irradiances. However, we have no idea what signals are involved, nor of their transduction pathway, nor of the mechanisms by which partitioning responds. Allocation to leaf area is clearly a subset of the more general problem: how is carbon partitioning regulated?

### The partitioning of carbon and dry weight
Partitioning is still one of the least understood areas of plant science, so much so that we have recently argued that even descriptions of it are frequently inadequate (Farrar & Gunn 1998). Some of the factors which may regulate partitioning are discussed below.

### The partial processes of partitioning
Some of the decisions which the plant takes to determine partitioning can be listed (Farrar & Gunn 1998; Farrar 1999); they help to define the partial processes of which it is composed. New assimilate can be stored in the leaf or exported (the

export fraction). Of that which is exported, a proportion goes up the plant to the shoot apex, and the rest down to the roots (gross partitioning fraction). Within a meristem, a proportion of the assimilate is used for growth (including the associated respiration) and the rest is stored. Within the shoot meristem, assimilate can be used to grow vegetative or reproductive sinks. Net partitioning of dry weight is the result of the transport and gross partitioning of carbon, and its subsequent metabolism. Each of these partial processes of partitioning is itself complex and is known to be a site of control (Farrar 1985, 1989; Farrar et al. 1994), although the evidence relating to induction of meristems is as yet weak. The problems are therefore to understand that control, and to see how the control of each individual step is integrated with control of the others.

The export fraction may be partly determined by the amount of readily transported sugar present in the leaf (B.E. Collis, C.J. Pollock & J.F. Farrar, unpublished data; Farrar 1999) but the control over phloem loading, perhaps localized at the sucrose transport protein, is not yet clear. Often only about half of the turgor in phloem is generated by sucrose itself, and so the loading of other solutes is critical for phloem transport.

Once inside the phloem, it is probably a network of turgor gradients connecting the sources and sinks which determines the direction of sucrose flux; we have no idea how they are controlled. Reviews by Farrar (1992, 1996) and a theoretical treatment of phloem partitioning (Minchin et al. 1993) suggest a number of relevant issues. First, the flux of solutes in the phloem is determined partly by events in the source which determine phloem loading and turgor pressure; partly by events in the sink which determine unloading and turgor there; and partly by properties of the transport path itself. Second, once there is more than one source or sink the system behaves in a way which is not intuitively obvious. For example, when one source is supplying two dissimilar sinks the proportioning of assimilate between the two sinks changes with the flux out of the source. Third, there need not be any simple link between the turgor established in the phloem (which will determine solute flux in the phloem) and metabolism in the tissue containing it; nor need there be a simple link between sugar loaded into the phloem and turgor generation because many other solutes contribute to turgor.

### Partitioning: regulated for resource acquisition?

The relative weight of shoot and root is both restored after an experimental perturbation (Brouwer 1962) and responds in an apparently adaptive manner to a changing environment (Wilson 1988). It could be regulated to achieve balanced acquisition of above- and below-ground resources (Farrar & Gunn 1998). It is widely believed that allocation and resource acquisition are related because relatively more root facilitates nutrient uptake at the expense of photosynthesis, and relatively more leaf facilitates photosynthesis at the expense of nutrient uptake, but this view neglects the physiological competence at resource acquisition of each unit of leaf or root. Such competence is determined both internally and by the environment (see below). It also neglects the fact that the ratio of shoot to root (S:R) usually changes with ontogeny. Perhaps functional leaf and fine root do indeed stay in a similar ratio, with support tissue accounting for differences in S:R. As the taproot of vegetative carrots develops, S:R falls dramatically while leaf and fibrous roots stay in proportion (O.A. Koroleva, A.D. Tomos & J.F. Farrar, unpublished data).

If it really is the ratio of resource-acquiring parts which is being regulated in partitioning, how does the plant achieve regulation? I will start with the classical pruning experiment of Brouwer (1962), where partial defoliation or de-rooting of bean plants resulted in the plant part which had been experimentally reduced growing faster than the intact part so that the control value of S:R was restored. This remarkable result means that the net partitioning fraction (the partitioning between sinks of that carbon which has been exported from a leaf, less subsequent respiration) is not conserved. It might imply that the plant has some means of measuring shoot and root weights, and adjusting allocation until these weights are in the correct ratio, but it is far more likely that the plant measures the internal status of one or more compounds—these will be called resource compounds—which reflect a balance between acquisition of a resource by the plant and its use. Several current views are described in a recent special issue of *Plant and Soil* (van der Werf & Lambers 1996), where a number of articles promote the role of 'resource compounds', such as sucrose and nitrate, which may be representative of the availability of the amount of resource, or of the balance between resource availability and use (Farrar 1996; Koch 1996, 1997; Scheible *et al.* 1997; Zhang & Forde 1998). There is a seductive logic in a resource being the signal for its own acquisition, and two such resource compounds could easily form the core of a system for achieving balanced shoot:root partitioning.

Critical tests of the idea that plants control an aspect of S:R which is relevant to resource acquisition, and an investigation of the mechanisms by which it does so, are desperately needed. It is important to distinguish between this approach and the closely related functional equilibrium hypothesis (Thornley 1976; Brouwer 1983), which is essentially descriptive and which does not distinguish between the export fraction and the partitioning fraction, and therefore is less precise than the questions demand. A simple experiment gives a result satisfyingly consistent with the idea that S:R reflects resource acquisition: if barley is grown at high nitrogen and partly defoliated, and at the same time transferred to a lower supply of N, it regrows to produce a plant with the S:R relations of a plant kept throughout at the low supply of nitrogen rather than the intact control maintained at high nitrogen (J.F. Farrar, unpublished data).

## Control of partitioning by sinks

It was suggested above that regulation of partitioning is partly determined by metabolism in sinks indirectly affecting the turgor in phloem within the sink. Sustained import into a sink cannot exceed the rate at which that sink can use assimilate in growth, respiration and storage. What might determine that rate for a single sink is considered here.

The growth rate of a sink depends on the number and size of meristems it possesses, the capacity for growth of each unit of meristem, and the extent to which prevailing conditions of environment and of supply of resources from elsewhere in the plant allow that capacity to be achieved. The control of meristem number is not understood, but both sugars and nitrate (or other nitrogenous compounds) may have a role. A simpler model for shoot apices is that the rate of cell division is set by availability of resources (including sugar; Francis 1998) and a new meristem is produced after a predetermined number of divisions. The capacity for each meristem to grow may also be set by supply of resources. When root tips are excised their metabolic rate falls rapidly; for example, their respiration is reduced, and their carbohydrate content falls. The succession of events following excision seems to be as follows. First, the roots deplete endogenous sugar pools, to the extent that respiration becomes limited largely by supply of substrate and its rate falls. Then, the low concentrations of sugars are associated with reduced expression of key genes encoding parts of the respiratory network and so the amounts of protein encoded by these genes fall. The fall may be a result of both natural turnover in the absence of synthesis and also increased degradation following the increased activity of proteases (Farrar & Williams 1991; Farrar 1999; B.E. Collis, P. Dwivedi & J.F. Farrar, unpublished data). Cell division, perhaps the most fundamental property of meristems, may itself be under the control of sugar supply (Francis 1998). Their metabolism can be restored by supplying them with sugars, but after a few hours of excision, the roots are no longer stimulated by sugar as a substrate. Rather, sugars now increase respiration and metabolism only after they have been supplied to the root for some hours and probably alter gene expression. In other words, there is a strong case that the number and capacity for growth of meristems is set by the status of one or more resource compounds.

The next question is, do sinks usually grow at their maximum possible rate, or at some lower rate? Frequently, the rate is lower. A series of experiments on barley roots (Farrar & Jones 1986; Bingham & Farrar 1988; Farrar & Williams 1991) showed that the growth and respiration rate of the roots on fast-growing intact plants could be increased by treatments which caused a sustained increase in assimilate inflow to the root. A similar conclusion comes from experiments on increased atmospheric $CO_2$ (Collis et al. 1996) and by analogy extends the argument to shoot apices. Why do sinks grow less quickly than they could? If their capacity to grow is set below the maximum because the status of key resource compounds such as sucrose is lower than that needed to induce maximal capacity, these compounds must be limiting. Similarly, it is easy to see how nitrogenous compounds must be plentiful in order to permit maximal growth of sinks, and that the consequence of their scarcity will be a reduction of sink growth rate whatever the rate of supply of assimilates. It will be argued below that it is an inevitable property of distributed control in a complex system that some of the mechanism of control does indeed lie in sinks.

## Determination of meristem number as a mechanism in the regulation of partitioning

Evidence is emerging that the number of meristems in a sink is crucial in regulating partitioning; this is a developmental effect, altering numbers of leaves, tillers and roots. The experiments described in this section have been designed and analysed allometrically, thus removing ontogenetic and size-dependent effects. In both barley and *Dactylis glomerata*, an increase in nitrate supply increased partitioning to root relative to shoot. In barley, there was no effect of nitrate on tiller production per unit shoot weight; the effect of nitrate was mainly on growth, not on development (K.B. Laghari & J.F. Farrar unpublished). By contrast, *D. glomerata* responded developmentally to nitrate: there were more tillers at high nitrogen because, whilst there was no increase in number of leaves on the main stem, a higher proportion of the possible tiller sites was filled, and the rate of appearance of leaves on the second tiller was increased (B.E. Collis, J.F. Farrar & C. Marshall, unpublished data). Plants which had been grown at low nitrogen and were then transferred to high nitrogen responded by altering the allocation of dry matter so that S:R allometry and root number per unit root dry weight changed towards the value of control plants at high nitrogen. Even if all new dry weight were allocated to the shoot and not the root (which is not what happens), it takes an appreciable amount of new growth to transform one S:R to another, and inevitably several days are needed to produce a transferred plant with the S:R appropriate for its new environment. More roots are produced per tiller and per unit root dry weight at high nitrogen, suggesting that development and partitioning are co-regulated.

Just what is detected to determine meristem number: the external concentration of resource, or some internal manifestation of it? Phosphate also alters morphology (Wilson 1987): in barley relatively more roots are produced at low P supply (J.F. Farrar, unpublished data). When paired ramets of *Agrostis stolonifera*, connected by a stolon, are grown hydroponically in low P concentrations, the length of root per unit root weight is increased. The signal for the P-sensitive alteration in morphology is unlikely to be the external concentration of Pi, because when just one of the pair of the ramets is at low, and the other at high, Pi, the length of root per unit root weight for the root in low Pi assumes a value intermediate between that of a ramet in high Pi, and that of a ramet connected to another at low Pi. A much better relationship exists between the root length:root weight ratio and the concentration of P within the shoot. Because this P will be in a number of chemically and anatomically distinct pools, it is presumably some discrete pool within it that represents the best candidate for a signal of P status (R. Solbe, C. Marshall & J.F. Farrar, unpublished data).

In conclusion, the deduction from limited data is that partitioning is controlled to regulate the relative amount of organs which acquire resources. The mechanism by which such regulation is achieved involves the detection of the internal status of 'resource compounds' such as sucrose, phosphate and nitrate. How size and density of resource acquiring machinery are interrelated is considered next.

## Coordinated regulation of dry matter partitioning and density of machinery for acquiring resources

It is not at all clear how these two components of whole-plant photosynthesis are mutually regulated: is there coordination between the mechanisms regulating each? Our inability to offer a good answer to this simple question is partly due to fashion: plant scientists interested in mechanism mainly work on parts of the plant rather than the whole organism. However, some real insights are coming indirectly from work on the effects of elevated $CO_2$ on plant growth and photosynthesis, and these may at least help to define an experimental system which will facilitate an approach to the problem.

Only experiments where roots have been physically unrestricted, and have had completely adequate supplies of nutrients and water, will be considered. Hydroponics provides the simplest technique for achieving this. Effects of treatment must also be unequivocally separated from those of ontogeny and allometry is a suitable separator. When very young plants are raised from seed in elevated atmospheric $CO_2$ ($C_a$) or when plants grown at 350 p.p.m. $CO_2$ are transferred into 700 p.p.m. $C_a$, the classical effect is a transient increase in relative growth rate (RGR), lasting about 7 days (Stulen et al. 1996; Baxter et al. 1997; D.L. Marriott & J.F. Farrar, unpublished data). The cause of the higher RGR is increased net photosynthesis per unit leaf area (Drake et al. 1997). The fact that the increase in RGR is transient could be caused by either, or both, a fall in photosynthesis per unit leaf area, or a reduction in leaf area ratio (LAR).

A number of possible signals may change partitioning or photosynthetic capacity. Typically the content of non-structural carbohydrate in source leaves rises, while the content of nutrients such as nitrogen falls (Drake et al. 1997). However partitioning of dry weight between organs is little affected (Farrar & Gunn 1996; Gunn & Farrar 1999), so the signals are having rather specific effects on source leaves. There is good evidence of feedforward effects of the higher sugar supply on sink metabolism; roots are heavier at elevated $CO_2$ as a result of the earlier and greater production of nodal roots and the greater rate of production of laterals on seminal roots (Farrar 1999).

Photosynthesis of leaf 2 of barley was initially higher when grown at elevated compared with atmospheric $CO_2$ (Plum et al. 1996), and only when the leaf was older was there downregulation of photosynthesis. The initial increase in the rate of photosynthesis of leaf 2 resulted in a greater accumulation of non-structural carbohydrate (Hibberd et al. 1996a). However, there is currently no firm evidence that this carbohydrate downregulated photosynthesis as the amount of Rubisco protein remained unchanged (Hibberd et al. 1996b). This suggests that a mechanism other than that involving sugar-mediated downregulation of gene expression is involved. Currently it must be concluded that while it is possible that photosynthetic rate per unit area and partitioning are co-regulated, there is no indication of how this might be achieved; certainly elevated $CO_2$ affects photosynthetic rate much more than partitioning.

A parallel question can be asked for the acquisition of nutrients: is there mutual regulation between the relative partitioning to roots and thus the amount of nutrient-acquiring organ, and the density of transporters per unit of root? Nutrients often occur in patches and/or in pulses of short duration. Roots grow rapidly through patches of high nutrients yet the growth responses of the plant are often slow. Can the plant respond more rapidly? The answer is yes, and the mechanism is to alter the capacity for uptake of nutrients. When *Dactylis glomerata* was switched from low to high relative addition rate (RAR) of nitrate, the maximum rate ($V_{max}$) for nitrate uptake per unit weight of root was halved within 24 hours, to a value typical of plants maintained at the high RAR. Thus two changes follow a switch from low to high RAR of nitrate: a rapid fall in $V_{max}$ for nitrate, and a change in carbon partitioning to favour the shoot which over several days brings S:R into line with plants grown at a high RAR of nitrate (B.E. Collis, C. Marshall & J.F. Farrar, unpublished data).

Phloem-mobile amino acids may control nitrate uptake (Touraine *et al.* 1994) by an unknown mechanism that might involve changing the activity of existing carriers, or changing the density of such carriers in the membrane. Certainly $V_{max}$ alters with RAR of nitrate in barley too (Larsson 1994). Clearly, resource compounds can affect both S:R partitioning and uptake kinetics, but whether the same compounds affect both, and how or if their effects on the two types of systems are integrated, remains unknown. We suggest a temporal distinction: physiology responds in the short term and morphology is a longer-term response. We would predict that plants in spatially and temporally heterogeneous environments have a plastic physiology while those in environments with a longer-term or coarser-grained heterogeneity respond morphologically.

In conclusion, a case can be made that partitioning is regulated to increase acquisition of resources, and that the mechanism involves the use of a variety of resource compounds as signals to trigger changes in metabolism and morphology within source leaves and sinks. However, we still have much to learn.

## Respiration

Half the carbon acquired by photosynthesis is lost as respiratory $CO_2$: this is the second of the three reasons for considering that post-photosynthetic processes are critical in the net acquisition of carbon. This number excludes carbon cycled during photorespiration: it refers only to dark respiration. It might be argued that respiration is loss, not gain, of carbon and thus quite irrelevant to its acquisition. I want to show you that this is not so, and then explore how respiration rate is determined.

### The relationship between the rates of growth and of respiration

The basic reason why so much carbon is respired has been understood for more than 20 years: it is the inevitable, unavoidable cost of synthesizing and maintaining the tissues which acquire resources (Beevers 1961; Amthor 1989) and of the net acquisition of nutrients (Clarkson 1998). Approximately, 0.45 g $CO_2$ is respired for every 1 g dry weight of plant tissue synthesized (if N and S are supplied as $NO_3$ and $SO_4$), and the cost of maintaining plant tissue is the respiration of about 1.5% of its weight in sucrose daily (Penning de Vries 1972).

Could the process be more efficient: could plants respire less and grow as fast? The isolation of fast-growing lines of ryegrass (Kraus *et al.* 1990) and tall fescue (Volenec *et al.* 1984) by selection for low rates of mature tissue respiration suggest that it might be. However, the reason for the success of these lines is not clear. In the case of ryegrass, it is not a result of low activity of the alternative oxidase (Kraus *et al.* 1990). It is odd to note that mature, non-growing leaf tissue has a respiration rate that correlates negatively with growth rate. This suggests that some process in leaves which consumes energy is inimical to the growth of the whole plant, possibly, but not certainly, beyond the reduced consumption of carbon. Is there a non-causal relationship? Other things being equal, high specific leaf area (SLA) is correlated with high RGR (both by definition, as discussed above, and in inter-species comparisons (Cornelissen *et al.* 1998; Poorter & van der Werf 1998)). Because respiration is to a first approximation a function of weight, not area, of tissue, fast-growing species would be expected to have lower rates of mature tissue respiration per unit area. However, there is a negative relationship between leaf respiration per unit weight and growth rate in *Lolium perenne* where the superiority of low-respiring lines is confined to growth in swards (Kraus *et al.* 1990).

To a first approximation, respiration energizes underlying processes efficiently. It is not wasteful, and so we can now ask: does respiration control the rate of processes which it energizes, or do the rates of these processes determine the rate of respiration?

### Short-term control of respiration rate

Many years of evidence from short-term experiments, mainly by biochemists, has shown that respiration rate is regulated by demand, via the turnover of adenylates and NAD(P)/NAD(P)H (Farrar & Williams 1991). The rate of electron transport increases when it is uncoupled from the supply of ADP, so the rate of release of ADP as a result of ATP hydrolysis coupled to energy-consuming processes was limiting respiration. (It will be argued below that this shows adenylate turnover to be a, not necessarily the, control of respiration rate, but all the evidence suggests that most control does lie with adenylate turnover.) The conclusion is refreshingly clear and robust: demand from growth, maintenance and ion uptake determines the rate of respiration. Neither the amount of respiratory machinery, nor the supply of assimilate as substrate for it, normally limits respiration or growth.

At once a possible problem appears: if the amount of respiratory machinery is not limiting, it may be in excess, and then needs to be maintained using carbon which otherwise could be used for growth. However in a number of systems, the capacity of the cytochrome pathway is just adequate to explain the respiratory flux (Farrar & Williams 1991). This is not coincidence; it is achieved by the long-term control (acclimation) of respiratory capacity to closely match the demand for respiratory energy.

## Acclimation of respiration rate

Respiration rate acclimates, that is it undergoes change in the amount of metabolic machinery in response to an altered environment (Farrar 1999), as does photosynthesis and sink metabolism. The best evidence for such acclimation comes from experiments where root respiration responds to both illumination or partial defoliation of the shoot, a response mediated by the supply of sugar from the shoot (Farrar & Williams 1991; B.E. Collis & J.F. Farrar, unpublished data). Sugar also regulates respiration in excised root tips and in plant cell cultures (Graham *et al.* 1994; Pollock & Farrar 1996; Farrar 1999). We know that, just as for photosynthesis, the expression of genes encoding key enzymes of sink metabolism (invertase, sucrose synthase and cytochrome oxidase) can be regulated by sugars or some component of sugar metabolism (Farrar & Williams 1991; Farrar 1999).

A second way of altering carbohydrate supply is to raise atmospheric $CO_2$ concentration. There is a debate about whether $CO_2$ can directly inhibit plant respiration (Drake *et al.* 1999), but there is clear evidence that there is acclimation to elevated $CO_2$. Even when data are expressed carefully, allowing for respiration changing with size, carbohydrate and nitrogen content, respiration is frequently lower in leaves and shoots of plants grown at elevated $CO_2$ (Amthor 1997; Drake *et al.* 1999). The most likely explanation is that the expression of genes such as cytochrome oxidase (*cox*II) has been reduced as a result of the higher sugar content of leaves in high $CO_2$, but critical evidence is needed. Respiration also acclimates to a change in temperature. While in the short term temperature alters respiration with a $Q_{10}$ of about 2, in the long term it has rather little effect (Lambers 1985; Semikhatova *et al.* 1992; Atkin & Lambers 1998; Farrar 1999). Low temperature results in increased expression of the gene encoding the alternative oxidase (Vanlerberghe & McIntosh 1992), but the signal transduction pathway is unknown.

To generalize, we know what causes respiration rate to change, but not always the means by which that change is accomplished. Clearly the simplest hypothesis to test is that the effect of environmental variables such as temperature is mediated by their effects on sugar status.

It cannot merely be respiration which is acclimating, but also many of the processes for which it is providing energy and carbon skeletons. Here respiration is providing a deep insight into the acclimatory process, and in particular the acclimation of sinks, which is much less studied than that of source leaves. The acclimation of respiration rate can therefore be seen in a more fundamental light: it is no less than the integral of how a tissue adjusts to a new internal or external environment. It is another aspect of the way in which the potential growth rate of a sink, or the capacity of a source leaf, is regulated. Evidence that the respiration rate of a root system acclimates is evidence that it is not only photosynthesis that determines plant growth.

## The control of growth rate

The third reason for implicating post-photosynthetic processes in regulating the acquisition of carbon is that the regulation of the rate of whole-plant growth is determined in part by each of the processes which constitute it. The strands of evidence which point to this conclusion are diverse, but underlying many of them is the core idea that growth can be treated as a flux of carbon compounds. At its simplest $CO_2$ is fixed in chloroplasts, and triose phosphate is exported from the chloroplast as the precursor for sucrose synthesis (or starch is synthesized in the chloroplast). Sucrose is loaded into phloem within the source leaf (or stored), and export results in the sucrose entering the hydrodynamic network of phloem in the whole plant. Flux of sucrose through the phloem results in partitioning of sucrose between sinks, and within the sinks sucrose is unloaded from phloem, moves to receiver cells, and enters metabolism where it is either used in biosynthesis or respired to $CO_2$ (or stored). This suite of processes can easily be drawn as a branched pathway and thus the control of growth can be considered as analogous to the control of flux along a metabolic pathway.

## Control analysis

The control of flux in metabolic pathways has been examined theoretically by metabolic control analysis, which was originally developed for glycolysis, but has been extended to include branched pathways (Fell 1997). Its main conclusion is easily stated: each step of a pathway exercises some degree of control over the flux through it. (The diffusion of $CO_2$ through stomata and layers of still air and water, and subsequent carboxylation within chloroplasts, is a familiar example where it is accepted that control of a pathway is distributed among each of its steps.) There are no single limiting factors; there is no 'law of the minimum'; rather, control is distributed throughout. However, control is not evenly distributed. The bulk of the control may be associated with one or a few key steps in the pathway. The task of metabolic control analysis is to measure experimentally just how much control is associated with each step of a pathway. Transgenic plants, particularly those over- or under-expressing an enzyme of interest, have become favoured tools for such investigations (Stitt 1996). Two conclusions of significance from work on photosynthesis are that significant control can be exercised by unexpected enzymes (such as those catalysing near-equilibrium reactions), and that environmental conditions (e.g. irradiance and nitrogen) which prevail during growth can have a profound effect on the distribution of control. Thus Rubisco exercises much more control when irradiance is high or nitrogen supply is low than when nitrogen supply is high or irradiance is low (Stitt 1996).

We are applying the concepts of metabolic control analysis to carbon flow in whole plants. The preliminary approach is to use a method which is analogous to, but simpler than, metabolic control analysis (Gunn & Farrar 1999): it simplifies a great deal, but hopefully not too much. Response coefficients *sensu* Jones and Lynn (1994) have been calculated for plants that have been partially defoliated. The argument is straightforward. If whole-plant growth is controlled solely by the pho-

**Table 2.1** Demonstration of partial sink limitation of growth using response coefficients. The two species were grown hydroponically at the two $CO_2$ concentrations and partially defoliated and their growth measured over the following 2 days. Response coefficients were calculated as the ratio of the proportional reduction in growth caused by partial defoliation/proportional reduction in leaf area by partial defoliation. From Farrar and Gunn (1996).

|                       | D. glomerata | | B. perennis | |
| --------------------- | ---- | ---- | ---- | ---- |
| $CO_2$ concentration  | 350  | 700  | 350  | 700  |
| Response coefficient  | 0.85 | 0.03 | 0.98 | 0.56 |

tosynthetic acquisition of carbon, then a reduction in photosynthesis should be matched by a proportional reduction in growth. The proportionality is calculated as a response coefficient. For two quite different species, it has been shown that growth is reduced less than would be expected when a significant proportion of the leaf area is removed, and the reduction of growth is less at 700 than at 350 p.p.m. $CO_2$ (Table 2.1). Our interpretation is that, before leaf area was removed, the growth of these plants was partially limited by their sinks (more precisely, by processes downstream from the source leaves), and that limitation by sinks was greater at 700 than at 350 p.p.m. $CO_2$. A greater limitation by sinks at elevated $CO_2$ is fully consistent with the regulatory effects of elevated $CO_2$ on photosynthesis described above.

### Additional evidence for distributed control of growth

With the benefit of hindsight, it is possible to reinterpret some existing literature to support the idea of distributed control of carbon flux. First, the transport system probably has a role in control. A long transport pathway reduces fluxes to sinks (Canny 1973; Cook & Evans 1978) and short-distance transport within sinks may present further constraints (Bret-Harte & Silk 1994; Patrick & Offler 1996).

Second, experiments on export of carbon from source leaves also demonstrate that the sinks to which they are attached alter the rate of export. Moorby and Jarman (1976) showed that the rate of export of [14]C from leaves of tomato was greater when the leaves were subtended by a truss of fruits, and greater still when those fruits were warmed. Several experiments since then have confirmed the sensitivity of export to events in sinks. Both these experiments, and those on the transport system, can be interpreted in terms of a model of phloem transport in which the metabolic activities at the ends of the phloem path are represented simply (Minchin et al. 1993). The essence of this model is that control of flux in the phloem is not confined to any single part of the plant, such as the turgor-generating source leaves, but is partly in source leaves, partly in the transport conduit and partly in sinks.

### Control of growth by sources and sinks

A number of experiments have been designed explicitly to address the problem of whether growth is limited by sources or by sinks. If control is indeed distributed around the plant, the very nature of such experiments means they will probably fail to demonstrate that fact. Let us say control is divided approximately equally between source leaves and postphotosynthetic processes. Then an experiment designed to detect, qualitatively, the presence of source limitation should find it; but an experiment designed to test for sink limitation would find it too. These experiments will therefore be most significant if they do not find evidence for limitation by the plant part which they examine, but are then less likely to have been reported. More generally, we must measure the degree of limitation, not simply look for qualitative evidence.

## Effects of environment on plant growth: a reinterpretation

Lastly it is possible to reconcile some effects of the environment on plants most easily if control of growth is distributed. For example, the growth of one strain of *Dactylis glomerata* under a fixed set of conditions can be increased both by increasing the ambient $CO_2$ concentration, and by raising the temperature by 4°C (Table 2.2). This effect is surely commonplace; we just happen to have data for this species. This simple observation can be interpreted as follows. $CO_2$ increases growth because photosynthesis is $CO_2$ limited and is one of the controllers of growth. Temperature increases growth because metabolism within sinks, and in particular cell division, is highly temperature sensitive in this range (Farrar & Gunn 1996; Francis 1998) whereas photosynthesis is not (Farrar 1988), and so an increased temperature relieves some of the sink limitation of growth. Both source and sink must therefore have been limiting growth together before the increases of $CO_2$ and of temperature were applied.

The critical experiment is one in which plants have been grown from seed in a constant set of conditions that ensure a low growth rate. The plants are then divided into groups each of which receives a more favourable regime by altering just one of those conditions (N, P, irradiance, etc.). The prediction is that the plants will respond to each of these by growing faster, because their growth was previously co-limited by all of them. Obviously if any resource were in large excess, then co-limitation by that factor is less likely to occur; the key prediction is that a single resource in low supply will not be the sole limit. We grew barley at low irradiance and nutrients, and 350 p.p.m. $CO_2$, and then transferred some plants to either high irradiance, or 700 p.p.m. $CO_2$, or high nutrients (Table 2.3). Both $CO_2$ and nutri-

**Table 2.2** $CO_2$ and temperature both increase plant growth. *Dactylis glomerata* and *Trifolium repens* were grown in controlled environments at 16 or 20°C and 350 or 700 p.p.m. $CO_2$. Data are expressed as a percentage of the lower temperature or $CO_2$ concentration. S. Gunn, S. Bailey & J.F. Farrar (unpublished).

|  | D. glomerata | T. repens |
| --- | --- | --- |
| $CO_2$ doubled | 86 | 76 |
| Temperature increased by 4°C | 47 | 40 |

**Table 2.3** Limitations to growth of barley seedlings. Plants were grown hydroponically in controlled environment cabinets at low nitrogen, low irradiance and 350 p.p.m. $CO_2$. Batches of plants were exposed to a greater supply of just one of these variables and their growth monitored over 7 days. J.F. Farrar & L. Thurlow (unpublished results).

| Treatment | Growth after applying treatment (% of the control) |
|---|---|
| Irradiance increased | 102 |
| $CO_2$ increased | 128 |
| Nutrients increased | 168 |
| N alone increased | 115 |

ents separately permitted a substantial increase in growth rate over the next 7 days, while nitrate alone and irradiance produced small increases. Although not performed with a switch in treatments, many old experiments can be reinterpreted as consistent with this argument. As one example, Blackman and Templeman (1940) grew *Agrostis tenuis* and *Festuca rubra* at combinations of irradiance and nitrogen supply; a plant was growing more slowly at 0.7 of daylight and low nitrogen than a plant where either nitrogen or irradiance was greater. Thus growth was probably limited by both factors simultaneously.

The picture that emerges, but clearly needs more rigorous experimental testing, is that of a plant which grows as it acquires and allocates carbon, and both the acquisition and allocation of carbon (and indeed all processes which contribute to the flux of carbon) partially control the total flux which constitutes growth. Further, small changes in internal or external conditions can alter the distribution of that control. It is pertinent to ask: is the redistribution of control programmed to ameliorate the effect of an external (or internal) change? It is also of course necessary to ask just how control is redistributed.

**What is the mechanistic basis of the distributed control of growth?**

The way in which control is distributed, and redistributed when conditions change, is not understood. Whilst it is implicit in the nature of linear and branched systems that control is distributed, the way in which it is distributed must reflect the properties of each part of the system. It shall be assumed that it is in some way under control, although it could be an emergent property of a complex system (Cheeseman *et al.* 1996).

First, there are encouraging parallels between the concepts of control analysis and what we know of regulation in the whole plant. Phloem is a network of hydraulically driven tubes, the turgor gradients in different parts of which depend on local metabolism; control of flow in the phloem must therefore be distributed. Sugars control metabolism in source leaves and in roots via their effects on gene expression. Because changes in carbohydrate metabolism in leaves will affect supply to roots, and changes in roots will feedback to leaves, control of carbohydrate metabolism is clearly distributed around the plant.

When the environment alters, the distribution of control changes; several examples have been cited above. The rate of whole-plant photosynthesis is altered in two quite different ways: the density of photosynthetic machinery per unit leaf area changes, and the leaf area per unit of plant weight changes. While the former can alter in both developing and mature leaves, the latter can only change significantly as a result of new growth being differently partitioned. We know that treatments such as elevated $CO_2$ can result in downregulation of the photosynthetic apparatus and, particularly in conditions where sinks may be the main limits to growth, expression of genes encoding photosynthetic proteins such as Rubisco and *cab* is downregulated (see above). Other things being equal, a reduction in the amount of photosynthetic protein will shift control towards photosynthesis and away from sinks (by definition, in terms of metabolic control analysis) and that shift will be of quantitative importance under those conditions where the protein concerned does indeed have a substantial role in control.

## An integrated view of the acquisition, partitioning and loss of carbon?

It has been argued that plants grow such that the control of growth is distributed between their sources and sinks for carbon, and control by environmental factors is similarly shared between a number of factors and is not attributable to one alone. The acquisition of carbon is regulated by the amount of leaf area relative to the total plant weight as well as by photosynthetic capacity per unit area of leaf, and both of these variables are themselves whole-plant properties and so are not defined solely by attributes of the source leaves themselves. Similarly, respiratory loss of carbon is mostly inevitably and tightly coupled to growth and resource acquisition, but these too are whole-plant properties. A *prime facie* case has also been made for resource compounds such as sucrose, nitrate and phosphate being the intermediaries in modulating whole-plant responses. An integrated view of the acquisition, partitioning and loss of carbon would therefore start with sucrose, its production, use and role in signalling.

## Acknowledgements

I would like to thank the BBSRC and NERC for financial support, and Bryan Collis and Sheila Gunn for formative discussions.

## References

Amthor, J.S. (1989). *Respiration and Crop Productivity*. Springer, Berlin.

Amthor, J.S. (1997). Plant respiratory responses to elevated carbon dioxide. In *Advances in Carbon Dioxide Effects Research* (Ed. by L.H. Allen & D.M. Olszyk). ASA Special Publication, Madison, W.I.

Atkin, O.K. & Lambers, H. (1998). Slow-growing alpine and fast-growing lowland species. In *Inherent Variation in Plant Growth* (Ed. by H. Lambers et al.), pp. 259–287. Backhuys, Leiden.

Baker, N.R., Ed. (1996). *Photosynthesis and the Environment*. Kluwer, Dordrecht.

Baxter, R., Ashenden, T.W. & Farrar, J.F. (1997).

Effect of elevated $CO_2$ and nutrient status on growth, dry matter partitioning and nutrient content of *Poa alpina*. *Journal of Experimental Botany*, **48**, 1477–1486.

Beevers, H. (1961). *Respiratory Metabolism in Plants*. Row Peterson, Evanston.

Bingham, I.J. & Farrar, J.F. (1988). Regulation of respiration in roots of barley. *Physiologia Plantarum*, **70**, 491–498.

Blackman, G.E. & Templeman, W.G. (1940). The interaction of light intensity and nitrogen in the growth and metabolism of grasses and clover. *Annals of Botany*, **4**, 533–588.

Bret-Harte, M.S. & Silk, W. (1994). Non-vascular, symplasmic diffusion of sucrose cannot satisfy the carbon demands of growth in the primary root tip of *Zea mays*. *Plant Physiology*, **105**, 19–32.

Brouwer, H. (1962). Nutritive influences on the distribution of dry matter in the plant. *Netherlands Journal of Agricultural Science*, **10**, 399–408.

Brouwer, H. (1983). Functional equilibrium: sense or nonsense? *Netherlands Journal of Agricultural Science*, **31**, 335–348.

Canny, M.J. (1973). *Phloem Translocation*. Cambridge University Press, Cambridge.

Cheeseman, J.M., Barreiro, R. & Lexa, M. (1996). Plant growth modelling and the integration of shoot and root activities without communicating messengers: opinion. *Plant and Soil*, **185**, 51–64.

Clarkson, D. (1998). Mechanisms for N-uptake and their running costs: is there scope for more efficiency? In *Inherent Variation in Plant Growth* (Ed. by H. Lambers *et al.*), pp. 221–236. Backhuys, Leiden.

Collis, B.E., Plum, S.A., Farrar, J.F. & Pollock, C.J. (1996). Root growth of barley at elevated $CO_2$. *Aspects of Applied Biology*, **45**, 181–185.

Cook, M.G. & Evans, L.T. (1978). Effect of relative sink size and distance of competing sinks on the distribution of assimilates in wheat. *Australian Journal of Plant Physiology*, **5**, 495–509.

Cornelissen, J.H.C., Castro-Diaz, P. & Carnelli, A.L. (1998). Variation in relative growth rate among woody species. In *Inherent Variation in Plant Growth* (Ed. by H. Lambers *et al.*), pp. 363–392. Backhuys, Leiden.

Drake, B.G., Gonzalez-Meler, M.A. & Long, S.P. (1997). More efficient plants: a consequence of rising atmospheric $CO_2$? *Annual Review of Plant Physiology and Plant Molecular Biology*, **48**, 609–639.

Drake, B.G., Berry, J., Bunce, J. *et al.* (1999). Does elevated atmospheric $CO_2$ inhibit mitochondrial respiration in green plants? *Plant, Cell and Environment*, **22**, 649–658.

Evans, J.R. (1996). Developmental constraints on photosynthesis. Effects of light and nitrogen. In *Photosynthesis and the Environment* (Ed. by N.R. Baker), pp. 281–304. Kluwer, Dordrecht.

Farrar, J.F. (1985). Fluxes of carbon in roots of barley plants. *New Phytologist*, **99**, 57–69.

Farrar, J.F. (1988). Temperature and the partitioning and translocation of carbon. In *Plants and Temperature* (Ed. by S.P. Long & F.I. Woodward), pp. 203–235. Company of Biologists, Cambridge.

Farrar, J.F. (1989). Fluxes and turnover of sucrose and fructans in healthy and diseased plants. *Journal of Plant Physiology*, **134**, 137–140.

Farrar, J.F. (1992). The whole plant: carbon partitioning during development. In *Carbon Partitioning Within and Between Organisms* (Ed. by C.J. Pollock, F.F. Farrar & A.J. Gordon), pp. 163–179. Bios Scientific, Oxford.

Farrar, J.F. (1996). Regulation of root weight ratio is mediated by sucrose: opinion. *Plant and Soil*, **185**, 13–19.

Farrar, J.F. (1999). Carbohydrate: where does it come from, where does it go?. In *Plant Carbohydrate Biochemistry* (Ed. by N. Kruger *et al.*) Bios Scientific, Oxford, pp. 17–46.

Farrar, J.F.& Gunn, S. (1996). Effects of temperature and atmospheric carbon dioxide concentration on source sink relations in the context of climate change. In *Photoassimilate Distribution in Plants and Crops* (Ed. E. Zamske & A.A. Sheffer), pp. 389–406. Dekker, New York.

Farrar, J.F., & Gunn, S. (1998). Allocation: allometry, acclimation—and alchemy? In *Inherent Variation in Plant Growth* (Ed. by H. Lambers *et al.*), pp. 183–198. Backhuys, Leiden.

Farrar, J.F. & Jones, C.L. (1986). Modification of respiration and carbohydrate status of barley roots by selective pruning. *New Phytologist*, **102**, 513–521.

Farrar, J.F. & Williams, J.H.H. (1991). Control of the rate of respiration in roots: compartmentation, demand and the supply of substrate. In

*Compartmentation of Plant Metabolism in Non-photosynthetic Tissues* (Ed. by M. Emes), pp. 167–188. Cambridge University Press, Cambridge.

Farrar, J.F., Minchin, P.E.H. & Thorpe, M.R. (1994). Carbon import into barley roots: stimulation by galactose. *Journal of Experimental Botany*, **45**, 17–22.

Fell, D. (1997). *Understanding the Control of Metabolism*. Portland Press, London.

Francis, D. (1998). The cell cycle and plant growth. In *Inherent Variation in Plant Growth* (Ed. by H. Lambers *et al.*), pp. 5–20. Backhuys, Leiden.

Graham, I.A., Denby, K.J. & Leaver, C.J. (1994). Carbon catabolite repression regulates glyoxylate cycle gene expression in cucumber. *Plant Cell*, **6**, 761–772.

Gunn, S. & Farrar, J.F. (1999). Partitioning of dry weight and leaf area within plants of three species grown at elevated $CO_2$. *Functional Ecology*, **13**, in press.

Hibberd, J.M., Richardson, P., Whitbread, R. & Farrar, J.F. (1996a). Effects of leaf age, basal meristem, and infection with powdery mildew on photosynthesis in barley grown in 700 μmol mol$^{-1}$ $CO_2$. *New Phytologist*, **134**, 317–325.

Hibberd, J.M., Whitbread, R. & Farrar, J.F. (1996b). Carbohydrate metabolism in source leaves of barley grown in 700 μl l$^{-1}$ $CO_2$ and infected with powdery mildew. *New Phytologist*, **133**, 659–671.

Jang, J.C. & Sheen, J. (1997). Sugar sensing in higher plants. *Trends in Plant Science*, **2**, 208–214.

Jones, H.G. & Lynn, J. (1994). Optimal allocation of assimilate in relation to resource limitation. In *Resource Capture by Crops* (Ed. by J.L. Monteith, R.K. Scott & M.H. Unsworth), pp. 234–256. Nottingham University Press, Nottingham.

Koch, K.E. (1996). Carbohydrate-modulated gene expression in plants. *Annual Review of Plant Physiology and Plant Molecular Biology*, **47**, 509–540.

Koch, K.E. (1997). Molecular cross-talk and the regulation of C- and N-responsive genes. In *A Molecular Approach to Primary Metabolism in Higher Plants* (Ed. by C.H. Foyer & W.P. Quick), pp. 105–124. Taylor and Francis, London.

Koroleva, O.A., Farrar, J.F., Tomos, A.D. & Pollock, C.J. (1997). Patterns of solute in individual mesophyll, bundle sheath and epidermal cells of barley leaves induced to accumulate carbohydrate. *New Phytologist*, **136**, 97–104.

Koroleva, O.A., Farrar, J.F., Tomos, A.D. & Pollock, C.J. (1998). Carbohydrates in individual cells of epidermis, mesophyll and bundle sheath in barley leaves with changed export or photosynthetic rate. *Plant Physiology*, **118**, 1525–1532.

Kraus, E., Wilson, D., Robson, M.J. & Pilbeam, C.J. (1990). Respiration: correlation with growth rate, and its quantitative significance for the net assimilation rate and biomass production. In *Causes and Consequences of Variation in Growth Rate and Productivity in Higher Plants* (Ed. by H. Lambers), pp. 187–198. SPB, The Hague.

Lambers, H. (1985). Respiration in intact plants and tissues. In *Encyclopedia of Plant Physiology*, 18, (Ed. by R. Douce & D.A. Day), pp. 418–473. Springer, Berlin.

Larsson, C.M. (1994). Responses of the nitrate uptake system to external nitrate availability. In *A Whole-Plant Perspective on Carbon–Nitrogen Interactions* (Ed. by J. Roy & E. Garnier), pp. 31–46. SPB, The Hague.

Lawlor, D.W. (1993). *Photosynthesis*, 2nd edn. Longman, Harlow.

Minchin, P.E.H., Thorpe, M.R. & Farrar, J.F. (1993). A simple mechanistic model of phloem transport which explains sink priority. *Journal of Experimental Botany*, **44**, 947–955.

Moorby, J. & Jarman, P.D. (1976). The use of compartmental analysis in the study of the movement of carbon through leaves. *Planta*, **122**, 155–168.

Patrick, J.W. & Offler, C.E. (1996). Post sieve-element transport of photoassimilate in sink regions. *Journal of Experimental Botany*, **47**, 1165–1177.

Paul, M. & Driscoll, P. (1997). Sugar repression of photosynthesis: the role of carbohydrates in signalling N deficiency through source: sink imbalance. *Plant, Cell and Environment*, **20**, 110–116.

Penning de Vries, F.W.T. (1972). Respiration and growth. In *Crop Processes in Controlled Environments* (Ed. by A.R. Rees), pp. 327–347. Academic, London.

Plum, S.A., Farrar, J.F. & Stirling, C. (1996). Carbon partitioning in barley following manipulation of source and sink. *Aspects of Applied Biology*, **45**, 177–180.

Pollock, C.J. & Farrar, J.F. (1996). Source-sink relations: the role of sucrose. In *Photosynthesis and the Environment* (Ed. by N. Baker), pp. 261–279. Kluwer, Dordrecht.

Poorter, H. & Garnier, E. (1996). Plant growth analysis: an evaluation of experimental design and computational methods. *Journal of Experimental Botany*, **47**, 1343–1351.

Poorter, H. & van der Werf, A. (1998). Is inherent variation in RGR determined by LAR at low irradiance and by NAR at high irradiance? A review of herbaceous species. In *Inherent Variation in Plant Growth* (Ed. by H. Lambers), pp. 309–336. Backhuys, Leiden.

Poorter, H., Gifford, R.M., Kriedmann, P.E. & Wong, S.C. (1992) A qualitative analysis of dark respiration and carbon content as factors in the growth response of plants to elevated $CO_2$. *Australian Journal of Botany*, **40**, 501–513.

Scheible, W.R., Lauerer, M., Schulze, E.D., Caboche, M. & Stitt, M. (1997). Accumulation of nitrate in the shoot acts as a signal to regulate shoot: root allocation in tobacco. *Plant Journal*, **11**, 671–691.

Semikhatova, O.A., Gerasimenko, T.V. & Ivanova, T.I. (1992). Photosynthesis, respiration and growth of plants in the Soviet arctic. In *Arctic Ecosystems in a Changing Climate* (Ed. by F.S. Chapin), pp. 169–192. Academic, New York.

Smeekens, S. (1998). Sugar regulation of gene expression in plants. *Current Opinion in Plant Biology*, **1**, 230–234.

Smeekens, S. & Rook, F. (1997). Sugar sensing and sugar-mediated signal transduction in plants. *Plant Physiology*, **115**, 7–13.

Stitt, M. (1991). Rising $CO_2$ levels and their potential significance for carbon flow in photosynthetic cells. *Plant, Cell and Environment*, **14**, 741–762.

Stitt, M. (1996). Metabolic regulation of photosynthesis. In *Photosynthesis and the Environment* (Ed. by N.R. Baker), pp. 151–190. Kluwer, Dordrecht.

Stitt, M. (1999). The first will be last and the last will be first: non-regulated enzymes call the tune. In *Plant Carbohydrate Biochemistry* (Ed. by N. Kruger *et al.*). Bios Scientific, Oxford, in press.

Stulen, I., Lambers, H. & van der Werf, A. (1996). Carbon use in root respiration as affected by elevated $CO_2$. *Plant and Soil*, **187**, 251–263.

Thornley (1976). *Mathematical Models in Plant Physiology*. Academic, London.

Touraine, B., Clarkson, D.T. & Muller, B. (1994). Regulation of nitrate uptake at the whole plant level. In *A Whole Plant Perspective on Carbon–Nitrogen Interactions* (Ed. by J. Roy & E. Garnier), pp. 11–30. SPB, The Hague.

Vanlerberghe, G.C. & McIntosh, L. (1992). Lower growth temperature increases alternative oxidase protein in tobacco. *Plant Physiology*, **100**, 115–119.

Volenec, J.J., Nelson, C.J. & Sleper, D.A. (1984). Influence of temperature on leaf dark respiration of tall fescue genotypes. *Crop Science*, **24**, 907–912.

van der Werf, A. & Lambers, H. (1996). Editorial: special issue on biomass partitioning to leaves and roots. *Plant and Soil*, **185**, ix–x.

Wilson, J.B. (1988). A review of evidence on the control of shoot: root ratio, in relation to models. *Annals of Botany*, **61**, 433–449.

Zhang, H. & Forde, B.G. (1998). An Arabidopsis MADS box gene that controls nutrient-induced changes in root architecture. *Science*, **279**, 407–409.

# Chapter 3

# Resource acquisition by plants: the role of crown architecture

*R.W. Pearcy[1] and F. Valladares[2]*

## Introduction

Physiological ecology has traditionally focused on organs such as leaves as units of function, in part because of their convenient scale of measurement, but obviously also because of their fundamental role in resource acquisition. Indeed, progress made in understanding gas-exchange responses of leaves to different environments and the linkage of these responses to underlying biochemical and stomatal mechanisms has been remarkable. At the same time, there has been recognition that understanding controls on leaf photosynthetic rate alone is rarely, if ever, explanatory of the ecological responses of species (e.g. Zipperlen & Press 1996). Although growth and net whole-plant carbon gain are inexorably linked, crop and whole-plant physiologists have often found no correlation or even a negative relationship between maximum leaf photosynthetic capacity ($A_{max}$) and crop yields (Gifford & Evans 1981). In wildland plants, the correlation between $A_{max}$ and relative growth rate (RGR) is also rather poor (Pereira 1994) and comparisons between fast- and slow-growing species reveal that leaf area ratio (LAR; leaf area/plant mass) is a better predictor of interspecific variation in RGR than is $A_{max}$ (Poorter *et al.* 1990; Kitajima 1994). However, these studies of RGR, of necessity, have focused on small plants growing at their maximum rate and with minimal self-shading. As plants increase in size, RGR decreases because changes in allocation reduce LAR. In addition, the increasing self-shading within the canopy reduces the mean realised photosynthetic rate. Self-shading creates an extremely heterogeneous light environment that influences the heterogeneity of leaf photosynthesis within the crown and the acclimation of these leaves to the contrasting light environments. Understanding how crown architecture influences this heterogeneity and the responses to it has been difficult because of the large number of interdependent characters that are involved in shaping it. Moreover, until recently, quantitative tools for evaluating the consequences of a particular architecture have not been available.

[1] *Section of Evolution and Ecology, Division of Biological Sciences, University of California, Davis, CA 95616, USA. E-mail: rwpearcy@ucdavis.edu*
[2] *Centro de Ciencias Medioambientales, CSIC, Serrano 115 dpdo, 28006 Madrid, Spain*

The role of canopy architecture in plant growth can be seen by examining two classic equations. The first (Evans 1972):

$$RGR = ULR \cdot LAR \qquad (1)$$

reveals that RGR is equal to the product of the unit leaf rate (ULR) and LAR. ULR is functionally equivalent to the whole-plant net photosynthetic rate per unit leaf area, including the effects of leaf display, while LAR on the other hand is a measure of the allocation to leaf area and hence the potential capacity for light capture of the plant. Architecture influences both ULR and LAR, the former through effects on the heterogeneity of the crown light environment and the latter through the biomass required for support and display. The second equation (Monteith 1977; Russell *et al.* 1989) is:

$$W = \int \varepsilon i Q \, dt, \qquad (2)$$

where W is the biomass accumulation, Q is the radiation incident on the top of the canopy, $\varepsilon$ is the ratio of dry matter produced to intercepted radiation, $i$ is the fraction of radiation intercepted and $t$ is time. Canopy structure is the primary determinant of $i$ but it also plays an important role in $\varepsilon$ because of the non-linear response of photosynthetic rate to light. These equations reveal the essential interdependence of photosynthetic properties of leaves and canopy structure in determining growth of plants.

This chapter will first consider how the multiple functions of, and the constraints imposed on, the form of plants influence crown architecture as an adaptation. It will then review some of the new models available for quantifying the role of crown architecture in the light interception and photosynthetic carbon gain of plants. Finally, the applications of a new simulation model to understanding light capture in deeply shaded forest understories and in habitats where light is present in excess will be presented.

## The adaptive biology of crown architecture

The function of plant crowns for light interception and photosynthesis has been approached from several different perspectives, including the scaling of photosynthesis from leaves to whole plants or stands and the morphological basis of light interception. Scaling of photosynthesis from leaves to individuals or stands requires knowledge of the heterogeneity of light environments created by the leaf angle distribution and self-shading patterns within the crown and then a coupling of this heterogeneity to the photosynthetic light response. Estimates of the distribution of light on the surfaces of leaves can be made by applying statistical models of light penetration following from the classical model of Monsi and Saeki (1953). Progress in understanding crown function of individual plants has depended on adapting this theory from plant stands to crowns of different shapes (Norman & Welles 1983; Wang & Jarvis 1990) or on the development of geometric models that array foliage in space according to branching patterns and shoot architecture.

Ecophysiological studies of the morphological basis of light interception and its consequences for photosynthesis have primarily focused on the orientation of leaves or cladodes as individual structures in isolation from other leaves (Ehleringer & Forseth 1980; Nobel 1980; Smith & Ullberg 1989). By contrast, ecophysiologists have given relatively little attention to the role of shoot and canopy structure in creating aggregations of leaves. In these aggregations, temporal and spatial patterns of self-shading become additional factors influencing the light environment. The classic study of Horn (1971) addressed how foliage should be arranged in layers of different photosynthetic properties in order to maximize photosynthetic carbon gain. However most studies of the role of crown structure and light interception have been undertaken from a functional morphology perspective to examine how specific characters such as branching angles or shoot phyllotaxies, influence radiation interception. These studies have been important in demonstrating that within a given architecture certain character values, such as the branching angles, can maximize the displayed leaf area (Honda & Fisher 1978). Early studies examined how the two-dimensional branching structure influenced displayed leaf area (Honda & Fisher 1978; Fisher 1986), but the models used apply when the light is received orthogonal to the horizontal plane of the model branching, and then only when the architecture can reasonably be represented in two dimensions. Studies of shoot structure carried out with relatively simple three-dimensional models incorporating more realistic solar radiation distributions reveal that the influence of a morphological character is interdependent on the other morphological characters specifying the architectural plan (Niklas 1988; Takenaka 1994). For example, in simulated shoot architectures with short internodes, light interception was strongly dependent on the phyllotactic spiral (Niklas 1988). However, lengthening the internodes, which allows better utilization of lateral light, could effectively compensate for inefficient spirals. These kinds of complex interactions between characters mean that quite different character combinations can result in functionally equivalent architectures in terms of light interception.

When viewed from an ecophysiological perspective, studies of the morphological basis of light interception by shoots such as those mentioned above have two limitations. First, most have focused on light interception capacity or efficiency rather than its consequences for photosynthesis, water loss or other functions. Second, most have focused on the architectural arrangement of the leaves in space, with minimal attention given to the costs of achieving this display. While light interception is the attribute most directly related to leaf display, it is important to emphasize that the function of this leaf display is photosynthesis. Thus, evaluation of the performance of a particular architecture is better made on the basis of its photosynthetic consequences rather than light interception. Because photosynthesis exhibits a non-linear, saturating response to light, maximizing light interception and maximizing photosynthesis are not necessarily the same, especially when potential reductions resulting from photoinhibition are considered or when reduced water-use efficiency may constrain long-term carbon gain. The architecture that maximizes carbon gain will therefore depend on the photosynthetic gas-exchange characteristics of the leaves. It follows then that the optimal architecture will be in part defined by these characteristics. Moreover, photosynthetic capacity is correlated with morphological factors such as the specific leaf area and leaf longevity, which in turn influence leaf display. Consequently, even at the morphological level it is difficult to divorce the photosynthetic characteristics of the leaves from their consequences for resource allocation and ultimately crown architecture.

In studies of adaptation, understanding the value of a particular character in a given environment is frequently undertaken by examining its costs and benefits. For crown architecture, this problem involves how resources can be utilized conservatively in order to maximize the return on investment, measured usually in terms of carbon balance. Natural selection should favour those architectures that maximize the returns through maximizing the ratio of benefits to costs. Optimality models incorporating this cost–benefit framework have frequently been used to establish a reference point against which observations in nature or experiment can be compared. The use of optimality as a conceptual framework for investigating adaptation has been criticized on a number of grounds (Gould & Lewontin 1979), but despite these criticisms it remains a major paradigm in evolutionary biology (Parker & Maynard-Smith 1990). The criticisms include the tendency to consider individual characters in isolation rather than viewing the organism as an integrated suite of characters. This criticism is particularly cogent for studies of optimality of crown architecture, because so many characters are involved in the architectural plan of a species. Among the array of possible character combinations, some will clearly be better, i.e. closer to optimal at this investment than others, but there may be several with equivalent optimal performances under the environment specified. Thus, the idea that a single 'best' architecture exists for a given environment is too simplistic and indeed not supported by the observed diversity of architectures present in most habitats. Nevertheless, we could expect some designs to emerge as better than others, and given the background of other characters, the investment in a particular character should be consistent with an optimal design. Testing this possibility requires methodology that allows for simultaneous variation in multiple characters, so that the 'best' among the possible phenotypes can be identified. Best in this case does not mean that the design actually has to be optimal, only that it be better than the alternatives.

Another factor influencing the application of optimality arguments to crown architecture is that the crowns of plants have multiple functions and design constraints (Fig. 3.1), therefore greatly increasing the complexity of the optimization problem (Farnsworth & Niklas 1995). Simulations of adaptive walks of crown design in fitness landscapes for simple forms similar to primitive plants (Fig. 3.2) revealed that as more functions are considered, more phenotypes representing local optima or maxima emerge (Niklas 1994). However, the apparent fitness value of these compromises between multiple functions was lower compared with phenotypes incorporating fewer functions. Architectural designs in a given environ-

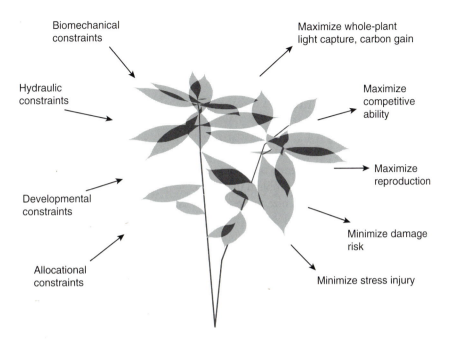

Biomechanical constraints

Maximize whole-plant light capture, carbon gain

Hydraulic constraints

Maximize competitive ability

Maximize reproduction

Developmental constraints

Minimize damage risk

Allocational constraints

Minimize stress injury

**Figure 3.1** Constraints (left side) and functions (right side) that influence crown architecture and determine the nature of the compromises involved in the evolutionary optimization of crown form.

ment should reflect different weightings of these functions, subject to the constraints involved in meeting the minimal design requirement of any single function. Thus, as the environment changes, there should be shifts in the composition of adaptive peaks on the landscape. The optimal design will also be constrained by a need for a sufficiency of allocation to provide adequate biomechanical support and hydraulic conductance. Because hydraulic and biomechanical design involves mostly supporting structures, allocation to these design constraints will restrict carbon allocation to leaves. Not all functions are strictly competing, however, because investment in hydraulic capacity in the stems may also increase biomechanical strength. Farnsworth and Van Gardingen (1995) found that the allometric relations among links in sitka spruce branches better fitted a minimum allocation based on biomechanical support rather than hydraulic requirements but, as both functions are correlated, an unequivocal conclusion about the limiting design requirement was not possible. Moreover, the possibility that the allocation followed some other criteria such as potential light interception could not be ruled out.

Substantial progress has been made in understanding various aspects of crown functions, such as the hydraulic architecture (Tyree & Ewers 1991; Zotz *et al.* 1998), biomechanics (Givnish 1986; Morgan & Cannell 1988; Niklas 1992) and competitive space exploitation (Küppers 1989). Although trade-offs among

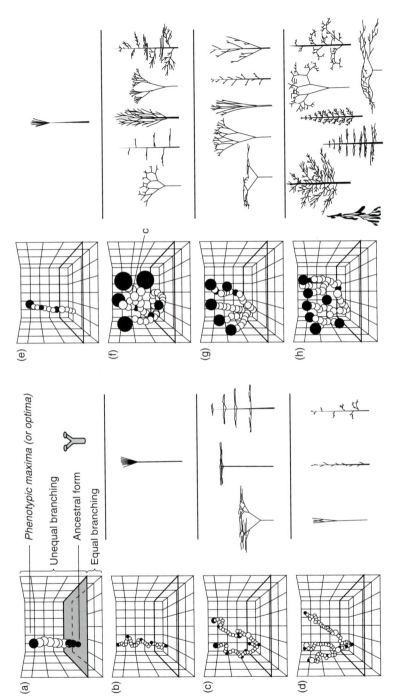

**Figure 3.2** Simulated adaptive walks through fitness landscapes for branching patterns involving selection for optimizing either single (b–d) or multiple tasks (e–h). The simulations were seeded with the simple shape resembling a Cooksonia-type organism (a) and then allowed to move in unfettered steps represented by spheres towards phenotypic maxima or optima, represented by the filled spheres. The branching architectures to the right depict the morphologies at the phenotypic maximas for each walk. The different landscapes are: (b) a reproductive success landscape; (c) a light interception landscape; (d) a mechanical stability landscape; (e) a mechanical stability and reproductive success landscape; (f) a light interception and mechanical stability landscape; (g) a light interception and reproductive success landscape; (h) a light interception, mechanical stability and reproductive success landscape. Adapted from Niklas (1994), figs 3 and 4.

these functions are sometimes discussed in general terms, the tendency has been to quantify only one function and then discuss trade-offs only in qualitative terms, if at all. Future research will need to examine these trade-offs in more quantitative detail.

## Models for simulating crown geometry and function

Because of the complex three-dimensional nature of crown geometry, mathematical simulation models are essential for teasing out important structural features that determine light interception and photosynthesis. Fortunately, the geometric principles are relatively straightforward, at least at a first approximation, so the principal limitation to modelling until recently has been insufficient computational power readily available to researchers. The dramatic increase in computational power in the past few years has now made it possible for these models to be available to ecophysiologists on their desktop microcomputers. Mechanistic leaf photosynthetic models are now quite complete and these can be coupled to architecture models. Coupling of leaf photosynthetic and crown architecture models allows exploration of the interactions between geometric placement of leaves, which determines light interception and self-shading, and leaf physiological properties. In addition, it allows a scaling of these properties up to whole-plant photosynthetic performance.

Models of crown architecture have utilized several different approaches depending on the overall objectives and the nature of the plants being simulated. The approaches can be categorized according to whether they are statistical or deterministic in nature with respect to the way that light capture is simulated. In addition, deterministic models differ in whether the architecture is elaborated by applying developmental rules to some starting point, such as a seedling, or is based on the geometric structure as derived from geometric relationships between morphological elements. The latter can be applied to simulate hypothetical shoot structures of defined characteristics, or if based on measurements, derived from actual plants to simulate light capture from these plants.

### Statistical models

Statistical approaches are based on the classical theory of Monsi and Saeki (Monsi & Saeki 1953; Monsi et al. 1973) that is now utilized in many crop models. The basic equation is:

$$P = \exp(-G\rho d) \tag{3}$$

which shows that the probability of beam penetration (P) decreases exponentially as the projected leaf area in the direction of the beam (G) increases. Both pathlength ($d$) and leaf area density ($\rho$) depend on beam direction and leaf angles. The theory assumes a random dispersion of foliage elements within the layer, horizontal homogeneity and independence between layers. Foliage dispersion is usually not random but the resulting errors are fairly small. Clumping of foliage around branches causes the model to overestimate the true radiation interception. Modifi-cations to account for factors such as non-random dispersions, upward reflection and scattering have been made in many models (Myneni *et al.* 1989; Baldocchi & Collineau 1994), but are beyond the scope of this discussion. Although developed for horizontally extensive crop canopies, statistical models can be applied to indi-vidual crowns if the shape can be approximated or measured. This can be as simple as a geometric shape such as a cone for a conifer crown (Kuuluvainen & Pukkala 1989) or it can be a much more complex mapping of foliage distributions (White-head *et al.* 1990). Roberts and Miller (1977) modelled crowns of Chilean mediter-ranean-climate shrubs by partitioning the canopy space into a cubic array of 64 cells and then determining the foliage content and geometry in each cell. The prob-ability of a beam that enters a cell also exits the cell (and hence enters a contiguous cell) is then calculated as above. Measurements showed a reasonably good agree-ment between measured and calculated downward radiation fluxes within the crown, supporting the general validity of the approach. Ryel *et al.* (1993) used con-centric subcylinders and layers to define cells of different foliage densities and char-acteristics for bunchgrasses and found good agreement between measured and simulated whole-plant photosynthesis (Fig. 3.3). Statistical models require a rela-tively large number of measurements to specify the cell properties for all the cells. Moreover, the foliage must fit the assumptions; that is, that the foliage elements must be relatively small and numerous within a cell.

### Deterministic models

Deterministic models recreate the actual geometric structure of the canopy, either by repeated application of developmental rules to an initial structure such as a

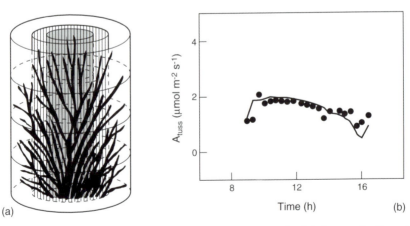

(a)    (b)

Time (h)

**Figure 3.3** (a) A schematic depiction of a grass tussock in the cylindrical array of cells used to define foliage properties in the model of Ryel *et al.* (1993). (b) Measured and simulated gas exchange of a tussock showing good agreement between model and measurement (Ryel *et al.* 1993).

seedling or from the geometry of canopy elements. Models based on developmental rules are dynamic, whereas those based on geometry alone are static and represent the architecture at only a single point in time. Developmentally based, three-dimensional models include the L-system models of Lindenmayer (Lindenmayer 1975; Prusinkiewicz *et al.* 1994) and the model AMAP developed by de Reffye (de Reffye 1988; de Reffye & Houllier 1997). Impressive, visually realistic simulations of large trees are possible with these types of models. Kurth (1994) and Room *et al.* (1996) have reviewed the properties and potentials of developmental rule-based models. The impetus for these models has been to understand how reiteration of a unit of morphogenesis, the metamer, defines the form of the plant, and indeed most have been developed as a result of collaborations between morphologists and computer scientists. By contrast, relatively little attention has been given to the functional aspects of the resulting crown form, although this may be changing (Room *et al.* 1996). For example, Mech and Prusinkiewicz (1996) have coupled L-system models to light absorption to simulate foraging by clonal plants and developmental responses of trees to competition. Models based on fractal geometry of individual tree forms as constrained by developmental rules have been used to simulate stand light interception and photosynthesis (Myneni 1991; Chen *et al.* 1994). The major limitation to use of developmental rule-based models is the specification for a species of the proper rules, especially those rules for plastic responses to the environment. Examples are the rules that prevent branches from growing into one another or that specify the differences between sun and shade branches. Progress in the application of these models to understanding the ecophysiology of crown form will also require a better coupling to models of resource capture and photosynthesis, as well as allocation and water transport.

Static geometric models recreate the crown structure at a given point in time and do not consider the developmental pathway that led to this point. Thus, they are not dynamic but they do have the potential to be applied sequentially to a developing shoot to evaluate whether an increment in growth is invested efficiently. They provide a picture in time of the architecture, which has the advantage of enabling a focus on the current structure and its relationship to physiological performance. It must also be remembered, however, that the current architecture also reflects the cumulative past history, including effects of damage or changes in the environment.

Several static geometric models have been developed having different objectives and capabilities. Niklas (1988) and Takenaka (1994) developed three-dimensional models for examining the roles of leaf shape, internode length and phyllotaxy for light interception. Both reconstruct two-dimensional images for different directions from the shoot geometry and then determine from the projected area the interception for light normal to the plane. Niklas simulated only interception of direct solar radiation interception, while Takenaka simulated only diffuse light interception. Although the geometries of the simulated shoots were rather simple and more stylistic than realistic, they did show that interactions between morphological parameters nevertheless resulted in complex variations in light interception. The effect of any one parameter clearly depended on the morphological context set by other characters and on the light environment.

Pearcy and Yang (1996) reported a model (Yplant) based on principles similar to those underlying the models of Niklas and Takenaka but capable of handling complex branched architectures and linked to leaf physiological simulations. Measurements from plants in the field provide the input for the crown geometry. Yplant simulates absorption of photon flux density (PFD) at any time, based on the crown geometry and the geometry and flux of direct and diffuse light as quantified from hemispherical photographs (Rich 1990) and standard equations (List 1971). Absorbed PFDs of the sunlit and shaded portions of each leaf are then used for simulation of their photosynthetic rates, which can then be summed over all leaves. Yplant has recently been revised and improved to run under Microsoft Windows 95/NT and to include energy balance simulations for each leaf, which then gives the transpiration rates and leaf temperatures. We have named this new version Y-plant to distinguish it from the earlier model.

The 'best' type of model depends on the particular application and the species being studied. Models based on developmental rules provide a path to the kind of dynamic optimization envisioned by Farnsworth and Niklas (1995), but do not yet have enough functional orientation to make them useful in ecophysiological studies. However, it is likely that this situation will change. Deterministic models such as Y-plant have been specifically developed for structure/function studies and are particularly useful for exploring how variation in the morphometry of the crown influences light capture and carbon gain. They are, however, limited to relatively small plants with about 500–1000 leaves as the practical upper limit. A partial way around this limitation for larger plants is to focus on branch architecture and to simulate the effects of shading by adjacent branches by grouping several around the target branch. Statistical models are useful for larger plants or plants with small leaves and dense foliage. These conditions make detailed geometric measurements difficult but they also better satisfy the assumptions of statistical models regarding foliage distributions. Statistical models are less suitable for studies of how variation in traits influences performance, because they are lumped into only a few parameters that are constant for a cell. A potentially promising avenue for further development is a nesting of deterministic and statistical approaches. Cell parameters in a statistically based model could be derived from deterministic models of shoot structure.

## Light capture and carbon gain in shaded environments

### Crown design

Selection for performance in the strongly limiting light conditions in shaded forest understories could be expected to place a high priority on maximizing light capture. Indeed, shade-tolerant understorey shrubs are noted for their plagiotropic branching and effectively monolayer crowns. Similarly, shade-tolerant herbs are often characterized by a few large leaves displayed in a way that minimizes self-shading. Although plagiotropy reduces self-shading along shoots, it also creates greater static loads than orthotropic shoots of the same mass, making plagiotropic shoots biomechanically less efficient (Morgan & Cannell 1988). Thus, the trade-off involves deploying the maximum leaf area with minimum self-shading while providing sufficient biomechanical support. Because damage by falling debris is common in understories, biomechanical sufficiency requires a margin of safety (Chazdon 1986) that must take into account risk *vs.* reward of survival allowing future reproductive output (Lerdau 1992).

Although the qualitative aspects of plant design for light capture in shaded environments are well known (Pickett & Kempf 1980), there have been relatively few quantitative examinations of the trade-offs involved. Givnish (1982) used a game-theoretic model involving trade-offs between investments in stem following biomechanical principles *vs.* allocation to leaf area to predict the optimal leaf height for forest herbs. The trade-offs suggest an evolutionary game, where greater stem investment that raises the leaves just above the competitors leads to more light capture and higher photosynthetic rates per unit area but less leaf area. There was a good general agreement between predicted and observed height along gradients of understorey herb density. Pearcy and Yang (1998) examined the morphometry of shoot architecture of the redwood-forest understorey herb *Adenocaulon bicolor*, using Y-plant simulations based on geometric measurements from plants in the field. This species is a rosette former and concurrent changes in leaf size, petiole length and petiole angle along the shoot axis act to minimize self-shading. Because of these morphometric changes, spaces between the larger, lower leaves were efficiently filled with smaller, upper leaves, resulting in only about 5% of the leaf area being self-shaded from any direction (Fig. 3.4). Through simulations in which the petiole length was varied from the values measured in the plants in the field, it was demonstrated that the efficiency of light absorption depended strongly on petiole length (Fig. 3.5a). Of course, long petioles that minimize self-shading are costly, so an optimal petiole length that balances the trade-off between the costs of the petiole *vs.* leaf area should be observed. Indeed, when this trade-off was simulated by reallocating biomass between leaf lamina and petiole, so that the total plant biomass remained constant, an optimum was found that closely matched the observed petiole lengths on the actual plants (Fig. 3.5b). Shorter petioles allowed for more leaf area but even more of it was self-shaded, resulting in smaller daily carbon gains. With longer petioles, the opposite trade-off occurred. By contrast, lengthening the internodes had, if anything, a slightly negative impact on light capture and carbon gain. Generally, longer internodes minimize within-shoot self-shading (Niklas 1988; Takenaka 1994) but the centripital leaf arrangement in *Adenocaulon* already minimized it. Moreover, longer internodes primarily enhance the use of lateral light but, in the understorey, most diffuse light originates near the zenith. Because *Adenocaulon* occurs in rather sparsely vegetated understories, height for competitive advantage as discussed by Givnish (1982) is unnecessary.

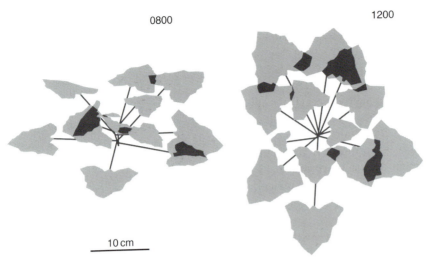

**Figure 3.4** Images created by Y-plant simulations of the crown architecture of an *Adenocaulon bicolor* plant as viewed at 08.00 and 12.00 hours from the vector of the solar beam.

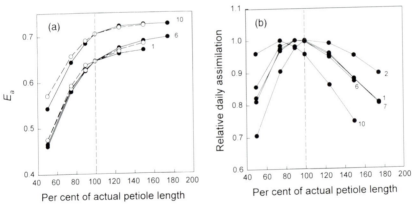

**Figure 3.5** Results of simulations showing the effect of varying the petiole lengths of *Adenocaulon bicolor* plants on light absorption efficiency ($Ea$) (a) and whole-plant daily assimilation (b). In (a) the filled symbols show simulations in which only petiole length was varied while the open symbols show simulations in which the accompanying change in biomass of the petioles was either added to or subtracted from the leaf biomass, causing the leaf sizes to be larger or smaller, respectively. In (b) the results of simulations in which petiole length and leaf area were varied simultaneously to maintain a constant mass are shown. The numbers next to the plots refer to the individual plants simulated. Adapted from Pearcy and Yang (1998).

These simulations show that it is possible to demonstrate that the architectural design of a plant conforms to an apparently optimal one without considering the multiple functions of the crown and the multiobjective optimization as discussed by Farnsworth and Niklas (1995). Of course, the architecture of *Adenocaulon* is relatively simple and determinate, which certainly facilitates simulating trade-offs. Moreover, light is strongly limiting, so a greater priority in terms of natural selection could be expected to be placed on maximizing light capture as compared with other crown functions. Finally, allocation to leaf area *vs.* other structures, such as petioles is orthogonal, making identification of optimal design easier than when design criteria are correlated, as is the case for investment in biomechanics and hydraulic conductance. It remains to be determined whether optimal design can be demonstrated in more complex architectures such as those of shrubs and trees.

## Performance and efficiency of different architectures

Tropical forest understories provide a diversity of life forms and, in some cases a diversity of architectures among closely related species that allow examination of the role of architecture in light capture and carbon gain. In the Neotropics, two diverse genera, *Piper* (Piperaceae) and *Psychotria* (Rubiaceae), provide striking examples of the congeneric diversity in architectural design that is possible. The *Piper* species studied by Field and colleagues (Chazdon & Field 1987; Chazdon *et al.* 1988; Field 1988; Fredeen & Field 1996) are all shrubs to small trees, which range from species that branch only infrequently to those that bifurcate at essentially every node. Leaf sizes and shapes of these species range from cordate and rounded (*Piper auritum*) with a diameter of almost 30 cm, to lanceolate with a length of 10–12 cm (*Piper aequale*). The gap species, such as *Piper auritum*, have monolayer, umbrella-shaped crowns with relatively little self-shading, whereas the understorey species have deeper crowns and greater self-shading. This seems counter to the expected pattern in sun and shade environments, but it can be explained by the role of leaf longevity as a component, and a constraint on, crown architecture. Rapid leaf turnover keeps the leaf area in high-light environments where carbon gain can be maximized and, because of recycling, it maximizes the efficiency of deployment of nitrogen resources (Field 1988). By contrast, a much greater leaf longevity is required in the shade to meet leaf costs (Williams *et al.* 1989), and because of new growth, older leaves tend to become shaded within the canopy.

The *Psychotria* species exhibit a similar diversity of architectures and leaf life-spans, and on Barro Colorado Island (BCI), Panama, they range from species occupying deeply shaded sites in moist ravines to species found principally in gaps on dry ridges. BCI experiences a long 4-month dry season, so drought, in addition to light, are likely to be major factors in habitat segregation between species. Application of Y-plant to 11 of the 21 species occurring on BCI revealed a strong convergence in efficiencies of light absorption (*Ea*; ratio of absorbed PFD to PFD incident on a horizontal surface in the same environment), with no apparent differences between species specializing on gap environments *vs.* those specializing on shaded forest understories (Fig. 3.6a) (R.W. Pearcy, F. Valladares, E. Lasso & S.T. Wright,

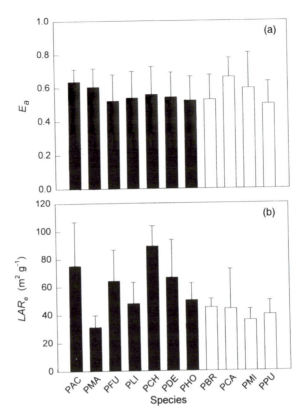

**Figure 3.6** The efficiency of light absorption (*Ea*) (a) and the effective leaf area ratio (b) for 11 *Psychotria* species from the understorey on Barro Colorado Island. The shaded bars are understorey species while the open bars are species typically associated with gaps.

unpublished results). This convergence is consistent with a trade-off among the repertoire of character variation inherent in *Psychotria* leading to functionally similar leaf displays. Similarly, Ackerly (1996) compared light-capture efficiencies of two neotropical gap species, *Trema micrantha* and *Cecropia obtusifolia*, and found a functional convergence despite large differences in leaf size and branching frequency. It is also important to recognize, however, that small variations in *Ea* can have large impacts on carbon gain in understories. Pearcy and Yang (1998) found that 10% variation in *Ea* among *Adenocaulon* plants resulted in 22% variation in daily carbon gain.

An alternative measure of light-capture efficiency is the efficiency of light absorption per unit of biomass. This ratio results in the units $m^2 g^{-1}$, that are the same as those of LAR that are extensively used in growth analysis. It therefore is a measure of how inefficiencies in leaf area display reduce the effectiveness of the investment in leaf area. Accordingly, we have named it the effective leaf area ratio

(LAR*e*). Among the *Psychotria* species, more variation was apparent in LAR*e* than in *E*a, with most gap species having lower values than most understorey species (Fig. 3.6b). The shade species with low LAR*e* such as *P. marginata* have large leaf masses per unit area, which may be important in increasing leaf longevity and for protection against herbivory. Smaller LAR*e* could also be due to greater investments required for hydraulic conductivity or biomechanical support, or more rapid leaf turnover leading to less leaf area per unit of stem. The gap species do have greater leaf-area-specific hydraulic conductivities than the understorey species (S.J. Wright & M.T. Tyree, personal communication), which might be explained by the relatively greater investment in stems in these species. Further work will be required, however, to understand the trade-offs responsible for the species differences in LAR*e*. We also found that LAR*e* is strongly dependent on plant size, as larger plants had a greater cumulative investment in stem biomass. Thus, comparisons need to be made on plants of equivalent size. Despite this complication, LAR*e* is a very useful measure of architectural efficiency because it incorporates both the leaf area display and the biomass investments required to achieve it.

## Architecture and stress avoidance in high-light environments

High-light environments, particularly those with other stresses such as drought or high temperature, impose an entirely different set of constraints on resource acquisition and utilization by plants. In these environments, a compromise must be reached between maximizing carbon gain and the avoidance of excessive temperatures, water loss and photoinhibition. Architecture plays an important role in this through regulating the receipt of solar radiation. The optimal condition will usually be when the leaves receive just below saturating PFDs on their surfaces but, because of the diurnal changes in solar zenith angle and azimuth, a compromise is necessary for most species. Leaf movements can function to maintain solar radiation on the surfaces nearer the optimum, but significant diurnal movement is possible only in species with pulvini at the petiole or, in compound leaves, at the base of each leaflet (Koller 1990). In these species, solar tracking can maximize carbon gain provided that they have a high photosynthetic capacity, allowing effective utilization of extra radiation (Mooney & Ehleringer 1978; Ehleringer & Forseth 1980). These same species will often exhibit paraheliotropic (solar-avoiding) movements that reduce solar radiation receipt when drought restricts photosynthesis (Forseth & Ehleringer 1980; Ludlow & Björkman 1984). Sometimes, upper leaves in canopies exhibit paraheliotropic movements while lower leaves simultaneously exhibit diheliotropic (solar-tracking) movements. It is possible that this may function to manage the tradeoffs at the level of the individual leaves and also to improve the distribution of solar radiation within the canopy. Species lacking pulvini are often capable of some leaf re-orientation through twisting of petioles, or through the irreversible operation of pulvinoids (Bell 1991), but these kinds of movements are clearly more important on a seasonal than on a daily basis.

For species that lack significant short-term leaf movements, shoot structure plays an important role in the trade-off between maximizing leaf carbon gain and minimizing stress effects. Valladares and Pearcy (1998, 1999) examined this trade-off in the sclerophyllous-leaved, evergreen chaparral shrub, *Heteromeles arbutifolia*. This species occurs both in open chaparral communities and in moderately shaded understorey sites of oak woodlands and broadleaved evergreen forests in California. A considerable plasticity of shoot architecture is expressed along this gradient, with shoots in the shade tending towards plagiotropy with a pseudo-distichous arrangement of nearly horizontal leaves (Fig. 3.7). By contrast, sun plants have orthotropic shoots with a spiral phyllotaxy of steeply inclined leaves (mean leaf angle = 71°). Further plasticity results from petiole twisting that moves individual leaves into more favourable microenvironments. There is evidence of leaf acclimation to these localized microenvironments around the shoot axis, because leaves differed in their structural arrangement of the mesophyll and in physiological responses to abaxial *vs.* adaxial illumination, depending on their orientation.

Application of Y-plant to *Heteromeles* shoots revealed how the contrasting architectures of sun and shade plants influence carbon gain and avoidance of stress. The architectures of sun shoots created highly heterogeneous leaf light environments which were well predicted by Y-plant (Fig. 3.7). Daily PFD in the shaded habitat was only 14% of that in the open habitat and shade shoots had only half the photosynthetic capacity of sun leaves. In spite of these large differences, the simulated daily carbon gain of shade shoots was 47% of that of sun shoots on a leaf area basis and 68% of that of sun shoots on a leaf mass basis. Although leaf sun/shade acclimation played a role, the convergence of carbon gains resulted principally from differences in architecture. In particular, the high $Ea$ for diffuse PFD of the shade shoots was important in enhancing their carbon gain. Shade shoots also had more than double the LARe of sun shoots. It should be noted, however, that the LARe of *Heteromeles* shade shoots was still much lower than those of plants from deeply shaded understories. This may possibly be a factor precluding *Heteromeles* from occurring in the most shaded understorey sites.

For *Heteromeles* sun shoots, their steep leaf angles are clearly a major determinant of how they capture and utilize solar radiation. The consequences of leaf angle were studied with Y-plant simulations in which the angles were varied (Fig. 3.8). As can be seen, the normally occurring leaf angles of 71° achieved a very efficient compromise. Carbon gain was only slightly reduced, while the time spent above light saturation that could lead to photoinhibition was strongly reduced. Moreover, simulated shoot water-use efficiencies were maximized by this angle and the maximum temperatures reached by the leaf surfaces that could lead to thermal damage were reduced. Shallower angles resulted in up to 15% more carbon gain but also led to the combination of high light and high leaf temperatures shown to exacerbate photoinhibition in *Heteromeles* (Valladares & Pearcy 1996). The real shoot architectures in which leaf angles varied around the mean of 71° actually performed better in all aspects, except minimizing maximum leaf temperatures,

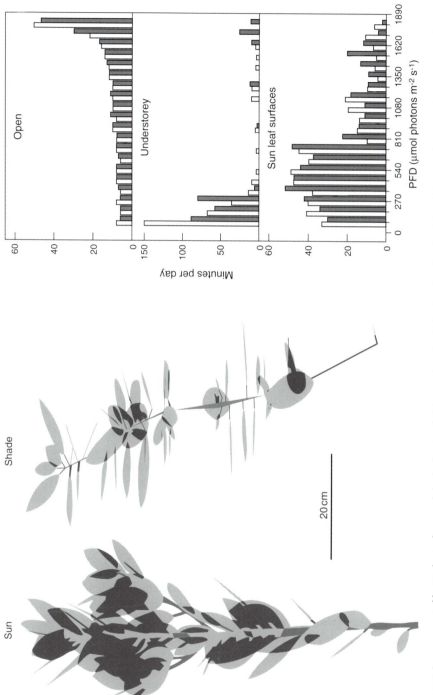

**Figure 3.7** Images created by Y-plant of sun and shade shoots of *Heteromeles arbutifolia* showing the plasticity of shoot architecture in response to the light environment. The dark shaded areas show leaf overlap and hence self-shading for light perpendicular to the view. Histograms on the right side show the frequency distribution for measured (solid bars) and predicted (open bars) PFD from Y-plant on horizontal surfaces in the open (top) and in the understorey (middle), and on the leaf surfaces of sun shoots (bottom).

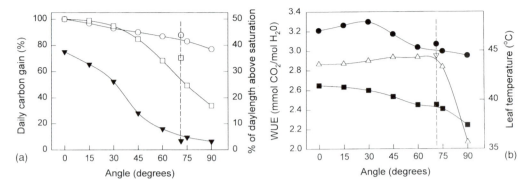

**Figure 3.8** (a) The effect of leaf angle on the simulated daily carbon gain in direct (□) and diffuse (○) PFD, and the percentage of the day that the PFD exceeds the light saturation point for photosynthesis (▼). (b) The effect of leaf angle on the the mean (■) and maximum (●) midday (11–15 hours) leaf temperatures and the daily water-use efficiencies (△). The vertical dashed lines give the mean leaf angle of the shoots and the symbols sitting on these lines are for simulations of sun shoots with their actual leaf angles. The values on the curves are from simulations of sun shoots done by setting the leaf angles of all leaves on a shoot equal to a specified value.

than did simulated shoots in which all the leaf angles were set to be the same. This indicates that the fine tuning accomplished by petiole twisting can be quite important for shoot function.

The architecture of vertical shoots of plants like *Heteromeles* also creates an important gradient in self-shading along their axis. Leaves of this species live for 3–4 years, by which time they have been overtopped by 50–90 younger leaves. About 50% of the shading of lower leaves is by leaves along the shoot axis, while the rest is accounted for by shading by adjacent shoots. Within-shoot shading on the other hand is most important for the upper leaves. Physiological gradients in leaf function caused by leaf ageing or sun/shade acclimation are seen in many canopies and *Heteromeles* is no exception. In *Heteromeles*, the decreasing photosynthetic capacity in the basipetal direction appeared to be related more to ageing than to acclimation. Redistribution of nitrogen associated with gradients of ageing and acclimation in dense herbaceous canopies has been shown to enhance carbon gain by as much as 20–40% compared to the case if nitrogen remained uniformly distributed (Hirose & Werger 1987; Evans 1993), though in more open tree or shrub canopies with lower total nitrogen concentrations the gains may be much less (Field 1983; Leuning *et al.* 1991). There was no evidence, however, for nitrogen movement from lower to upper leaves in *Heteromeles*, because the nitrogen per unit leaf area was constant along the shoot. Moreover, simulations showed that the changes in photosynthetic capacity along the shoot had little effect on whole-shoot carbon gain. This is in contrast to the findings of Meister *et al.* (1987), who found that sun/shade acclimation was important in increasing canopy carbon gain of *Quercus coccifera*. The difference may be accounted for by the denser canopies of *Q. coccifera*.

## Concluding remarks

The complexity of crown organization of plants has clearly been an inhibitory factor in research progress in understanding its role in the ecological success of species. However, an understanding of crown function in displaying leaves and avoiding stress is essential in understanding the productive structure of plants, particularly those of saplings or larger size plants where significant self-shading may occur. Recently developed models such as Y-plant provide tools for assessing the heterogeneity of light environments within crowns and its consequences for leaf and whole-plant photosynthesis, but in the future more dynamic modelling approaches will be needed. This effort will require collaboration between plant morphologists and ecophysiologists, so that the salient features from both disciplinary areas can be included in the models. Incorporation of the developmental plasticity exhibited by plants into these models is particularly important. In addition, a better understanding of the integration of the multiple functions of crowns is needed. For example, at the leaf level, the integration between photosynthesis and transpiration is well appreciated, but at the whole-plant level, much less is understood about the interrelationships between the hydraulic architecture and the light-capture properties of the crowns of plants.

## Acknowledgements

Support of NSF grant IBN-9604424 is gratefully acknowledged. We thank M. Allen and R.S. Loomis for helpful comments on the manuscript.

## References

Ackerly, D.D. (1996). Canopy structure and dynamics: integration of growth processes in tropical pioneer trees. In *Tropical Forest Plant Ecophysiology* (Ed. by S.S. Mulkey, R.D. Chazdon & A.P. Smith), pp. 619–658. Chapman & Hall, New York.

Baldocchi, D. & Collineau, S. (1994). The physical nature of solar radiation in heterogeneous canopies: spatial and temporal attributes. In *Exploitation of Environmental Heterogeneity by Plants: Ecophysiological Processes Above- and Belowground* (Ed. by M.M. Caldwell & R.W. Pearcy), pp. 21–72. Academic Press, San Diego.

Bell, A.D. (1991). *Plant Form: an Illustrated Guide to Flowering Plant Morphology.* Oxford University Press, Oxford.

Chazdon, R. (1986). The costs of leaf support in understory palms: economy vs. safety. *American Naturalist*, **127**, 9–30.

Chazdon, R.L. & Field, C.B. (1987). Determinants of photosynthetic capacity in six rainforest *Piper* species. *Oecologia*, **73**, 222–230.

Chazdon, R.L., Williams, K. & Field, C.B. (1988). Interactions between crown structure and light environment in five rain forest *Piper* species. *American Journal of Botany*, **75**, 1459–1471.

Chen, S.G., Ceulemans, R. & Impens, I. (1994). A fractal-based *Populus* canopy structure model for the calculation of light interception. *Forest Ecology and Management*, **69**, 97–110.

Ehleringer, J. & Forseth, I. (1980). Solar tracking by plants. *Science*, **210**, 1094–1098.

Evans, G.C. (1972). *The Quantitative Analysis of Plant Growth.* Blackwell Scientific Publications, Oxford.

Evans, J.R. (1993). Photosynthetic acclimation and nitrogen partitioning within a lucerne canopy. II. Stability through time and comparison with a theoretical optimum. *Australian Journal of Plant Physiology*, **20**, 69–82.

Farnsworth, K.D. & Niklas, K.J. (1995). Theories of optimization, form and function in branching architecture in plants. *Functional Ecology*, **9**, 355–363.

Farnsworth, K.D. & Van Gardingen, P.R. (1995). Allometric analysis of sitka spruce branches: mechanical versus hydraulic design principles. *Trees: Structure and Function*, **10**, 1–12.

Field, C.B. (1983). Allocating leaf nitrogen for the maximization of carbon gain: leaf age as a control on the allocation program. *Oecologia*, **56**, 341–347.

Field, C.B. (1988). On the role of photosynthetic responses in constraining the habitat distribution of rainforest plants. *Australian Journal of Plant Physiology*, **15**, 343–358.

Fisher, J.B. (1986). Branching patterns and angles in trees. In *On the Economy of Plant Form and Function* (Ed. by T.J. Givnish), pp. 493–523. Cambridge University Press, Cambridge.

Forseth, I.N. & Ehleringer, J.R. (1980). Solar tracking response to drought in a desert annual. *Oecologia*, **44**, 159–163.

Fredeen, A.L. & Field, C.B. (1996). Ecophysiological constraints on the distribution of *Piper* species. In *Tropical Forest Plant Ecophysiology* (Ed. by S.S. Mulkey, R.L. Chazdon & A.P. Smith), pp. 597–618. Chapman & Hall, New York.

Gifford, R.M. & Evans, L.T. (1981). Photosynthesis, carbon partitioning, and yield. *Annual Review of Plant Physiology*, **32**, 485–509.

Givnish, T.J. (1982). On the adaptive significance of leaf height in forest herbs. *American Naturalist*, **120**, 353–381.

Givnish, T.J. (1986). Biomechanical constraints on crown geometry in forest herbs. In *On the Economy of Plant Form and Function* (Ed. by T.J. Givnish), pp. 525–583. Cambridge University Press, Cambridge.

Gould, S.J. & Lewontin, R.C. (1979). The spandrels of San Marco and the Panglossian paradigm: a critique of the adaptationist programme. *Proceedings of the Royal Society of London. Series B*, **205**, 581–598.

Hirose, T. & Werger, M.A. (1987). Maximizing daily canopy photosynthesis with respect to the leaf nitrogen pattern in the canopy. *Oecologia*, **72**, 520–526.

Honda, H. & Fisher, J.B. (1978). Tree branch angle: maximizing effective leaf area. *Science*, **199**, 888–889.

Horn, H.S. (1971). *The Adaptive Geometry of Trees*. Princeton University Press, Princeton, NJ.

Kitajima, K. (1994). Relative importance of photosynthetic traits and allocation patterns as correlates of seedling shade tolerance of 13 tropical trees. *Oecologia*, **98**, 419–428.

Koller, D. (1990). Light-driven leaf movements. *Plant, Cell and Environment*, **13**, 615–632.

Küppers, M. (1989). Ecological significance of above-ground architectural patterns in woody plants: a question of cost benefit relationships. *Trends in Ecology and Evolution*, **4**, 375–379.

Kurth, W. (1994). Morphological models of plant growth: possibilities and ecological relevance. *Ecological Modelling*, **75/76**, 299–308.

Kuuluvainen, T. & Pukkala, T. (1989). Simulation of within-tree and between-tree shading of direct radiation in a forest canopy — effect of crown shape and sun elevation. *Ecological Modelling*, **49**, 89–100.

Lerdau, M. (1992). Future discounts and resource allocation in plants. *Functional Ecology*, **6**, 371–375.

Leuning, R., Wang, Y.P. & Cromer, R.N. (1991). Model simulations of spatial distributions and daily totals of photosynthesis in *Eucalyptus grandis* canopies. *Oecologia*, **88**, 494–503.

Lindenmayer (1975). Developmental algorithms for multicellular organisms: a survey of L-systems. *Journal of Theoretical Biology*, **54**, 3–22.

List, R.J. (1971). *Smithsonian Meteorological Tables*. Smithsonian Institution Press, Washington, DC.

Ludlow, M.M. & Björkman, O. (1984). Paraheliotropic leaf movement in *Siratro* as a protective mechanism against drought-induced damage to primary photosynthetic reactions: damage by excessive light and heat. *Planta*, **161**, 505–518.

Mech, R. & Prusinkiewicz, P. (1996). Visual models of plants interacting with their environment. In *SIGGRAPH 96 Conference Proceedings*, pp. 397–410. Association for Computing Machinery, New York.

Meister, H.P., Caldwell, M.M., Tenhunen, J.D. & Lange, O.L. (1987). Ecological implications of sun/shade differentiation in sclerophyllous canopies: assessment by canopy modelling. In *Plant Response to Stress: A Functional Analysis in*

*Mediterranean Ecosystems* (Ed. by J.D. Tenhunen, F.M. Catarino, O.L. Lange & W.C. Oechel), pp. 401–411. Springer-Verlag, Berlin.

Monsi, M. & Saeki, S. (1953). Uber den lichtfaktor in den pflanzengesellschaften und seine Bedeutung fur de stoffproduktion. *Japan Journal of Botany*, **14**, 22–52.

Monsi, M., Uchijima, Z. & Oikawa, T. (1973). Structure of foliage canopies and photosynthesis. *Annual Review of Ecology and Systematics*, **4**, 301–328.

Monteith, J.L. (1977). Climate and the efficiency of crop production in Britain. *Philosophical Transactions of the Royal Society of London, Series B*, **281**, 277–294.

Mooney, H.A. & Ehleringer, J.R. (1978). The carbon gain benefits of solar tracking in a desert annual. *Plant, Cell and Environment*, **1**, 307–311.

Morgan, J. & Cannell, M.G.R. (1988). Support costs of different branch designs: effects of position, number, and deflection of laterals. *Tree Physiology*, **4**, 303–313.

Myneni, R.B. (1991). Modeling radiative transfer and photosynthesis in three-dimensional vegetation canopies. *Agricultural and Forest Meteorology*, **55**, 323–344.

Myneni, R.B., Ross, J. & Asrar, G. (1989). A review on the theory of photon transport in leaf canopies. *Agricultural and Forest Meteorology*, **45**, 1–153.

Niklas, K.J. (1988). The role of phyllotactic pattern as a 'developmental constraint' on the interception of light by leaf surfaces. *Evolution*, **42**, 1–16.

Niklas, K.J. (1992). *Plant Biomechanics*. University of Chicago Press, Chicago.

Niklas, K.J. (1994). Morphological evolution through complex domains of fitness. *Proceedings of the National Academy of Sciences USA*, **91**, 6772–6779.

Nobel, P.S. (1980). Interception of photosynthetically active radiation by cacti of different morphology. *Oecologia*, **45**, 160–166.

Norman, J.M. & Welles, J.M. (1983). Radiative transfer in an array of canopies. *Agronomy Journal*, **75**, 481–488.

Parker, G.A. & Maynard-Smith, J. (1990). Optimality theory in evolutionary biology. *Nature*, **348**, 27–33.

Pearcy, R.W. & Yang, W. (1996). A three-dimensional crown architecture model for assessment of light capture and carbon gain by understory plants. *Oecologia*, **108**, 1–12.

Pearcy, R.W. & Yang, W. (1998). The functional morphology of light capture and carbon gain in the redwood forest understorey plant *Adenocaulon bicolor* Hook. *Functional Ecology*, **12**, 543–552.

Pereira, J.S. (1994). Gas exchange and growth. In *Ecophysiology of Photosynthesis* (Ed. by E.-D. Schulze & M.M. Caldwell), pp. 147–181. Springer-Verlag, Berlin.

Pickett, S.T.A. & Kempf, J.S. (1980). Branching patterns in forest shrubs and understory trees in relation to habitat. *New Phytologist*, **86**, 219–228.

Poorter, H., Remkes, C. & Lambers, H. (1990). Carbon and nitrogen economy of 24 wild species differing in relative growth rate. *Plant Physiology*, **94**, 621–627.

Prusinkiewicz, P.W., Remphrey, W.R., Davidson, C.G. & Hammel, M.S. (1994). Modeling the architecture of expanding *Fraxinus pennsylvanica* shoots using L-systems. *Canadian Journal of Botany*, **72**, 701–714.

de Reffye, P. (1988). Plant models faithful to botanical structure and development. *Computer Graphics*, **22**, 151–158.

de Reffye, P. & Houllier, F. (1997). Modelling plant growth and architecture: some recent advances and applications to agronomy and forestry. *Current Science*, **73**, 984–992.

Rich, P. (1990). Characterizing plant canopies with hemispherical photographs. *Remote Sensing Reviews*, **5**, 13–29.

Roberts, S.W. & Miller, P.C. (1977). Interception of solar radiation as affected by canopy organization in two mediterranean shrubs. *Oecologia Plantarum*, **12**, 273–290.

Room, P., Hanan, J. & Prusinkiewicz, P. (1996). Virtual plants: new perspectives for ecologists, pathologists and agricultural scientists. *Trends in Plant Science*, **1**, 33–38.

Russell, G., Jarvis, P.G. & Monteith, J.L. (1989). Absorption of radiation by plant canopies and stand growth. In *Plant Canopies: Their Growth, Form and Function* (Ed. by G. Russell, B. Marshall & P.G. Jarvis), pp. 21–39. Cambridge University Press, Cambridge.

Ryel, R.J., Beyschlag, W. & Caldwell, M.M. (1993). Foliage orientation and carbon gain in two tussock grasses as assessed with a new whole-plant gas-exchange model. *Functional Ecology*, 7, 115–124.

Smith, M. & Ullberg, D. (1989). Effect of leaf angle and orientation on photosynthesis and water relations in *Silphium terebunthinaceum*. *American Journal of Botany*, 76, 1714–1719.

Takenaka, A. (1994). Effects of leaf blade narrowness and petiole length on the light capture efficiency of a shoot. *Ecological Research*, 9, 109–114.

Tyree, M.T. & Ewers, F.W. (1991). The hydraulic architecture of trees and other woody plants. *New Phytologist*, 119, 345–360.

Valladares, F. & Pearcy, R.W. (1996). Interactions between water stress, sun-shade acclimation, heat tolerance and photoinhibition in the sclerophyll *Heteromeles arbutifolia*. *Plant, Cell and Environment*, 20, 25–36.

Valladares, F. & Pearcy, R.W. (1998). The functional ecology of shoot architecture in sun and shade plants of *Heteromeles arbutifolia* M. Roem, a California chaparral shrub. *Oecologia*, 114, 1–10.

Valladares, F. & Pearcy, R.W. (1999). The geometry of light interception by shoots of *Heteromeles arbutifolia* M. Roem: morphological and physio-logical consequences for individual leaves. *Oecologia*, in press.

Wang, Y.P. & Jarvis, P.G. (1990). Influence of crown structural properties on PAR absorption, photosynthesis, and transpiration in sitka spruce: application of a model (MAESTRO). *Tree Physiology*, 7, 297–316.

Whitehead, D., Grace, J.C. & Godfrey, M.J.S. (1990). Architectural distribution of foliage in individual *Pinus radiata* d. Don. crowns and the effects of clumping on radiation interception. *Tree Physiology*, 7, 135–155.

Williams, K., Field, C.B. & Mooney, H.A. (1989). Relationships among leaf construction costs, leaf longevity, and light environment in rainforest species of the genus *Piper*. *American Naturalist*, 133, 198–211.

Zipperlen, S.W. & Press, M.C. (1996). Photosynthesis in relation to growth and seedling ecology of two dipterocarp rain forest tree species. *Journal of Ecology*, 84, 863–876.

Zotz, G., Tyree, M.T., Patino, S. & Carlton, M.R. (1998). Hydraulic architecture and water use of selected species from a lower montane forest in Panama. *Trees: Structure and Function*, 12, 302–309.

# Chapter 4

# Plant responses to small perturbations in soil water status

*W.J. Davies[1] and D.J.G. Gowing[2]*

## Introduction

It is well known that plants modify their biochemistry, physiology, growth and development in response to a reduction in soil water availability. Their sensitivity to soil drying is not so widely appreciated however. Henson *et al.* (1989) demonstrated that plants are able to modify their physiology and growth rate in response to soil drying when moisture tensions exceed just 5 kPa or 0.5 m (Fig. 4.1). In most soils, a supply of water would still be freely available at such a tension. There is therefore a conceptual difficulty in understanding why plants should begin to reduce growth in such relatively unstressed conditions. The possibility that such mild stress may indeed have a significant ecological role has been observed under experimental conditions (Ellenberg 1953) and in the field (Gowing & Spoor 1998).

The classic experiments of Ellenberg (1953) clearly demonstrate the effect of mild soil drying. A number of grass species were grown in monoculture on a water-table gradient and their shoot biomass recorded. All species exhibited greatest productivity when the root zone moisture tension was in the range 3–5 kPa. At lower tensions, productivity declined as a result of waterlogging and soil anoxia, while in drier soils biomass was also lowered, pointing to a limiting factor connected to soil drying. The experiment, when repeated in mixed culture to observe competition effects, yielded much wider variation in optimal water regimes.

In the field, the water regime of a species-rich grassland on moist peaty soils in the Somerset Levels was modelled and the distribution patterns of common grassland species were related to a long-term measure of soil moisture status. These grasslands are in an area of controlled drainage and have a constantly high water-table, thus they are never exposed to severe drought stress. A clear pattern emerges if microsites throughout the area are ranked with respect to the number of weeks in which the moisture tension in the root zone exceeds 0.5 m and then superimposed on the frequency of a species such as *Carex nigra* (Fig. 4.2a). Meanwhile, a more drought-tolerant species, *Dactylis glomerata*, showed the opposite response (Fig.

[1] *Department of Biological Sciences, Lancaster, University, Bailrigg, Lancaster, LA1 4YQ, UK. E-mail: w.davies@lancaster.ac.uk.*
[2] *Institute of Water and Environment, Cranfield University, Silsoe, Bedfordshire, MK45 4DT, UK. E-mail: d.gowing@cranfield.ac.uk*

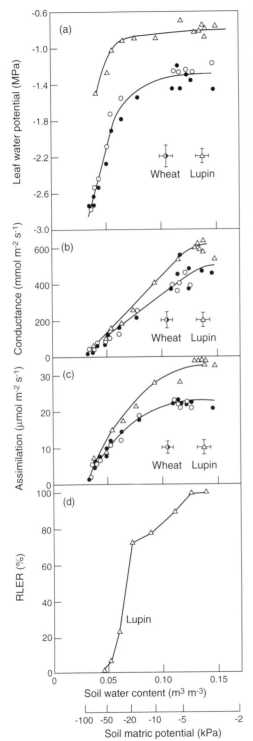

**Figure 4.1** Leaf water potential (a), leaf conductance (b), net $CO_2$ assimilation rate (c) and relative leaf extension rate (RLER) (d) as a function of bulk soil water content for lupin ($\triangle$) and wheat cultivars Gamenya ($\bullet$) and Warigal ($\bigcirc$). The equivalent values of soil matric potential are also shown. Data are the means for 10 lupin or 4 wheat leaves. Bars in (a), (b) and (c) indicate $\pm$ pooled SE for lupin or wheat. In (d), RLER is expressed as a percentage of the controls. Curves were fitted by eye. After Henson *et al.* (1989).

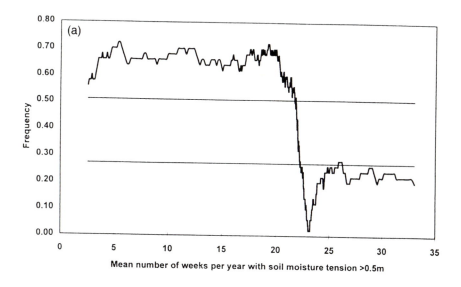

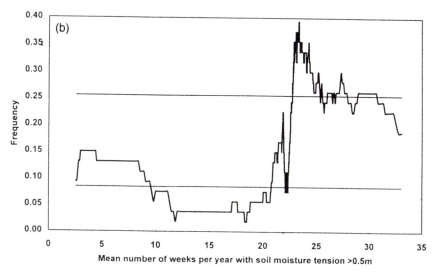

**Figure 4.2** Frequency of occurrence of *Carex nigra* (a) and *Dactylis glomerata* (b) in randomly placed quadrats from species-rich wet grassland (Somerset, England). The 900 quadrats used are ranked to form the abscissa according to the mean number of weeks per year in which the soil matric potential of the surface layer exceeds 5 kPa. These data are based on the results of a hydrological model run for a 15-year period (1980–94). The plot shows species frequency as a moving average within a subset of 50 quadrats. Horizontal lines represent 95% confidence limits around the mean frequency of the species.

4.2b). The impedance of oxygen diffusion within the root zone might be expected to explain such patterns, but a parallel analysis of root-zone saturation gives less explanatory power (Gowing et al. 1997). Nevertheless, the apparent sensitivity of some species to mild soil drying may not necessarily be due to restricted water uptake, but could rather be a response to reduced diffusability of soluble nutrients, increased rate of nitrogen mineralization or to increased soil shear strength retarding root elongation.

Ellenberg demonstrated that mild soil drying could control plant competition, reinforcing the idea that the effect illustrated in Fig. 4.2(a) is not a result purely of the absence of waterlogging. Ellenberg's work was extended to observe the interaction of water regime and nitrogen supply (Ellenberg 1954), which proved to be complex. It is not therefore a straightforward task to partition the direct hydraulic effects of mild soil drying from a possible reduction in nutrient supply. Indeed, the two environmental stresses may be sensed in the root and communicated to the shoot by a common signalling mechanism (Palmer et al. 1996) and therefore they may need to be considered together rather than separately. The increasing shear strength of drying soil may also be sensed and communicated in a similar way (Tardieu et al. 1992).

Whatever the mechanism of stress imposition, field and controlled environment studies show that mild soil drying can exert an impact on plant-community composition. It is interesting to speculate therefore whether the ability of a plant to sense the onset of this mild soil drying and to respond appropriately plays a role in determining interspecific competition within natural or semi-natural communities. To improve our understanding of this area, the physiological adaptations that underlie the observed species distribution patterns with respect to soil moisture need to be investigated. Distinguishing the relative importance of the three potentially limiting factors, water supply, nutrient supply and soil strength, would be a first step. It is only comparatively recently that physiologists have started to try to understand the effects of very mild soil drying. Previously, most attention had focused on changes in physiology and growth that accompany large variation in leaf water status (desiccation stress). Most molecular investigations of the effects of drought concentrate on desiccation-related phenomena or on responses to what are often massive doses of the plant stress hormone abscisic acid (ABA). These responses are important, but they are not the only important drought responses and clearly such stress cannot develop when the soil water status changes by only 5 kPa. If growth, development and gas exchange are modified before plant water status changes in response to soil drying, then clearly a great deal is being missed by concentrating on desiccation phenomena. Even when variation in soil water status is very much larger than 5 kPa, the mechanistic basis of responses to such changes does not always involve the much-discussed effects of variation in shoot–water relations. This chapter considers the mechanistic basis of subtle modifications in plant response that have traditionally received much less attention than plant responses to desiccation stress.

## Cell water relations and growth

The most commonly discussed explanation for a drought stress effect on plants is that the drying of the soil and/or the air will result in a reduction in the water potential of key parts of the plant. This may or may not be accompanied by a reduction in shoot water content or turgor. For the control of many processes, it is the regulation of cell turgor that is the key variable and it is undeniable that a reduction in turgor will eventually reduce growth and restrict gas exchange. If turgor is reduced sufficiently, then potentially severe biochemical lesions will develop, perhaps because of the development of damaging concentrations of key inorganic ions (Kaiser 1987).

In many plants, shoot turgors are sustained even though water potentials decline. Thus, critical plant parts can remain for longer at a more favourable water status and avoid lower, potentially lethal, water contents. This postponement of desiccation injury is accomplished by the accumulation of so-called compatible solutes. Conventional plant breeding programmes have increased yield under drought by selection for osmotic adjustment (Morgan 1984), and there is currently much interest in inserting into crop plants genes that result in the accumulation of certain compatible solutes (e.g. Holmberg & Bulow 1998). When compared with non-transgenic counterparts, such transgenic plants often show enhanced root and even shoot growth under drought (Kavi Kishor *et al.* 1995), but it is not always clear why this is the case (Blum *et al.* 1996). Many resulting transgenic plants are very different from plants that will produce economic yields under drought conditions and, indeed, it is difficult to see how engineering greater desiccation resistance in isolation can help in this regard. This is because crop yield is inextricably linked to water availability and high yields are only obtained when water is relatively freely available (Passioura 1977). Desiccation resistance is of more interest to ecologists as this may keep the plant alive under extreme stress, allowing the life cycle to be completed, however long this process takes. In less harsh environments, the ability to maintain shoot turgor and leaf area production differs between species and may exert important controls on the relative abundance of species in natural communities.

There is considerable uncertainty over why vegetative growth and reproductive development of plants are so highly sensitive to drought stress, and crop physiologists and ecologists alike have an interest in understanding why this is the case. It is clear that both cell production and cell expansion can be influenced by mild water deficit (Lecoeur *et al.* 1995), but much more attention has been given to the modification of cell expansion, presumably because a role for turgor variation is easier to understand in this context. Nevertheless, the growth of shoots is often restricted as soil dries, even when shoot turgor is completely maintained. Indeed, shoot turgor may be maintained *because* shoot growth is restricted as soil dries (Kuang *et al.* 1990). In addition to this, there is some controversy over the sensitivity of shoot growth to small reductions in turgor. The traditional view is that any small reduction in cell turgor will reduce growth, and above a threshold turgor for growth, increases in turgor will result in growth stimulation. The Lockhart equation, first developed for unicellular organisms (Lockhart 1965), describes this relationship and analysis of this kind has now been extended to multicellular organisms (McDonald & Davies 1996). More recently, careful analysis of the turgor/growth relations of the unicellular alga *Chara* suggests that above a turgor threshold, growth rate varies independently of turgor and that beyond this, growth is probably regulated by the changing properties of cell walls (Zhu & Boyer 1992). This is an important observation and similar studies on multicellular organisms have yet to be undertaken.

There is a difficulty in measuring the water relations and growth of exactly the same cells in a multicellular organism, where not all of the cells will be growing at the same rate and indeed some cells may not be growing at all. Although the micro-pressure probe can be used to monitor the turgor of individual cells, the growth of only quite large groups of cells can be measured. Spollen *et al.* (1993) have shown how easy it is to infer Lockhart-like behaviour from a group of cells, an increasing proportion of which stop growing as water availability and turgor decline, while the remaining cells continue to grow at the maximum rate. Analysis of root growth under drought by Sharp's group has also shown how important it can be to describe in detail the spatial distribution of growth through an organ, if we are to understand the regulation of growth of that organ as a whole.

While reduced soil water availability can limit shoot growth, even when turgor is completely maintained, root growth is generally much less sensitive to cell dehydration (Spollen *et al.* 1993), an observation that explains the increase in root: shoot ratio which is commonly recorded when plants grow in drying soil. Maize roots growing at low water potential show sustained growth close to the root tip and this growth maintenance correlates well with increased proline deposition in this part of the growing zone (Ober & Sharp 1994). Proline accumulation contributes to the maintenance of some cell turgor in this zone, although turgor maintenance is not complete (Spollen & Sharp 1991).

The maintenance of growth rate close to the root tip at reduced cell turgors suggests some modification of cell wall properties in this zone. Lower growth rates of roots at low water potentials (even though growth rates close to the tip are sustained) are explained by a shortening of the growing zone as the water potential of the growing medium declines, together with a reduction in the relative growth rate of cells as they reach their maximum rate of growth. Where water supplies at depth are available when the surface soil becomes dry, maintenance or even promotion of root elongation in drying soil has a clear competitive benefit when neighbours, whose root growth ceases, are forced to become dormant or die (Sharp & Davies 1985).

## Cell wall properties

The effects of soil drying on the properties of growing cell walls should be considered with respect to both wall-loosening and wall-stiffening enzymes (Thompson *et al.* 1997). In the literature, there is interest in the roles played by two putative wall-loosening enzymes, xyloglucan endoglycosylase (XET) (Fry *et al.* 1992) and a class of enzymes termed the expansins (McQueen-Mason 1995). Cell walls may be stiffened through an upregulation of peroxidase enzymes (MacAdam *et al.* 1992). There are now reports that soil water deficits will influence the activity of all these enzymes and the assumption is that, as a result, growth will change. However, there is not always a clear spatial or temporal relationship between the change in enzyme activity and the change in growth rate (Palmer & Davies 1996). In the primary root growth system used by Sharp's group to investigate the effects of low water potential, maintenance of growth close to the root tip is apparently a function of upregulation of the activity of both XET and expansins (Wu *et al.* 1994, 1996). The resultant wall loosening allows the maintenance of growth, even though turgor is reduced in these cells.

A recent paper by Fleming *et al.* (1997) shows very elegantly that expansins applied to apical meristems of tomato can exert a significant influence on the induction of leaf primordia. This is an interesting observation in the light of the commonly observed reduction in leaf production when plants are subjected to drought. The activity of expansins in leaves is highly pH dependent (McQueen-Mason 1995; see below) and it is interesting to speculate that drought may limit leaf production via a pH effect on expansin activity.

The identification of clones encoding XET (Arrowsmith & de Silva 1995), expansins (Shcherban *et al.* 1995) and peroxidase (Lagrimini *et al.* 1987) mean that molecular techniques can now be used to manipulate enzyme activity. Investigation of the effects of soil drying on the growth of plants, where activities of these enzymes have been up- or down-regulated, will be a useful way of elucidating the cellular and molecular effects of drought on growth processes.

Antisense tobacco plants deficient in an anionic peroxidase (Lagrimini *et al.* 1997a) have been shown to display altered growth characteristics. While these plants had an apparently normal morphology, both plant height and leaf thickness were greater than for non-transformed counterparts. This particular peroxidase enzyme is not normally found in the roots of tobacco plants, but overexpression, in contrast to downregulation, resulted in plants that had a normal appearance above ground but that wilted on reaching maturity (Lagrimini *et al.* 1997b). Wilting occurred despite the presence of functional stomata and a normal vascular anatomy and physiology. The transgenics, however, differed from wild-type plants in having considerably less root mass, which resulted from an inhibition of root branching. It seems clear that although enzymes may have predictable effects on cell wall properties and changes in enzyme activity may be correlated with changes in growth rate, intervention to manipulate the expression of particular genes can promote complex and not entirely predictable effects at the whole-plant level.

## Stomatal behaviour and gas exchange: hydraulic and non-hydraulic regulation?

As with the regulation of growth in droughted plants, the traditional explanation for drought-induced regulation of gas exchange is a response to a change in shoot–water relations. In the past 15 years several ingenious techniques have been used to break the link between soil drying and modified shoot–water relations to demonstrate the existence of an additional factor or factors limiting gas exchange of plants in drying soil. Using one such technique, the root pressure vessel (Passioura & Munns 1984), Gollan et al. (1992) demonstrated that stomata of sunflower plants closed as the soil dried, irrespective of whether or not pressure was applied to roots to counteract the increasing soil suction with drought. Pressure was applied such that the shoots of plants in soil at low water potential exhibited water relationships that were comparable with those of well-watered plants. Despite this, stomata still closed in response to the soil drying treatment. More recently, Comstock and Mencuccini (1998) used the same technique to restore the shoot water potential of seedlings of the woody desert shrub *Hymenoclea salsola* as the soil dried and demonstrated that stomata re-opened in response to this treatment. This suggests that drought-induced limitations in gas exchange in this species are mostly hydraulic and raises the possibility that the relative importance of chemical and hydraulic signalling differs between species and genotypes and may, for example, differ between woody and herbaceous plants.

Additional evidence for non-hydraulic regulation of gas exchange is provided by an early study reported by Turner et al. (1985). These authors showed that the relationship between stomatal conductance and shoot water status was highly dependent on the method used to generate variation. When stomatal conductances of plants subjected to a range of different drying treatments were plotted against soil water status, a single tightly defined relationship was observed, suggesting that some measure of soil moisture status was regulating the behaviour of stomata.

In a similar, more recent experiment, Tardieu et al. (1996) generated relationships between stomatal conductance and xylem ABA concentration from sunflower plants grown in a range of different environments and vapour-pressure deficits. Despite the fact that plants were grown in contrasting environments these relationships were remarkably similar, suggesting that the plant hormone may act as a regulator of stomatal behaviour, overriding even the influence of variation in the water relations of the leaf. There are now many studies reported in the literature which suggest that ABA and other chemical regulators may provide a drought-induced non-hydraulic influence on gas exchange and growth. The nature of this control can be extremely subtle and finely tuned to variation in soil moisture supply. The ecological implications of the different mechanisms of stomatal control require further investigation.

## Non-hydraulic influences on gas exchange and growth: what are the regulators?

Almost since the identification of ABA as one of the major plant growth regulators or hormones (Addicott *et al.* 1968), it has been apparent that this compound has a potent effect on stomata (Mittelheuser & van Steveninck 1969; Jones & Mansfield 1970). This, combined with a general acceptance of its activity as an inhibitor of shoot growth and the early observation that synthesis was stimulated by dehydration of cells (Wright 1977), has led many to propose a role for ABA as a general stress hormone. It has been suggested that when cell dehydration occurs, extra ABA will reach sites of action within the leaf, restricting the development of transpiring area and, thus, the rate of water loss per unit leaf area. Over the years, this hypothesis has been refined with the suggestion that ABA arriving in the leaf in the transpiration stream can be an important regulating influence (Loveys 1984). It has been proposed that the concentration of ABA in the xylem stream can provide the shoots with a 'measure' of the water available in the soil, thereby allowing the plant to regulate its physiology, growth and development as a function of water availability (Zhang & Davies 1989; Tardieu & Davies 1993).

Tardieu *et al.* (1992) have considered the nature of the information content of the root to shoot signal and have argued that it may reflect the access that roots have to soil water. This will be a function of both the water status and structure of the soil, the latter influencing the distribution of roots. In soil with a high mechanical impedance, roots will clump and soil drying may therefore be localized. Signal development will reflect this, even though the bulk of the soil may still be relatively well supplied with water. Other authors have suggested that roots may respond to changes in soil strength as the soil water status changes (Passioura 1988). Jones (1989) and Davies & Zhang (1991) have noted the desirability of the long-term regulation of development by such a signal rather than by a signal such as leaf turgor that can vary from minute to minute as evaporative demand changes. The root signal will become more intense as soil dries, perhaps over a season. In tall plants this signal will take days to reach the shoots but such a time delay would not be inconsistent with the time scale of the regulation of plant development. The more dynamic regulation of stomatal behaviour will require the involvement of climatic influences and these effects are considered below.

Many xylem-borne regulators of growth and development have been considered in the literature as candidates of chemical 'measures' of soil drying, but none has received the sustained attention accorded to ABA. This is because of the potent influence of this compound on growth and stomatal behaviour and because of the potentially substantial accumulation of this compound in response to stress. Despite this, it has been difficult to make an unequivocal case for the involvement of ABA in the plant's stress response. Some of these issues are discussed below.

There is little doubt that soil drying often results in an increase in the ABA concentration in the xylem stream of plants and it seems likely that a reasonable proportion of this extra ABA is root sourced (Wolf *et al.* 1990). In order to argue for root to shoot signalling it is necessary to show that this increase in xylem ABA concentration is the result of extra ABA loading into the xylem and not just the result of changes in the concentration of the molecule as the transpiration flux decreases (Jackson 1993). Schurr (1998) discusses ways of doing this and Jarvis and Davies (1997) have also provided a novel means of demonstrating that concentration changes are not merely a result of changes in flux. Others have calculated the delivery of ABA molecules to the shoot by multiplying the transpiration rate by the concentration of ABA in the xylem and shown substantial enhancement of ABA as a result of soil drying. The interaction between the root and the drying soil will cause some extra ABA synthesis in the root but it will also cause enhanced re-circulation of ABA arriving in the root from the shoot.

Having established that soil drying can increase the delivery of ABA to shoots, it is important to determine whether this extra ABA is enough to bring about the observed closure of stomata and limitation of leaf growth. This question has received a lot of attention (Munns & King 1988; Munns 1989) but is difficult to answer without very close control and monitoring of experimental conditions.

It now seems likely that even well-watered plants contain enough ABA to close stomata and limit leaf growth, if the ABA can get to the sites of action in the leaf. This is an important point in the context of the results reported above in the Introduction. We suggest that the chemical signalling mechanism present in the well-watered plants can be utilized to promote growth and gas-exchange responses to extremely subtle changes in water availability. No extra synthesis of signalling molecules is required. In order to address this issue we need to define a mass balance of the hormone in a compartment adjacent to the sites of action on the guard cell (and the site of action for leaf growth) and recognize that variation in any one of the components of this mass balance will affect the sensitivity of the plant to the hormone. Most workers consider that the binding sites for ABA action on guard cells are on the apoplastic face of the guard cells (Hartung 1983), but recent evidence suggests that there may also be some intracellular binding of ABA in the guard cells (Allan et al. 1994). This will be considered later in this chapter. There is much less information on binding sites for hormonal regulation of leaf growth other than a general acceptance that the epidermis of the leaf must play an important role.

The various components of the mass balance are as follows. There will be a flux of ABA into the apoplastic compartment outside the guard cells from the xylem and perhaps from other leaf cells, although a modelling approach has suggested that this may not be the case (Slovik et al. 1995). ABA will also flow out of the compartment and will be broken down. It is very difficult to quantify all of these variables (particularly the apoplastic concentration of ABA outside guard cells) and each can influence the apparent sensitivity of stomata to ABA arriving in the transpiration stream. The flux of ABA into the compartment will be a function of the concentration of the hormone in the transpiration stream and of the transpiration flux. Gowing et al. (1993) have shown that this can be an important variable in the control of stomatal behaviour.

It is clear that a large proportion of the ABA arriving in the transpiration stream never reaches the site of action in the leaf, because so much hormone is delivered to the shoot over the day that, were it to penetrate to the guard cells, the stomata would never be open. The extent to which penetration occurs will be determined largely by the pH of different compartments of the leaf. When xylem sap pH is relatively acid (as it is in well-watered plants) most ABA arriving in the leaf will be sequestered away from sites of action for regulation of stomatal behaviour and leaf growth. At more alkaline sap pH values, a greater proportion of xylem sap ABA will get to sites of action in the leaf (Hartung *et al.* 1998).

It is well known that soil drying can result in the alkalinization of the xylem sap (Fig. 4.3) and Wilkinson *et al.* (1998) have shown that when this occurs a very low concentration of ABA is enough to close stomata substantially (Fig. 4.4). This can be equivalent to the ABA present in the xylem stream of a plant that is well supplied with water. These experiments illustrate one of the attractions of ABA as a chemical signal of soil drying, i.e. that subtle control of leaf physiology is possible with relatively small variations in the delivery of the chemical regulator to its site of action.

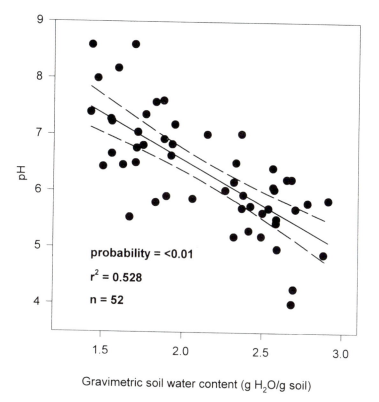

**Figure 4.3** The effect of gravimetric soil water content on the pH of xylem sap expressed within 2 minutes from 30-mm shoot stumps of 6- to 9-week-old wild-type tomato plants. Linear regression with 95% confidence limits is shown. After Wilkinson *et al.* (1998).

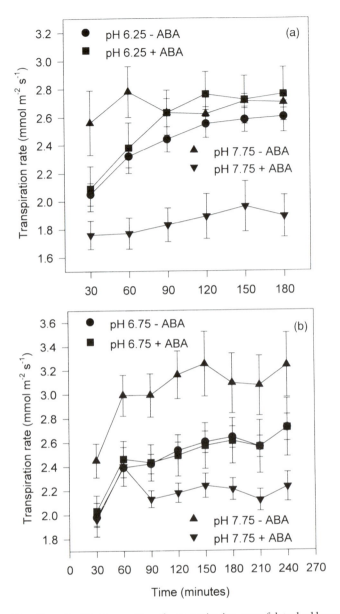

**Figure 4.4** The effect of artificial sap pH on the transpiration rate of detached leaves of *Lycopersicon esculentum* in the light in the presence and absence of 0.03 μM (+)-ABA. Leaves were preincubated in artificial sap for 1 hour in the dark. It was assumed the endogenous ABA content of the leaves used in experiment (a) was higher than that in the leaves used in experiment (b). Transpiration rates were calculated for each leaf every 30 minutes and expressed as a mean ($n=4$–6). After Wilkinson *et al.* (1998).

**Table 4.1**  pH changes that occur in plant xylem or apoplastic sap under various conditions. After Wilkinson *et al.* (1998).

| Species | Source of sap | pH change | Conditions of change | Reference |
|---|---|---|---|---|
| *Phaseolus coccineus* | Root xylem | 6.5–7.0 | Wet soil—2-day drought | Hartung & Radin (1989) |
| *Anastatica hierochuntica* | Root xylem Shoot xylem | 6.5–7.1 6.5–7.6 | Wet soil—dry soil | Hartung & Radin (1989) |
| *Helianthus annuus* | Shoot xylem | 5.8/6.6–7.1 | High soil water— below 0.13 g g⁻¹ | Gollan *et al.* (1992) |
| *Commelina communis* | Shoot xylem | 6.0–6.5/6.7 | Wet soil—4/5 day drought | Wilkinson & Davies (1997) |
| *Lycopersicon esculentum* | Root xylem at 0.1 MPa | 6.13–6.74 | Drained soil— flooded soil | Else (1996) |
| *Ricinus communis* | Shoot xylem | 6.0–6.6 | End of night—end of day | Schurr & Schulze (1995) |
| *Helianthus annuus* | Leaf apoplast | 5.7–6.4 | Light—dark (+2.5 mM nitrate) | Hoffman & Kosegarten (1995) |
| *Samanea pulvinus* | Extensor apoplast | 6.2–6.7 | White light—dark | Lee & Satter (1989) |
| *Robinia wood* | Shoot xylem | 6.0–5.0–5.5 | Jan—April/May —Nov/Dec | Fromard *et al.* (1995) |
| *Actinidea chinensis* | Shoot xylem | 5.3–6.2 | Spring—rest of year | Ferguson *et al.* (1983) |
| *Betula pendula* | Shoot xylem | 7.5/8.0–5.7 | Rest of year— catkin bud break | Sauter & Ambrosius (1986) |
| *Hordeum vulgare* | Leaf apoplast | 6.6–7.3 | Control—brown rust infected | Tetlow & Farrar (1993) |
| *Helianthus annuus* | Leaf apoplast | 6.8–7.4 | Low nitrate—high nitrate | Dannel *et al.* (1995) |
| *Helianthus annuus* | Leaf apoplast | 6.0–6.5 6.2–7.0 | $NH_4NO_3$—$NO_3^-$ $NH_4NO_3$—$HCO_3^-$ | Mengel *et al.* (1994) |

It seems likely that the pH of the various compartments of the leaf can be sensitive to a wide variety of environmental variables (Table 4.1), thus linking the root to shoot signalling system to variations in the climatic environment. This has led to the hypothesis that soil drying or some other change in edaphic conditions will increase both the delivery of ABA to shoots and the pH of the xylem sap. The latter will ensure that more of the ABA signal penetrates to sites of action for stomatal control or growth within the leaf. Variation in pH will therefore modify the apparent sensitivity of the leaf to the root-sourced ABA signal.

Climatic factors can also modify the sensitivity of the plant to ABA through a variation in pH, thus extra delivery of ABA from roots to shoots is not necessary to close stomata if climatic conditions are conducive to dehydration of the shoots. This is important, because the roots of trees can be tens of metres from the stomata and sap flux, particularly in conifers, can be very slow. Thus, it is hypothesized that the ABA concentration provides the shoot with a 'measure' of soil water availability, but whether or not the stomata respond depends upon other factors (climatic and edaphic) that influence the degree of penetration of the hormone to sites of action. In many cases this will occur via a change in the pH relations of the leaf. It is well known that water deficit (Tardieu & Davies 1992), the nutrient relations of the soil (Radin *et al.* 1982), temperature (Dodd & Davies 1994) and vapour pressure deficit can all change the apparent sensitivity of the leaf to ABA and each of these variables could act through modification of pH (Table 4.1). The sensitizing of stomata to xylem ABA by alkaline pH is consistent with the importance of outward-facing receptors for the hormone in the plasma membrane. When the xylem sap is more acid, ABA will move into all cells, presumably including the guard cells, and yet the response of the stomata to the delivery of the hormone under these conditions is only very restricted, arguing against the importance of intracellular receptors.

Interestingly, Bacon *et al.* (1998) have recently shown that leaf growth of barley is also sensitive to pH signals (Fig. 4.5) and that these responses are ABA based. In a direct parallel with the regulation of stomatal behaviour, alkalinization of xylem sap will limit leaf growth, but only if a low background concentration of ABA is present in xylem sap. Alkaline sap sensitizes leaf growth to very low concentrations of ABA or very mild soil drying. The effects of pH on the ABA signalling process raise an interesting possibility that plants that exhibit a very low sensitivity to ABA may do so because the xylem pH has not become more alkaline with soil drying. If this were the case then even moderate increases in ABA concentration might not penetrate to sites of action in the leaf. We can start to see how variation in the nutrient status of the soil may modify the plant's sensitivity to ABA and therefore to soil drying. Alternatively, a drought-induced reduction in nutrient supply to shoots (Turner 1986) may modify the plant's ABA responses and thus, as suggested in the Introduction, contribute to the plant's drought-sensing capability. Schurr and Schulze (1995, 1996) have shown that there are significant differences between species in the effects of soil drying on hormone and nutrient fluxes to shoots. This is also true for changes in pH. This suggests that we are likely to find significant interspecific variation in drought signalling and that this variation may form the basis of differences in competitive potential.

The difficulty in establishing the adequacy or otherwise of an ABA signal is exacerbated by the difficulty in accurately assessing the ABA dose arriving in the leaf. This is not just a case of matching flows of sampled sap with those in the intact plant so that deliveries are accurately assessed, but also of ensuring that wounding effects on hormone concentrations in sample sap are minimized. Else *et al.* (1994) have shown that such effects can be substantial but can be minimized if suitable delays are introduced between cutting plant tissue and sampling sap. Correira and Pereira (1994) have also considered these effects and have proposed a method of accounting for wounding effects.

It is worth emphasizing at this point that most of the responses that we have discussed are responses to very subtle variation in soil water availability. These changes will increase the delivery of ABA to the site of action in the leaf (perhaps through a change in sap pH), but they do not necessarily result in an accumulation of the hormone. It is important to differentiate between effects brought about by

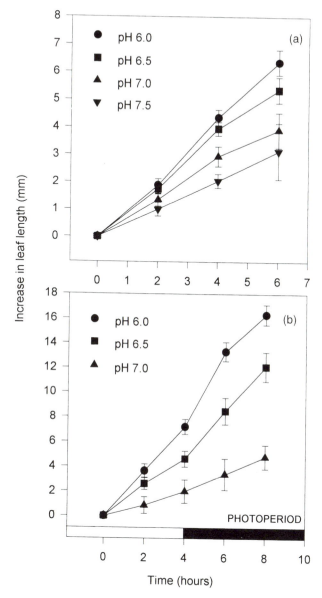

**Figure 4.5** The effect of feeding artificial sap buffered to different pH values to de-rooted barley seedlings via the subcrown internode on the elongation rate of the third leaf. Each point is the mean of 10 determinations of leaf extension rate (mm h$^{-1}$±SE). (a) In the light for 6 hours at 21–24°C. (b) In the light for 4 hours at 28–30°C then in the dark for a further 4 hours at 20°C. After Bacon *et al.* (1998).

increased hormone delivery and effects that could be a response to substantial accumulations of the hormone. It has always been difficult to understand why plants accumulate so much more ABA then is apparently required for the modification of growth and physiology (Jones & Davies 1991). It should be emphasized that, at least in shoots, the accumulation of ABA to what can be very high concentrations may be responsible for a different class of drought responses to those discussed in this chapter. These may, for example, require the modification of gene expression for the synthesis of compatible solutes or the synthesis of other compounds that can protect against desiccation injury. Studies by Spollen and Sharp (1994) and Spollen *et al.* (1996) have suggested that one of the roles of ABA in the stressed plant may be to counteract the effects of increased ethylene synthesis. Ethylene may act as a growth inhibitor in roots and shoots so that under certain circumstances ABA accumulation can promote growth (Mulholland *et al.* 1996a,b). It seems likely that the growth-inhibiting effects described above, caused by subtle modifications of ABA delivery, in what is effectively a well-watered plant will be very different from the effects of large increases in ABA concentration against a backgound of lowered water potential (and increased ethylene production).

In conclusion, it seems that under many circumstances there is excellent evidence of root to shoot signalling of the effects of soil drying and evidence of circumstances where these signals can control leaf growth and gas exchange. One of the best pieces of evidence in the literature is still the experiment of Gowing *et al.* (1990). Here, young apple trees had roots that were split between two containers, the soil in one of which was allowed to dry. This treatment resulted in a limitation in leaf growth that was alleviated by removing from the plant the roots in contact with drying soil. This result was interpreted as the removal of the source of an inhibitor of leaf growth but this inhibitor was not identified. This has also been the case in several other experiments. In other studies, authors have ruled out the involvement of chemical regulators, but in many of these studies no measurements of chemical regulators are made. The experiments of Tardieu's group have shown how easy it is to attribute control to a hydraulic effect when in fact the case for chemical control is more compelling. There is an important need for more comprehensive investigation of the effects of soil drying where both chemical and hydraulic influences are fully quantified. This is not an easy task because, as we have seen, there are many potential chemical regulators of growth and gas exchange, several of which may play an important role in the modification of growth and development.

## Interactions between hydraulic and chemical signals

Tardieu and Davies (1993) have suggested that chemical and hydraulic influences can interact with mild water deficits that on their own do not affect growth and development. This interaction can enhance the influence of a given dose of ABA. Our discussion of the distribution patterns of ABA (above) shows that it is likely that the drought-induced increase in sensitivity to ABA is brought about by increased penetration of ABA to the sites of action within the leaf. The interaction described by Tardieu and Davies (1993) contributes to the regulation of stomatal behaviour in maize plants, but also to the regulation of the water status of this plant. Tardieu and Simmoneau (1998) have described this type of behaviour as 'isohydric'. However, many plants show 'anisohydric' behaviour, and here water status is not regulated. Tardieu *et al.* (1996) have investigated the stomatal control in one of these plants (sunflower) and, interestingly, found no interaction between ABA and plant water status in the regulation of stomatal behaviour. It would be interesting to know whether the pH of the xylem sap of these plants varied significantly with soil drying. It is possible that without the increase in stomatal sensitivity to soil drying provided by the ABA–water status interaction, the plant is unable to effectively regulate its water status (Tardieu 1993).

Another excellent example of an apparent interaction between chemical and hydraulic influences on growth is provided by the work on maize root growth by Sharp's group. Here, growth of roots is sustained even at quite low water potentials and Saab *et al.* (1990) have suggested that this response requires the presence of ABA. Application of the carotenoid synthesis inhibitor fluridone increased the sensitivity of root growth to low soil water potential. Fluridone had no influence on growth at high water potential. These observations underline the possibility that ABA has a general stress physiological role within plants at low water potential. The combination of restriction of transpiration rate per unit leaf area, restriction of leaf area development and maintenance of root growth and presumably, therefore, maintenance of water uptake as the soil dries, will be expected to maintain turgor within the plant as water supply is restricted. It should be stressed, however, that the regulation of root growth described by Sharp's group does apparently involve a response to an accumulation of ABA. This is in contrast to the regulation of shoot growth considered above. One might speculate that the requirement for dynamic, rapidly reversible responses is less in the root than in the shoot, at least when soil drying is relatively mild. Substantial accumulation of ABA in shoots will also modify growth patterns, but as we have suggested this type of control may also be different to the more dynamic control described above.

Although Sharp's work on maize is the most comprehensive analysis of root growth at low water potential, it is important to note that all of this work has been conducted in vermiculite held at different water potentials. In drying soil, soil strength will increase greatly as water potential falls but this does not occur in vermiculite. Increased soil strength will often result in a fattening and shortening of roots which is a very different response to the water potential response. It is important to understand the interplay of these influences if we are to explain the below-ground responses of plants rooted in soil.

Wilkinson *et al.* (1998) have suggested that ABA may also play a role in modifying the water relations of the well-watered plant. It has been reported above that a variety of changes in the edaphic environment can lead to an alkalinization of xylem sap (Table 4.1). Without the presence of ABA in the sap this change will open stomata to a degree that is greater than that set by the prevailing climatic conditions. Such a change could lead to excessive water loss and to the development of potentially damaging water deficits.

## Scaling-up and conclusions

Much of the work discussed above has been performed with single plants or even with detached plant parts in the laboratory. To place these results properly in an agricultural or an ecological context, the evidence for a role for chemical signalling in the regulation of the functioning of crops and natural vegetation in the field must be considered. There have been comparatively few field-based studies that have properly considered a role for chemical regulators in the control of gas exchange, growth and development of plants in communities. One exception is the continuing work of Tardieu's group on field-grown maize and sunflower plants (described above). In another study, Wartinger et al. (1991) defined seasonal courses of gas exchange and xylem sap ABA concentration in almond trees under desert conditions. These authors suggested that the chemical regulator may have set the range of maximum conductances for the trees during the first part of the seasonal drying cycle. Later in the year when soil water content was lower, conductance was apparently determined by other factors. Fussader et al. (1992) report data from the same study which suggest an ameliorating influence of xylem cytokinin concentration on the ABA effect on stomatal behaviour. The results of the other detailed field study by Tardieu's group (see above) provide a good case for the involvement of ABA in the regulation of stomatal behaviour of maize and sunflower plants in the field. Here, the relationship between stomatal conductance and xylem ABA concentration is conserved across a range of environmental conditions, from leaf to leaf, through the crop canopy and from day to day, as drought stress develops. This conservation provides strong evidence that this hormone acts as a driving variable for stomatal behaviour in the field.

This chapter presents the evidence that plants have a well-developed mechanism or mechanisms for sensing soil moisture availability and are able to respond dynamically to episodes of relatively mild soil drying. It would appear that different species employ different control measures. For example, the isohydric and anisohydric species described by Tardieu and Simmoneau (1998) show very different control mechanisms and we have suggested that woody species may demonstrate a greater role for hydraulic signals than do herbaceous ones. In spite of this wealth of physiological information, the ecological significance of a plant's ability to respond to drying soil has been relatively little explored. Jones (1980) proposed a classification of responses which labelled species as 'optimists' or 'pessimists', depending on whether they used all the resources available to them in the expectation of more arriving (in the case of water, usually in the form of rain) or whether they modified their growth and physiology to conserve current resources and to control their demand for future resources in the expectation that such replenishment would not be forthcoming.

This classification of 'expectation' could perhaps be related to the life strategy classification propounded by Grime (1979). Competitors could be regarded as optimists in this regard while both stress tolerator and ruderal strategies correspond to a pessimistic viewpoint. It would appear that the control of water use via root-to-shoot communication would be a fundamentally conservative response, which would be expected to be more highly developed in species with stress-tolerant strategies compared with those with competitive strategies.

## References

Addicott, F.T., Lyon, J.L., Ohkuma, K. *et al.* (1968). Abscisic acid: a new name for abscisin II (dormin). *Science*, **159**, 1493.

Allan, A.C., Fricker, M.D., Ward, J.L., Beale, M.H. & Trewavas, A.J. (1994). Two transduction pathways mediate rapid effects of abscisic acid in *Commelina* guard cells. *Plant Cell*, **6**, 1319–1328.

Arrowsmith, D.A. & de Silva, J. (1995). Characterisation of two tomato fruit-expressed cDNAs encoding xyloglucan *endo*-transglycosylase. *Plant Molecular Biology*, **28**, 391–403.

Bacon, M.A., Wilkinson, S. & Davies, W.J. (1998). pH-regulated leaf cell expansion in droughted plants is abscisic acid-dependent. *Plant Physiology*, **118**, 1507–1515.

Blum, A., Munns, R., Passioura, J.B. & Turner, N.C. (1996). Genetically engineered plants resistant to soil drying and salt stress: how to interpret osmotic relations. *Plant Physiology*, **110**, 1051–1053.

Comstock, J. & Mencuccini, M. (1998). Control of stomatal conductance by leaf water potential in *Hymenoclea salsola* (T and G), a desert shrub. *Plant, Cell and Environment*, **21**, 1029–1038.

Correira, M.J. & Pereira, J.S. (1994). Abscisic acid in apoplastic sap can account for the restriction of leaf conductance of white lupins during moderate soil drying and after re-watering. *Plant, Cell and Environment*, **17**, 845–852.

Dannel, F., Pfeffer, H. & Marschner, H. (1995). Isolation of apoplasmic fluid from sunflower leaves and its use for studies on influence of nitrogen supply on apoplasmic pH. *Journal of Plant Physiology*, **146**, 273–278.

Davies, W.J. & Zhang, J. (1991). Root signals and the regulation of growth and development of plants in drying soil. *Annual Review of Plant Physiology and Molecular Biology*, **42**, 55–76.

Dodd, I.C. & Davies, W.J. (1994). Leaf growth responses to ABA are temperature-dependent. *Journal of Experimental Botany*, **45**, 903–907.

Ellenberg, H. (1953). Physiologisches und ökologisches Verhalten derselben Pflanzarten. *Berichte der Deutschen Botanischen Gesellschaft*, **65**, 351–362.

Ellenberg, H. (1954). Über einige Fortschritte der Kausalen Vegetationskundle. *Vegetatio*, **5**, 199–211.

Else, M.A. (1996). *Xylem-borne messages in the regulation of shoot responses to soil flooding.* PhD Thesis. Lancaster University, UK.

Else, M.A., Davies, W.J., Whitford, P.N., Hall, K.C. & Jackson, M.B. (1994). Concentrations of abscisic acid and other solutes in xylem sap from root systems of tomato and caster-oil plants are distorted by wounding and variable sap flow rates. *Journal of Experimental Botany*, **45**, 317–324.

Ferguson, A.R., Eiseman, J.A. & Leonard, J.A. (1983). Xylem sap from *Actinidia chinensis*: seasonal changes in composition. *Annals of Botany*, **51**, 823–833.

Fleming, A.J., McQueen-Mason, S., Mandel, T. & Kuhlemeier, C. (1997). Induction of leaf primordia by the cell wall protein expansin. *Science*, **276**, 1415–1418.

Fromard, L., Babin, V., Fleurat-Lessard, P., Froment, J.-C., Serrano, R. & Bonnemain, J.-L. (1995). Control of vascular sap pH by the vessel-associated cells in woody species. *Plant Physiology*, **108**, 913–918.

Fry, S.C., Smith, R.C., Renwick, K.F., Martin, D.J., Hodge, S.K. & Matthews, K.J. (1992). Xyloglucanendotransglycosylase, a new wall-loosening enzyme activity from plants. *Biochemical Journal*, **282**, 821–828.

Fussader, A., Wartinger, A., Hartung, W., Schulze, E.-D. & Heilmeier, H. (1992). Cytokinins in the xylem sap of desert grown almond (*Prunus dulcis*) trees: daily courses and their possible interactions with abscisic acid and leaf conductance. *New Phytologist*, **122**, 45–52.

Gollan, T., Schurr, U. & Schulze, E.-D. (1992). Stomatal responses to drying soil in relation to changes in xylem sap composition of *Helianthus annuus* I. The concentration of cations, anions and amino acids in and pH of the xylem sap. *Plant, Cell and Environment*, **15**, 551–559.

Gowing, D.J.G., Davies, W.J. & Jones, H.G. (1990). A positive root-sourced signal as an indicator of soil drying in apple, *Malus × domestica* Bork. *Journal of Experimental Botany*, **41**, 1535–1540.

Gowing, D.J.G. & Spoor, G. (1998). The effect of water table depth on the distribution of plant species in lowland wet grassland. In *UK Floodplains* (Ed. by R. Bailey, P. Jose & B. Sherwood), pp. 185–196. Westbury, Otley.

Gowing, D.J.G., Davies, W.J., Trejo, C.L. & Jones, H.G. (1993). Xylem-transported chemical signals and the regulation of plant growth and physiology. *Philosophical Transactions of the Royal Society of London, Series B*, **341**, 41–47.

Gowing, D.J.G., Gilbert, J.C., Youngs, E.G. & Spoor, G. (1997). *Water Regime Requirements of the Native Flora — With Particular Reference to ESAs.* Report to Ministry of Agriculture, Fisheries and Food, London. Project BD0209, Cranfield University, UK.

Grime, J.P. (1979). *Plant Strategies and Vegetation Processes.* Wiley, Chichester.

Hartung, W. (1983). The site of action of abscisic acid at the guard cell plasmalemma of *Valerianella locusta. Plant, Cell and Environment*, **6**, 427–428.

Hartung, W. & Radin, J.W. (1989). Abscisic acid in the mesophyll apoplast and in the root xylem sap of water stressed plants: the significance of pH gradients. *Current Topics in Plant Biochemistry and Physiology*, **8**, 110–124.

Hartung, W., Wilkinson, S. & Davies, W.J. (1998). Factors that regulate abscisic acid concentrations at the primary site of action at the guard cell. *Journal of Experimental Botany*, **49**, 361–368.

Henson, I.E., Jenson, C.R. & Turner, N.C. (1989). Leaf gas exchange and water relations of lupins and wheat. I. Shoot responses to soil water deficits. *Australian Journal of Plant Physiology*, **16**, 401–413.

Hoffman, B. & Kosegarten, H. (1995). FITC-dextran for measuring apoplast pH and apoplastic pH gradients between various cell types in sunflower leaves. *Physiologia Plantarum*, **95**, 327–335.

Holmberg, N. & Bulow, L. (1998). Improving stress tolerance of plants by gene transfer. *Trends in Plant Science*, **3**, 61–66.

Jackson, M.B. (1993). Are plant hormones involved in root to shoot communication? *Advances in Botanical Research*, **19**, 104–187.

Jarvis, A.J. & Davies, W.J. (1997). Whole plant water flux and the regulation of water loss in *Cedrella odorata. Plant, Cell and Environment*, **20**, 521–527.

Jones, H.G. (1980). Interaction and integration of adaptive responses to water stress: the implications of an unpredictable environment. In *Adaptation of Plants to Water and High Temperature Stress* (Ed. by N.C. Turner & P.J. Kramer), pp. 353–365. Wiley, New York.

Jones, H.G. (1989). Control of growth and stomatal behaviour at the whole plant level: effects of soil drying. In *Importance of Root to Shoot Communication in the Response to Environmental Stress* (Ed. by W.J. Davies & B. Jeffcoat), pp. 81–93. BPGRG Monograph 21, British Plant Growth Regulator Group, Bristol.

Jones, H.G. & Davies, W.J. (1991). A perspective on ABA research in the 1990s. In *Abscisic Acid* (Ed. by W.J. Davies & H.G. Jones), pp. 1–4. Bios Scientific, Oxford.

Jones, R.J. & Mansfield, T.A. (1970). Suppression of stomatal opening in leaves treated with abscisic acid. *Journal of Experimental Botany*, **21**, 714–719.

Kaiser, W.M. (1987). Effect of water deficit on photosynthetic capacity. *Physiologia Plantarum*, **71**, 142–149.

Kavi Kishor, P.B., Hong, Z., Miao, G., Hu, C. & Verma, D.P. (1995). Over-expression of pyrroline-5-carboxylate synthetase increases proline production and confers osmotolerance in transgenic plants. *Plant Physiology*, **108**, 1387–1394.

Kuang, J.B., Turner, N.C. & Henson, I.E. (1990). Influence of xylem water potential on leaf elongation and osmotic adjustment of wheat and lupin. *Journal of Experimental Botany*, **41**, 217–221.

Lagrimini, L.M., Burkhart, W.A., Moyer, M.B. & Rothstein, S. (1987). Molecular cloning of complementary, DNA encoding the lignin-forming peroxidase from tobacco: molecular analysis, tissue specific expression. *Proceedings of the National Acadamy of Sciences USA*, **84,** 7542–7546.

Lagrimini, L.M., Gingas, V., Finger, F., Rothstein, S. & Liu, T.-T.Y. (1997a). Characterisation of antisense transformed plants deficient in tobacco anionic peroxidase. *Plant Physiology*, **114,** 1187–1196.

Lagrimini, L.M., Joly, R.J., Dunlap, J.R. & Liu, T.-T.Y. (1997b). The consequence of peroxidase overexpression in transgenic plants on root growth and development. *Plant Molecular Biology*, **33,** 887–895.

Lecoeur, J., Wery, J., Turc, O. & Tardieu, F. (1995). Expansion of pea leaves subjected to short water deficit: cell number and cell size are sensitive to stress at different periods of leaf development. *Journal of Experimental Botany*, **46,** 1093–1101.

Lee, Y. & Satter, R.L. (1989). Effects of white, blue, red light and darkness on pH of the apoplast in the *Samanea* pulvinus. *Planta*, **178,** 31–40.

Lockhart, J.A. (1965). An analysis of irreversible plant cell elongation. *Journal of Theoretical Biology*, **8,** 264–275.

Loveys, B.R. (1984). Diurnal changes in water relations and abscisic acid in field-grown *Vitis vinifera* cultivars. III The influence of xylem-derived abscisic acid on leaf gas exchange. *New Phytologist*, **98,** 563–573.

MacAdam, J.W., Nelson, C.J. & Sharp, R.E. (1992). Peroxidase activity in the leaf elongation zone of tall fescue. 1. Spatial distribution of ionically bound peroxidase activity in genotypes differing in length of the elongation zone. *Plant Physiology*, **99,** 872–878.

McDonald, A.J.C., Davies, W.J. (1996). Keeping in touch: responses of the whole plant to deficits in water and nitrogen supply. *Advances in Botanical Research*, **22,** 228–300.

McQueen-Mason, S. (1995). Expansins and cell wall expansion. *Journal of Experimental Botany*, **46,** 1639–1650.

Mengel, K., Planker, R. & Hoffman, B. (1994). Relationship between leaf apoplast pH and iron chlorosis of sunflower (*Helianthus anuus* L.). *Journal of Plant Nutrition*, **17,** 1053–1065.

Mittelheuser, C.J. & van Steveninck, R.F.M. (1969). Stomatal closure and inhibition of transpiration induced by (RS)—abscisic acid. *Nature*, **221,** 281–282.

Morgan, J.M. (1984). Osmoregulation and water stress in higher plants. *Annual Review of Plant Physiology*, **35,** 299–319.

Mulholland, B.J., Black, C.R., Taylor, I.B., Roberts, J.A. & Lenton, J.R. (1996a). Effect of soil compaction on barley (*Hordeum vulgare* L.). 1 Possible role for ABA as a root-sourced chemical signal. *Journal of Experimental Botany*, **47,** 539–549.

Mulholland, B.J., Taylor, I.B., Black, C.R. & Roberts, J.A. (1996b). Effect of soil compaction on barley (*Hordeum vulgare* L.). 2 Are increasing xylem sap ABA concentrations involved in maintaining leaf expansion in compacted soils? *Journal of Experimental Botany*, **47,** 552–556.

Munns, R. (1989). Chemical signals moving from roots to shoots: the case against ABA, In *Importance of Root to Shoot Communication in the Response to Environmental Stress* (Ed. by W.J. Davies & B. Jeffcoat), pp. 175–184. BPGRG Monograph 21, British Plant Growth Regulator Group, Bristol.

Munns, R. & King, R.W. (1988). Abscisic acid is not the only stomatal inhibitor in the transpiration stream. *Plant Physiology*, **88,** 703–708.

Ober, E.S. & Sharp, R.E. (1994). Proline accumulation in maize (*Zea mays* L.) primary roots at low water potentials. 1. Requirement for increased levels of abscisic acid. *Plant Physiology*, **105,** 981–987.

Palmer, S.J., Berridge, D.M., McDonald, A.J.S. & Davies, W.J. (1996). Control of leaf expansion in sunflower (*Helianthus annuus* L) by nitrogen nutrition. *Journal of Experimental Botany*, **47,** 359–368.

Palmer, S.J. & Davies, W.J. (1996). An analysis of the relative elemental growth rate, epidermal cell size and xyloglucan endotransglycosylase activity through the growing zone of aging maize leaves. *Journal of Experimental Botany*, **47,** 339–347.

Passioura, J.B. (1977). Grain yield, harvest index and water use of wheat. *Journal of Australian Institute of Agricultural Science*, **43,** 117–121.

Passioura, J.B. (1988). Root signals control leaf expansion in wheat seedlings growing in drying soil. *Australian Journal of Plant Physiology*, **15,** 687–693.

Passioura, J.B. & Munns, R. (1984). Hydraulic resistance of plants II. Effects of rooting medium and time of day in barley and lupins. *Australian Journal of Plant Physiology*, **11**, 341–350.

Radin, J.W., Parker, L.L. & Guinn, G. (1982). Water relations of cotton plants under nitrogen deficiency. V. Environmental control of abscisic acid accumulation and stomatal sensitivity to abscisic acid. *Plant Physiology*, **70**, 1066–1070.

Saab, I.N., Sharp, R.E., Pritchard, J. & Voetberg, G.S. (1990). Increased endogenous ABA maintains primary root growth and inhibits shoot growth of maize seedlings at low water potentials. *Plant Physiology*, **93**, 1329–1336.

Sauter, J.J. & Ambrosius, T. (1986). Changes in the partitioning of carbohydrates in the wood during bud break in *Betula pendula* Roth. *Journal of Plant Physiology*, **124**, 31–43.

Schurr, U. (1998). Xylem sap sampling—new approaches to an old topic. *Trends in Plant Science*, **3**, 293–298.

Schurr, U. & Schulze, E.-D. (1995). The concentration of xylem sap constituents in root exudate, and in sap from intact, transpiring castor bean plants (*Ricinus communis* L). *Plant, Cell and Environment*, **18**, 409–420.

Schurr, U. & Schulze, E.-D. (1996). Effect of drought on nutrient transport and ABA transport in *Ricinus communis*. *Plant, Cell and Environment*, **19**, 665–674.

Sharp, R.E. & Davies, W.J. (1985). Root growth and water uptake by maize plants in drying soil. *Journal of Experimental Botany*, **36**, 1441–1456.

Shcherban, T.Y., Shi, J., Durachko, D.M. *et al.* (1995). Molecular cloning, sequence analysis of expansins—a highly conserved multi-gene family of proteins that mediate cell wall extension in plants. *Proceedings of the National Acadamy of Sciences USA*, **92**, 9245–9249.

Slovik, S., Daeter, W. & Hartung, W. (1995). Compartmental redistribution and long-distance transport of ABA in plants influenced by environmental changes in the rhizosphere—a biomathematical model. *Journal of Experimental Botany*, **46**, 881–894.

Spollen, W.G. & Sharp, R.E. (1991). Spatial distribution of turgor and root growth at low water potentials. *Plant Physiology*, **96**, 438–443.

Spollen, W.G. & Sharp, R.E. (1994). Role of ABA in root growth maintenance at low water potentials

involves regulation of ethylene synthesis or responsiveness. *Plant Physiology*, **105S**, 25.

Spollen, W.G., Sharp, R.E., Saab, I.N. & Wu, Y. (1993). Regulation of cell expansion in roots and shoots at low water potentials. In *Water Deficits: Plant Responses from Cell to Community* (Ed. by J.A.C. Smith & H. Griffiths), pp. 37–52. Bios Scientific, Oxford.

Spollen, W.G., LeNoble, M.E. & Sharp, R.E. (1996). Regulation of root ethylene production by accumulation of endogenous ABA at low water potentials. *Plant Physiology*, **114S**, 446.

Tardieu, F. (1993). Will progress in understanding root–soil relations substantially alter water flux models? *Philosophical Transactions of the Royal Society of London Series B*, **341**, 57–66.

Tardieu, F. & Davies, W.J. (1992). Stomatal response to ABA is a function of current plant water status. *Plant Physiology*, **98**, 540–545.

Tardieu, F. & Davies, W.J. (1993). Integration of hydraulic and chemical signalling in the control of stomatal conductance and water status of droughted plants. *Plant, Cell and Environment*, **16**, 341–350.

Tardieu, F. & Simmoneau, T. (1998). Variability among species of stomatal control under fluctuating soil water status and evaporative demand: modelling isohydric and anisohydric behaviours. *Journal of Experimental Botany*, **49**, 419–432.

Tardieu, F., Zhang, J. & Davies, W.J. (1992). What information is conveyed by an ABA signal from maize roots in drying field soil? *Plant, Cell and Environment*, **15**, 185–192.

Tardieu, F., Lafarge, T. & Simmoneau, T. (1996). Stomatal control by fed or endogenous xylem ABA in sunflower: interpretation of observed correlations between leaf water potential and stomatal conductance in anisohydric species. *Plant, Cell and Environment*, **19**, 75–84.

Tetlow, I.J. & Farrar, J.F. (1993). Apoplastic sugar concentration and pH in barley leaves infected with brown rust. *Journal of Experimental Botany*, **44**, 926–936.

Thompson, D.S., Wilkinson, S., Bacon, M.A. & Davies, W.J. (1997). Multiple signals and mechanisms that regulate leaf growth and stomatal behaviour during water deficit. *Physiologia Plantarum*, **100**, 303–313.

Turner, N.C. (1986). Crop water deficits: a decade

of progress. *Advances in Agronomy*, **39**, 1–51.

Turner, N.C., Schulze, E.-D. & Gollan, T. (1985). The responses of stomata and leaf gas exchange to vapour pressure deficit and soil water content. II. In the mesophytic herbaceous species *Helianthus annuus*. *Oecologia*, **65**, 348–355.

Wartinger, A., Heilmeier, H., Hartung, W. & Schulze, E.-D. (1991). Daily and seasonal courses of leaf conductance and abscisic acid in xylem sap of almond trees (*Prunus dulcis*) under desert conditions. *New Phytologist*, **116**, 581–587.

Wilkinson, S. & Davies, W.J. (1997). Xylem sap pH increase: a drought signal received at the apoplastic face of the guard cell which involves saturable ABA uptake by the epidermal symplast. *Plant Physiology*, **113**, 559–573.

Wilkinson, S., Corlett, J.E., Oger, L. & Davies, W.J. (1998). Effects of xylem pH on transpiration from wild-type and *flacca* tomato leaves. A vital role for abscisic acid in preventing excessive water loss even from well watered plants. *Plant Physiology*, **117**, 703–710.

Wolf, O., Jeschke, W.D. & Hartung, W. (1990). Long distance transport of abscisic acid in salt stressed *Lupinus albus* plants. *Journal of Experimental Botany*, **41**, 581–587.

Wright, S.T.C. (1977). The relationship between leaf water potential and levels of abscisic acid and ethylene in excised wheat leaves. *Planta*, **134**, 183–189.

Wu, Y.J., Spollen, W.G., Sharp, R.E., Hetherington, P.R. & Fry, S.C. (1994). Root growth maintenance at low water potentials—increased activity of xyloglucan endotransglycosylase and its possible regulation by ABA. *Plant Physiology*, **106**, 607–615.

Wu, Y.J., Sharp, R.E., Durachko, D.M. & Cosgrove, D.J. (1996). Growth maintenance of the primary maize root at low water potentials involves increases in cell wall extension properties, expansin activity and wall susceptibility to expansins. *Plant Physiology*, **111**, 765–772.

Zhang, J. & Davies, W.J. (1989). Abscisic acid produced in dehydrating roots may enable the plant to measure the water status of the soil. *Plant, Cell and Environment*, **12**, 73–81.

Zhu, G.L. & Boyer, J.S. (1992). Enlargement in *Chara* studied with a turgor clamp—growth-rate is not determined by turgor. *Plant Physiology*, **100**, 2071–2080.

# Chapter 5

# Evolution and ecology of plant mineral nutrition

*G.R. Stewart[1] and S. Schmidt[2]*

## In the beginning

Astronomers appear to have discovered that universes have an ecology and indeed a metabolism (Rees 1997) and no doubt our universe has a distinctive physiological ecology. The task here, however, is to address the physiological ecology of plants. Traditionally, this has deployed two approaches: the *in vivo*, in which physiological processes were studied at the whole-plant or organ level, in field and laboratory, and the *in vitro*, in which cellular and subcellular processes were dissected in order to understand the mechanisms underlying physiological responses. Bringing us closer to the astronomy fraternity there is a third approach, the *in silico,* in which computer simulation and graphic modelling provide tools for the study of past and future plants and processes. This contribution attempts to look back to the past, to the origins of life, in order to understand the selectivity in the use of mineral elements and their roles in extant physiological ecology.

What do we know of the conditions present on early Earth that shaped the origin and evolution of life and determined the selection and roles of the mineral elements that are now essential for plant growth? The emergence of life may have occurred in a fairly short interval, microfossils are reported in 3.5-Gyr rock (Schopf & Walter 1983) and there is evidence of biogenic carbon isotope fractionation in 3.8-Gyr rock (Schidlowski 1988). The early Earth on which life appeared was very different to the planet we inhabit today. Life originated not in a quiescent, favourable environment but rather a violent, impact-ridden one, characterized by enormous tides, intense volcanic activity and heavy meteoritic bombardment (Sleep *et al.* 1989; Chyba 1993). Current thinking is that this early Earth had an atmosphere rich in carbon dioxide and nitrogen (Walker 1986), rather than one of methane and ammonia — conditions rather unfavourable for cooking up prebiotic soups.

## Metals and early evolutionary steps

The dominant biogenic elements that are found in extant organisms, hydrogen,

[1] *Faculty of Science, University of Western Australia, Nedlands, WA 6907, Australia. E-mail: gstewart@science.uwa.edu.au.*
[2] *Department of Botany, University of Queensland, Brisbane, QLD 4072, Australia. E-mail: s.schmidt@botany.uq.edu.au.*

carbon, nitrogen, oxygen and phosphorus, formed by cosmic nucleosynthesis, are those that are the most common in the Universe. They are mostly elements whose roles in biological processes are irreplaceable. Life on planet Earth is a carbon-based chemistry; possible replacement elements, silicon, boron, phosphorus or germanium, have the potential to form a much less rich array of stable compounds and linkages with themselves. Hydrogen and oxygen are essential for that often limiting resource, water. Nitrogen is essential for protein and nucleic acids. Phosphorus, a much less abundant element, is required for nucleic acids and high-energy compounds. None of these has any obvious, alternative replacement elements.

However, when other essential elements of extant organisms are considered, their biological significance is clear but their irreplaceability is not. It has been argued (Frausto da Silva & Williams 1991) that they were selected on the basis of their abundance and availability. Thus, calcium, magnesium, potassium and sulphur were selected in preference to strontium, beryllium, rubidium and selenium simply because they were more abundant in the environment that life originated in.

The classic theory of a heterotrophic origin of life, as expressed in a number of papers by both Oparin and Haldane, is inconsistent with the environmental conditions now thought to have been prevalent on early Earth 4.5 Gyr ago — the atmosphere above a liquid water ocean comprised $CO_2+CO$ ($10^6$ Pa) and $N_2$ ($10^5$ Pa) (Maden 1995). Certainly this would have been a world with an environment unfavourable for organic synthesis. The heterotrophic schema requires the prebiotic synthesis of the basic amino acids (arginine and lysine), various coenzymes and other metabolites whose abiotic synthesis is as yet to be convincingly demonstrated. Molecular studies are consistent with a universal phylogeny and identify the earliest living organisms as hyperthermophilic autotrophs (Barns *et al.* 1996). The soup theory gives scant attention to the evolution of mineral nutrition and offers little explanation of the role of metals in metabolism. In fact, minerals may have played a key role in the earliest stages of molecular evolution. Suggested roles for minerals include the synthesis of amino acids and sugar-like molecules, adsorption of organic molecules, catalytic condensation of peptides and oligonucleotides, as a reductant for the early fixation of carbon dioxide and as primitive genetic material.

The first influential hypothesis that minerals were involved in the origin of life came from Bernal (1951), who suggested alumino-silicate (clay) minerals could have served as adsorbents for amino acids and sugars, acting to concentrate them. Bernal remained a soup stirrer but provided a mechanism for concentrating the broth of otherwise rather dilute reactants. Cairns-Smith (1982) developed a more radical hypothesis for the role for clay minerals in the so-called 'clay world'. Clay crystals were suggested to act as a prebiotic informational surface upon which a primitive metabolism started. Following on from the suggestions of Hartman (1975) of an early oxidative citric acid cycle, and Racker (1965) and Hartman (1975) of thioester energy preceding phosphate bond energy, Wächtershäuser

(1988, 1992) put forward an alternative, speculative but intriguing hypothesis of a chemoautotrophic origin for life. This highlights the pivotal role of iron and other catalytic metals in the evolution of carbon-fixing organisms (Wächtershäuser 1992). Similarly Russell and coworkers have invoked FeS 'membranes' laced with nickel, as a means for creating an environment for the development of biochemical reactions in the Hadean ocean (Russell *et al.* 1994; Russell & Hall 1997).

In the pyrite world the energy source for early biochemical reactions is postulated to be the reducing power of FeS/$H_2$S with the oxidative formation of pyrite and reduced organic compounds:

$$FeS + HS^- \rightarrow FeS_2 + H^+ + 2e^- \tag{1}$$

Early metabolism is suggested as having developed on the pyrite surface. Organic intermediates with anionic groups were able to bond to the cationic surface of pyrite and transition-metal sulphides. It is postulated that initially pyrite formation drove an archaic version of the reductive citric acid cycle, and from this the central biosynthetic pathways developed. In this scheme pyrite formation is the driving force for thioester activation, thioesters being the forerunner of high-energy phosphate compounds. In this pyrite world, enzymes and nucleic acids are viewed as products, not the precursors to archaic biochemistry.

## Sulphur and selenium

Wächtershäuser and Russell accord sulphur a key function in early metabolism. However selenium has similar redox properties similar to those of sulphur and could have had a catalytic function in an archaic pyrite-driven metabolism. In extant organisms selenocysteine is present in a small number of enzymes, including the iron-nickel-sulphur hydrogenases, formate dehydrogenase, glutathione peroxidase and glycine reductase (Stadtman 1990; Baker *et al.* 1998). The formyl-methanofuran dehydrogenase of the hyperthermophilic archaeon, *Methanopyrus kandleri,* has an active site selenocysteine (see below) and is a tungsten enzyme (Vorholt *et al.* 1997). There appears to be a higher frequency of selenocysteine residues in proteins of archaeons, suggesting that the present-day distinction between sulphur and selenium may have been less in archaic metabolism and that through evolution there has been progressive discrimination against selenium in favour of sulphur.

For higher plants in general, selenium is toxic, its toxicity arising from substitution of sulphur by selenium in essential bio-organic molecules. A small number of higher plants are recognized by their capacity to accumulate large amounts of selenium: 2–4% of plant dry weight (Stewart & Larher 1980). Such plants sequester selenium primarily as methylselenocysteine. Even though these so-called primary indicator plants accumulate large amounts of selenium, there is little evidence of an essential metabolic role for selenium-compounds in accumulating species.

## The copper anomaly

The proposed involvement of metals that take on a biocatalytic function in the pyrite world arises from the solubility of their sulphides: $Mn^{2+} > Fe^{2+} > Ni^{2+} > Co^{2+} > Zn^{2+}$. Under acid conditions they would have been available to play catalytic functions in a surface metabolism. At a later evolutionary stage they become universal biocatalysts. It is striking that many of the other transition metals ($Cd^{2+}$, $Pb^{2+}$, $Ag^{2+}$, $Hg^{2+}$, $Sn^{2+}$, $As^{3+}$, $Sb^{3+}$, $Bi^{3+}$) have insoluble sulphides and for extant organisms many of these metals are extremely toxic (Wächtershäuser 1992).

$Cu^{2+}$ has an anomalous position in this hypothesis because its sulphide is insoluble, yet for extant aerobic organisms it is an essential trace element. Copper is known to be essential as a biocatalyst for plant growth, although it can become toxic at high concentrations. All biological functions involve copper in the monovalent form, either alone or in the form of the $Cu^{II}/Cu^{I}$ redox pair. Copper has unique chemical characteristics in its two relevant oxidation states, +I and +II. Copper enzymes have almost exclusive functions in the metabolism of $O_2$ and N/O compounds and in association with oxidizing organic and inorganic radicals. Our understanding of copper metabolism has been enhanced by the discovery of copper chaperone proteins that deliver copper to copper-containing proteins. These have been identified in diverse organisms (Culotta et al. 1997).

It is striking that the requirement for copper is widespread among aerobic organisms but absent in anaerobes (Huheey 1988). The lack of a copper requirement in anaerobes is consistent with the insolubility of CuS. Moreover the high potential necessary for oxidation to the +II state suggests that copper bioavailability would have been very limited prior to the photosynthetic generation of an oxidizing environment. In this scheme of things copper would have became bioavailable 2–3 Myr ago. Copper can therefore be regarded as a 'modern' bioelement (Kaim & Rall 1996).

## Zinc, nickel and tungsten

The requirement for zinc is universal among living organisms but we can well ask why zinc has been selected for specific biochemical functions in preference to other metals? The electrostatic properties of zinc are similar to those of magnesium, copper and nickel and, to a lesser extent, calcium. Zinc is a strong Lewis acid (a good electron acceptor) and in this it is similar to copper and nickel. Unlike copper and nickel it does not have variable valence. Its closest similarity is with cadmium, a metal toxic to many, if not all, living organisms. However, Lee and Morel (1995) have shown that cadmium can alleviate growth limitation imposed by suboptimal concentrations of zinc in a variety of marine phytoplankton. It may simply be that the preference of living organisms for zinc over related metals reflects its greater availability in the biosphere. Did then increasing specificity for zinc lead to increasing toxicity of cadmium? In at least one instance we know that the biochemical function of zinc can in fact be replaced by cobalt. Cobalt can substitute for zinc in the carbonic anhydrase of the marine diatom *Thalassiosira weissflogii* (Lee & Morel

1996). The identification of zinc transfer mechanisms between thioneins and zinc proteins (Jacob *et al.* 1998) may help in understanding the biochemical basis of zinc specificity.

Many of the zinc-containing enzymes hydrolyse esters or amides, and include peptidases, proteases and 3′ nucleotidases. In addition, zinc is a component of some superoxide dismutases, many dehydrogenases, carbonic anhydrase, RNA polymerase, reverse transcriptase and the zinc finger proteins. The latter include zincDNA-binding proteins. Zinc is a catalyst in RNA synthesis and hydrolysis, an anomaly that has implications for evolutionary hypotheses invoking ribo-organisms that utilized RNA as both genetic material and catalyst (Lazcano *et al.* 1988).

In plants (and other higher organisms) nickel is only in found in urease with 2 nickel atoms per molecule. In marked contrast, anaerobic bacteria, particularly the archaeon methanogens, have nickel dehydrogenases and other nickel-containing enzymes (Hausinger 1994). This suggests that nickel was more widely used in archaic organisms than it is today. In contrast to zinc, nickel and cobalt have acid-soluble sulphides and would have been able to participate in catalytic functions on pyrite surfaces. Later in the evolution of metalloenzymes we can speculate that some of the catalytic functions of nickel may have been taken over by zinc. The nickel enzymes of archaeons may be in one sense, living 'fossil' enzymes.

It is interesting that many archaeons also have a number of tungsten-containing proteins (Johnson *et al.* 1996). Neither molybdenum nor vanadium can replace tungsten in the catalytically active forms of aldehyde ferredoxin oxidoreductase, formaldehyde ferredoxin oxidoreductase and glyceraldehyde-3-phosphate oxido-reductase from the archaeon *Pyrococcus furiosus* (Mukund & Adams 1996).

It is tempting to speculate that what appear to be the unusual features of metal-lobiochemistry of some archaeons, particularly the hyperthermophiles, reflect remnants of an archaic, less selective mineral nutrition. Alternatively it might simply be that their unusual metal requirements reflect their evolution in a 'mineral environment' rather different to that of most other organisms.

## Archaic nitrogen metabolism

Was nitrogen essential for the origin of life and if so in what form was it available? All theories of a heterotrophic origin for life assume an archaic chemistry characterized by a rich diversity of nitrogenous compounds. Current views of the early atmosphere, however, suggest that dinitrogen is likely to have been the available source of nitrogen for the earliest biochemical systems (see above). There are few obvious sources of ammonium in an anaerobic, prebiotic earth. Fe(II) reduction of nitrate and nitrite has been suggested, providing a source of ammonia on early Earth (Summers & Chang 1993). Another possible source could have been the release of $NH_3/NH_4^+$ during degassing of the Earth's mantle, particularly if this occurred in hydrothermal vents. Interestingly, studies of the growth requirements of modern archaeons, including hyperthermophilic species isolated from marine hydrothermal systems, indicate that many are heterotrophic with respect to nitrogen use. They require complex amino acid mixtures, peptides or proteins (Duffaud *et al.* 1998; Gonzalez *et al.* 1998; Keller *et al.* 1998). The evolutionary implications of such studies depend very much on the evolutionary position assigned to hyperthermophilic archaeons. It is likely that the ancestor to the Archaea was a hyperthermophile, consistent with the 'hot' origin of life hypothesis and the supposed primitiveness of this group. If the hyperthermophilic archaeons do resemble early life forms that evolved in habitats akin to the deep-sea vents of the mid-ocean rifts, then the implication is that heterotrophic nitrogen use may have preceded dinitrogen fixation. However, archaeon hyperthermophily may not be a primitive characteristic but rather an elaborate reductive adaptation, essential for survival at high temperatures (Forterre 1995). Furthermore, it has been argued that prokaryotes and eukaryotes evolved from a common mesophilic ancestor (Forterre 1995). This evolutionary scenario would be consistent with the view that autotrophic nitrogen assimilation preceded heterotrophic nitrogen nutrition.

In the context of Wächtershäuser's chemo-autotrophic hypothesis it is significant that many of the enzymes of nitrogen assimilation are iron-sulphur proteins or are dependent on reduced ferredoxin as a reductant. In the archaic world nitrogen fixation could have been driven by pyrite formation. Extant nitrogenases are Fe-S or Fe-S-Mo or Fe-S-V enzymes, moreover the Fe-Mo co-factor is polyanionic, and is able to bind to a pyrite surface. Homocitrate or citrate is a co-factor of Mo-nitrogenases, suggesting a relationship between nitrogen fixation and the reductive citrate cycle (Wächtershäuser 1992). The extant ferredoxins and other iron–sulphur proteins can be seen as having evolved from archaic pyrite-driven reactions.

Eukaryotic nitrate reductases have three highly conserved binding domains for flavin adenine dinucleotide (FAD), haem and the molybdenum co-factor and use pyridine nucleotide electron donors (Zhou & Kleinhofs 1996). By contrast, nitrite reductases from photosynthetic organisms have a sirohaem prosthetic group and a 4Fe-4S cluster at the active centre and they utilize ferredoxin (Kleinhofs & Warner 1990). The nitrite reductase of fungi uses NADPH as the electron donor.

It is suggested (Wächtershäuser 1992) that the pyrite-forming reducing agent $FeS/H_2S$ was the reductant for an archaic glutamate pathway:

$$-CO- + NH_3 + FeS + H_2S \rightarrow -CHNH_2- + FeS_2 + H_2O \qquad (2)$$

Later this was replaced by the glutamate synthase cycle. Glutamate synthase of cyanobacteria, algae and plants is an iron-sulphur-ferredoxin-dependent enzyme (Stewart *et al.* 1980). This is suggestive of an early evolution of the glutamate synthase cycle. The extant 'alternative' ammonia assimilatory route is catalysed by the Zn-dependent glutamate dehydrogenase.

Kinetic studies have shown that extant glutamate synthases are capable of both a glutamine-dependent and ammonia-dependent synthesis of glutamate (Stewart *et al.* 1980), albeit that the rate with ammonia is less than 8% that with glutamine. In an early 'glutamate synthase cycle' the ability to use ammonia as a substrate may have been advantageous in the absence of a route via glutamate dehydrogenase and in a hyperthermal environment of unstable amides.

Glutamine serves as a $H_2N$ source for the synthesis of purines, pyrimidines, folates, histidine, tryptophane and asparagine (Lea *et al.* 1990). These amide-transfer enzymes from eukaryotes can substitute ammonia for glutamine (Mei & Zalkin 1989). One such group of amide-transfer enzymes, the carbomyltransferases, have attracted much attention for studies of molecular evolution and it is striking that the ornithine carbomyltransferase from the hyperthermophilic archaeon, *P. furiosus,* is ammonia rather than glutamine dependent (Legrain *et al.* 1995). This is possibly indicative that glutamine dependence is a more recent biochemical characteristic.

In the pyrite world biocatalytic metals would have played a key role in the prebiotic synthesis of amino acids and peptides. Consistent with this, amino acid synthesis via a reductive amination of alpha-oxo acids with the concomitant formation of pyrite has been shown (Hafenbrandl *et al.* 1995). Furthermore, it has been demonstrated that there is the potential to synthesize amide (peptide) bonds via a reductive acetylation of amino acids driven by pyrite formation (Keller *et al.* 1994) and, recently, peptide formation using co-precipitated (Ni, Fe)S and CO together with $H_2S$ has been observed (Huber & Wächtershäuser 1998).

## Evolution of the nitrogen cycle

For extant plants there is an almost universal reliance on the intermediates of the nitrogen cycle and it is generally assumed that they predominantly assimilate ammonium and nitrate as their principal nitrogen sources. In the archaic world dinitrogen is likely to have been the most stable and most abundant form of nitrogen. As discussed earlier, nitrogen fixation is of great antiquity, there is a high homology of the nitrogenase complex and the process is strictly anaerobic. The stromatolites found today at Shark Bay in Western Australia are formed by living descendants of the reef-forming cyanobacteria that produced the 3500-Myr stromatolites from Northernwestern Australia. Some of these cyanobacteria are nitrogen fixing so that their ancestors, as well as being the earliest known photosynthetic organisms, may have also fixed nitrogen.

It would seem probable that fixed inorganic nitrogen may have been quite scarce in the archaic world until after the evolution of nitrogen-fixing organisms such as the cyanobacteria. A wide range of organisms, anaerobes as well as aerobes, mediate the release of nitrogen from the protein of dead organisms through ammonification. However the formation of nitrate from ammonium involves two oxidation steps, indicating that nitrification must have evolved after the atmosphere became oxygen enriched. Denitrification was probably the last process of the cycle to emerge. It is likely to have evolved independently several times given that denitrifying organisms are diverse and include archaeons and Eubacteria (Zumpft 1992). The bacterial denitrifiers have nitrite reductases that are copper dependent and, as discussed above, copper enzymes are indicative of an oxygen-rich atmosphere.

The initial colonizers of terrestrial ecosystems were most probably bacteria, algae, lichens and fungi. Unequivocal evidence for multicellular land plants in and prior to the Mid Silurian is sparse. However spore tetrads have been recovered from various mid-Ordovician sites, although their exact relationship to land plants is uncertain. Plant megafossils and spores recovered from the Early Silurian suggest early land plants comparable with extant liverworts (Gray & Shear 1992). Palaeoenvironmental reconstructions of this time imply these early land plants occupied a humid coastal habitat (Moore *et al.* 1994).

Early land plants were, most probably, closely related to the Charophyceae and probably had a freshwater origin. Gene sequencing of 18S RNA and mitochondrial DNA is consistent with a monophyletic origin (Kenrick & Crane 1997). The Barag-wanathian flora of Victoria (Australia) is of late Silurian age and here the lycopod-like *Baragwanathia* occurs together with Rhyniophytes and a Zosterophyll (Dettmann 1994). For these early plants transition from an aqueous to a gaseous habitat meant exposure to new environmental conditions that necessitated structural and physiological changes. In particular they would have needed to obtain nutrients from substrate. Root-like structures are almost recognizable in Late Silurian palaeosols and were certainly present by the Early Devonian (Kenrick & Crane 1997). *Rhynia* had a fungal associate but its mineral nutrition is little understood.

The extant species *Psilotum nudum*, although now regarded as being more closely related to fern–seed plant lineage than the clubmoss or rhyniophyte precursors of modern vascular plants (Pryer *et al.* 1995), can still be regarded as possessing many of the features that would have characterized the earliest land plants. Its rhizomes have rhizoidal hairs that are active in nutrient absorption. The rhizome is colonized by endomycorrhizal fungi. Studies with this species show that it has an inducible nitrate reductase (N. Robinson, S. Schmidt & G.R. Stewart, unpublished data). Extant mosses and liverworts from diverse habitats are active in nitrate utilization (Fig. 5.1) and it can be speculated that the ancestors of these and other early land plants probably were also able to assimilate nitrate. The development of more advanced land plants and the expansion of terrestrial biomass probably proceeded concomitantly with increased nitrogen cycling.

In the early terrestrial world the first land plants would have encountered a 'soil' environment with a nitrogen cycle limited by carbon availability. These plants had poorly developed root systems. Simon *et al.* (1993) have argued that endomycorrhization was critical for the exploitation of terrestrial nutrient sources. In a low biomass ecosystem opportunities for fungal colonization may have been limited. Even if endomycorrhizas were common, we suggest that these simple land plants would have had specific nutrient transport systems inherited from their aquatic ancestors. The past few years have seen remarkable progress in our understanding of the molecular mechanisms of nutrient uptake systems. A high-affinity transporter for ammonium has been identified in root cells of rice (von Wirén *et al.*

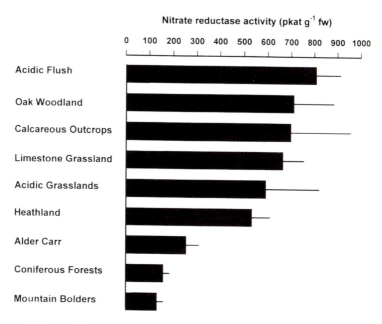

**Figure 5.1** Nitrate reductase activity of bryophytes from different habitats. Samples were collected from various sites in England, Scotland and Wales. Averages of 6–10 species per site and standard deviations are shown.

1997) and root hairs of tomato (Lauter *et al.* 1996). High-affinity nitrate transporters, belonging to the major facilitator superfamily, have been cloned from higher plants (Quesada *et al.* 1997) and there are several reports describing the nitrate transporters of higher plants (Wang & Crawford 1996; Amarasinghe *et al.* 1998; Krapp *et al.* 1998).

## Nitrogen and the evolution of the Australian flora

Nutrients and the specialized mechanisms for their acquisition have had a long and critical role in determining the ecophysiology of Australian plants. Australia is an old continent and all major landscape elements have existed for as long as 90 Myr. Many of its ancient landscapes have been exposed to the processes of weathering and soil formation for long periods of stable tectonic conditions. Consequently, a major factor in the evolution of the flora has been the much leached and largely infertile soils. The substrate for much of Australian vegetation since at least the Early Tertiary has been highly leached acid soils. Although what is regarded as 'typical' Australian vegetation is descended from the ancient Gondwanan flora, its history is a post-Jurassic phenomenon; few pre-Cretaceous plants had a major impact on the extant flora (Hill 1994).

The preangiosperm flora suggests dominance of araucarians and podocarps in the overstorey associated with pteridosperms, cycads, cycadeoids and cryptogams

**Table 5.1** Foliar nitrate reductase activity (pkat g$^{-1}$ fresh weight) and leaf nitrogen content (per cent of dry weight) of Pteridophytes and Gymnosperms. Samples were collected from Southeast Queensland and assayed as described in Stewart *et al.* (1986).

| Species | Nitrate reductase activity (pkat g$^{-1}$ fw) | Foliar nitrogen (%) |
|---|---|---|
| *Blechnum cartilagineum* | 40–80 | 0.8–1.0 |
| *Cyathea woollsiana* | 50–60 | 1.0–1.2 |
| *Platycerium bifurcatum* | 50–70 | 0.9–1.1 |
| *Cycas media* | 20–40 | 1.0–1.2 |
| *Macrozamia miquelli* | 25–45 | 0.9–1.2 |
| *Agathis robusta* | 25–45 | 0.7–0.9 |
| *Araucaria bidwillii* | 12–34 | 0.7–1.0 |
| *Araucaria cunninghamii* | 16–35 | 0.6–0.9 |
| *Callitris columellaris* | 30–40 | 0.8–0.9 |
| *Podocarpus elatus* | 10–20 | 0.5–0.7 |
| *Podocarpus spinulosus* | 0–10 | ND |

ND = not determined.

in the understorey of Early Cretaceous vegetation (Dettmann 1994; Douglas 1994). Modern representatives of these groups tend to have a low capacity for leaf nitrate utilization and they have low leaf concentrations of nitrogen (Table 5.1; Smirnoff *et al.* 1984; Stewart *et al.* 1986). Presumably the only vascular components with a potential for symbiotic nitrogen fixation would have been those of the Cycadophyta. No other extant and presumably no extinct gymnosperms fix(ed) dinitrogen and, to our knowledge, have no associative nitrogen fixers.

Angiosperms arrived in Australia during Late Barremian to Early Aptian times. The earliest forms were magnoloid and probably occupied lakeside and riverine areas (Dettmann 1994). Consistent with what we regard today as the relative inefficiency of the magnoloid root system and the characteristics of similar modern habitats, they are most likely to have colonized sites of relatively high nutrient availability.

During the Late Cenomanian, environmental patterns underwent a marked change, particularly in Southeast Australia, namely a shift from riparian and aquatic communities to dry zone associations. In the Late Cenomanian–Early Turonian there are further changes and the appearance of early proteaceous species (Dettmann 1994). What is then seen is the appearance of species characteristic of sclerophyllous communities on nutrient-deficient soils. This process probably accelerated during the Tertiary where prolonged leaching of soils occurred. The evolution of sclerophylly in the Australian vegetation has been ascribed to the development of low soil fertility (Barlow 1981).

What can be inferred from these early angiosperm taxa, present as fossils, regarding their nitrogen nutrition? Our studies of nitrate assimilation in extant species suggest that the capacity to assimilate nitrate is very much a taxonomic character, and families with a high or low capacity to reduce nitrate can be recognized (Smirnoff *et al.* 1984; Turnbull *et al.* 1996). Almost all members of families

**Table 5.2** Foliar nitrate reductase activity of flowering plant families. Plants were sampled from undisturbed sites in Queensland, Northern Territory and Western Australia. Leaf tissue was analysed as described in Stewart *et al.* (1986). Average activity (±SD) is shown.

| Plant family | Nitrate reductase activity (pkat g$^{-1}$ fw) |
| --- | --- |
| Amaranthaceae | 1115 ± 50 |
| Chenopodiaceae | 1117 ± 190 |
| Epacridaceae | <50 |
| Ericaceae | <50 |
| Moraceae | 800 ± 280 |
| Myrtaceae | 100 ± 120 |
| Polygonaceae | 2200 ± 300 |
| Proteaceae | <50 |
| Urticaeae | 2100 ± 200 |

such as the Proteaceae, Epacridaceae, Ericaceae and Myrtaceae generally exhibit low levels of root and shoot nitrate reductase, irrespective of the availability of nitrate (Table 5.2). By contrast, members of the Amaranthaceae, Chenopodiaceae, Moraceae, Polygonaceae and Urticaceae generally grow in nitrogen-rich conditions and have the potential to express high levels of leaf nitrate reductase. Mostly those species with a high potential for nitrate assimilation also have high leaf nitrogen contents, while those with a low nitrate assimilation potential have low leaf nitrogen. The angiosperm families that appear in the Cretaceous include Epacridaceae (low leaf nitrogen), Fagaceae (high leaf nitrogen), Gunneraceae (potential for nitrogen fixation) and Proteaceae (low leaf nitrogen). This would suggest that there was considerable heterogeneity in the nitrogen sources exploited by the component species of the Cretaceous vegetation of Australia.

Few, if any, Early Tertiary floras of Australia have close resemblance to modern Australian vegetation (Macphail *et al.* 1994). As already discussed, the period is generally characterized by prolonged leaching of nutrients from soils. The Droseraceae and Loranthaceae, families with specialized modes of nitrogen nutrition (carnivorous and parasitic, respectively), appear in the Early Eocene and Middle Oligocene, respectively (Table 5.3). The Santalaceae, another family of parasitic genera, appears in the Late Eocene of Australia. Another group with specialized nitrogen nutrition, the nitrogen-fixing Papilionoideae and *Acacia*, appear in the Late Eocene and Middle Oligocene, respectively. The appearance of families with specialized modes of nitrogen nutrition is consistent with increasing oligotrophy. However the Amaranthaceae, Chenopodiaceae and Polygonaceae, families characterized by a high potential for nitrate assimilation, also appear in the Oligocene and Early Miocene, respectively.

It is argued that the biota of Quaternary Australia had to adapt to the most dramatic environmental changes since the evolution of the major groups in the Mesozoic (Hope 1994). This was a period of seasonally dry conditions with widespread drought inland, reflected by the dominance of the Asteraceae, Casuarinaceae and Poaceae. Arguably, the arrival of humans 40 000 years ago was the most

**Table 5.3** Possible nitrogen-acquisition strategies of Tertiary flowering plant families based on strategies of extant species. Fossil data is taken from Macphail *et al.* (1994). Heterotrophic nitrogen use is suggested to indicate assimilates of amino acids or protein, often via a mycorrhizal associate. High- and low-nitrogen (N) strategies refer to known leaf nitrogen contents of extant species in the appropriate family.

| First appearance in Southeastern Australia | Family | Nitrogen acquisition strategy |
|---|---|---|
| Oligocene | Amaranthaceae | $NO_3^-$ assimilating, high N strategy |
| | Chenopodiaceae | $NO_3^-$ assimilating, high N strategy |
| Middle Oligocene | Mimosaceae (*Acacia*) | $N_2$ fixing, high N strategy |
| Late Eocene | Caesalpinoidae | $N_2$ fixing (?), high N strategy |
| | Proteaceae (*Hakea*, *Grevillea*) | Non-mycorrhizal—cluster roots— heterotrophic(?), low N strategy |
| Middle Eocene | Epacridaceae | Mycorrhizal—heterotrophic, low N strategy |
| | Loranthaceae | Heterotrophic—parasitic, low N strategy |
| Early Eocene | Droseraceae | Heterotrophic—insectivorous |
| | Myrtaceae | Ectomycorrhizal—heterotrophic, low N strategy |
| | Restionaceae | Non-mycorrhizal, low N strategy |
| | Santalaceae | Parasitic—heterotrophic, low N strategy |
| Late Paleocene | Polygonaceae | $NO_3^-$ assimilating, high N strategy |
| Paleocene | Casuarinaceae | $N_2$ fixing (actinorrhizal), low N strategy |

seminal event. Massive floristic changes occurred through human activity, including increased fire and faunal extinctions (mostly grazers and browsers). Human occupation of practically all available niches, including montane and arid regions, was complete by 35 000 years ago (Dodson 1992). Climate change coupled with the change in fire regime and the loss of the megafauna had dramatic effects on the vegetation (Flannery 1990), resulting in what we recognize as the present-day vegetation. Nutritionally, one outcome of this is likely to have been a further decrease in soil fertility. The most recent event of the Quaternary was of course the arrival of agriculture and with it what Hope (1994) has described as 'a major laboratory of evolution of adaptation to disturbance'.

## Nitrogen source utilization by extant higher plants

Needless to say an adequate supply of nitrogen is a prerequisite for plant growth. However the amount required and the form utilized is quite variable among species. Current evidence suggests higher plants can assimilate various forms of soil nitrogen: dissolved organic nitrogen, the mineralization product ammonium and its nitrification product nitrate. Eviner and Chapin (1997) have suggested that there is increasing evidence that species differ quite markedly in their preference for particular nitrogen sources and that these preferences reflect the prevalent form of available nitrogen. Much of the evidence supporting these suggestions is derived from studies of arctic and boreal plant species. The extent to which these findings can be generalized to plants in other ecosystems is uncertain.

We have investigated the availability and utilization of nitrogen by plants of diverse Australian ecosystems. Such investigations have employed a suite of techniques to elucidate nitrogen source utilization and include measurements of the substrate-inducible nitrate reductase; the composition of xylem sap nitrogenous compounds; the $^{15}$N natural abundance of plant tissues; and the uptake and metabolism of $^{15}$N-labelled nitrogen sources.

## Variation in nitrate reductase activity

Leaf nitrate reductase activity is variable both between and within plant communities. As a generalization, nitrate reductase activities are highest in sites of natural or man-made disturbance. These would include plants of regrowth vegetation following windthrow (Stewart *et al.* 1988), fire (Stewart *et al.* 1993) and animal disturbance (Schmidt *et al.* 1998). Most of the exotic weeds invading natural ecosystems tend to have high nitrate reductase activities (Schmidt *et al.* 1998).

Although plant communities differ with respect to their utilization of nitrate (as judged by relative nitrate reductase activities), there are quite marked differences in apparent nitrate utilization within some of these communities. In the tropical *Eucalyptus* woodlands of northern Australia nitrate reductase activities are generally low; however, some species, such as *Ficus,* have activities at least an order of magnitude higher than the average for the community as a whole. In the case of *Ficus* this high nitrate reductase activity was associated with much greater levels of soil nitrate (Schmidt *et al.* 1998). In general these high-activity species belong to plant families characterized by a high potential for nitrate reduction (Turnbull *et al.* 1996). These findings and those that have been obtained on other plant communities (Stewart *et al.* 1992) suggest that within many communities there is considerable variation among species in the extent to which they are active in nitrate assimilation. This could reflect that the component species have different nitrogen requirements or that within a particular habitat they exploit different nitrogen sources.

## Xylem sap nitrogen

In general, a close correspondence between the proportion of nitrogen present as nitrate in the xylem sap and nitrate reductase activity has been found (Stewart *et al.* 1993; Schmidt & Stewart 1997, 1998; Schmidt *et al.* 1998). Nitrate is more or less absent from the xylem sap of plants of the coastal heathland of southeast Queensland and little or no leaf nitrate reductase is measurable in their leaves. Root nitrate reductase activity in these species is also low (Schmidt & Stewart 1997). In other communities there is much variability in the proportion of nitrate present in xylem sap (Erskine *et al.* 1996; Schmidt & Stewart 1998). Again, this suggests variation among species in these communities in nitrate utilization. As discussed above it could indicate that species have differing nitrogen requirements or that they exploit different nitrogen sources.

## Uptake and metabolism of $^{15}$N-labelled substrates

The potential of selected species from diverse communities to take up and incorporate the $^{15}$N-labelled substrates ammonium, nitrate and glycine has been tested (Fig. 5.2). Glycine has been detected as a common, soil amino acid (Abuarghub & Read 1988; Kielland 1994; Atkin 1996; Schmidt & Stewart 1997; Näsholm et al. 1998) and was chosen here as a model with which to determine plant use of soluble,

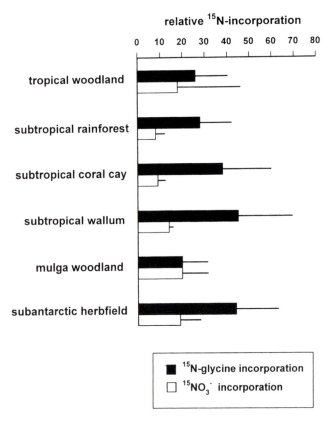

**Figure 5.2** Average incorporation of ammonium, glycine and nitrate by species in Australian plant communities. Four to nine species per community were analysed and community averages and standard deviations are shown. Field-excavated roots were incubated for 4 hours in 1 mol m$^{-3}$ $^{15}$N-labelled ammonium, glycine or nitrate (>98% atom excess). Values represent the average of three replicates per species and the incorporation of glycine and nitrate is shown relative to ammonium (100%). Experimental procedures were as described by Schmidt and Stewart (1997). Data are adapted from Schmidt and Stewart (1999).

soil organic soil nitrogen. Without exception highest rates of [15]N incorporation were observed when ammonium was the substrate. In most communities rates of [15]N incorporation from glycine exceeded those from nitrate.

These results indicate that within a particular community species exhibit marked differences in their capacity to take up and assimilate ammonium, nitrate and glycine. The results are quite different to those published for some arctic species, where a much greater uptake of glycine relative to ammonia has been observed (Chapin *et al.* 1993; Raab *et al.* 1996). Our results show that the capacity of roots to incorporate glycine is widespread among species in plant communities, ranging from subantarctic herbfield to tropical savanna woodland. This indicates that glycine is readily taken up not only by plants growing in soils of low temperatures and/or low rates of N mineralization, but also by species from subtropical and tropical environments. In species from most communities glycine incorporation exceeded nitrate incorporation, suggesting that glycine and possibly other amino acids are an important nitrogen source in all the communities we studied. In mulga woodland, where many species appear to use nitrate as their major nitrogen source (Erskine *et al.* 1996), roots incorporated equal or higher amounts of nitrate than glycine. By contrast, species from the subtropical heathland incorporated on average threefold higher amounts of glycine than nitrate. This may reflect the higher availability of amino acids compared with nitrate in this community (Schmidt & Stewart 1997). Such observations support the hypothesis that species differ in their preference for specific nitrogen sources and that these preferences reflect the availability of different nitrogen sources in particular habitats (Kielland 1994; Eviner & Chapin 1997).

### [15]N natural abundance

In recent years there have been a number of studies surveying the variation in [15]N natural abundance of various plant communities. Analysis of the $\delta$[15]N signatures of plants has the potential to be a powerful integrating tool for probing plant–soil nitrogen relations in natural and managed ecosystems (Chapter 21). However, the processes that determine $\delta$[15]N signatures at the whole-plant level are poorly understood and interpretations of $\delta$[15]N are frequently empirical (Handley & Raven 1992). Fractionation in the content of [15]N in plant tissue may occur as a consequence of biological and physicochemical processes (Handley & Raven 1992; Högberg 1997).

The $\delta$[15]N of nitrogen available to plants in soils may be influenced by soil age (Vitousek *et al.* 1989), disturbance (Peoples *et al.* 1991), the frequency of fire (Schulze *et al.* 1991), flooding (Yoneyama 1994) and soil moisture (Selles *et al.* 1986; Garten 1993; Sutherland *et al.* 1993). Soil $\delta$[15]N may be influenced by the presence of nitrogen-fixing species or nitrogen-rich plant parts in the soil profile and by point sources of isotopically distinct nitrogen such as animal excreta (Mizutani *et al.* 1986; Erskine *et al.* 1998).

Plants relying primarily on $NH_4^+$-N are commonly reported to be [15]N depleted (Nadelhoffer & Fry 1994). Either the signature of source or assimilatory fractionation or internal plant partitioning, or a combination of these factors may explain this. Plant $\delta^{15}N$ may be influenced by the type and extent of mycorrhizal association formed (Högberg 1990). Depleted $\delta^{15}N$ values in many temperate woody perennials have been attributed to the dependence of such species on ectomycorrhizas (Nadelhoffer & Fry 1994). Experimental work involving vesicular-arbuscular (VA) mycorrhizal *Ricinus communis* has found that the presence/absence of a fungal symbiont changed the $\delta^{15}N$ value of the host by as much as 2‰ (Handley *et al.* 1993). In a parallel experiment, ectomycorrhizal seedlings of *Eucalyptus globulus* had $\delta^{15}N$ values not significantly different from non-mycorrhizal seedlings (Handley *et al.* 1993).

The activity of mycorrhizal fungi has the potential to influence plant $\delta^{15}N$ in a number of ways. They can assimilate and transport large amounts of nitrogen and enhance nitrogen uptake by host plants (Arnebrant *et al.* 1993; Frey & Schlepp 1993). In addition, ectomycorrhizal fungi may enable host plants to utilize organic forms of nitrogen (Abuzinadah & Read 1986a,b, 1988; Abuzinadah *et al.* 1986; Finlay *et al.* 1992) and mycelial activity may be concentrated in pockets rich in organic nitrogen (Read 1991). Our studies with *Eucalyptus* (Turnbull *et al.* 1995) have shown that colonization of *E. maculata* and *E. grandis* roots with ectomycorrhizal fungi enables them to utilize amino acids and protein nitrogen sources that cannot be accessed by non-mycorrhizal seedlings. Mycorrhizal infection may therefore broaden the nitrogen source of the host to include a range of inorganic and organic compounds with potentially distinct $\delta^{15}N$ signatures.

While a number of field studies have attempted to relate mycorrhizal infection and nitrogen source to plant $\delta^{15}N$ (Högberg 1990; Pate *et al.* 1993; Högberg & Alexander 1995), few studies have addressed this issue experimentally (e.g. Handley *et al.* 1993). A number of studies have reported distinct patterns and differences in nitrogen isotope signatures between plant communities and within the species of those communities. The significance of such results is somewhat controversial. Many such studies are based on what we would regard as rather young communities (postglacial) and subject to very significant and poorly quantified anthropogenic inputs of nitrogen.

The variation in $\delta^{15}N$ in species of diverse Australian plant communities has been examined and several general points have emerged from these investigations. There are differences in community average values; the most $^{15}N$-enriched values are associated with sites where there is significant input of animal-derived nitrogen, such as the penguin colonies on Macquarie Island (Erskine *et al.* 1998) and the noddy colonies on Heron Island (Fig. 5.3). The most depleted $^{15}N$ values are found at remote sites such as in the alpine and subalpine vegetation of Tasmania and the feldmark vegetation of Macquarie Island. The range of values within these sites was largest in the alpine/subalpine vegetation and smallest in the guano-dominated vegetation of Heron Island. Within all of the communities that we have studied there are consistent differences in $\delta^{15}N$ of component species. Irrespective of the community average value there are groups of species that always tend to have higher values and those that always tend to have lower $\delta^{15}N$ values. Of particular

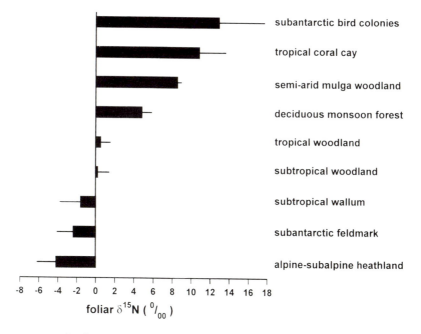

**Figure 5.3** Foliar δ15N signatures of species from Australian plant communities. Plants were sampled and analysed as described in Erskine *et al.* (1998). Community averages and standard deviations are shown.

interest are those species that consistently have the highest δ15N values. These are almost always species belonging to plant families that are regarded as being non-mycorrhizal. The best example of this in Australia is the Proteaceae, possibly the only family of woody plants that is consistently non-mycorrhizal. In the alpine/subalpine vegetation, the tropical *Eucalyptus* woodland, coastal subtropical heathlands and semi-arid mulga woodland members of the Proteaceae have δ15N values higher than the community average. Other families that have higher δ15N values are the generally non-mycorrhizal Cyperaceae and Restionaceae. Comparable results have been obtained for subarctic communities where mycorrhizal species were found to be depleted in 15N compared with those that were non-mycorrhizal (Michelsen *et al.* 1996; Nadelhoffer *et al.* 1996). Similar results for European and African plant communities report that ectomycorrhizal species tend to be 15N-depleted (Högberg 1990, 1997). One explanation of these findings is that mycorrhizal species utilize organic nitrogen in the litter and that this nitrogen source is depleted relative to soil mineral nitrogen (Michelsen *et al.* 1996). This explanation would be consistent with the notion that species differ in their preference for particular nitrogen sources and that these preferences reflect the prevalent form of available nitrogen (Eviner & Chapin 1997). We were unable to relate the nitrogen isotope signature of soil nitrogen fractions to those of the component species of a wallum heathland (Schmidt & Stewart 1997). Rather, we showed that the fine roots of ericoid and ectomycorrhizal species were enriched in $^{15}N$ and suggested that there was fractionation at the point of nitrogen transfer from fungal to plant partner with the plant receiving $^{15}N$-depleted nitrogen. It can be hypothesized that species differences in nitrogen isotope signatures relate more to physiological differences in root functioning rather than being indicative of differences in the exploitation of soil nitrogen sources.

## Concluding remarks

In early terrestrial communities it is suggested that the nitrogen cycle was rudimentary and that nitrification may have been quite limited. If this was so, nitrogen sources other than nitrate may have been assimilated. Continued dependence on heterotrophic nitrogen sources is present in many low-fertility plant communities. Reliance on mycorrhizal associates for nitrogen assimilation may well be a less advanced and ancient nutritional strategy. There are fossil arbuscular mycorrhizas dating from 400 Myr ago and ectomycorrhizas from about 50 Myr. Ecto- and ericoid-mycorrhizal associations are suggested to have special roles in higher plant nutrition. Some gymnosperms, and flowering plants such as the Myrtaceae and *Nothofagus* (Fagaceae), are mycorrhizal and have been components of the Australian vegetation since Late Cretaceous times.

Many of the ericoid- and ecto-mycorrhizal species have low concentrations of nitrogen in their leaves. This could be a less advanced characteristic. Many of the ancient plant families that are important components of the present-day (Australian) flora, Myrtaceae and Proteaceae, tend to be low nitrogen species and to be less active in nitrate assimilation. Furthermore, oligotrophic species may themselves contribute to limitations on the availability of nitrogen, particularly in the form of nitrate. Many Myrtaceae accumulate large concentrations of terpenoids in their leaves. Compounds such as cineole are highly flammable, increasing the likelihood of fire with a consequent decrease in nitrogen availability. In addition, the release of terpenoids from the leaf litter could inhibit nitrification. These oligotrophic communities, exemplified by the Western Australian Kwongan, have high biodiversity and the component species exhibit physiological diversity with respect to nitrogen acquisition strategies (Lamont 1982). These are communities with a high occurrence of heterotrophic nitrogen assimilation.

By contrast, members of the Mimosaceae appear to have highly diverse nitrogen-use strategies: dinitrogen fixation; access to organic nitrogen via mycorrhizal associates; and ability to assimilate nitrate and ammonium. This apparent catholicism in nitrogen use may be essential to maintain their high nitrogen strategy, which enables them to have the fast growth rates necessary for early sexual reproduction that would otherwise be threatened by fire. In most of the sites that we have studied, ruderal herb and tree species with higher nitrate reductase activities and with nitrate in xylem or leaves generally had higher than average leaf nitrogen contents. McKey (1994) has argued that high foliar nitrogen concentrations in legumes have enabled them to employ distinct life strategies, leading to the evolution of deciduous trees and ephemeral species. Such species are able to maximize growth during favourable conditions, because they have a high photosynthetic potential as a consequence of their high leaf nitrogen. The ruderal herbs and deciduous trees studied here have the 'high nitrogen strategy' and this may enable them to restrict photosynthetic activity to favourable periods, obviating the impact of water limitation and fire danger during the dry season. It is striking that these high-nitrogen-strategy species tend to accumulate nitrogen-based osmolytes such as proline and betaines (Erskine *et al.* 1996). By contrast, the low-nitrogen-strategy species use non-nitrogen osmoticants (Poljakoff-Mayber *et al.* 1987; Erskine *et al.* 1996). The high-nitrogen-strategy species also synthesize nitrogen-rich compounds, including non-protein amino acids, toxic peptides and proteins and alkaloids for defence against herbivore predation. Many of those pursuing the low nitrogen strategy tend to use phenylpropanoids and other non-nitrogen compounds.

Nitrate reduction, particularly in leaves, may thus be viewed as a more advanced character among flowering plants. Certainly ferns and gymnosperms, representing less advanced terrestrial plants, tend to have a low potential for leaf nitrate reduction (Smirnoff *et al.* 1984; Stewart *et al.* 1986). It can be argued that when nitrate assimilation is restricted to the root there are limitations on total nitrogen gain. Leaf reduction of nitrate increases considerably the potential for acquiring nitrogen. However, leaf nitrate reduction does incur some potential disadvantages. While hydroxyl ions resulting from nitrate reduction can be neutralized by oxalate synthesis and storage or the retranslocation of organic acids to the root system, the continuous accumulation of organic acids in nitrate-assimilating leaves may present problems of osmotic and acid–base balance (Raven 1986). The ruderal or deciduous habit is well suited to leaf nitrate assimilation, because these problems can be minimized through the regular disposal of leaves.

It is widely accepted that in the Northern Hemisphere there has been a general nitrogen eutrophication (Pearson & Stewart 1993 and references therein). This has led to increased tissue nitrogen concentrations (Pitcairn *et al.* 1995). These changes in nitrogen availability are likely to favour the spread of high-nitrogen-strategy species, a decrease in heterotrophic nitrogen use with a concomitant decline in biodiversity. Welcome to the autotrophic world of the weed.

## References

Abuarghub, S.M. & Read, D.J. (1988). The biology of mycorrhiza in the Ericaceae. XII. Quantitative analysis of individual 'free' amino acids in relation to time and depth of the soil profile. *New Phytologist*, **108**, 433–441.

Abuzinadah, R.A. & Read, D.J. (1986a). The role of proteins in the nitrogen nutrition of ectomycorrhizal plants. I. Utilisation of peptides and pro- teins by ectomycorrhizal fungi. *New Phytologist*, **103**, 481–493.

Abuzinadah, R.A. & Read, D.J. (1986b). The role of proteins in the nitrogen nutrition of ectomycorrhizal plants. III. Protein utilisation by *Betula*, *Picea* and *Pinus* in mycorrhizal association with *Hebeloma crustuliniforme*. *New Phytologist*, **103**, 507–514.

Abuzinadah, R.A. & Read, D.J. (1988). Amino acids as nitrogen sources for ectomycorrhizal fungi: utilisation of individual amino acids. *Transactions of the British Mycological Society*, **91**, 473–479.

Abuzinadah, R.A., Finlay, R.D. & Read, D.J. (1986). The role of proteins in the nitrogen nutrition of ectomycorrhizal plants. II. Utilisation of protein by mycorrhizal plants of *Pinus contorta*. *New Phytologist*, **103**, 495–506.

Amarasinghe, B.H.R.R., deBruxelles, G.L., Braddon, M., Onyeocha, I., Forde, B.G. & Udvardi, M.K. (1998). Regulation of GmNRT2 expression and nitrate transport activity in roots of soybean (*Glycine max*). *Planta*, **206**, 44–52.

Arnebrant, K., Ek, H., Finlay, R.D. & Söderström, B. (1993). Nitrogen translocation between *Alnus glutinosa* (L.) Gaertn. seedlings inoculated with *Frankia* sp. & *Pinus contorta* Doug. ex Loud seedlings connected by a common ectomycorrhizal mycelium. *New Phytologist*, **124**, 231–242.

Atkin, O.K. (1996). Reassessing the nitrogen relations of Arctic plants: a mini review. *Plant, Cell and Environment*, **19**, 695–704.

Baker, R.D., Baker, S.S. & Rao, R. (1998). Selenium deficiency in tissue culture, implications for oxidative metabolism. *Journal of Pediatric Gastroenterology and Nutrition*, **27**, 387–392.

Barlow, B. (1981). The Australian flora: its origin and evolution. *Flora of Australia,* Vol 1, pp. 25–75. Griffin Press, Netley, South Australia.

Barns, S.M., Delwiche, C.F., Palmer, J.D. & Pace, N.R. (1996). Perspective of archaeal diversity, thermophily and monophyly from environmental rRNA sequences. *Proceedings of the National Academy of Sciences USA*, **93**, 9188–9193.

Bernal, J.D. (1951). *The Physical Basis of Life*. Routledge Kegan Paul, London.

Cairns-Smith, A.G. (1982). *Genetic Takeover and the Mineral Origin of Life*. Cambridge University Press, Cambridge.

Chapin, F.S. III, Moilanen, L. & Kielland, K. (1993). Preferential use of organic nitrogen for growth by a nonmycorrhizal arctic sedge. *Nature*, **361**, 150–153.

Chyba, C.F. (1993). The violent environment of the origin of life: progress and uncertainties. *Geochimica et Cosmochimica Acta*, **57**, 3351–3358.

Culotta, V.C., Klomp, L.W.J., Strain, J., Caserno, R.L.B., Krems, B. & Gitlin, J.D. (1997). The copper chaperone for superoxide dismutase. *Journal of Biological Chemistry*, **272**, 23469–23472.

Dettmann, M.E. (1994). Cretaceous vegetation: the microfossil record. In *History of the Australian Vegetation* (Ed. by R.S. Hill), pp. 143–170. Cambridge University Press, Cambridge.

Dodson, J.R. (1992). *The Native Lands — Prehistory and Environmental Change in the South West Pacific*. Longman Cheshire, Melbourne.

Douglas, J.G. (1994). Cretaceous vegetation: the macrofossil record. In *History of the Australian Vegetation* (Ed. by R.S. Hill), pp. 171–189. Cambridge University Press, Cambridge.

Duffaud, G.D., Dhennezel, O.B., Peek, A.S., Reysenbach, A.L. & Kelly, R.M. (1998). Isolation and characterization of *Thermococcus barossii* sp. nov. a hyperthermophilic archaeon isolated from a hydrothermal vent flange formation. *Systematic and Applied Microbiology*, **21**, 40–49.

Erskine, P.D., Stewart, G.R., Schmidt, S., Turnbull, M.H., Unkovich, M. & Pate, J.S. (1996). Water availability — a physiological constraint on nitrate utilization in plants of Australian semi-arid mulga woodlands. *Plant, Cell and Environment*, **19**, 1149–1159.

Erskine, P.D., Bergstrom, D.M., Schmidt, S., Stewart, G.R., Tweedie, C.E. & Shaw, J.D. (1998). Subantarctic Macquarie Island — a model ecosystem for studying animal derived nitrogen sources using $^{15}$N natural abundance. *Oecologia*, **117**, 187–193.

Eviner, V.T. & Chapin, F.S. III (1997). Nitrogen cycle: plant–microbial interactions. *Nature*, **385**, 26–27.

Finlay, R.D., Frostegård, C. & Sonnerfeldt, A.-M. (1992). Utilisation of organic and inorganic nitrogen sources by ectomycorrhizal fungi in pure culture and in symbiosis with *Pinus contorta* Dougl. ex Loud. *New Phytologist*, **120**, 105–115.

Flannery, T.M. (1990). Pleistocene faunal loss: implications of the aftershock for Australia's past and future. *Archaeology in Oceania*, **25**, 45–67.

Forterre, P. (1995). Looking for the most primitive organism(s) on earth today — the state of the art. *Planetary and Space Science*, **43**, 167–177.

Frausto da Silva, J.J.R. & Williams, R.J.P. (1991). *The*

*Biological Chemistry of the Elements*. Clarendon Press, Oxford.

Frey, B. & Schlepp, H. (1993). Acquisition of nitrogen by external hyphae of arbuscular mycorrhizal fungi associated with *Zea mays* L. *New Phytologist*, **124**, 221–230.

Garten, C.T. Jr (1993). Variation in foliar $^{15}$N abundance and the availability of soil nitrogen on Walker Branch watershed. *Ecology*, **74**, 2098–2113.

Gonzalez, J.M., Masuchi, Y., Robb, F.T. *et al.* (1998). *Pyrococcus horikoshii* sp. nov. a hyperthermophilic archaeon from a hydrothermal vent at the Okinawa trough. *Extremophiles*, **2**, 123–130.

Gray, J. & Shear, W. (1992). Early life on land. *American Scientist*, **80**, 444–456.

Hafenbrandl, D., Keller, M., Wächtershäuser, G. & Stetter, K.O. (1995). Primordial amino acids by reductive amination of alpha-oxo acids in conjunction with the oxidative formation of pyrite. *Tetrahedron Letters*, **36**, 5179–5182.

Handley, L.L. & Raven, J.A. (1992). The use of natural abundance of nitrogen isotopes in plant physiology and ecology. *Plant, Cell and Environment*, **15**, 965–985.

Handley, L.L., Daft, M.J., Wilson, J., Scrimgeour, C.M., Ingelby, K. & Sattar, M.A. (1993). Effects of the ecto- and va-mycorrhizal fungi *Hydnagium* and *Glomus clarum* on the $\delta^{15}$N and $\delta^{13}$C values of *Eucalyptus globulus* and *Ricinus communis*. *Plant, Cell and Environment*, **16**, 375–382.

Hartman, H. (1975). Speculations on the origin and evolution of metabolism. *Journal of Molecular Evolution*, **4**, 359–370.

Hausinger, R.P. (1994). Nickel enzymes in microbes. *Science of the Total Environment*, **148**, 157–166.

Hill, R.S. (1994). The Australian fossil plant record: an introduction. In *History of the Australian Vegetation* (Ed. by R.S. Hill), pp. 1–4. Cambridge University Press, Cambridge.

Högberg, P. (1990). $^{15}$N natural abundance as a possible marker of the ectomycorrhizal habit of trees in mixed African woodlands. *New Phytologist*, **115**, 483–486.

Högberg, P. (1997). $^{15}$N natural abundance in soil–plant systems. *New Phytologist*, **137**, 179–203.

Högberg, P. & Alexander, I.J. (1995). Roles of root symbioses in African woodland forest: evidence from $^{15}$N abundance and foliar analysis. *Journal of Ecology*, **83**, 217–224.

Hope, G.S. (1994). Quaternary vegetation. In *History of the Australian Vegetation* (Ed. by R.S. Hill), pp. 368–389. Cambridge University Press, Cambridge.

Huber, C. & Wächtershäuser, G. (1998). Peptides by activation of amino acids with Co on (NiFe) surfaces—implications for the origin of life. *Science*, **281**, 670–672.

Huheey, J.E. (1988). *Anorganische Chemie*. Walter de Gruyter, Berlin.

Jacob, C., Maret, W. & Vallee, B.L. (1998). Control of zinc transfer between thionein, metallothionein, and zinc proteins. *Proceedings of the National Academy of Sciences USA*, **95**, 3489–3494.

Johnson, M.K., Rees, D.C. & Adams, M.W.W.W. (1996). Tungstoenzymes. *Chemical Reviews*, **96**, 2817–2839.

Kaim, W. & Rall, J. (1996). Copper—a 'modern' bioelement. *Angewandte Chemie International Edition in English*, **35**, 43–60.

Keller, M., Blochl, E., Wächtershäuser, G. & Stetter, K.O. (1994). Formation of amide bonds without a condensation agent and implications for origin of life. *Nature*, **368**, 836–838.

Keller, D.R., Hafenbradl, D., Braun, F.J., Rachel, R., Burggraf, S. & Stetter, K.O. (1998). *Thermococcus acidaminovorans* sp. nov., a new hyperthermophilic alkalophilic archaeon growing on amino acids. *Extremophiles*, **2**, 109–114.

Kenrick, P. & Crane, P.R. (1997). The origin and early evolution of plants on land. *Nature*, **389**, 33–39.

Kielland, K. (1994). Amino acids absorption by arctic plants: implications for plant nutrition and nitrogen cycling. *Ecology*, **75**, 2373–2383.

Kleinhofs, A. & Warner, R.L. (1990). Advances in nitrate assimilation. In *Biochemistry of Plants*, Vol. 16 (Ed. by B.J. Miflin & P.J. Lea), pp. 89–120. Academic, New York.

Krapp, A., Fraisier, V., Scheible, W.R. *et al.* (1998). Expression studies of Nrt2: 1Np, a putative high-affinity nitrate transporter: evidence for its role in nitrate uptake. *Plant Journal*, **14**, 723–731.

Lamont, B. (1982). Mechanisms for enhancing nutrient uptake in plants with particular reference to mediterranean South Africa and Western Australia. *The Botanical Review*, **48**, 597–689.

Lauter, F.R., Ninnemann, O., Bucher, M., Riesmeier, J.W. & Frommer, W.B. (1996). Preferential expression of an ammonium transporter and of two putative nitrate transporters in root hairs of tomato. *Proceedings of the National Academy of Sciences USA*, **93**, 8139–8144.

Lazcano, A., Guerro, R., Marguilis, L. & Oro, J. (1988). The evolutionary transition from RNA to DNA in early cells. *Journal of Molecular Evolution*, **27**, 283–290.

Lea, P.J., Robinson, S.A. & Stewart, G.R. (1990). The enzymology and metabolism of glutamine, glutamate and asparagine. In *Biochemistry of Plants*, Vol. 16 (Ed. by B.J. Miflin & P.J. Lea), pp. 121–160. Academic, New York.

Lee, D. & Morel, F.M.M. (1995). Replacement of zinc by cadmium in marine phytoplankton. *Marine Ecology—Progress in Series*, **127**, 305–309.

Lee, D. & Morel, F.M.M. (1996). *In vivo* replacement of zinc by cobalt in carbonic anhydrase of a marine diatom. *Limnology and Oceanography*, **41**, 573–577.

Legrain, C., Villeret, V., Rooveres, M., Gigot, D. *et al.* (1995). Biochemical characterisation of ornithine carbomyltransferase from *Pyrococcus furiosus*. *European Journal of Biochemistry*, **247**, 1046–1055.

McKey, D. (1994). Legumes and nitrogen: the evolutionary ecology of a nitrogen-demanding lifestyle. In *Advances in Legume Systematics* (Ed. by J.I. Sprent & D. McKey), pp. 221–228. Royal Botanic Gardens, Kew.

Macphail, M.K., Alley, N.F., Truswell, E.M. & Sluiter, I.R.K. (1994). Early Tertiary vegetation: evidence from spores and pollen. In *History of the Australian Vegetation* (Ed. by R.S. Hill), pp. 189–261. Cambridge University Press, Cambridge.

Maden, B.E.H. (1995). No soup for starters? Autotrophy and the origins of metabolism. *Trends in Biochemical Science*, **20**, 337–341.

Mei, B. & Zalkin, H. (1989). A cysteine-histidine-aspartate catalytic triad is involved in glutamine amide transfer functions in *pur*F-type glutamine transferases. *Journal of Biological Chemistry*, **264**, 16613–16619.

Michelsen, A., Schmidt, I.K., Jonasson, S., Quarmby, C. & Sleep, D. (1996). Leaf [15]N abundance of subarctic plants provides evidence that ericoid, ectomycorrhizal and non and arbuscular mycorrhizal species access different sources of soil nitrogen. *Oecologia*, **105**, 53–63.

Mizutani, H., Hasegawa, H. & Wada, E. (1986). High nitrogen isotope ratio for soils of seabird rookeries. *Biogeochemistry*, **2**, 221–247.

Moore, G.T., Jacobson, S.R., Ross, C.A. & Hays, D.N. (1994). A palaeoclimate simulation of the Wenlockian (late early Silurian) world using a general circulation model with implications for land plant palaeoecology. *Palaeogeography, Palaeoclimatology, Palaeoecology*, **110**, 115–144.

Mukand, S. & Adams, M.W. (1996). Molybdenum and vanadium do not replace tungsten in the catalytically active forms of three tungstoenzymes in the hyperthermophilic archaeon *Pyrococcus furiosus*. *Journal of Bacteriology*, **178**, 163–167.

Nadelhoffer, K.J. & Fry, B. (1994). Nitrogen isotope studies in forest ecosystems. In *Stable Isotopes in Ecology and Environmental Science* (Ed. by K. Lathja & R.H. Michener), pp. 22–44. Blackwell Scientific Publications, London.

Nadelhoffer, K.J., Shaver, G., Fry, B., Giblin, A., Johnson, L. & McKane, R. (1996). [15]N natural abundance and N use by tundra plants. *Oecologia*, **107**, 386–394.

Näsholm, T., Ekblad, A., Nordin, A., Giesler, R., Högberg, M. & Högberg, P. (1998). Boreal forest plants take up organic nitrogen. *Nature*, **392**, 914–916.

Pate, J.S., Stewart, G.R. & Unkovich, M. (1993). [15]N natural abundance of plant and soil components of a *Banksia* woodland ecosystem in relation to nitrate utilization, life form, mycorrhizal status and $N_2$-fixing abilities of component species. *Plant, Cell and Environment*, **16**, 365–373.

Pearson, J. & Stewart, G.R. (1993). Tansley review No 56. The deposition of atmospheric ammonia and its effects on plants. *New Phytologist*, **125**, 283–305.

Peoples, M.B., Bergensen, F.J., Turner, G.L. *et al.* (1991). Use of the natural enrichment of [15]N in soil mineral N for the measurement of symbiotic $N_2$ fixation. In *Stable Isotopes in Plant Nutrition, Soil Fertility and Environmental Studies*, pp. 431–440. International Atomic Energy Agency/FAO, Vienna.

Pitcairn, C.E.R., Fowler, D. & Grace, J. (1995). Deposition of fixed atmospheric nitrogen and foliar nitrogen content of bryophytes and *Calluna vulgaris* (L.) Hull. *Environmental Pollution*, **88**, 193–205.

Poljakoff-Mayber, A., Symon, D.E., Jones, G.P., Naidu, B.P. & Paleg, L.G. (1987). Nitrogenous compatible solutes in native South Australian plants. *Australian Journal of Plant Physiology*, **14**, 341–350.

Pryer, K.M., Smith, A.R. & Skog, J.E. (1995). Phylogenetic relationships of extant ferns based on evidence from morphology and rbcL sequences. *American Fern Journal*, **85**, 205–282.

Quesada, A., Krapp, A., Trueman, L.J. *et al.* (1997). PCR-identification of a *Nicotiana plumbaginifolia* cDNA homologous to the high affinity nitrate transporters of the *crnA* family. *Plant Molecular Biology*, **34**, 265–274.

Raab, T.K., Lipson, D.A. & Monson, R.K. (1996). Nonmycorrhizal uptake of amino acids by roots of the alpine sedge *Kobresia myosuroides*: implications for the alpine nitrogen cycle. *Oecologia*, **108**, 488–494.

Racker, E. (1965). *Mechanisms in Bioenergetics*. Academic, New York.

Raven, J.A. (1986). Biochemical disposal of excess $H^+$ in growing plants? *New Phytologist*, **104**, 175–206.

Read, D.J. (1991). Mycorrhizas in ecosystems. *Experientia*, **47**, 376–391.

Rees, M. (1997). *Before the Beginning — Our Universe and Others*. Simon and Schuster, London.

Russell, M.J. & Hall, J. (1997). The emergence of life from iron monosulphide bubbles at a submarine hydrothermal redox and pH front. *Journal of the Geological Society*, **154**, 377–402.

Russell, M.J., Daniel, R.M., Hall, J. & Sherringham, J.A. (1994). A hydrothermally precipitated catalytic iron sulphide membrane as a first step toward life. *Journal of Molecular Evolution*, **39**, 231–243.

Schidlowski, M.A. (1988). 3800-million-year isotope record of life from carbon in sedimentary rocks. *Nature*, **333**, 313–318.

Schmidt, S. & Stewart, G.R. (1997). Waterlogging and fire impacts on nitrogen availability and utilization in a subtropical wet heathland (wallum). *Plant, Cell and Environment*, **20**, 1231–1241.

Schmidt, S. & Stewart, G.R. (1998). Transport, storage and mobilization of nitrogen by trees and shrubs in the wet/dry tropics of northern Australia. *Tree Physiology*, **18**, 403–410.

Schmidt, S. & Stewart, G.R. (1999) Glycine metabolism by plant roots and its occurrence in Australian plant communities. *Australian Journal of Plant Physiology*, **26**, 253–264.

Schmidt, S., Stewart, G.R., Turnbull, M.H., Erskine, P.D. & Ashwath, N. (1998). Nitrogen relations of natural and disturbed plant communities in tropical Australia. *Oecologia*, **117**, 95–105.

Schopf, J.W. & Walter, M.R. (1983). Archaean microfossils: new evidence of ancient microbes. In *Earth's Earliest Biosphere* (Ed. by J.W. Schopf), pp. 214–239. Princeton University Press, Princeton.

Schulze, E.-D., Gebauer, G., Ziegler, H. & Lange, O.L. (1991). Estimates of nitrogen fixation by trees on an aridity gradient in Namibia. *Oecologia*, **88**, 451–455.

Selles, F., Karamanos, R.E. & Kachanoski, R.G. (1986). The spatial variability of nitrogen-15 and its relation to the variability of other soil properties. *Soil Science Society of America Journal*, **50**, 105–110.

Simon, L., Bousquet, J., Lveque, C. & Lalonde, M. (1993). Origin and diversification of endomycorrhizal fungi with vascular plants. *Nature*, **363**, 67–69.

Sleep, N.H., Zahnle, K.J., Kasting, J.F. & Morowitz, H.J. (1989). Annihilation of ecosystems by large asteroid impacts on the early earth. *Nature*, **342**, 139–142.

Smirnoff, N., Todd, P. & Stewart, G.R. (1984). The occurrence of nitrate reduction in the leaves of woody plants. *Annals of Botany*, **54**, 363–374.

Stadtman, T.C. (1990). Selenium biochemistry. *Annual Review of Biochemistry*, **59**, 111–127.

Stewart, G.R. & Larher, F. (1980). Nitrogen and environmental stress. In *Biochemistry of Plants*, Vol. 5 (Ed. by B.J. Miflin), pp. 609–635. Academic, New York.

Stewart, G.R., Mann, A.F. & Fentem, P.A. (1980). Enzymes of glutamate formation: glutamate dehydrogenase, glutamine synthetase and glutamate synthase. In *Biochemistry of Plants*, Vol. 5 (Ed. by B.J. Miflin), pp. 271–328. Academic, New York.

Stewart, G.R., Popp, M., Holzapfel, I., Stewart, J.A. & Dickie-Eskew, A. (1986). Localisation of nitrate reduction in ferns and relationship to environment and physiological characteristics. *New Phytologist*, **104**, 373–384.

Stewart, G.R., Hegarty, E.E. & Specht, R.L. (1988). Inorganic nitrogen assimilation in plants of Australian rainforest communities. *Physiologia Plantarum*, **74**, 26–33.

Stewart, G.R., Joly, C.A. & Smirnoff, N. (1992). Partitioning of organic nitrogen assimilation between roots and shoots of cerrado and forest trees of contrasting plant communities of South East Brazil. *Oecologia*, **91**, 511–517.

Stewart, G.R., Pate, J.S. & Unkovitch, M. (1993). Characteristics of nitrogen assimilation of plants from fire prone Mediterranean-type vegetation. *Plant, Cell and Environment*, **16**, 351–363.

Summers, D.P. & Chang, S. (1993). Prebiotic ammonia from reduction of nitrite by iron (II). *Nature*, **365**, 630–633.

Sutherland, R.A., van Kessel, C., Farrell, R.E. & Pennock, D.J. (1993). Landscape-scale variations in plant and soil nitrogen-15 natural abundance. *Soil Science Society of America Journal*, **57**, 169–178.

Turnbull, M.H., Goodall, R. & Stewart, G.R. (1995). The impact of mycorrhization on nitrogen source utilisation in *Eucalyptus grandis* and *Eucalyptus maculata*. *Plant, Cell and Environment*, **18**, 1386–1394.

Turnbull, M.H., Schmidt, S., Erskine, P.D., Richards, S. & Stewart, G.R. (1996). Root adaptation and nitrogen source acquisition in natural ecosystems. *Tree Physiology*, **16**, 941–948.

Vitousek, P.M., Shearer, G. & Kohl, D.H. (1989). Foliar [15]N natural abundance in Hawaiian rainforest: patterns and possible mechanisms. *Oecologia*, **78**, 383–388.

Vorholt, J.A., Vaupel, M. & Thauer, R.K. (1997). A selenium-dependent and a selenium-independent formylmethanfuran dehydrogenase and their transcriptional regulation in the hyperthermophilic *Methanopyrus kandleri*. *Molecular Microbiology*, **23**, 1033–1042.

Wächtershäuser, G. (1988). Pyrite formation, the first energy source for life: a hypothesis. *Systematic Applied Microbiology*, **10**, 207–210.

Wächtershäuser, G. (1992). Groundworks for an evolutionary biochemistry: the iron-sulphur world. *Progress in Biophysics and Molecular Biology*, **58**, 85–201.

Walker, J.C.G. (1986). Carbon dioxide on the early earth. *Origins of Life*, **16**, 117–127.

Wang, R.C. & Crawford, N.M. (1996). Genetic identification of a gene involved in constitutive, high-affinity nitrate transport in higher plants. *Proceedings of the National Academy of Sciences USA*, **93**, 9297–9301.

von Wirén, N., Bergfeld, A., Ninnemann, O. & Frommer, W.B. (1997). High affinity ammonium transporter from rice (*Oryza sativa* cv. Nipponbare) (OsAMT1–1). *Plant Molecular Biology*, **35**, 681.

Yoneyama, T. (1994). Nitrogen metabolism and fractionation of nitrogen isotopes in plants. In *Stable Isotopes in the Biosphere* (Ed. by E. Wada, T. Yoneyama, M. Minagawa, T. Ando & B.D. Fry), pp. 92–102. Kyoto University Press, Kyoto.

Zhou, J. & Kleinhofs, A. (1996). Molecular evolution of nitrate reductase genes. *Journal of Molecular Evolution*, **42**, 432–460.

Zumpft, W.G. (1992). The prokaryotes. In *A Handbook on the Biology of Bacteria: Ecophysiology, Isolation Applications* (Ed. by A. Ballow, H.G. Truper, M. Dworkin, W. Harder & K.H. Schleifer), pp. 554–582. Springer, New York.

# Chapter 6

# Roots as dynamic systems: the developmental ecology of roots and root systems

*A.H. Fitter*

## Introduction

Roots perform two primary functions: anchorage and resource acquisition. Other root functions, such as synthesis and storage, are secondary, arising because roots existed and not *vice versa*. Anchorage depends largely on properties of the roots nearest the stem base (Ennos 1989) and is most significant as a root function in tall plants, notably trees. The ability to acquire resources, however, is a fundamental requirement for all plants, and is determined by four features of root systems: their architecture; their interactions with microorganisms; their ability to influence the rhizosphere by exudation; and their systems for transporting materials across membranes. Only the last two of these are instantly recognizable as 'physiological ecology', but the others are certainly as important. For a root to acquire nutrient ions, for example, the ion must be adjacent to the root surface. That may happen because: (i) transpiration induces mass flow of water and ions to the root surface (convective flux); (ii) uptake has reduced the concentration at the rhizoplane and created a concentration gradient (or depletion zone) down which diffusion occurs (diffusive flux); (iii) exudates from the root, such as citrate or ionophores, have altered the solubility of ions in the rhizosphere; (iv) a mycorrhizal fungal symbiont has, via its hyphae, transported the ion to the root surface from beyond the depletion zone surrounding the root (see Chapter 7); and (v) a lateral root has grown out from the parent root or the root has itself extended into undepleted soil.

Much research on physiological ecology of roots has focused on the transport of ions across cell membranes. Considerable advances are being made in this field at a molecular level, with the identification, for example, of a family of phosphate transporters in a range of species, with each species having several transporters varying in affinity for phosphate ions. It is already possible to identify which transporters are active at any given time, where in a root system they are active, and possibly how active each transporter is. For example, Liu *et al.* (1998) cloned two phosphate transporters (PT) from the legume *Medicago truncatula*, which encoded proteins that were very similar to PTs from other plants (*Arabidopsis thaliana*, *Solanum tuberosum*) and fungi (*Saccharomyces*, *Neurospora*). The function of one

*Department of Biology, University of York, York, YO10 5DD, UK. E-mail: afh10york.ac.uk*

of the proteins was demonstrated by a complementation study with a yeast PT mutant (*pho84*). The power of these techniques is considerable: using Northern blot analysis to detect transcript levels, Liu *et al.* (1998) showed that transcription was clearly suppressed by increasing the phosphate concentration at which the plants were grown over a range from 0 to 5 mM. The transporters were expressed in roots only and not in leaves, and even more significantly they were downregulated when the plants were colonized by the mycorrhizal fungus *Glomus versiforme*, which would have increased plant phosphate uptake. Using such techniques, it is now possible to undertake studies to measure the molecular plasticity of uptake systems, and rapid advances are likely.

Molecular biology is also providing new tools to investigate another of the major components of root function, namely the development of roots and the consequent architecture of root systems. This chapter focuses on what might be termed 'developmental ecology' because a number of ecologically exciting questions can be approached by that route. Developmental ecology of roots is significant for two main reasons. First, it is often the case that the movement of ions through soil to the root surface is the rate-limiting step in the overall uptake process, rather than the movement of ions across membranes. Therefore the distance between individual roots within a root system determines its effectiveness at exploiting soil, especially for slowly diffusing ions such as phosphate. This does not imply that transport across membranes is an unimportant phenomenon: there has presumably been long-sustained and powerful selection pressure on plants to possess ion acquisition systems that can effectively obtain ions at the rhizoplane in competition with other soil organisms. Second, the costs to a plant of maintaining uptake systems of high affinity and ability to move ions across cell membranes are likely to be small relative to those of constructing and maintaining roots (Eissenstat & Yanai 1997). A mismatch between uptake ability and environment can be quickly and cheaply rectified by the destruction of existing transporters and the synthesis of new, more appropriate ones; by contrast, the growth of roots that are in the wrong place or of the wrong morphology can only be rectified by the sacrifice of the investment in those roots and the construction of new roots.

There is no suggestion here that physiological responses to the environment, which cover more than just the operation of transport mechanisms to the environment, are unimportant. Plants can alter the solubility in the rhizosphere of relatively insoluble ions such as phosphate by the secretion of ions such as citrate (Gardner *et al.* 1983; Grierson 1992). In some species, such as lupins (Fabaceae), some sedges (Cyperaceae) and Proteaceae, this seems to be a key element in their phosphate acquisition strategy, notably where it is combined with the morphological development of cluster roots (Lamont 1993; Dinkelaker *et al.* 1995). It appears that the cluster roots provide the ability to flood the rhizosphere with chelating agents or ions that will exchange for nutrient ions, and so overwhelm both soil chemistry and any microbial antagonists.

Some plants certainly respond to variations in nutrient concentration by altering transport abilities of roots (Drew & Saker 1975; Black *et al.* 1994). In natural environments, it is likely that there will be short-term fluctuations (i.e. on a time scale of hours rather than days) in ion concentrations at the root surface, and it may be that roots can track these by the selective synthesis of appropriate transporter molecules. It may also be the case that there will be important variation in the ability of species to do this, presumably because there will be costs associated with such syntheses that are only affordable by certain species; in other words, there will be trade-offs that cannot as yet be quantified. It is, however, certain that such trade-offs exist for the morphological responses of roots, and here the ecological basis is at least partly understood, and potentially of great importance in defining functional groups, community composition and ultimately ecosystem function.

## Responses to heterogeneity

Root systems have developmental plans. When grown under uniform conditions (e.g. on an agar plate) they produce predictable and often recognizable architectures, which can be altered in equally predictable ways by specific changes in the environment (e.g. phosphate concentration). When roots develop in soil, the architecture alters, because their development is plastic, and soil is an exceptionally heterogeneous environment, both temporally and spatially.

The classic work on root responses to spatially heterogeneous environments was performed by Malcolm Drew and co-workers in the 1970s (e.g. Drew 1975; Drew & Saker 1975; Drew *et al.* 1975), although this was far from the first demonstration of the effect (see Fitter 1987). The important development was the construction of an apparatus which allowed the part of the root system that experienced a distinct nutrient concentration to be defined precisely; the publication of dramatic pictures (e.g. Fig. 6.1) of proliferation of roots in zones enriched in nitrate, ammonium or phosphate (but not potassium) probably enhanced the impact of this work. One consequence was that it came to be assumed that a proliferation response was what *all* root systems did when they encountered a nutrient-rich patch. An ingenious experiment by Campbell *et al.* (1991) showed that the response was far from uniform across species. Species that were competitively inferior in a mixture allocated far more of their *new* root growth to a nutrient-rich area than the competitive dominants. The latter, however, were much larger and still had absolutely more root in the rich patches than the subordinate species. Campbell *et al.* (1991) referred to this as a contrast between the scale and the precision of proliferation.

An assumption in all this work is the intuitively obvious one that it is beneficial to a plant to put roots into nutrient-rich zones rather than elsewhere. In fact, this assumption may not always be true. Whether it is will depend on a number of factors, including the characteristics of the patch (its size, duration, intensity, e.g. Fitter 1994) and the number and identity (e.g. type of plant species and microbial populations) of competitors. One problem with the literature on proliferation is that very few workers have demonstrated that the proliferation they observed resulted in increased nutrient capture. Van Vuuren *et al.* (1996) showed that wheat

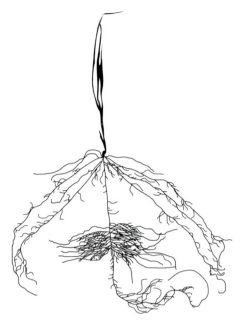

**Figure 6.1** Lateral root proliferation by a barley plant grown in a nutrient solution containing 0.01 mM nitrate, except for the central section which received 1.0 mM nitrate. Courtesy of Dr M.C. Drew.

roots proliferated in a patch created by incorporating grass leaves into soil, a more realistic situation than those in many experiments. However, this proliferation did not occur until 3 weeks after the roots encountered the patch, by which time the plant had absorbed almost all the nitrogen available in the patch. This led Robinson (1996) to ask, 'why do plants bother?' It might have been that wheat was unusual in this respect, but Hodge *et al.* (1998) confirmed this result for five grass species which differed markedly in their ability to proliferate in response to an organic patch: uptake of nitrogen from the patch was unrelated to the degree of root proliferation. Similarly Fransen *et al.* (1998), working with a different set of five grass species, found no relationship between proliferation response and the ability to acquire nitrogen or phosphorus from heterogeneous supplies.

The explanation for this apparent paradox is probably that all these experiments have been performed on single plants in pots. When plants compete for nutrients in a patch, it is much more likely that proliferation by some individuals will lead to increased nutrient capture (although whether that will outweigh the costs of proliferation, however they are to be measured, is less certain), and offers the basis for an economic argument that may explain why species differ so much in their proliferation responses.

Root systems can only respond to spatially and temporally heterogeneous nutrient supplies because of their dynamic nature. The principal process that leads to differential growth of roots in various parts of the root system is variation in the initiation of new meristems. Morphological plasticity therefore implies control of the processes of root production and, probably, root death. We need therefore to understand root demography.

## The demography of roots

Because they are harder to observe, our understanding of the demography of roots is rudimentary compared with that of leaves. A whole vocabulary exists to describe leaf demography—evergreen, wintergreen, deciduous—which has never been invented for roots. Yet root systems exhibit exactly the same patterns: there are 'everwhite' root systems that maintain absorbing roots (the functional equivalent of leaves) at all times, and deciduous systems that shed all (or at least the bulk) of their absorbing roots during unfavourable periods, notably during dry seasons. With leaves, we must differentiate the maintenance of a leaf canopy from the longevity of individual leaves: both pasture grasses and trees such as conifers are evergreen, but the grasses achieve that by continuous production of short-lived leaves, whereas the conifers do so by the seasonal production of long-lived leaves. Grass root systems appear to behave similarly to their leaf systems: half-lives of cohorts of grass roots can be as short as 10 days for roots produced in spring and summer, but much longer for those produced in late summer (Fitter *et al.* 1998), a pattern identical to that found for grass leaves (Harper 1977). In these root cohorts, it is common to find that a very small number of individual roots are very long lived so that a standard demographic plot of the fraction of surviving roots against time is often biphasic, with an initial exponential decline (from which a half-life can be calculated) and a second phase during which little or no mortality occurs (Fig. 6.2). It is reasonable to suppose that these few long-lived roots form the structural root system, analogous to the branch structure of a tree above ground, implying that these may be important differences in demography between above- and below-ground morphologies of grasses.

Compilations of half-life data for roots (Eissenstat & Yanai 1997; Black *et al.* 1998; Fig. 6.3) show that roots are generally surprisingly short lived, although the selection of species so far studied is strongly biased to forest and fruit trees and agricultural crops. A problem with these data is that various authors have used a range of criteria for root death (disappearance, colour change, loss of cortex, etc.). Nevertheless, 70% of all the studies reported to date produced root half-lives of less than 100 days, and 46% were less than 60 days. However, these data may be unrepresentative of types of species that have not been well studied, and even of some that have. For example, annual crops might be expected to show rapid root turnover, yet two studies on wheat (*Triticum aestivum*) found very low levels of mortality. Gibbs and Reid (1992), who measured root demography in horizontal minirhizotrons in the field, and van Vuuren *et al.* (1997), who worked with single plants in long tubes in a growth chamber, both found almost no mortality before maturity (135 days after sowing in the former case). It is well known that root

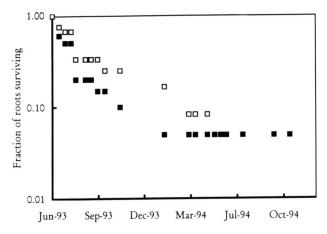

**Figure 6.2** Demography of a single cohort of roots under a *Nardus stricta* grassland on peaty soil taken from Great Dun Fell (UK) and grown at ambient (open symbols) or elevated (solid symbols) $CO_2$ in solardomes. Elevated $CO_2$ represented ambient $+ 250\,\mu\text{mol}$ $\text{mol}^{-1}$. From Fitter *et al.* (1997).

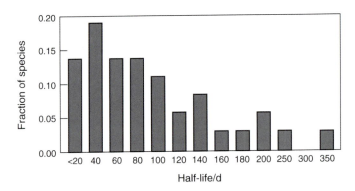

**Figure 6.3** Published estimates of root half-life for a range of species. From various sources; see text.

biomass in wheat declines at maturity, which clearly demonstrates that considerable root mortality occurs then, but it seems that the wheat root system is genuinely deciduous, undergoing synchronous mortality at the end of the growing season. A comparable demography is found in bulbous plants and possibly in many geophytes: bluebell (*Hyacinthoides nonscripta*) has a strictly annual root system, losing its roots entirely in mid-summer in Britain (Merryweather & Fitter 1995), and although no proper demographic studies have been carried out, direct observations show that there is effectively no root mortality prior to this synchronized death. These considerations suggest that it may be possible to recognize distinct categories of root system demography, dependent upon the synchronicity of root

**Table 6.1** A possible classification of root system demography, with examples of plant groups possessing given combinations of demographic characteristics, and analogies to leaf demography shown in *italic* type.

|  | Production | |
| --- | --- | --- |
|  | *Synchronous* | *Continuous* |
| **Mortality** | *Strict deciduous* | *Deciduous* |
| *Synchronous* | Geophytes (Merryweather & Fitter 1995) | Annuals, herbaceous perennials, some trees (Gibbs & Reid 1992) |
| *Continuous* | *Seasonal evergreen* | *Productive evergreen* |
|  | Temperate trees (Hendrick & Pregitzer 1992; Berntson & Bazzaz 1996) | Grasses, tropical trees (Fitter *et al.* 1998) |

birth and death (Table 6.1), which suggests that root and leaf demographies are generally closely linked.

Variation in root demography has large consequences in studies of ecosystem productivity. Estimates of below-ground productivity have traditionally been made by sequential destructive harvests; Hendrick and Pregitzer (1992) showed that such methods would have underestimated root production by sugar maple *Acer saccharum* by more that 50% compared with minirhizotron methods, because they cannot account for roots that die between measurement periods. Overestimation is also possible (Fairley & Alexander 1985). Minirhizotrons (which are simply glass tubes inserted in soil) can be used to quantify below-ground productivity: Fitter *et al.* (1997) found good agreement between estimates of production from minirhizotrons and from gas exchange measurements.

## How do roots die?

Leaf demography is to a great extent controlled by leaf senescence, a process under plant control. By contrast, little is known about root death. The short life-spans of most roots show that it is a common occurrence, but there is no evidence of a senescence process comparable to that in leaves. In rhizotron studies, roots often disappear completely between observation events separated by only a few days, which might suggest that herbivory or pathogen attack is a major cause of death (e.g. Kosola *et al.* 1995); other roots can be observed turning brown and losing their cortex, although apparently the stele can remain functional for long periods in such cases.

Rapid turnover of fine roots is, at first sight, an expensive process. However, although fine roots may be more nutrient rich than the structural roots in a woody plant root system, they comprise a very small fraction of the total root system. For example, Pregitzer *et al.* (1997) showed that high-order laterals in the woody root systems of *Acer saccharum* have higher nitrogen concentrations (Fig. 6.4). This is an important finding, because it implies that fine roots are more expensive to con-

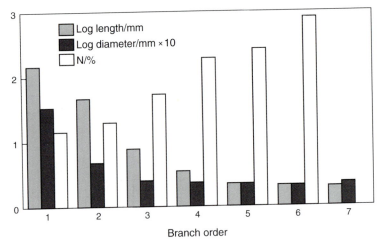

**Figure 6.4** The relationship between total root length, mean root diameter and root nitrogen concentration for roots of *Acer saccharum* of different developmental order. No data were available for nitrogen concentration for order 7 roots. From Pregitzer *et al.* (1997).

struct and maintain than the structural roots in terms of a critical nutrient; even if the nitrogen in fine roots is recycled, it still represents an opportunity cost. However, the highest-order laterals represent a tiny fraction of the total root system biomass: in *A. saccharum* 99% of the total root nitrogen was in the lowest-order laterals, which were large-diameter roots (mean 3.27 mm). This equation is simply driven by root diameter: coarse roots have such a greater biomass than fine roots that they dominate all such budgets. However, the nitrogen costs of root systems will be very different for plants such as grasses, where there is much less distinction between structural and 'feeder' roots, and where root turnover must represent a potentially large loss of nitrogen, in the absence of any resorption mechanism for which the evidence is as yet limited (Persson 1979; Ferrier & Alexander 1991).

Root death is, however, certainly partly under plant control, and there is abundant evidence that root longevity is affected by environmental factors (both biotic and abiotic) that impact more generally on plants and on root systems. For example, longevity of poplar roots was much shorter when they were fertilized with nitrogen (Pregitzer *et al.* 1995) and longevity of *Lolium perenne* roots was very much shorter at 27°C than at 15 or 21°C (Forbes *et al.* 1997; Fig. 6.5). In the latter case, nearly 70% of roots lived longer than 35 days at 15°C, compared with 40% at 21°C and 16% at 27°C. At the highest temperature, over 60% of roots lived less than 21 days. Similarly, longevity of roots (mainly of *Nardus stricta* and *Juncus squarrosus*) in a peat soil was significantly reduced by growing the plants *in situ* in soil monoliths at elevated (ambient $+250\,\mu\mathrm{mol\,mol^{-1}}$) as opposed to ambient atmospheric $CO_2$ concentration (Fitter *et al.* 1997). Some of these responses may have

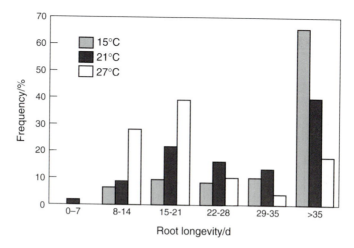

**Figure 6.5** Mean root longevity (time to death) of roots of *Lolium perenne* grown at three temperatures, 15, 21 or 27°C, for 86 days. From Forbes *et al.* (1997).

been caused by changes in pathogen or herbivore activity, but it seems much more likely that they represent evidence for plant control over root longevity, presumably brought about by the regulation of assimilate supply to individual roots. One relevant point is that the C:N ratio of plants was increased in elevated $CO_2$, as is typically found in such conditions, which might determine the duration of root activity.

The role of biotic factors other than predation awaits clarification. Hooker *et al.* (1995) found that the half-life of a cohort of roots of *Populus generosa* cv *interamericana* was about 50 days when non-mycorrhizal, but less than half that when colonized by the arbuscular mycorrhizal fungus *Glomus mosseae*. They suggested that this might be caused by changes in root architecture which they had previously reported (Hooker *et al.* 1992), with mycorrhizal plants having more short-lived, high-order laterals. An alternative is that the higher carbon demand of a mycorrhizal root might either initiate a root senescence process more rapidly through effects on carbon:nutrient ratios or extend root life-span, because the fungal carbon sink acts to maintain the flow of assimilate to the root, at least in ectomycorrhizal roots (Cairney & Alexander 1992). There is an urgent need for experimental studies on the control of the timing of root death.

## What controls root production?

Root growth has been extensively studied under controlled conditions, and the basic physiology is well understood. Very few studies, however, have determined the controls on root growth in the field, where most environmental factors (especially temperature, nutrient supply and, in many soils, water availability) are likely to be suboptimal. More importantly, all these factors fluctuate significantly, on a daily basis in surface soils where most production occurs, and on longer time scales throughout the soil profile.

## Environmental heterogeneity

Understanding root production requires an appreciation of the modular nature of root systems. An environmental factor could act in two ways on roots: it might increase the elongation rate of existing meristems, or it might increase the rate at which meristems are produced. These two parameters are linked, but not absolutely so. Changes in the overall architecture of root systems arise because they can be disconnected, with an emphasis on lateral production in some situations and on elongation in others: plants grown at high nutrient supply rates tend to have more dichotomous architecture (Fitter & Stickland 1991; Taub & Goldberg 1996), which means that they have produced more lateral meristems of high developmental order. The dramatic proliferation responses seen where nutrients are supplied locally to roots (Drew 1975) demonstrate changes in lateral meristem initiation even more markedly. Conversely, many physiological studies concentrate on rates of elongation as indicators of root responsiveness. Nevertheless, the responses are connected: as elongation proceeds, the numbers of lateral meristems will normally increase in parallel.

Soil nutrient concentration affects both elongation and lateral initiation. For example, in *Arabidopsis thaliana*, high nitrate concentration has no effect on the elongation rate of the main axis but inhibits the lateral elongation rate; by contrast, high phosphate increases both axial and lateral elongation rates (L. Williamson, unpublished data). Zhang and Forde (1998) have identified a gene homologous to MADS-box transcription factors that is involved in the response to nitrate by lateral roots in *A. thaliana*, but not in the response of the apparently separate, shoot-controlled system that results in reduced elongation of the main axis. When the expression of this gene, *ANR1*, was blocked by antisense effects, plants were unable to show the proliferation response to locally applied nitrate. This is the first report of a gene that is directly involved in lateral root proliferation.

Zhang and Forde's (1998) study offers an understanding of the mechanisms by which plant development responds to nutrient ion concentrations. Because local proliferation is a common response to heterogeneous supply of nitrate, ammonium and phosphate, but not to potassium (Drew 1975; Robinson 1994), one potential mechanism would be that proliferation is triggered by metabolism: of those ions only potassium is not metabolized on entry. However, Zhang and Forde (1998) found that mutants of *A. thaliana* that were deficient in nitrate reductase, and therefore poorly able to metabolize nitrate, could proliferate lateral roots in response to heterogeneous nitrate supply as well as could wild-type plants. This must mean that nitrate acts as at least part of the signal pathway. If this is so, then responses to other ions may be equally specific.

## Temperature and radiation flux

Responses of roots to other environmental factors (temperature, oxygen, light) are likely to display more complex control pathways. Because of the direct impact of temperature on metabolic processes, it is common to find that growth responses relate directly to temperature, especially in controlled conditions. There are numerous experiments in which root growth has been shown to be a simple function of temperature. Most of these have either been undertaken in growth chambers, where environmental conditions are optimal, or on crop plants in fertilized, irrigated fields, where again environmental limitations are minimized. However, there is now increasing evidence that, at least for wild plants growing in natural environments, root growth can acclimate to temperature variation to such an extent that it becomes almost independent of temperature, at least at low temperatures, as illustrated in the following examples.

Fundamental to all discussion of temperature effects is the concept of optimal temperature, typically measured by exposing plants to a range of temperatures for short time periods. Optimal temperatures for root growth vary greatly between species: in crop plants they can be as low as 5°C for *Avena sativa* and as high as 33°C for *Gossypium sativum* (McMichael & Burke 1996), although most species have values between 20 and 30°C. However, the measurement of an optimal temperature requires environmental conditions that are not experienced by plants in the field, namely a constant temperature. The physiological literature on temperature responses suffers from a lack of appreciation of the importance of acclimation and of measurements of plant function under natural temperature regimes.

We have measured the effects of temperature on root growth in a series of experiments at Great Dun Fell, in the northern Pennines (UK), which was the base for an experimental programme conducted under the UK Natural Environment Research Council's TIGER programme (Terrestrial Initiative in Global Environmental Research). Root dynamics were measured in minirhizotrons at four sites along an altitudinal gradient from the summit of Great Dun Fell (845 m), through sites at 600 m (High Carle Band) and 480 m (Sink Beck), to a lowland site at 170 m (Newton Rigg), which represents a gradient in mean annual temperature of 4.5°C. Turf dominated by the grass *Festuca ovina* was transplanted from the summit site to all the other sites, and root birth and death rates were measured in minirhizotrons over a 2-year period (Fitter *et al.* 1998). Because an altitudinal gradient also demonstrates variation in a range of environmental factors other than temperature (notably rainfall), the effect of temperature was also measured directly by use of a soil-warming grid at the summit site, which uniformly elevated soil temperature by 2.8°C above ambient, at a depth of 2 cm (Ineson *et al.* 1998).

Cumulative root production (i.e. the total number of roots produced in a season) was greatest at the lowest and warmest sites, as expected, but root birth rates (i.e. daily appearance of new roots) did not differ among the sites. The differences in total production were a result of differences in the length of the growing season, defined as the period during which there was net positive root production.

This raises the question as to what environmental factor does determine root production rate, and this was resolved by stepwise regression of production rate against a range of environmental factors measured on site. In these regressions, mean radiation flux over the preceding 10 days consistently emerged as the controlling variable (Fig. 6.6a). Exactly the same result emerged from a similar analysis of data from the soil-warming experiment. In this case, there were no confounding variables such as those introduced by the use of an altitudinal gradient, and yet although heating the soil almost doubled cumulative root production and mean

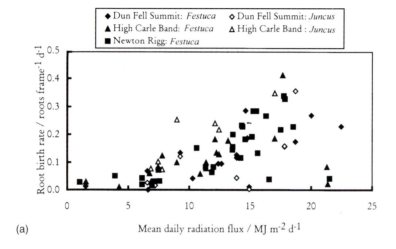

(a)

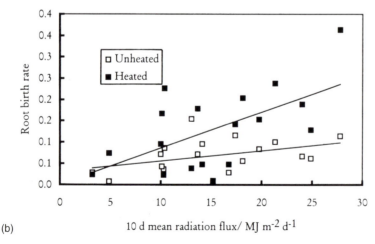

(b)

**Figure 6.6** Relationship between root birth rate and irradiance averaged over the previous 10 days, in minirhizotrons in: (a) two different grassland sites at a range of altitudes on Great Dun Fell, UK; (b) heated (+3°C) and unheated plots of a *Festuca ovina* grassland at the summit of Great Dun Fell (845 m). Data from Fitter *et al.* (1998, 1999).

root birth rate, there was no relationship between birth rate and soil temperature. Again, radiation flux averaged over the preceding 10 days was the environmental factor that explained most of the variation (Fig. 6.6b). There is an apparent paradox in the data from the heating experiment, because root birth rate was higher in the heated plots than in the controls, but was unrelated to temperature. The explanation for this is almost certainly that heating the soil increased nitrogen mineralization (Ineson *et al.* 1998) and so made more nitrogen available for plant growth.

These results suggest that root growth under these field conditions is determined by the availability of photosynthate, and that roots can acclimate to variation in soil temperature. This phenomenon has been shown previously: Aguirrezabal *et al.* (1994) found the same with *Helianthus annuus* under field conditions, as did Gregory (1986) with *Pennisetum typhoides*. We were able to obtain direct evidence that acclimation was responsible from measurements of respiration rate *in situ*. In August 1992, tillers of one of the dominant species at Great Dun Fell, the grass *Festuca ovina*, were transplanted into $4 \times 15$ cm plastic tubes, overwintered in a glasshouse and embedded into soil at all four altitudes in May 1993. The tillers were then removed at various times in 1994 and 1995, and measurements made of the respiration rates of extracted roots in a field laboratory at three temperatures (3, 9 and 16°C) in order to measure the temperature sensitivity of respiration. Roots of three different species (*Festuca ovina, Nardus stricta* and *Juncus squarrosus*) were also collected from undisturbed vegetation on four occasions between July 1994 and May 1995, and measurements were made of the respiration rates of the roots at the soil temperature of collection. It was expected that respiration rates would be a simple function of temperature and would vary with altitude, season and other temperature-related factors. However, once again, the factor that best described variation in root respiration rate was radiation flux, in this case averaged over the previous 2 days, showing that respiration rate had acclimated to the prevailing temperature, even down to soil temperatures below 5°C. It is unsurprising that growth, a process involving the integration of many physiological processes, should respond to an environmental factor over a longer time-scale than respiration, which is only one of the component processes.

It seems then that roots of some species can acclimate successfully to very low temperatures and continue to be active when soil temperature is only a few degrees above freezing. In those conditions, they are highly dependent on the availability of resources from photosynthesis, and hence on photosynthetically active radiation (PAR) flux. It is well established that root growth is sensitive to PAR: growth of roots of *Pseudotsuga menziesii* in spring is a simple linear function of PAR flux (van den Driessche 1987), and similar effects have been recorded in deciduous trees (*Acer saccharum*: Webb 1976) and cereals (wheat: Bingham & Stevenson 1993). Root respiration is also recorded as being able to acclimate to temperature, within 4 days of a 10°C change in temperature in citrus (Bryla *et al.* 1997); whole-plant respiration acclimated within 2 days in wheat (Gifford 1995).

It is easy to see what adaptive benefits plants might gain from a relative independence of root growth and metabolism from temperature. In temperate climates, at least, root growth tends to occur early in the year, when nutrient availability is high and before leaf growth has resulted in high transpirational demand. Species that produce roots early may be able to acquire nutrients that are only briefly available. Most importantly, this temperature independence may enable species to react more precisely to soil factors that directly affect their performance, notably the availability of nutrients in local patches. Precisely such a response has been shown by Bilbrough and Caldwell (1995): they showed that the grass *Agropyron deserto-rum* was unable to respond to locally available nutrients when shaded, simply because of the reduced availability of photosynthate.

If root growth is largely uncoupled from temperature in many ecosystems, this will have important implications for global carbon cycle models—in many of these models temperature is an important driver both of productivity and respiratory carbon loss. It is certainly the case that acclimation responses have not been taken sufficiently into account in these models (Ryan 1991).

## Conclusions

The production of roots consumes a large fraction of plant resources. A fundamental aspect of plant adaptation therefore is an ability to position root growth in both time and space in such a way as to maximize the effectiveness of individual root elements in acquiring resources. The developmental programme of plant root systems is exceptionally plastic, in comparison with almost all other plant organ systems. Root system architecture can be directly modified by variation in nutrient ion concentration, with a high degree of spatial precision. In the case of nitrate, at least, it appears that it is nitrate itself that acts as the stimulus to root production. By contrast, the developmental control of root production by temperature is indirect, to the extent that root systems appear to be able to acclimate so as to maintain production even at very low temperatures, as long as fluxes of photosynthetically active radiation permit continuing photosynthesis and ensure the availability of carbon resources. Again, the adaptive significance of this is apparent, although the mechanisms of acclimation are little understood. A fuller understanding of the developmental ecology of plant root systems will require study of both mechanisms underlying acclimation and of the control of root death, an event which is probably also determined by the supply of carbon resources.

## Acknowledgements

I am very grateful to A. Hodge, D. Robinson and I. Alexander for detailed comments on a draft and to the many collaborators who have been responsible for much of the work reported here, especially: A. Hodge, J. Graves, J. Merryweather, G. Self, D. Bogie, L. Williamson, T. Brown, O. Leyser and S. Ribrioux at York; D. Robinson, B. Griffiths, M. van Vuuren and J. Stewart at the Scottish Crop Research

Institute, Dundee; P. Ineson, D. Benham and K. Taylor at ITE Merlewood; and J. Wolfenden and T. Mansfield at Lancaster.

## References

Aguirrezabal, L.A.N., Deleens, E. & Tardieu, F. (1994). Root elongation rate is accounted for by intercepted PPFD and source–sink relations in field and laboratory grown sunflower. *Plant, Cell and Environment,* **17,** 443–450.

Berntson, G.M. & Bazzaz, F.A. (1996). The allometry of root production and loss in seedlings of *Acer rubrum* (Aceraceae) and *Betula papyrifera* (Betulaceae): implications for root dynamics in elevated G$_2$. *American Journal for Botany,* **831,** 608–616.

Bilbrough, C.J. & Caldwell, M.M. (1995). The effects of shading and N-status on root proliferation in nutrient patches by the perennial grass *Agropyron desertorum* in the field. *Oecologia,* **103,** 10–16.

Bingham, I.J. & Stevenson, E.A. (1993). Control of root growth: effects of carbohydrates on the extension, branching and rate of respiration of different fractions of wheat roots. *Physiologia Plantarum,* **88,** 149–158.

Black, K.E., Harbron, C.G., Franklin, M., Atkinson, D. & Hooker, J.E. (1998). Differences in root longevity of some tree species. *Tree Physiology,* **18,** 259–264.

Black, R.A., Richards, J.H. & Manwaring, J.H. (1994). Nutrient-uptake from enriched soil microsites by 3 Great-Basin perennials. *Ecology,* **75,** 110–122.

Bryla, D.R., Bouma, T.J. & Eissenstat, D.M. (1997). Root respiration in citrus acclimates to temperature and slows during drought. *Plant, Cell and Environment,* **20,** 1411–1420.

Cairney, J.W.G. & Alexander, I. (1992). A study of aging of Spruce [*Picea sitchensis* (Bony.) Carr] ectomycorrhizas. II. Carbohydrate allocation in aging *Picea sitchensis/Tylospora fibrillosa* (Burt.) Dark ectomycorrhizas. *New Phytologist,* **122,** 153–158.

Campbell, B.D., Grime, J.P. & Mackey, J.M.L. (1991). A trade-off between scale and precision in resource foraging. *Oecologia,* **87,** 532–538.

Dinkelaker, B., Hengeler, C. & Marschner, H. (1995). Distribution and function of proteoid roots and other root clusters. *Acta Botanica,* **108,** 183–200.

Drew, M.C. (1975). Comparison of the effects of a localised supply of phosphate, nitrate, ammonium and potassium on the growth of the seminal root system, and the shoot, in barley. *New Phytologist,* **75,** 479–490.

Drew, M.C. & Saker, L.R. (1975). Nutrient supply and the growth of the seminal root system in barley. II. Localized, compensatory increases in lateral growth and rates of nitrate uptake when nitrate supply is restricted to only part of the root system. *Journal of Experimental Botany,* **26,** 79–90.

Drew, M.C., Saker, L.R. & Ashley, T.W. (1975). Nutrient supply and the growth of the seminal root system in barley. I. *Journal of Experimental Botany,* **24,** 1189–1202.

van den Driessche, R. (1987). Importance of current photosynthate to new growth in planted conifer seedlings. *Canadian Journal of Forest Research,* **17,** 776–782.

Eissenstat, D.M. & Yanai, R.D. (1997). The ecology of root lifespan. *Advances in Ecological Research,* **27,** 1–60.

Ennos, A.R. (1989). The mechanics of anchorage in seedlings of sunflower, *Helianthus annuus* L. *New Phytologist,* **113,** 185–192.

Fairley, R.I. & Alexander, I. (1985). Methods of calculating fine-root production in forests. In *Ecological Interactions in Soil* (Ed. by A.H. Fitter), pp. 37–42. British Ecological Society special publication Number 4, Blackwell Scientific Publications, Oxford.

Ferrier, R.C. & Alexander, I. (1991). Internal redistribution of N in Sitka spruce seedlings with partly elongated root systems. *Forest Science,* **37,** 860–870.

Fitter, A.H. (1987). An architectural approach to the comparative ecology of plant root systems. *New Phytologist,* **106**(Suppl.), 61–77.

Fitter, A.H. (1994). Architecture and biomass allocation as components of the plastic response of root systems to soil heterogeneity. In *Exploitation of Environmental Heterogeneity by Plants* (Ed. by M.M. Caldwell & R.W. Pearcy), pp. 305–323. Academic, New York.

Fitter, A.H. & Stickland, T.R. (1991). Architectural analysis of plant root systems II. Influence of nutrient supply on architecture in contrasting plant species. *New Phytologist*, **118**, 383–389.

Fitter, A.H., Graves, J.D., Wolfenden, J. *et al.* (1997). Root production and turnover and carbon budgets of two contrasting grasslands under ambient and elevated atmospheric carbon dioxide concentrations. *New Phytologist*, **137**, 247–255.

Fitter, A.H., Graves, J.D., Self, G.K., Brown, T.K., Bogie, D. & Taylor, K. (1998). Root production, turnover and respiration under two grassland types along an altitudinal gradient: influence of temperature and solar radiation. *Oecologia*, **114**, 20–30.

Fitter, A.H., Self, G.K., Brown, T.K. *et al.* (1999) Root production and turnover in an upland grassland subjected to artificial soil warming responds to radiation flux and nutrients, not temperature. *Oecologia*, in press.

Forbes, P.J., Black, K.E. & Hooker, J.E. (1997). Temperature-induced alteration to root longevity in *Lolium perenne*. *Plant and Soil*, **190**, 87–90.

Fransen, B., de Kroon, H. & Berndse, F. (1998). Root morphological plasticity and nutrient acquisition by perennial grass species from habitats of different nutrient availability. *Oecologia*, **115**, 351–358.

Gardner, W.K., Barber, D.A. & Parbery, D.G. (1983). The acquisition of phosphorus by *Lupinus albus* L. III. The probable mechanism by which phosphorus movement in the soil/root interface is enhanced. *Plant and Soil*, **70**, 107–124.

Gibbs, R.J. & Reid, J.B. (1992). Comparison between net and gross root production by winter wheat and by perennial ryegrass. *New Zealand Journal of Crop and Horticultural Science*, **20**, 483–487.

Gifford, R.M. (1995). Whole plant respiration and photosynthesis of wheat under increased $CO_2$ concentration and temperature: long-term vs short-term distinctions for modelling. *Global Change Biology*, **1**, 385–396.

Gregory, P.J. (1986). Response to temperature in a stand of pearl millet (*Pennisetum typhoides* S. & H.). VIII. Root growth. *Journal of Experimental Botany*, **37**, 379–388.

Grierson, P.F. (1992). Organic acids in the rhizosphere of *Banksia integrifolia* L. *Plant and Soil*, **144**, 259–265.

Harper, J.L. (1977). *The Population Biology of Plants*. Academic, London.

Hendrick, R.L. & Pregitzer, K.S. (1992). The demography of fine roots in a northern hardwood forest. *Ecology*, **73**, 1094–1104.

Hodge, A., Stewart, J., Robinson, D., Griffiths, B.S. & Fitter, A.H. (1998). Root proliferation, soil fauna and plant nitrogen capture from nutrient-rich patches in soil. *New Phytologist*, **139**, 479–494.

Hooker, J.E., Monroe, M. & Atkinson, D. (1992). Vesicular-arbusclar mycorrhizal fungi induced alteration in poplar root system morphology. *Plant and Soil*, **145**, 207–214.

Hooker, J.E., Black, K.E., Perry, R.L. & Atkinson, D. (1995). Arbuscular mycorrhizal fungi induced alteration to root longevity in poplar. *Plant and Soil*, **172**, 327–329.

Ineson, P., Benham, D.G., Poskitt, J., Harrison, A.F., Taylor, K. & Woods, C. (1998). Effects of climate change on nitrogen dynamics in upland soils. II. A soil warming study. *Global Change Biology*, **4**, 153–161.

Kosola, K.R., Eissenstat, D.M. & Graham, J.H. (1995). Root demography of mature citrus trees: the influence of *Phytophthora nicotianae*. *Plant and Soil*, **171**, 283–288.

Lamont, B.B. (1993). Why are hairy root clusters so abundant in the most nutrient-impoverished soils of Australia? *Plant and Soil*, **156**, 269–272.

Liu, H., Trieu, A.T., Blaylock, L.A. & Harrison, M.J. (1998). Cloning and characterization of two phosphate transporters from *Medicago truncatula* roots: regulation in response to phosphate and to colonization by arbuscular mycorrhizal (AM) fungi. *Molecular Plant–Microbe Interactions*, **11**, 14–22.

McMichael, B.L. & Burke, J.J. (1996). Temperature effects on root growth. In *Plant Roots: the Hidden*

*Half,* 2nd edn (Ed. by Y. Waisel, A. Eshel & U. Kafkafi), pp. 383–396. Dekker, New York.

Merryweather, J.W. & Fitter, A.H. (1995). Arbuscular mycorrhiza and phosphorus as controlling factors in the life history of *Hyacinthoides non-scripta* (L.) Chouard ex Rothm. *New Phytologist*, **129**, 629–636.

Persson, H. (1979). Fine root production, mortality and decomposition in forest ecosystems. *Vegetatio*, **41**, 101–107.

Pregitzer, K.S., Zak, D.R., Curtis, P.S., Kubiske, M.E., Teeri, J.A. & Vogel, C.S. (1995). Atmospheric $CO_2$, soil nitrogen and turnover of fine roots. *New Phytologist*, **129**, 579–585.

Pregitzer, K.S., Kubiske, M.E., Yu, C.K. & Hendrick, R.L. (1997). Relationships among root branch order, carbon and nitrogen in four temperate species. *Oecologia*, **111**, 302–308.

Robinson, D. (1994). The responses of plants to non-uniform supplies of nutrients. *New Phytologist*, **127**, 635–674.

Robinson, D. (1996). Resource capture by localized root proliferation: why do plants bother? *Annals of Botany*, **77**, 179–185.

Ryan, M.G. (1991). Effects of climate change on plant respiration. *Ecological Applications*, **1**, 157–167.

Taub, D.R. & Goldberg, D. (1996). Root system topology of plants from habitats differing in soil resource availability. *Functional Ecology*, **10**, 258–264.

van Vuuren, M.M.I., Robinson, D. & Griffiths, B.S. (1996). Nutrient inflow and root proliferation during the exploitation of a temporally and spatially discrete sources of nitrogen in soil. *Plant and Soil*, **178**, 185–192.

van Vuuren, M.M.I., Robinson, D., Fitter, A.H., Chasalow, S.D., Williamson, L. & Raven, J.A. (1997). Effects of elevated atmospheric $CO_2$ and soil water availability on root biomass, root length and N, P and K uptake by wheat. *New Phytologist*, **135**, 455–465.

Webb, P. (1976). Root growth in *Acer saccharum* Marsh: effects of light intensity and photoperiod on root elongation rates. *Botanical Gazette*, **137**, 211–217.

Zhang, H. & Forde, B.G. (1998). An *Arabidopsis* MADS box gene that controls nutrient-induced changes in root architecture. *Science*, **279**, 407–409.

# Chapter 7

# The ecophysiology of mycorrhizal symbioses with special reference to impacts upon plant fitness

*D.J. Read*

## Introduction

When plants were first emerging into terrestrial environments their access to soil-borne resources was constrained by the geometrical inadequacy of their root systems. Fossil evidence for the presence of fungal colonists producing tree-like or 'arbuscular' branches in root cortical cells of these first land plants suggests that these constraints favoured partnerships at an early evolutionary stage. Of the heterotrophs present in these primitive environments mycelial fungi, with their potential to produce extensively branched extremely fine hyphal systems of enormous surface area, would be the best placed physically to fulfil an absorptive role. What, however, were the factors which determined that relationships between the partners were of a balanced rather than of a pathogenic kind? Vanderplank (1978) has pointed out that whereas in parasitic associations mutation towards resistance in the host would be an advantage, any such resistance to fungi that confer essential benefits would be selectively eliminated when plants lacking such advantages perished.

The force of such arguments is strengthened by comparative analysis of fossil and present-day evidence, the latter coming from both molecular and organismic sources. Striking parallels exist between the 'arbuscular' structures seen in Devonian fossils and those produced by zygomycetes of the order Glomales in the roots of extant plants. These give the name 'arbuscular' to this apparently oldest and now most widespread type of mycorrhizal association. Molecular methods have confirmed glomalean fungi as being a monophyletic group that originated at about the same time as that proposed by palaeontologists for the origin of land plants, which lies between 460 and 350 million years BP (Simon *et al.* 1993).

From these original colonists of the land a diverse flora has arisen in which mycorrhizas, both of the original arbuscular kind and of more recently evolved types, are extensively represented (Fig. 7.1). It is true that only a small proportion

*Department of Animal and Plant Sciences, University of Sheffield, Sheffield, S10 2TN, UK. E-mail: d.j.read@sheffield.ac.uk*

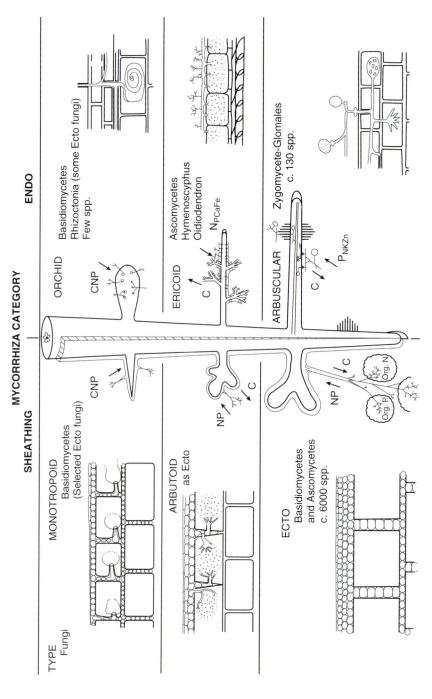

**Figure 7.1** The structural and major nutrient pathways of the six recognized types of mycorrhiza. Two basic categories are designated, one in which the root surface is sheathed in a fungal mantle (SHEATHING), and one lacking a mantle but in which hyphae proliferate internally (ENDO). The defining structures of each type are fungal pegs (MONOTROPOID), Hartig net and intracellular penetration (ARBUTOID — also ectendo), Hartig net, mantle, external mycelial network (ECTO), peloton (ORCHID), hyphal complexes in hair roots (ERICOID) and arbuscules or hyphal coils (ARBUSCULAR). C, carbon; N, nitrogen; P, phosphorus. *Arrows* indicate direction of flow: light, fungus; black, plant tissue. From Read (1988a) with permission.

of the roots of the world's terrestrial plant species have been examined—perhaps around 3% (Trappe 1987). However, based upon those that have been studied it can be concluded that about 95% of present-day species belong to families that are characteristically mycorrhizal. Emerging from such analyses of 'occurrence' of mycorrhizas is the recognition that the symbiotic condition is 'the norm' in nature except in a small number of plant families such as Cruciferae, Caryophyllaceae, Polygonaceae and Juncaceae, and under highly fertile conditions.

While knowledge of the generality of the 'occurrence' of the symbiosis has been gained and may indeed represent the main achievement of the first century of research on mycorrhiza, much less progress has been made towards understanding of 'function'. Early work stressed a rather simplistic view of function in which access to nutrients, in particular phosphorus, was conventionally seen to be of predominant importance. It has become increasingly recognized that species, communities and ecosystems respond to differing extents and in different ways to the presence of the mycorrhizal symbiosis and we are moving towards identification of the nature and extent of these responses under relevant ecological circumstances. If definition of structures and charting of occurrence of the symbiosis were the achievements of the first hundred years of mycorrhizal research, elucidation of their contributions to ecosystem function will be the goal of the next century.

The divergence between the more conventional view of mycorrhizas as nutrient-absorbing organs and the emerging perception of their wider biological contribution is highlighted by recent definitions of the symbiosis. Trappe *et al.* (1996) defined mycorrhizas as 'dual organs of absorption formed when symbiotic fungi inhabit healthy organs of most terrestrial plants ...'. Similarly, Smith and Read (1997) described a mycorrhiza as 'a symbiosis in which an external mycelium of a fungus supplies soil derived nutrients to plant roots', a definition which placed emphasis upon the importance, often overlooked in the past, of the foraging role of hyphal networks.

Pressures to broaden the definition come from two directions. First, there is the recognition that nutrient acquisition, and the biomass gain which is its consequence, while being useful measures for agriculturalists and foresters are, at best, only indirectly related to success in natural ecosystems. Second, there has been increasing awareness of broader impacts of the symbiosis upon the physiological ecology of both partners. Among these there are effects, examined in this paper, upon photosynthetic activity of the plant, upon the response to plant pathogens and upon resistance to environmental toxins.

In combination these attributes contribute to determine what in Darwinian terms can be described as the 'fitness' of partners involved in the symbiosis here, defined as 'the ability of parents to contribute descendants to subsequent generations' fitness'. This is the driving force in natural selection. Viewed from this perspective it has been suggested that mycorrhizas be defined as 'Structures in which a symbiotic union between fungi and plant roots leads to increases of fitness in one or both partners' (Read 1998a). An important consequence of acceptance of this definition would be recognition of the need to design experiments targeting

135

fitness-related aspects of both symbionts in the association. Here those aspects of particular interest to the physiological ecology of plants, rather than of fungi, will be emphasized. Experimental evaluation of 'fitness' of the fungal components of arbuscular mycorrhizas (AM) and ectomycorrhiza (ECM) may in any event be unnecessary, because a large literature documents the dependence of these organisms upon their autotrophic partners for the carbon necessary to sustain growth, reproduction and hence fitness (Smith & Read 1997).

Only a relatively small literature directly addresses the question of 'fitness' in mycorrhizal plants, although often a tacit assumption of fitness gain is made by reference to the mycorrhizal symbiosis as being a 'mutualism'. According to classifications originating with De Bary (1887), and subsequently accepted by Starr (1975), Lewin (1982) and Lewis (1985), a symbiotic relationship can be considered to be mutualistic only if the fitness of both of the associating individuals is greater when they are together than when apart. Because short-term increases of nutrient uptake or growth do not directly increase fitness they cannot provide confirmation that the relationship is mutualistic.

The main factors determining fitness, as defined above, fall into two categories: those enabling the organism to reach the reproductive state, these being termed 'indirect' (Table 7.1), and those which, through their effects upon quality and quantity of the reproductive propagules themselves, contribute in a 'direct' manner. Among the indirect factors clearly the ability of a new generation to establish itself in the community, 'regeneration', is a key feature, while physiological vigour in the vegetative phase of the life cycle, together with enhanced resistance to various environmental stresses, are evidently of potential importance. To identify these as factors requiring analysis is not to imply that their effects will be independent of one another. Clearly, for example, seed quality might influence the success of a germinating seedling in the regeneration niche, as might an enhanced ability to acquire nutrients. The latter could indeed be the basis for enhanced resistance to disease in some circumstances. Whatever the nature of the interactions between the factors, the advantage of the fitness-based analysis of mycorrhizal function is that it requires evaluation of each of them and in so doing promises to bring the evolutionary significance of the symbiosis into sharper focus. This chapter provides an overview of the contribution of the mycorrhizal symbiosis to plant fitness using the framework established in Table 7.1.

**Table 7.1** Factors influencing the fitness of plant and fungal partners in the mycorrhizal symbiosis.

| Indirect influences | Direct influences |
| --- | --- |
| Ability to regenerate | The life cycle |
| Vigour in the vegetative condition | Fecundity (seed number and |
|   Physiological state |   quality) |
|   Ability to acquire resources | |
| Resistance to pathogens | |

## Indirect influences of mycorrhizas upon fitness

### Ability to regenerate

Whereas many of the turf-forming species of grasslands are clonal perennials which reproduce largely by tillering, the recruitment of numerous annual and biennial forbs into the community depends upon establishment from seed. In a seminal paper Grubb (1977) pointed out that the conditions prevailing in the 'regeneration niche', through their effects upon establishment of seedlings, determined community structure. It is now recognized that mycorrhizal fungi play a key role in facilitating successful establishment.

Hartnett *et al.* (1994) investigated the influence of AM fungi upon the regeneration of six tall-grass prairie species. Seed of each species was sown either into undisturbed prairie soil or into the same soil from which the fungal symbionts had been removed by benomyl application. Interestingly, seedling emergence of $C_4$ species such as *Andropogon gerardii*, shown to be very responsive to colonization in terms of vegetative growth and tiller production, was not influenced by removal of inoculum. By contrast, the $C_3$ grasses *Elymus canadensis*, *Koeleria cristata* and the forb *Liatris aspera*, all considered to be unresponsive in biomass terms, each showed significantly lower emergence if AM inoculum was removed.

Two studies (Grime *et al.* 1987; van der Heijden *et al.* 1998) of reconstructed communities representative of calcareous and old field grasslands have demonstrated that through their effects upon survivorship of component plants, AM fungi can determine the structure and diversity of these types of ecosystems. Grime *et al.* (1987) showed that seedlings of a number of forbs typically present in naturally occurring calcareous grasslands failed to survive in the absence of their fungal symbionts, while the turf-forming grass *Festuca ovina*, although normally colonized by the fungi, was relatively insensitive to their removal. In the presence of AM symbionts the forbs showed strong interspecific differences in the extent of their responsiveness to the fungi, but the overall effect of their presence was to increase plant diversity.

The work of van der Heijden *et al.* (1998) was an advance because it involved manipulation of fungal as well as plant diversity. Each of a set of four *Glomus* species was first shown to produce a distinctive response in each of a range of plant species. Thus, while a given fungal symbiont could promote the growth of one species it had little effect upon another. Such effects are initially difficult to interpret because, as a consequence of the early evolutionary elimination of resistance referred to earlier, all of these plant species would normally be colonized by all of the fungi in nature. Indeed, molecular analyses have confirmed that the roots of any AM-compatible plants in nature may be simultaneously occupied by several species, even genera, of these fungi (Clapp *et al.* 1995; Helgason *et al.* 1998). In these circumstances the observation of van der Heijden *et al.* (1998) that there are interspecific differences in their responses to a given fungus, can best be interpreted in terms of differences in functional compatibility between individual plant–fungus combinations (Read 1998b). These effects challenge the view that the physiological basis of plant response to colonization by AM fungi is similar in all taxa and call for a more sophisticated analysis of relationships between partners of defined genomes.

The experiment of van der Heijden *et al.* (1998) demonstrated that plant biodiversity, nutrient capture and productivity can be significantly increased by increasing the species richness of the AM fungal community. Progressively greater numbers of AM fungal taxa (from non-mycorrhizal to 14 species) were added to a standardized mix of old field species. It was shown that both the plant biodiversity as measured by Simpson's diversity index (Fig. 7.2a) and productivity above and below ground (Fig. 7.2b,c) increased with an increasing AM fungal species richness. These observations are again consistent with a role for functional compatibility, because progressive addition of fungal species will increase the chance of achieving optimal combinations of symbionts. Increasing AM fungal biodiversity resulted in the production of a greater length of fungal hyphae (Fig. 7.2d), a more effective capture of soil phosphorus (Fig. 7.2e) and increases in phosphorus content of the plant tissue (Fig. 7.2f). This demonstration of the interconnectedness of plant and fungal biodiversity in nutrient-poor ecosystems has important conservational implications because there is increasing evidence that intensive management of soils has an adverse effect upon the diversity of AM fungi (Helgason *et al.* 1998).

The recognition that mycorrhizal fungi influence plant community structure by exercising different levels of benefit is itself an important advance, but also of great interest is the observation that some species can be excluded from the community by antagonistic effects of the fungus. It has been known for some time that members of certain plant families, notably the Brassicaceae, Caryophyllaceae, Chenopodiaceae and Polygonaceae, which characteristically occur as ruderals of disturbed ground, are not normally colonized by AM fungi. However, most of the emphasis in research has been placed upon mechanisms adopted by the plant to resist colonization. If, conversely, the fungus is acting as an antagonist of those plant species with which it is incompatible, its role must be seen from a new perspective in which it has the ability to determine recruitment into, as well as performance within, the community.

In order to examine the impact of AM fungi upon the establishment phase of the plant life cycle, chambers were specially designed to enable simulation of regeneration niches either with or without the presence of AM mycelia (Francis & Read 1994, 1995). Seeds of a range of species representing families which are known to be either compatible or incompatible with AM fungi were then sown, at the time of radicle emergence, into both types of environment. There were clearly distinguishable responses of the two categories of plants. Whereas compatible species grew rapidly and became colonized in the niches containing AM hyphae, the reverse was true of seedlings of incompatible species, many of which died within days of sowing into these compartments and before colonization of their roots had occurred. Species of this latter category flourished only in the absence of AM mycelia.

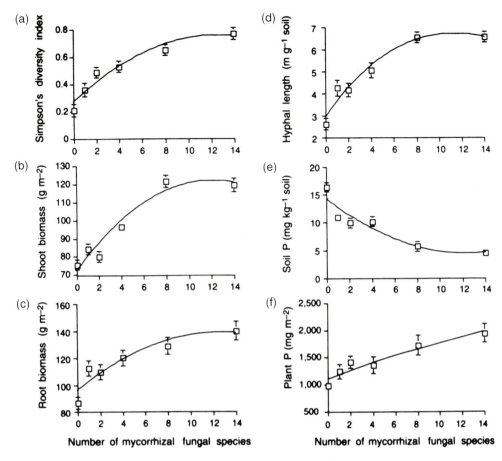

**Figure 7.2** The effect of AM fungal species richness on different parameters. (a) Simpson's diversity index (fitted curve is $y=0.0271+0.077x-0.003x^2$; $r^2=0.63$; $P\leq0.0001$); (b) shoot biomass ($y=0.334x^2+8.129x+72.754$; $r^2=0.69$; $P\leq0.0001$); (c) root biomass ($y=0.265x^2+6.772x+96.141$; $r^2=0.55$; $P\leq0.0001$); (d) length of external mycorrhizal hyphae in soil ($y=0.001x^3-0.046x^2+0.765x+2.979$; $r^2=0.60$; $P\leq0.0001$); (e) soil phosphorus concentration ($y=0.065x^2-1.593x+14.252$; $r^2=0.67$; $P\leq0.0001$); (f) total plant phosphorus content (linear relationship; $y=61.537x+1156.281$; $r^2=0.48$; $P\leq0.001$), in macrocosms simulating North American old field ecosystems (expt 2). Squares represent mean values ($\pm$SEM). From van der Heijden *et al.* (1998).

To date most analyses of plant–fungus incompatibility in the AM mycorrhizal system have adopted the 'plant centric' view that non-host plants are 'resistant' to colonization. Antifungal compounds have been sought (Schreiner & Koide 1993a) and a role for resistance genes (Anderson 1988) postulated. In Brassicaceae products of glucosinolate hydrolysis, notably isothiocyanates, have been identified (Schreiner & Koide 1993b) and shown to have properties inhibitory to AM fungi.

However, no such compounds have been found in other 'non-host' families. The results of Francis and Read (1995) suggest a pro-active role for the mycorrhizal fungus, which, independent of any colonization process exercises the power selectively to determine success in the regeneration niche. The important difference between this mycocentric view and the plant-centric notion of resistance is that whereas the latter sees the fitness of the plant as being unaffected by AM fungi, the former identifies a direct challenge upon this fitness and so envisages a role for the fungi as determinants of recruitment into, and composition of, plant communities. In short, the mycocentric view is that ruderals are ruderals because they must avoid networks of AM mycorrhizal mycelia. Conversely, the stability of closed communities made up of species that are compatible with AM fungi depends on the sharing of resources facilitated by interconnecting mycelial systems.

In nature ruderals, as would be predicted by this view, dominate early stages of succession in which, as a result of preceding disturbance events, AM inoculum is absent or fragmented. The introduction of propagules of AM-compatible plants and of the fungus itself would enable the production of the first mycelial networks, a progressive reduction of the ability of ruderals to regenerate and eventually their elimination. Successional sequences of this kind can be observed in maritime sand dune ecosystems.

## Vigour in the vegetative condition

### Physiological state

When mycorrhizal and non-mycorrhizal plants of a given species are compared under similar soil conditions the superior ability of the colonized plant to forage for nutrients, in particular nitrogen and phosphorus, normally enables them to accumulate greater amounts of these elements in their tissues and eventually to produce greater biomass. These types of response are conventionally seen as the 'benefits' of mycorrhizal colonization, while the 'costs' are described in terms of the 'carbon' required of the plant to sustain the foraging mycelium of the fungus. However, these nutritional differences mask other, perhaps more fundamental, effects of colonization upon the physiology of the host plant. In order to expose such effects it is necessary to eliminate the differences in nutrient status between the two categories of plant. Elimination of these nutritional impacts, which can be achieved by supplementing the mineral supply to uncolonized individuals in such a way as to produce similar nitrogen and phosphorous concentrations in both categories of plant, enables these effects to be exposed.

Using this approach Wright et al. (1998a,b) compared the impacts of mycorrhizal colonization upon the carbon economy of clover (*Trifolium repens*). They showed, using nutritionally matched plants, that diurnal rates of photosynthetic carbon gain in mycorrhizal plants were significantly higher than those of uncolonized individuals. Rather than being allocated to plant biomass the additional carbon appeared in the form of larger pools of sucrose, glucose and fructose in the roots of the mycorrhizal plants. Further analyses suggested that these sugars were utilized for the synthesis of lipids and the fungal sugar trehalose, which would contribute to sustenance of the large extraradical mycelium. The increased allocation of carbon to mycorrhizal roots was associated with a stimulation of activities of cell wall and cytoplasmic invertases and of sucrose synthase in these systems. These enzymes have been shown to be involved in the regulation of photosynthetic carbon metabolism in a number of plant–fungus associations, by providing increases of 'sink' strength. They have not, however, previously been considered to provide a mechanism for enhancement of carbon gain by mycorrhizal plants.

An interesting feature of the carbon budget of mycorrhizal relative to non-mycorrhizal plants calculated by Wright *et al.* (1998b) was that the net increase of carbon acquisition arising from colonization was approximately 15%. This value is similar to those reported as representing the respiratory 'cost' of mycorrhizal fungi in a number of AM associations (Pang & Paul 1980; Paul & Kucey 1981; Kucey & Paul 1982; Snellgrove *et al.* 1982; Koch & Johnson 1984; Harris *et al.* 1985; Douds *et al.* 1988; Wang *et al.* 1989). The implication is that the stimulatory effects of sink strength are sufficient to eliminate the 'costs' of the symbiosis. If this is in fact generally the case it would further help to explain the almost universal presence of the symbiosis as an effective mechanism for nutrient capture in the plant kingdom. Stimulatory effects of mycorrhizal colonization upon photosynthetic activity of nutritionally matched plants have also been observed in ectomycorrhizal plants.

*Ability to acquire resources*

It has been pointed out (see e.g. Harley 1989) that while the allocation of carbon to the production of an extensive mycelial network may constitute a cost to the plant, in geometric terms any such investment represents an efficient use of resource. On the conservative assumption that the dry weight per unit volume of root, root hairs or hyphae are similar and that the radius of a root is $200\,\mu m$, a root hair $20\,\mu m$ and a mycorrhizal hypha $2\,\mu m$, then the surface area of a given length of each structure per unit is equal to:

$$\frac{2\pi RL}{\pi R^2 L} = \frac{2}{R} \tag{1}$$

Hence the surface area obtained per unit volume of root is $0.01\,\mu m^2$, for the root hair $0.1\,\mu m^2$ and for the hypha $1\,\mu m^2$. Investment in hyphae expressed per unit area is thus 100 times more efficient than in a root and 10 times more efficient than in a root hair. When, to these purely arithmetic considerations, are added the fact that root hairs are unbranched and of limited growth whereas hyphae are branched, show extensive growth and in many cases also possess an elaborate array of nutrient-mobilizing capabilities not seen in roots, the selective advantages arising from preferential allocation of carbon to these structures become evident.

Direct measurements of length of the hyphal elements that make up these mycelial systems confirm that they are indeed extensive. In AM systems reported hyphal lengths per unit of colonized root length and weight of associated soil are in the ranges $3–3000\,m\,m^{-1}$ and $3–20\,m\,g^{-1}$. The mycelial systems of ECM plants are even more extensive and elaborate. Equivalent values cover the ranges 30–8000 m m$^{-1}$ colonized root length and 10–200 mg$^{-1}$ soil (Smith & Read 1997). In ECM plants the mycorrhizal lateral roots are of finite growth, and ensheathed by a mantle of fungal mycelium from which a complex network of hyphae extend over considerable distances into soil. Much has been learned about these networks by the use of microcosms, enabling non-destructive visual and radiographic analysis of their development. These have transformed our view of the function of the ECM symbiosis. In early studies (e.g. Hatch 1937) it was proposed that the benefits of mycorrhizal colonization in plants such as *Pinus* arose through the stimulation of root branching. Much emphasis subsequently upon absorptive capabilities of excised roots further encouraged the view that the root rather than the extraradical mycelium was the critical absorptive structure. This was despite the demonstration (Melin & Nilsson 1950, 1953) of the absorptive functions of the mycelial systems.

Analyses of ECM systems developed on natural organic substrates of low nutrient availability demonstrate that the soil is explored by an extensive hyphal front extending at rates between 2 and 4 mm d$^{-1}$ through soil. The region behind the foraging front and the mycorrhizal mantle is sparsely occupied by linear hyphal aggregates, rhizomorphs, which in some species of fungi can show considerable internal differentiation (Duddridge *et al.* 1980; Agerer 1991) and which provide pathways of transport between root and hyphal tips.

It is a feature of ECM systems that the mycorrhizal roots often proliferate preferentially in air pockets or, as in tuberculate systems (Smith & Read 1997), under an impermeable rind of fungal tissues where they have no direct contact with soil. Even in those cases where intimate contact between the mantle of the mycorrhizal root and the substrate can be demonstrated, it is now known that the contact zone can be hydrophobic. What emerges from studies of entire systems is that the absorptive functions are localized not at the mantle surface but at the distal parts of the hyphal systems often at considerable distances from the ensheathed lateral root. Some of the dynamics of these processes have been revealed by feeding $^{14}CO_2$ to mycorrhizal plants in observation chambers (Plate 7.1a, facing page 146). Photosynthate rapidly flows, by way of roots and rhizomorphs, to the advancing hyphal front. This allocation apparently sustains two functions, one being to facilitate 'capture' by the fungus of new carbon sources, in the form of uncolonized roots, the other being the location of spatially restricted zones of mineral enrichment. The dynamic nature of these foraging processes is revealed first by 'real time' analysis of carbon allocation after pulse labelling with $^{14}C$ (Plate 7.1b) and secondly by observation of responses of mycelial systems after they encounter resource-rich zones.

Localized zones of enrichment representative of those likely to be encountered by fungal mycelium foraging through forest soil in nature can be recreated in observation chambers. This approach has been used (Bending & Read 1995a) to examine the response of the advancing hyphal front of the ECM fungus *Suillus bovinus* to the presence of pine residues freshly collected from the fermentation horizon (FH) of a pine forest soil. The residues were added as discrete 3-cm² blocks to otherwise nutritionally uniform peat. These studies show that dense proliferation of mycelium occurs in these blocks immediately following contact by the hyphae (Plate 7.1c), and that within 21 days of the initial contact the entire volume of added organic matter can be intensively occupied by the fungus. Analysis of allocation of ¹⁴C, fed as $^{14}CO_2$ to the shoots of mycorrhizal *Pinus sylvestris* seedlings, shows that these zones of intensive exploitation are major sinks for carbon (Plate 7.1d). However, the period of time over which carbon allocation takes place is again short, emphasizing the rapid and dynamic nature of the fungal responses to increased resource availability. The entire system thus shows considerable morphological plasticity with the ability throughout the zone of occupation to respond dynamically to changing patterns of resource input.

Analyses of the macronutrient status of these organic resources before and after colonization by the ectomycorrhizal mycelium have confirmed that during the periods of occupation nitrogen and phosphorus are exported leaving residues of reduced quality. In both quantitative and qualitative terms, the nature of the events observed in the experiments using introduced resource materials, are very similar to those reported from field studies of soils occupied by mycelial 'mats'.

What are the mechanisms involved in these processes of nutrient mobilization? One possibility, of course, is that non-mycorrhizal microbial communities are mineralizing these organic residues and that ectomycorrhizal proliferation simply enables an effective 'mopping up' of the released mineral ions. A number of observations suggest, however, that these fungi carry out direct attack upon the substrates. Although there are large inter- and intra-specific differences in abilities of ECM fungi to metabolize complex organic polymers, extensive studies (see Leake & Read 1997) have shown that a wide range of extracellular enzymes can be deployed by these organisms. Perhaps more importantly, measurements of enzyme activities within introduced blocks of natural substrates (Bending & Read 1995a,b) and in hyphal mats (Griffith & Caldwell 1992) have demonstrated that exploitation of soil residues by these organisms is associated with significant increases of activity of those enzymes, particularly acid proteinase and acid phosphatase which are likely to be involved in direct attack upon nitrogen- and phosphorus-containing organic polymers. An additional attribute of the armoury of some of these fungi is that they also produce enzymes such as polyphenol oxidase and peroxidase, which provide the potential for attack on those structural components of litter residue that can 'mask' the nutrients elements (Leake & Read 1997).

In combination the physical and biochemical attributes of the ectomycorrhizal mycelium provide the fungal symbiont with a far greater potential for effective exploitation of complex nutritional environments than could be achieved by roots. The effective physical response to contact with resource-enriched zones is based on the ability of individual hyphal tips to branch rapidly, and repeatedly, over periods of hours. By contrast, more complex processes of information transfer and translation seen in the organization of root meristem activity will inevitably lead to much longer response times.

In the dynamic environment of the soil where nutrient inputs take the form of localized pulses, the importance of rapid responses to potential resource availability cannot be overemphasized. When to these attributes are added the biochemical plasticity which enables direct attack upon complex substrates, the advantage to be gained from allocation of carbon to a fungal symbiont would seem to be large indeed. It is not surprising therefore that in environments characterized by seasonal and patchy inputs of complex organic residues selection has favoured plants that are capable of forming symbioses with mycelial heterotrophs.

While there is some evidence for increased branching of AM fungal hyphae after contacting organic residues, there is, as yet, little indication that these fungi have the ability to mobilize nutrients contained in such substrates.

## Enhanced resistance to pathogens as a basis for increased fitness

Emphasis upon the nutritional roles of mycorrhizal fungi has led to a relative neglect of their contributions to other fitness-related aspects of plant biology. Analyses of the interactions between these benign associates of the root and other groups of microorganisms has suffered particular neglect. However, there is increasing awareness that occupancy of the root by mycorrhizal fungi can confer resistance to pathogenic attack and the possibility of using these symbionts in biocontrol programmes is now emerging.

### Arbuscular mycorrhizal fungi

It has been demonstrated that diseases arising from attack by *Fusarium* (Caron *et al.* 1986; Jalali & Jalali 1991; Dugassa *et al.* 1996), *Pythium* (Kaye *et al.* 1984; Rosendahl & Rosendahl 1990), *Phytophthora* (Davis & Menge 1980; Guillemin *et al.* 1994; Cordier *et al.* 1996; Trotta *et al.* 1996) and *Verticillium* (Bååth & Hayman 1983; Liu 1995) can be reduced in intensity where plants are previously colonized by AM fungi. Unfortunately, these studies have, for the most part, been carried out under artificial or controlled environment conditions. Hence, we still know little of the role played by AM colonization in natural communities of plants. However, recent work (Newsham *et al.* 1994, 1995) is illuminating in that for the first time it investigates the interactive effects of the presence of an AM fungus (*Glomus*) and a pathogen (*Fusarium*) on the fitness of a field-grown population of a grass, *Vulpia ciliata*. These studies demonstrate that through their antagonistic effects upon the pathogen, which has been shown (Carey *et al.* 1995) to reduce fecundity of the grass species, the AM symbionts can directly increase the fitness of the plant without influencing its mineral nutrient status. Clearly, more analyses under natural conditions of nutrient supply, climate and using realistic levels of inoculum potential are highly desirable.

### Ectomycorrhizal fungi

It has been pointed out (Marx 1969, 1973) that the presence of the thick fungal mantle over the tissues of the ectomycorrhizal root could be expected to provide a physical barrier to penetration by pathogens. However, more subtle effects of the presence of ECM fungi are suggested by recent work. Duchesne *et al.* (1988a,b), for example, showed that the mere presence of *Paxillus involutus* in the rhizosphere of *Pinus resinosa* was sufficient to reduce the pathogenicity of *Fusarium oxysporum*. In this work seedling survival, and hence plant fitness, was markedly enhanced even without mycorrhiza formation. The basis of such effects are not known, although antifungal compounds have been isolated from pure cultures of some ECM fungi (Kope *et al.* 1991). The work carried out to date on biocontrol by ECM fungi has been restricted to laboratory conditions and so suffers from the same limitations as that involving AM fungi. There is an urgent need to investigate epidemiology under natural conditions.

## Direct influences upon fitness

### Effects through an entire life cycle

One of the few studies designed specifically to examine the impact of mycorrhizal colonization upon fitness of plants growing through an entire life cycle under field conditions was recently carried out using the bulbiferous vernal geophyte 'bluebell' (*Hyacinthoides nonscripta*) (Merryweather & Fitter 1995a,b). It was first shown, using plants provided with nutrients under laboratory conditions, that in the absence of mycorrhizal fungi; bluebell was unable to take up sufficient phosphorus to restock the new bulb to the previous season's level. When mycorrhizal colonization was eliminated in the field by the application of benomyl over two growing seasons (Merryweather & Fitter 1996), leaf phosphorus was strongly reduced relative to levels in untreated plants. During the first growing season concentrations of phosphorus in flowers and seeds of non-mycorrhizal plants appeared to be unaffected, suggesting that essential reproductive units were being protected, although by the second season these structures were adversely affected. By the final harvest the reductions of phosphorus concentration in bulbs of treated plants, at 42%, was greater than for any other plant part. While there was again the suggestion that some processes, notably seed production, were being selectively favoured in terms of phosphorus allocation at the expense of others, all tissues showed significant reduction in their phosphorus concentrations. These studies show that bluebell is obligately dependent upon mycorrhizal colonization for the construction of its bulb and hence for survival in nature.

The most clear-cut case of fitness-related dependence upon mycorrhizal colonization is seen in the Orchidaceae. In this, the largest of all plant families, seeds are reduced to consist only of minute embryos. In the absence of storage products their germination, development and the advance of the plant to adulthood all depend on the import of carbon and minerals from exogenous sources by mycorrhizal hyphae. The evidence suggests that in orchid mycorrhiza carbon moves exclusively from the fungus to the plant, this polarity being the reverse of that seen in all but the monotropoid type of the symbiosis (Fig. 7.1). The effectiveness of this 'reverse' carbon flow process is demonstrated by the fact that in a considerable number of orchids, perhaps as many as 20% of the *c.* 1700 species in the family, plants are completely achlorophyllous throughout their life cycles. One of the additional fitness-related advantages of this so-called 'mycoheterotrophic' mode of nutrition (Leake 1994) is that plants are able to complete their life cycles in deeply shaded environments free of competition from autotrophs.

Because most of the fungi that colonize autotrophic orchid species are known to have considerable saprotrophic capabilities, it has become the convention to think of soil organic polymers as the ultimate source of the carbon which sustains the development of orchids, at least to the above-ground stage. This may indeed be true of those orchids forming mycorrhizas with symbionts of the *Rhizoctonia* type. Recent findings indicate, however, that a number of orchid species, particularly those which retain the mycoheterotrophic mode of nutrition into adulthood, are associated with fungi that are, at the same time, forming ectomycorrhizas with the canopy forming forest trees. Among these, species of *Corallorhiza* and *Cephalanthera* have been shown to have typical orchid-type mycorrhizas formed by fungi of russulacean and thelephoroid affinities that are the ectomycorrhizal associates of the adjacent tree species (Taylor & Bruns 1997). It is possible to hypothesize that the fitness of these orchids is increased because these symbionts are attached to more reliable supplies of carbon than those available to species of *Rhizoctonia*. However, the mechanisms whereby compatibilities between partners are switched and interspecific carbon flows are facilitated remain a matter of speculation. Whatever the mechanisms, they are not unique to the Orchidaceae, because parallel examples of 'cross-dressing' are seen in the unrelated achlorophyllous family Monotropaceae where ectomycorrhizal associates of trees, for example in the genera *Rhizopogon, Suillus* and *Russula*, form monotropoid mycorrhiza (Cullings *et al.* 1996) upon which, again, the plants are entirely dependent for carbon.

## Fecundity — seed number and quality

It is evident from the bluebell studies that reproductive propagules may be protected from deficiency under conditions of phosphorus stress. Among all of the fitness-related attributes of the individual organisms, its reproductive traits are of the greatest direct importance. It is therefore surprising that until recently little attention has been paid to the influence of mycorrhizal colonization upon such a vital aspect of plant biology. Following the observation (Jensen 1983) that seed of barley grown in the mycorrhizal condition contained greater amounts of phosphorus than that produced from non-mycorrhizal plants, Koide and his group have demonstrated a number of direct impacts of mycorrhizal colonization upon seed number, quality and vigour of offspring. Seed production was greater in mycorrhizal plants of oat (Koide *et al.* 1988) and tomato (Bryla & Koide 1990) than in their non-mycorrhizal counterparts, and the quality of these seeds, in terms of their nutrient composition, was improved. In wild oats, Koide and Lu (1992) found a mean phosphorus content of 50 µg seed in mycorrhizal compared with only 32 µg seed in non-mycorrhizal plants. The difference was largely caused by increases in the phytate-phosphorus in seeds from mycorrhizal plants.

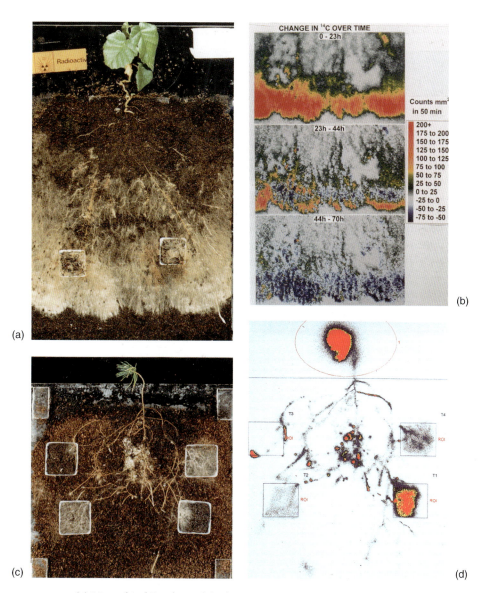

**Plate 7.1** (a) Mycorrhizal *Betula pendula* plant grown in association with the fungus *Paxillus involutus*. The fungal mycelium spreads from the colonized roots across the growth medium as a fan with a dense foraging front of hyphal tips. (b) Sequential real-time analysis of the distribution of carbon following feeding of the plant with $^{14}CO_2$ reveals selective allocation (red colour) of $^{14}C$ to the foraging front. Loss of activity over time is attributable to a combination of respiratory release and onward dispersion into the soil community. Photographs courtesy of J.R. Leake and D. Donnelly. (c) Plant of *Pinus sylvestris* colonized by the fungus *Suillus bovinus* in an observation chamber in which trays of plant litter have been added to the homogeneous peat. Selective intensive foraging is seen in the relatively resource-rich litter trays. (d) Pulse labelling of the plant with $^{14}CO_2$ followed by real-time scanning of the system demonstrates selective allocation of carbon to those resource patches which are being most actively colonized. Leake *et al.* (in press).

[Facing p. 146]

When seed of *Abutilon* was collected from mycorrhizal and non-mycorrhizal parents and allowed to germinate in the absence of AM fungi, plants originating from mycorrhizal parents were more vigorous than those from their non-mycorrhizal counterparts, the greater vigour being apparently a result of increased length and branching of roots (Koide & Lu 1995). A conclusion which might readily be drawn from these observations is that the vigour of the mycorrhiza-derived seedling is attributable to greater seed phosphorus reserves combined with a superior ability to scavenge for nutrients. However, two sets of observations cast doubt upon a simple nutrient-based explanation of the phenomenon. Lewis and Koide (1990) showed that when seed phosphorus content was raised, by the manipulation of phosphorus availability in nutrient cultures, there was no effect on offspring vigour. Much smaller effects of phosphorus content arising from mycorrhizal colonization did, however, significantly increase vigour. The second observation concerns the possibility of enhanced nutrient scavenging associated with superior root architecture. When mycorrhizal offspring with this characteristic were grown in nutrient-free water culture they still produced seedlings of greater biomass. The possibility remains that the type of phosphorus allocated to seeds of mycorrhizal plants differs. Whatever the basis of the positive effect on mycorrhizal plant offspring, it is clearly likely to contribute to their ability to overcome the critical early threats to survivorship posed by the regeneration niche (see above).

It is important in this context that the phenomenon of enhanced offspring vigour has now been recorded under field (Carey *et al.* 1992; Stanley *et al.* 1993) as well as under laboratory conditions. Stanley *et al.* (1993) showed that mycorrhizal colonization led to increases of yield and fecundity in *Abutilon* plants grown in field plots. With or without supplementation of plant-available phosphorus, seed weight and mean seed phosphorus contents and concentration were significantly higher in mycorrhizal than in non-mycorrhizal plants (Fig. 7.3). No significant differences between nitrogen contents of the two sets of plants was observed. Importantly, the superior provisioning of seeds from mycorrhizal plants enabled significantly greater recruitment of seedlings in the following year.

## Conclusion

Greater emphasis upon 'fitness' as a defining factor in analysis of the mycorrhizal condition will provide a deeper understanding of the significance of this almost universal symbiosis. It will enable experimental approaches to be targeted at questions which are of greater relevance to the success of the individuals involved in the symbiotic partnership, and lead to a more accurate evaluation of their contribution to the wider communities of which they are a part. It is increasingly clear that these fundamental attributes of the symbiosis can only be adequately evaluated against the background of the complexity prevailing in natural environments. As a result, the design of experiments required to decipher the interactions will themselves need to become more intricate. They will take into account the fact that mycorrhizal fungi and their plant partners are normally present as mixtures

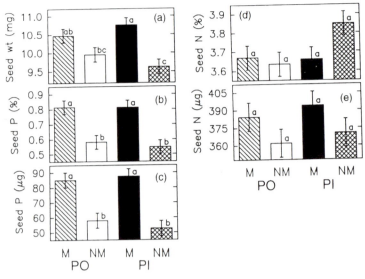

**Figure 7.3** (a) Mean seed weight of *Abutilon theophrasti* grown in the mycorrhizal (M) and non-mycorrhizal (NM) condition with field (PO) and supplementary (PI) phosphorus availability. (b) Phosphorus concentration of above seeds (%). (c) Mean seed phosphorus content. (d) Mean seed nitrogen concentration. (e) Mean seed nitrogen content. From Stanley *et al.* (1993), with permission.

of symbiotic genomes which are interacting with other, often non-symbiotic, members of the community. A further level of complexity of which we are increasingly aware, and one which may have led to selection favouring biological diversity, is found in the physicochemical environment which supports the interacting symbionts. Recognition that the natural substrates which contain essential nutrients for plants are often polymeric and accessible only to selected heterotrophs is providing new insights into the basis of the symbiotic relationship.

Undoubtedly the wider importance of mycorrhizas lies in the contribution they make to ecosystems. Only by identifying the impacts of the symbiosis on the fitness of plant and mycobiont shall we be in a position to fully understand the role of the symbiosis in the context of ecosystem function.

## References

Agerer, R. (1991). Characterisation of ectomycorrhiza. *Methods in Microbiology*, **23**, 25–73.

Anderson, A.J. (1988). Mycorrhizae–host specificity and recognition. *Phytopathology*, **78**, 375–378.

Bååth, E. & Hayman, D.S. (1983). Plant growth responses to vesicular-arbuscular mycorrhiza. XIV. Interactions with *Verticillium* wilt on tomato plants. *New Phytologist*, **95**, 419–426.

Bending, G.D. & Read, D.J. (1995a). The structure and function of the vegetative mycelium of ectomycorrhizal plants. V. The foraging behaviour of ectomycorrhizal mycelium and the translocation of nutrients from exploited organic matter. *New Phytologist*, **130**, 400–409.

Bending, C.D. & Read, D.J. (1995b). The structure and function of the vegetative mycelium of ecto-mycorrhizal plants. VI. Activities of nutrient mobilising enzymes in birch litter colonised by *Paxillus involutus* (Fr.) *New Phytologist*, **130**, 411–417.

Bryla, D.R. & Koide, R.T. (1990). Role of mycor-rhizal infection in the growth and reproduction of wild vs. cultivated plants. II. Eight wild acces-sions and two cultivars of *Lycopersicon esculen-tum* Mill. *Oecologia*, **84**, 82–92.

Carey, P.D., Fitter, A.H. & Watkinson, A.R. (1992). A field study using the fungicide benomyl to inves-tigate the effect of mycorrhizal fungi on plant fitness. *Oecologia*, **90**, 550–555.

Carey, P.D., Watkinson, A.R. & Gerard, F.F.O. (1995). The determinants of the distribution and abundance of the winter annual grass *Vulpia ciliata* ssp. *ambigua*. *Journal of Ecology*, **83**, 177–187.

Caron, M., Fortin, J.A. & Richard, C. (1986). Effect of *Glomus intraradices* on infection by *Fusarium oxysporum*. f. sp. *radicis-lycopersici* in tomatoes over 12-week period. *Canadian Journal of Botany*, **64**, 552–556.

Clapp, J.P., Young, J.P.W., Merryweather, J. & Fitter, A.H. (1995). Diversity of fungal symbionts in arbuscular mycorrhizal from a natural commu-nity. *New Phytologist*, **130**, 259–265.

Cordier, C., Gianinazzi, S. & Gianinazzi-Pearson, V. (1996). Colonisation patterns of root tissues by *Phytolophthora nicotianae* var. *parasitica* related to reduced disease in mycorrhizal tomato. *Plant and Soil*, **185**, 223–232.

Cullings, K.W., Szaro, T.M. & Bruns, T.D. (1996). Evolution of extreme specialisation with a lineage of ectomycorrhizal parasites. *Nature*, **379**, 63–66.

Davis, R.M. & Menge, J.A. (1980). Influence of *Glomus fasciculatus* and soil phosphorus on *Phytophthora* root rot of citrus. *Phytopathology*, **70**, 447–452.

De Bary, A. (1887). *Comparative Morphology and Biology of the Fungi, Mycetozoa and Bacteria.* [English translation of 1884 edn.] Clarendon Press, Oxford.

Douds, D.D., Johnson, C.R. & Koch, K.E. (1988).

Carbon cost of the fungal symbiont relative to net leaf P accumulation in a split-root VA mycorrhizal symbiosis. *Plant Physiology*, **86**, 491–496.

Duchesne, L.C., Peterson, R.L. & Ellis, B.E. (1988a). Interaction between the ectomycorrhizal fungus *Paxillus involutus* and *Pinus resinosa* induces resistance to *Fusarium oxysporum*. *Canadian Journal of Botany*, **66**, 558–562.

Duchesne, L.C., Peterson, R.L. & Ellis, B.E. (1988b). Pine root exudate stimulates antibiotic synthesis by the ectomycorrhizal fungus *Paxillus involutus*. *New Phytologist*, **108**, 470–476.

Duddridge, J.A., Malibari, A. & Read, D.J. (1980). Structure and function of mycorrhizal rhi-zomorphs with special reference to their role in water transport. *Nature*, **287**, 834–836.

Dugassa, G.D., von Alten, A. & Schönbeck, F. (1996). Effects of arbuscular mycorrhiza (AM) on health of *Linum usitatissimum* L. infected by fungal pathogens. *Plant and Soil*, **185**, 173–182.

Francis, R. & Read, D.J. (1994). The contribution of mycorrhizal fungi to the determination of plant community structure. *Plant and Soil*, **159**, 11–25.

Francis, R. & Read, D.J. (1995). Mutualism and antagonism in the mycorrhizal symbiosis, with special reference to impacts on plant community structure. *Canadian Journal of Botany*, **73**(Suppl. 1), 1301–1309.

Griffiths, R.P. & Caldwell, B.A. (1992). Mycorrhizal mat communities in forest soils. In *Mycorrhizal Mat Communities in Forest Soils* (Ed. by D.J. Read, D.H. Lewis, A.H. Fitter & I.J. Alexander), pp. 98–105. CAB International, Wallingford.

Grime, J.P., Mackey, J.M.L., Hillier, S.H. & Read, D.J. (1987). Floristic diversity in a model system using experimental microcosms. *Nature*, **328**, 420–422.

Grubb, P.J. (1977). The maintenance of species richness in plant communities: the importance of the regeneration niche. *Biological Reviews*, **52**, 107–145.

Guillemin, J.P., Gianinazzi, S., Gianinazzi-Pearson, V. & Marchal, J. (1994). Contribution of arbus-cular mycorrhizas to biological protection of micropropagated pineapple [*Ananas comosus* (L.) Merr] against *P. cinnamomi* rands. *Agricul-tural Science Finland*, **3**, 241–251.

Harley, J.L. (1989). The significance of mycorrhiza. *Mycological Research*, **92**, 129–139.

Harris, D., Pacovsky, R.S. & Paul, E.A. (1985). Carbon economy of soybean–*Rhizobium*–*Glomus* associations. *New Phytologist*, **101**, 427–440.

Hartnett, D.C., Samenus, R.J., Fischer, L.E. & Hetrick, B.A. (1994). Plant demographic responses to mycorrhizal symbiosis in tallgrass prairie. *Oecologia*, **99**, 21–26.

Hatch, A.B. (1937). The physical basis of mycotrophy in the genus *Pinus*. *Black Rock Forest Bulletin*, **6**, 168.

van der Heijden, M.G.A., Klironomos, J.N., Ursic, M. *et al.* (1998). Mycorrhizal fungal diversity determines plant biodiversity, ecosystem variability and productivity. *Nature*, **396**, 69–72.

Helgason, T., Daniell, T.J., Husband, R., Fitter, A.H. & Young, J.P.Y. (1998). Ploughing up the wood-wide web. *Nature*, **394**, 431.

Jalali, B.L. & Jalali, I. (1991). Mycorrhiza in plant disease control. In *Handbook of Applied Mycology*, Vol. 1: *Soil and Plants* (Ed. by D.K. Arora, B. Rai, K.G. Mukherjee & D. Knudsen), pp. 131–154. Dekker, New York.

Jensen, A. (1983). The effect of indigenous vesicular-arbuscular mycorrhizal fungi on nutrient uptake and growth of barley in two Danish soils. *Plant and Soil*, **70**, 155–163.

Kaye, J.W., Pfleger, F.L. & Stewart, E.L. (1984). Interaction of *Glomus fasciculatum* and *Pythium ultimum* on greenhouse-grown poinsettia. *Canadian Journal of Botany*, **62**, 1575–1579.

Koch, K.E. & Johnson, C.R. (1984). Photosynthate partitioning in split-root seedlings with mycorrhizal and nonmycorrhizal root systems. *Plant Physiology*, **75**, 26–30.

Koide, R.T. & Lu, X. (1992). Mycorrhizal infection of wild oats: parental effects on offspring nutrient dynamics, growth and reproduction. In *Mycorrhizas in Ecosystems* (Ed. by D.J. Read, D.H. Lewis, A.H. Fitter & I.J. Alexander), pp. 55–58. CAB International, Wallingford.

Koide, R.T. & Lu, X. (1995). On the cause of offspring superiority conferred by mycorrhizal infection of *Abutilon theophrasti*. *New Phytologist*, **131**, 435–441.

Koide, R.T., Li, M., Lewis, J. & Irby, C. (1988). Role of mycorrhizal infection in the growth and reproduction of wild vs. cultivated plants.

I. Wild vs. cultivated oats. *Oecologia*, **77**, 537–542.

Kope, H.H., Tsantrizos, Y.S., Fortin, J.A. & Ogilvie, K.K. (1991). *p*-Hydroxybenzoylformic acid and (R)-(−)-*p*-hydroxymandelic acid, two antifungal compounds isolated from the liquid culture of the ectomycorrhizal fungus *Pisolithus arhizus*. *Canadian Journal of Microbiology*, **37**, 258–264.

Kucey, R.M.N. & Paul, E.A. (1982). Carbon flow photosynthesis, and $N_2$ fixation in mycorrhizal and nodulated faba beans (*Vicia faba* L.). *Soil Biology and Biochemistry*, **14**, 407–412.

Leake, J.R. (1994). The biology of myco-heterotrophic 'saprophytic' plants. *New Phytologist*, **127**, 171–216.

Leake, J.R. & Read, D.J. (1997). Mycorrhizal fungi in terrestrial habitats. In *The Mycota*, Vol. 5 (Ed. by D.T. Wicklow & B. Söderström), pp. 282–301. Springer, Berlin.

Leake, J.R., Donnelly, D.P., Saunders, E.M., Boddy, L. & Read, D.J. Rates and quantities of carbon flux to ectomycorrhizal mycelium following pulse labelling of *Pinus syvestris* L seedlings: effects of litter patches on interaction with wood-decomposed fungus. *Tree Physiology* (in press).

Lewin, R.A. (1982). Symbiosis and parasitism—definitions and evaluations. *Bioscience*, **32**, 254–259.

Lewis, D.H. (1985). Symbiosis and mutalism: crisp concepts and soggy semantics. In *The Biology of Mutualism Ecology and Evolution* (Ed. by D.H. Boucher), pp. 29–39. Croom Helm, London.

Lewis, J.D. & Koide, R.T. (1990). Phosphorous supply, mycorrhizal infection and plant offspring vigour. *Functional Ecology*, **4**, 695–702.

Liu, R.J. (1995). Effect of vesicular-arbuscular mycorrhizal fungi on *Verticilum* wilts of cotton. *Mycorrhiza*, **5**, 293–297.

Marx, D.H. (1969). The influence of ectotrophic ectomycorrhizal fungi on the resistance of pine roots to pathogen infections. I. Antagonism of mycorrhizal fungi to pathogenic fungi and soil bacteria. *Phytopathology*, **59**, 153–163.

Marx, D.M. (1973). Mycorrhizae and feeder root disease. In *Ectomycorrhizae: their ecology and physiology* (Ed. by G.C. Marks & T.T. Kozlowski), pp. 351–382. Academic, New York.

Melin, E. & Nilsson, H. (1950). Transfer of radioactive phosphorus to pine seedlings by means of mycorrhizal hyphae. *Physiologia Plantarum*, **3**, 88–92.

Melin, E. & Nilsson, H. (1953). Transfer of labelled nitrogen from glutamic acid to pine seedlings through the mycelium of *Boletus variegatus* (SW) Fr. *Nature*, **171**, 434.

Merryweather, J. & Fitter, A. (1995a). Phosphorus and carbon budgets: mycorrhizal constriction in the obligately mycorrhizal *Hyacinthoides non-scripta* (L.) Chouard ex Rothm. under natural conditions. *New Phytologist*, **129**, 619–627.

Merryweather, J. & Fitter, A. (1995b). Arbuscular mycorrhiza and phosphorus as controlling factors in the life history of the obligately mycorrhizal *Hyacinthoides non-scripta* (L.) Chouard ex Rothm. *New Phytologist*, **129**, 629–636.

Merryweather, J. & Fitter, A. (1996). Phosphorus nutrition of an obligately mycorrhizal plant treated with the fungicide benomyl in the field. *New Phytologist*, **132**, 307–311.

Newsham, K.K., Fitter, A.H. & Watkinson, A.R. (1994). Root pathogenic and arbuscular mycorrhizal fungi determine fecundity of asymptomatic plants in the field. *Journal of Ecology*, **82**, 805–814.

Newsham, K.K., Fitter, A.H. & Merryweather, J.W. (1995). Multifunctionality and biodiversity in arbuscular mycorrhizas. *Tree*, **10**, 407–411.

Pang, P.C. & Paul, E.A. (1980). Effects of vesicular-arbuscular mycorrhiza on $^{14}$C and $^{15}$N distribution in nodulated baba beans. *Canadian Journal of Soil Science*, **60**, 241–250.

Paul, E.A. & Kucey, R.M.N. (1981).Carbon flow in plant microbial associations. *Science*, **213**, 473–474.

Read, D.J. (1998a). Mycorrhiza—the state of the art. In *Mycorrhiza—Structure, Function, Molecular Biology and Biotechnology*, 2nd edn (Ed. by A. Varma & B. Hock), pp. 3–34. Springer, Berlin.

Read, D.J. (1998b). Plants on the web. *Nature*, **396**, 22–23.

Rosendahl, C.N. & Rosendahl, S. (1990). The role of vesicular-arbuscular mycorrhiza in controlling damping-off and disease reduction in cucumber caused by *Pythium ultimum. Symbiosis*, **9**, 363–366.

Schreiner, R.P. & Koide, R.T. (1993a). Antifungal compounds from the roots of mycotrophic and non-mycotrophic plant species. *New Phytologist*, **123**, 99–105.

Schreiner, R.P. & Koide, R.T. (1993b). Mustards, mustard oils and mycorrhizas. *New Phytologist*, **123**, 107–113.

Simon, L., Bousquet, J., Levesque, R.C. & Lalonde, M. (1993). Origin and diversification of endomycorrhizal fungi and coincidence with vascular land plants. *Nature*, **363**, 67–69.

Smith, S.E. & Read, D.J. (1997). *Mycorrhizal Symbiosis*, 2nd edn. Academic, London.

Snellgrove, R.C., Splittstoesser, W.E., Stribley, D.P. & Tinker, P.B. (1982). The distribution of carbon and the demand of the fungal symbiont in leek plants with vesicular-arbuscular mycorrhizas. *New Phytologist*, **92**, 75–87.

Stanley, M.R., Koide, R.T. & Shumway, D.L. (1993). Mycorrhizal symbiosis increases growth, reproduction and recruitment of *Abutilon theophrastic* Medic in the field. *Oecologia*, **94**, 30–35.

Starr, M.P. (1975). A generalized scheme for classifying organismic associations. *Symposium of Social Experimental Biology*, **29**, 1–20.

Taylor, D.L. & Bruns, T.D. (1997). Independent, specialized invasions of ectomycorrhizal mutualism by two nonphotosynthetic orchids. *Proceedings of the National Academy of Sciences USA*, **94**, 4510–4515.

Trappe, J.M. (1987). Phylogenetic and ecological aspects of mycotrophy in the angiosperms from an evolutionary standpoint. In *Ecophysiology of VA Mycorrhizal Plants* (Ed. by G.R. Safir), pp. 5–25. CRC Press, Boca Raton.

Trappe, J.M., (1996). What is a mycorrhiza? In *Proceedings of the 4th European Symposium on Mycorrhizae, Granada, Spain*, Vol. 8, pp. 3–9, EC Report EUR 16728.

Trotta, A., Varese, G.C., Gnavi, E., Fusconi, A., Sampo, S. & Berta, G. (1996). Interactions between the soilborne root pathogen *Phytophthora nicotianae* var. *parasitica* and the arbuscular mycorrhizal fungus *Glomus mosseae* in tomato plants. *Plant and Soil*, **185**, 199–209.

Vanderplank, J.E. (1978). *Genetic and Molecular Basis of Plant Pathogenesis*. Springer-Verlag, Berlin.

Wang, G.M., Coleman, D.C., Freckman, D.W., Dyer, M.I., McNaughton, S.J., Acra, M.A. & Goeschl, J.D. (1989). Carbon partitioning patterns of mycorrhizal versus non-mycorrhizal plants: real-time dynamic measurements using $^{11}CO_2$. *New Phytologist*, **112**, 489–493.

Wright, D.P., Scholes, J.D. & Read, D.J. (1998a). Effects of VA mycorrhizal colonisation on photosynthesis and biomass production of *Trifolium repens* L. *Plant, Cell and Environment*, **21**, 209–216.

Wright, D.P., Read, D.J. & Scholes, J.D. (1998b). Mycorrhizal sink strength influences whole plant carbon balance of *Trifolium repens* L. *Plant, Cell and Environment*, **21**, 881–891.

# Chapter 8

# Measuring symbiotic nitrogen fixation: case studies of natural and agricultural ecosystems in a Western Australian setting

*J.S. Pate and M.J. Unkovich*

## Introduction

Whereas photosynthetic assimilation of $CO_2$ and accompanying transpiration of plants can be examined relatively simply in the field by cuvette-based gas exchanges of foliage, comparable assessments of symbiotic nitrogen fixation in species and ecosystems are much more difficult and in certain respects unreliable. Ruling out measurements based on $^{15}N_2$ feeding as economically non-viable under field conditions, acetylene ($C_2H_2$) reduction assays and $^{15}N$-based natural abundance (NA) techniques remain as possible candidates for usage. However, both have methodological problems (Peoples & Herridge 1990; Hansen 1994; Handley & Scrimgeour 1997; Högberg 1997).

Rather than reviewing yet again recent literature concerning the assessment of nitrogen fixation in ecosystems, this chapter examines a relatively limited number of studies conducted by ourselves and colleagues in Western Australia. By parading apparent successes and failures, including instances where the NA and $C_2H_2$ reduction methods have been used on the same experimental subjects, it is hoped to pinpoint circumstances where reproducible results are likely to be generated and how failures in application of the techniques can be avoided.

Our mainstay when using $C_2H_2$ reduction assays has been to calibrate the technique against the plant species and field conditions under which it is being applied, whether by means of short-term paired assays of $^{15}N_2$ fixation and $C_2H_2$ reduction of freshly detached symbiotic structures, or longer-term glasshouse-based comparisons of cumulative $C_2H_2$ reduction and plant nitrogen gain in the absence of inorganic nitrogen sources. For NA methodologies, special attention is paid to proper choice of reference species and to possible complications resulting from isotope discrimination during uptake and assimilation of nitrogen by components of an ecosystem. Featuring prominently are case studies of annual crop and pasture legumes cultured on the low-organic-matter, deep sands of certain agricultural regions of Western Australia. These apparently provide an ideal setting for the

*Botany Department and Centre for Legumes in Mediterranean Agriculture, The University of Western Australia, Nedlands, WA 6907, Australia. E-mail: unkmj@cyllene.uwa.edu.au*

application of the NA technique. A special feature of this chapter is advocacy of visual assessments of presence and quality of symbiotic organs, preferably using the same material on which $C_2H_2$ reduction or NA assays are being performed.

This chapter deals first with the biology of the systems examined, followed by case studies involving first the $C_2H_2$-based assays and then the NA method. Application of the latter technique to agricultural or natural systems then sets the scene for a study in which $N_2$-fixation inputs and related cycling of nitrogen in pristine, phosphorus-limited bushland are compared with those in adjacent cleared land, where the application of superphosphate has promoted substantially increased inputs of legume-fixed nitrogen. A short concluding section suggests avenues for further study, particularly broader-scale evaluations of $N_2$ fixation against total nitrogen cycling of ecosystems.

## Assessments of symbiotic status by observations on, and recovery of, root nodules and comparable structures

Any investigation of symbiotic nitrogen fixation in a given ecosystem should first list the putative nitrogen fixing species present and identify which of these bear symbiotic organs. Unfortunately, this outwardly simple approach may be difficult to follow, especially where species inhabit intractable soils of high rock or clay content.

Shallow-rooted crop and pasture legumes (e.g. *Trifolium subterraneum*, *Pisum sativum* and *Lupinus angustifolius*) on agricultural systems in Southwestern Australia will be dealt with first. Because of the deep sands at the sites, seasonal cycles of nodulation are relatively easily studied by progressive sampling, thus enabling nodule masses and numbers to be compared with increases in plant biomass and nitrogen content (Unkovich & Pate 1998).

Proportions of young to old nodules, the presence or absence of leghaemoglobin in nodules and mean nodule size are usually found to vary characteristically through a season, thus providing a semi-quantitative measure of the extent to which symbiosis is contributing to plant nitrogen status. Because legumes exhibit an enormous capacity to compensate for poor nodulation by developing nodules of extremely large size, mass of nodules per unit plant weight offers the most definitive index of symbiotic potential. Then, by making similar measurements on fully symbiotic plants raised under controlled environment conditions in the absence of inorganic nitrogen, an indication of whether field-grown material is likely to be achieving a similar level of nodulation may be gained; otherwise, partial reliance on soil nitrogen is implicated.

However, matching field performance against that of fully symbiotic cultures is inherently difficult. Field recoveries of nodules are likely to be incomplete, especially in the case of old plants nodulated mostly on minor roots. Drought-induced shedding of nodules may also occur and, if further nodules form once the soil re-wets, contributions from these will be underestimated unless plants are sampled frequently.

The perennial woody species that we have studied include a native cycad (*Macrozamia riedlei*), a number of forest understorey legumes (*Acacia* spp., *Bossiaea* spp.), the water-logging tolerant legumes *Aeschynomene indica*, *Viminaria juncea* and *Sesbania* sp., and a range of tree and shrub legumes. An exotic species also studied is tagasaste (*Chamaecytisus proliferus*), a deep-rooted shrubby legume from the Canary Islands which is currently planted on deep sand ecosystems of Western Australia as summer fodder for sheep and cattle. This species carries the additional potential to combat rising water-tables in agricultural landscapes of the region currently devoted to shallow-rooted herbaceous species (Pate & Dawson 1999).

*Macrozamia riedlei* develops cyanobacteria-containing coralloid roots which are easily recoverable from upper parts of its root stock (Plate 8.1a,b, facing page 162) (Halliday & Pate 1976; Pate 1976). The annual legumes, *Aeschynomene indica* and *Sesbania* sp., both from wetlands of Northwestern Australia, are equally amenable to study as they have nodules confined to submerged lower stems (*Aeschynomene* (Plate 8.1c)) or lateral roots (*Sesbania*). *Viminaria juncea*, a shrub of winter wetlands of Southern Australia, forms pneumatophores on its roots and develops mid-cortical layers of aerenchyma throughout immersed regions of its stem, roots and nodules (Plate 8.1d; Walker *et al.* 1983; Walker & Pate 1986). The pneumatophores and aerenchyma collectively ventilate flooded parts of the plant. Superficially located nodule-bearing roots of *Viminaria* are easily extricated from the water body or mud during the wet season, when symbiotic activity is most prominent (Walker *et al.* 1983). Other woody $N_2$-fixing species have proven more difficult to study. For example, understorey shrub legumes of the genera *Acacia* and *Bossiaea* typically develop scattered nodules on long lateral feeding roots, thus requiring greater excavation to recover nodules, particularly on older plants.

In the case of larger, deep-rooted species (e.g. tagasaste and native species of the genera *Allocasuarina*, *Casuarina*, *Jacksonia*, *Daviesia* and *Paraserianthes*), the sheer size and complexity of root systems make it impossible to harvest all nodules, even when using mechanically aided excavations extending to 3 m or more in diameter and depth. Furthermore, nodulation may be strictly seasonal, so frequent sampling is required to follow a nodulation cycle. Certain tap-rooted genera (*Jacksonia*, *Daviesia* and tagasaste) develop two distinct populations of nodules, one an ephemeral set initiated each wet season (Plate 8.1e) on lateral feeding roots, the other consisting of elongated and branched perennial nodules attached to tap roots at 2–5 m depth (Plate 8.1f).

## Measurements of $N_2$-fixing performance using calibrated acetylene reduction assays

Our first study of this nature involved the cycad *Macrozamia riedlei*, growing in *Banksia* woodland near Perth (Halliday & Pate 1976). Allometric plots of plant mass against coralloid root mass (Fig. 8.1a) and stem diameter (Fig. 8.1b) were first

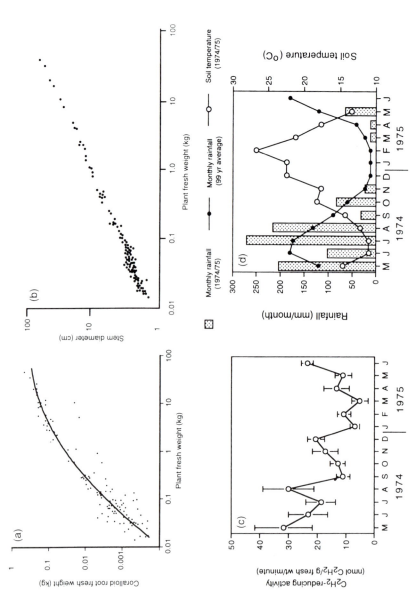

**Figure 8.1** Assessment of symbiotic nitrogen fixation by the native Southwestern Australian *Macrozamia riedlei* using the $C_2H_2$ reduction technique. (a) Allometric relationships (log×log plot) between coralloid root fresh weight per plant and plant fresh weight across a full size range of plants of the species. (b) Allometric plot (log×log plot) of stem diameter against fresh weight, thus enabling plant weights and coralloid root weights (data from (a)) to be predicted non-destructively in the field. (c) Mean $C_2H_2$-reducing activity of coralloid roots assayed in the field during the 14-month study period (1974–75). (d) Data for rainfall for the same period. All data from Halliday and Pate (1976).

**Table 8.1** Amounts of symbiotic nitrogen fixation by populations of the cycad *Macrozamia riedlei* in *Banksia* woodland near Perth, Western Australia. From Halliday and Pate (1976).

|  | Population 1 | Population 2 |
|---|---|---|
| Annual fixation |  |  |
| (a) Plant density ha$^{-1}$ | 1590 | 1270 |
| (b) Biomass per plant (kg) | 20.8 | 34.1 |
| (c) Biomass of plants (kg ha$^{-1}$; a × b) | 33 050 | 43 290 |
| (d) Biomass of coralloid roots (kg ha$^{-1}$) | 499 | 495 |
| (e) Annual return of fixed N (kg ha$^{-1}$ yr$^{-1}$) | 18.8 | 18.6 |
| Potential of N$_2$ fixation as N source for plant growth |  |  |
| Coralloid root N (kg ha$^{-1}$) | 2.80 | 2.78 |
| Whole-plant N (kg ha$^{-1}$) | 155 | 203 |
| Annual return of fixed N (kg ha$^{-1}$ yr$^{-1}$) | 18.8 | 18.6 |
| Doubling time for coralloid root N (weeks) | 7.7 | 7.8 |
| Doubling time for whole-plant N (yr) | 8.2 | 10.9 |

constructed to predict plant biomass and symbiotic potential of the two study populations.

Calibrations of C$_2$H$_2$ reduction against $^{15}$N$_2$ fixation, using freshly harvested coralloid roots, indicated that 5.8 mol C$_2$H$_2$ were reduced per mol N$_2$ fixed. C$_2$H$_2$ reduction rates per unit coralloid root mass proved to be appreciably higher in winter than in summer (Fig. 8.1) and, using data for coralloid root mass per plant, annual fixation rates were estimated as 37.6 g N kg$^{-1}$ of coralloid root. Then, by incorporating information on population densities, mean biomass and plant total nitrogen contents, fixation inputs of the two populations were estimated at 18.8 and 18.6 kg N ha$^{-1}$ yr$^{-1}$ (Table 8.1). Fixation rates of this order were considered sufficient to double total plant nitrogen every 8–11 years (Table 8.1).

A second case study (Monk *et al.* 1981) on *Acacia pulchella* in *Banksia* woodlands near Perth showed individuals of the species to nodulate, grow and accumulate nitrogen over a life-span of up to 14 years after fire (Fig. 8.2a–c). Study over the first four seasons after recruitment showed that each season's nodule initiation followed the first rains of autumn (Fig. 8.2b) and that maxima in C$_2$H$_2$ reduction rate and nodule mass per plant coincided with late spring or early summer (Fig. 8.2e). Nodules rarely survived the summer.

Calibrations of cumulative C$_2$H$_2$ reduction against plant nitrogen accumulation in inorganic nitrogen-free cultures (Fig. 8.2d) indicated that 2.3 mol C$_2$H$_2$ were reduced per mol N$_2$ fixed. With a knowledge of active nodule mass and plant nitrogen increments during the study seasons (Fig. 8.2), it was estimated that only 8% of the first season's nitrogen gain had been derived from fixation, in contrast to 45 and 68% in the second and fourth season, respectively (Table 8.2). Incorporating data on declines in plant density with population age, a mean annual turnover of 3.9 kg N ha$^{-1}$ was estimated for dense stands of the species, with approximately half of this derived from fixation. The bulk of the annual return of nitrogen came from the

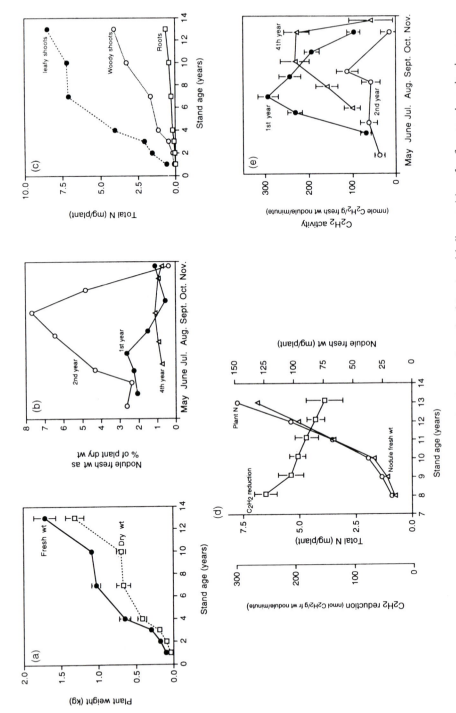

**Figure 8.2** Growth, nitrogen relationships and nodulation of various aged stands of *Acacia pulchella* recruiting after fire at various sites in open *Banksia* woodland around Perth, Western Australia. (a) Mean fresh and dry weights of plants with age of stand. (b) Intensity of nodulation of three age groups of plants expressed in terms of nodule fresh weight as percentage of total plant dry weight. (c) Nitrogen accumulation per plant in stands of different age. (d) Calibration curve for cumulative rate of $C_2H_2$-reducing activity by nodules against nitrogen increment of plants grown in –N pot culture. (e) Seasonal courses of $C_2H_2$-reducing activity for nodules of the same age groups of plants given in (b). All data from Monk *et al.* (1981).

**Table 8.2** Estimates of proportional dependence of a population of second year plants of *Acacia pulchella* on symbiotic nitrogen fixation, based on comparisons of acetylene-reducing performance of nodules and observed increments in plant total nitrogen. Data from Monk *et al.* (1981).

| Interval of growth (mth) | Nodule mass (g FW plant$^{-1}$) | $C_2H_2$ reduction (nmol g FW$^{-1}$ nodules$^{-1}$ min$^{-1}$) | Predicted $N_2$ fixed (mg plant$^{-1}$) | Increment in total N (mg plant$^{-1}$) | Plant N derived from fixation (%) |
|---|---|---|---|---|---|
| 13–14 | 0.15 | 48 | 4.0 | 79 | 5 |
| 14–15 | 0.32 | 63 | 10.8 | 33 | 33 |
| 15–16 | 0.58 | 62 | 19.9 | 29 | 68 |
| 16–17 | 0.99 | 61 | 33.4 | 45 | 75 |
| 17–18 | 1.04 | 114 | 63.5 | 89 | 71 |
| 18–19 | 0.11 | 50 | 3.0 | 21 | 15 |

Note: Estimates assume a calibration value of 2.26 mol $C_2H_2$ per mol $N_2$ fixed (see Fig. 8.2(d)).

death of the plant, with the remainder from shed litter and seed. Poor symbiotic performance in the first season was attributed to inhibition of nodulation by nitrate released from biomass killed during the fire (see also Stewart *et al.* 1993).

A further study using calibrated $C_2H_2$ reduction assays was undertaken on two understorey shrub legumes (*Acacia pulchella* and *A. alata*) common in the postfire succession of jarrah (*Eucalyptus marginata*) forest (Hansen & Pate 1987a; Hansen *et al.* 1987a). Comparisons of masses and $C_2H_2$ reducing performances of nodules against plant nitrogen accumulation in nitrogen-free sand culture indicated calibration values of 1.60 and 3.58 mol $C_2H_2$ reduced per mol $N_2$ fixed for *A. pulchella* and *A. alata*, respectively. Seasonal patterns of $C_2H_2$ reduction (data for *A. alata* shown in Fig. 8.3) showed peak $C_2H_2$ reduction activity in July–September and senescence of most nodules by mid-summer (Fig. 8.3; Hansen & Pate 1987a). *A. pulchella* was concluded to be dependent on $N_2$ fixation for 37% of its nitrogen in the first year but only for 9% in the second season of growth. Comparable values for *A. alata* were 29% and 2%, respectively. Both species exhibited substantial increments in total nitrogen after achieving seasonal peaks in nodule mass and $C_2H_2$ reducing performance (Fig. 8.3), thereby indicating recourse to soil nitrogen in late season.

A parallel study on fully symbiotic, glasshouse-cultured plants of *A. pulchella* and *A. alata*, enjoying unrestricted access to water and all nutrients except nitrogen, showed up to 100-fold greater increases in total plant nitrogen over their first year of growth than their counterparts in natural forest (Hansen & Pate 1987a). Applications of various nutrient solution supplements to plants in native habitat indicated that slow growth was a result of restricted availability of water and phosphorus (Hansen & Pate 1987a).

We conclude from our studies in jarrah forest that symbiotically active understorey legumes are likely to be highly competitive against non-legumes in the first few seasons after a fire. This is because most above-ground nitrogen capital will have been lost to the atmosphere during the fire, whereas phosphorus availability

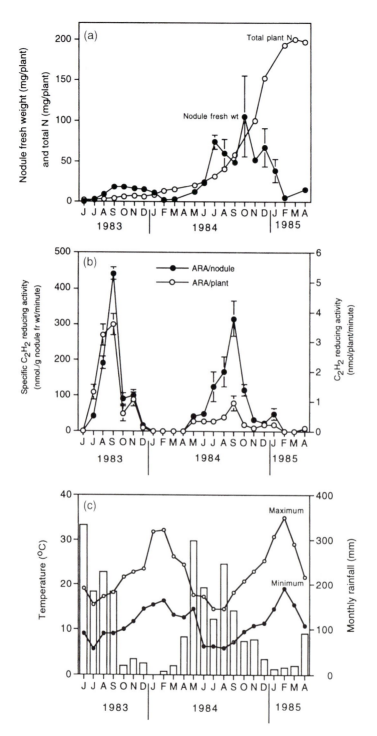

**Figure 8.3** Nodule mass, plant nitrogen accumulation (a) and acetylene-reducing activity of nodule mass (b) of plants of *Acacia alata* recruiting after fire in jarrah forest, Western Australia. (c) Rainfall and temperature data for the 2-year study. Data from Hansen and Pate (1987a).

will typically increase as a result of the deposition of ash. Later, as this phosphorus becomes progressively immobilized, plant growth becomes limited by phosphorus, whereupon the ability to access the element through mycorrhizal agencies will become paramount for all species. The fixation of nitrogen should carry little advantage under the latter circumstances and might well incur penalties from extra requirements for photosynthate and phosphorus to support the symbionts.

### Use of the $^{15}$N natural abundance technique to measure proportional dependencies on N$_2$ fixation and amounts of N$_2$ fixed

The $^{15}$N natural abundance (NA) method for assessing symbiotic N$_2$ fixation exploits small differences in proportions of the stable isotopes of nitrogen ($^{14}$N and $^{15}$N) in different biotic and abiotic components of a system, expressed as $\delta^{15}$N values (Chapter 21). Soil nitrogen pools are typically enriched in $^{15}$N ($d$-values positive) compared with atmospheric N$_2$ ($\delta^{15}$N$=$O), thus enabling one to estimate dependence on symbiotic fixation by comparing the $\delta^{15}$N value of a supposedly N$_2$-fixing plant with that of adjacent non-fixing 'reference' species relying on soil nitrogen (Shearer & Kohl 1986; Ledgard & Peoples 1988; Peoples *et al.* 1989; Unkovich *et al.* 1994).

The formula regularly used to calculate the proportion of plant nitrogen derived from N$_2$ fixation (%Ndfa) is:

$$\%\text{Ndfa} = \frac{\delta^{15}\text{N of reference plant} - \delta^{15}\text{N of N}_2\text{-fixing association}}{\delta^{15}\text{N of reference plant} - B} \qquad (1)$$

where $B$ is the $\delta^{15}$N value of the legume growing solely on atmospheric N$_2$.

Successful use under Southwestern Australian agricultural conditions has been estimated to require a difference in $\delta^{15}$N value between reference plant and air that is at least 10 times the precision of measurement of $\delta^{15}$N (Unkovich *et al.* 1994). Thus, under the seemingly ideal conditions in which we have operated in the deep sand ecosystems of Western Australia, a differential of only 2‰ is capable of estimating %Ndfa with ±6% accuracy (Unkovich *et al.* 1994). In a broader context, across Australian agricultural soils ranging in $\delta^{15}$N from 2.5 to 6.8‰, NA methodologies have been applied with supposedly similar levels of confidence (Ledgard *et al.* 1984; Pate *et al.* 1994; Unkovich *et al.* 1997). The unusually great homogeneity in NA enrichment displayed by many Australian agricultural systems probably reflects crop rotation practices, the small resident pools of soil nitrogen present, a generally low usage of fertilizer nitrogen and the likelihood of nitrogen mostly coming from the mineralization of recent legume residues.

Even where site variations in $\delta^{15}$N are unusually small, strict criteria must be met before the technique can be applied to full effect. First, the rooting morphologies and phenologies of reference species should match closely those of the legume and reference plants should be accessing forms of soil inorganic nitrogen similar to those utilized by the study legume.

Second, no direct transfer of newly fixed nitrogen should occur from legume to reference species, because this would have the possible effect of lowering the $\delta^{15}N$ of the latter, and thereby lead to underestimations of %Ndfa. Pate *et al.* (1994) have accordingly advocated a pair-wise sampling procedure in which each reference plant is harvested close to, but outside, the rooting catchment of its designated partner legume or any other legumes.

Third, dicotyledonous herbs have proven to be better reference species than grasses under Western Australian conditions (Sanford *et al.* 1994). Ideal reference material would, of course, be non-nodulating genotypes of a study legume, such as used for soybean by Bergersen *et al.* (1985) and Herridge *et al.* (1990). Alternatively, uninoculated plots of a legume can be used as references in study sites lacking relevant effective nodule bacteria (Pate *et al.* 1994; see Unkovich *et al.* 1997).

Finally, special consideration is required when obtaining $B$-values for legumes totally reliant on fixed nitrogen. Above-ground biomass of shoots of such plants is usually depleted in $^{15}N$ because legume nodules are generally enriched in $^{15}N$ ($\delta^{15}N$ in the range +4 to +10‰) compared with atmospheric $N_2$. Both host and bacterium influence the extent of such nodule discriminations (Shearer & Kohl 1986; Ledgard 1989; Unkovich & Pate 1998), so rhizobia likely to nodulate the legume in the field should be employed when assessing $B$-values in pot culture. Furthermore, estimates of %Ndfa should use $B$-values for symbiotically dependent plants of the same age as those being examined in the field.

## Case studies on annual pasture and grain legumes

Unkovich *et al.* (1997) reviewed NA-based studies on annual legumes in Australian mediterranean-type agriculture and the principal findings are summarized below.

1 Values for %Ndfa, seasonal nitrogen fixation and post-harvest nitrogen balances of crops of narrow-leaved lupin (*Lupinus angustifolius*), field pea (*Pisum sativum*) and chickpea (*Cicer arietinum*) indicated wide variability in symbiotic effectiveness among and within species (Table 8.3). The estimates for net nitrogen balance are given in terms of nitrogen in unharvested above-ground residues minus nitrogen removed as harvested seed. Unfortunately, below-ground biomass has generally not been considered, although this may represent up to a third or possibly much more of the peak nitrogen of above-ground crop biomass (Russell & Fillery 1996) and comprise an important source of mineralizable nitrogen to crops in the subsequent season (McNeill *et al.* 1998).

2 Extending the approach described above to legume–cereal rotations at Wagga, New South Wales, Armstrong *et al.* (1997) demonstrated considerable differences between legumes in nitrogen carryover to subsequent cereals. At their study site, for example, only lupins (*Lupinus albus* and *L. angustifolius*) effected sufficient returns of nitrogen to elicit a reasonable yield from a following wheat crop.

3 Application of the NA technique across pastures of southern coastal Western Australia by Sanford *et al.* (1994) indicated wide variation in %Ndfa between soil types and management practices. Mean %Ndfa values were 72% for subterranean

(a,b)

(c)

(d)

(e)

**Plate 8.1** (a) Upper part of root system and lower stem of a 20-year-old plant of the cycad *Macrozamia reidlei* showing clusters of apogeotropic coralloid roots attached to the root crown, and contractile roots which pull the stem base up to 1 m below ground. The tap root shown is 6 cm in diameter. (b) Longitudinal section through the apical region of part of the coralloid roots shown in (a). Note the dark midcortical zone harbouring the $N_2$-fixing cyanobacteria (*Nostoc* sp.). The excrescences on the surface of the root are lenticels. The root shown is 7 mm in diameter. (c) Nitrogen-fixing nodules on the submerged stem of the aquatic legume *Aeschynomene indica*. The dark coloration of the nodules derives from a chlorophyllous outer cortex overlying the haemoglobin pigmentation of the underlying bacterial tissue. The stem shown is 9 mm in diameter. (d) Nodulated root stystem of partly submerged plant of the shrub legume *Viminaria juncea* showing proliferation of aerenchyma in roots, nodules and lower stem. Two pneumatophores are visible to the lower left of the picture. These projected above water level when the plant was in its natural habitat. The base of the stem shown is 11 mm in diameter. (e) Part of the matted upper lateral root system of a tree of tagasaste (*Chamaecystus proliferus*) harvested in the wet season. The region shown is 30 cm across and was photographed *in situ* after blowing away fine sand from between the roots. A number of small ephemeral nodules are visible. (f) Near terminal region of a tap root of tagasaste showing a cluster of elongated perennial nodules. The root portion shown was located at 3.5 m below the soil surface and the nodulated zone shown is 70 mm in length. Note the much larger nodules in (f) than in (e).

(f)

*[Facing p. 162]*

**Table 8.3** Nitrogen economies of annual legume crops in southern Australia studied using the [15]N natural abundance technique. Data from Unkovich *et al.* (1997).

|  | Narrow-leaved lupin (*Lupinus angustifolius*) | Field pea (*Pisum sativum*) | Chickpea (*Cicer arietinum*) |
|---|---|---|---|
| Peak N in above-ground biomass (kg ha$^{-1}$) |  |  |  |
| Mean | 214 | 122 | 115 |
| Range | 53–322 | 29–197 | 30–283 |
| Dependence on N$_2$ fixation (%Ndfa) |  |  |  |
| Mean | 77 | 68 | 60 |
| Range | 29–97 | 31–95 | 37–86 |
| N$_2$ fixed (kg ha$^{-1}$) |  |  |  |
| Mean | 165 | 83 | 70 |
| Range | 30–283 | 26–183 | 43–124 |
| Net N balance (±) for season (kg ha$^{-1}$) |  |  |  |
| Mean | +68 | +14 | +8 |
| Range | −41 to +141 | −32 to +96 | −41 to +56 |

clover (*Trifolium subterraneum*), 71% for medics (*Medicago* spp.), 81% for trefoils (*Lotus* spp.), 76% for serradella (*Ornithopus compressus*) and 69% for balansa clover (*Trifolium balansae*). Subterranean clover, the commonest legume, showed unsatisfactorily low %Ndfa values (0–65%) at almost one-third of 184 sites investigated, and at almost half of these aluminium toxicity was indicated.

**4** Sanford *et al.* (1995) employed the NA technique to compare %Ndfa and cumulative yields of dry matter (DM) and nitrogen in grazed and ungrazed subterranean clover pasture, and demonstrated greater total herbage yields per season in grazed (11.8 t DM ha$^{-1}$) than ungrazed plots (7.8 t DM ha$^{-1}$). Associated inputs of fixed nitrogen by clover were 188 and 103 kg N ha$^{-1}$, respectively, and symbiotic dependence of the clover in both classes of plot increased with seasonal declines in soil NO$_3$ concentration. In another study, Sanford *et al.* (1995) grew clover with varying proportions of companion grass (*Lolium* spp.) or capeweed (*Arctotheca calendula*) and found that the grass was much more effective in driving the clover towards high dependency on N$_2$ fixation than was the broadleaved capeweed.

**5** Further exploitation of δ[15]N signals was used to assess possible nitrogen transfer from clover to associated ryegrass, using nodulated subterranean clover grown in single rows bordered on each side at 10, 20, 30 or 60 cm distance by rows of ryegrass (J.S. Pate, P. Sanford and M.J. Unkovich, unpublished). δ[15]N values of the ryegrass (Table 8.4) were then found to be lowered by an extent proportional to the proximity of the grass to the N$_2$-fixing clover (see also Pate *et al.* 1994).

**6** Build up of nitrate during the summer and after wetting of soils in autumn is a commonly accepted cause of poor symbiosis of seedling pasture and grain legumes (Unkovich *et al.* 1997). Unkovich and Pate (1998) evaluated such effects by comparing nitrogen accumulation of sand-cultured subterranean clover (cv. Trikkala) partnered with the recommended inoculum (WU95) or with each of a series of mixed native rhizobia. Inocula of the latter were prepared as dilute suspensions of

**Table 8.4** Evidence for below-ground transfer of fixed nitrogen of low $\delta^{15}N$ value from clover to ryegrass, where ryegrass was grown in lines at given distances from the clover with bare ground in between the clover and ryegrass rows. Mean $\delta^{15}N$ values (‰), with standard errors in parentheses, are reported from four replicates. Values of $\pm 5.9$‰ for grass sampled 60 cm from the clover would be expected to be representative of full subsistence on soil nitrogen, implying that grass 10 cm away from the clover (4.4‰) might be gaining 29% of its nitrogen from the clover.

| | Ryegrass $\delta^{15}N$: distance from clover | | | |
|---|---|---|---|---|
| Clover $\delta^{15}N$ | 10 | 20 | 30 | 60 |
| 0.77 (0.17) | 4.42 (0.64) | 5.17 (0.59) | 5.42 (0.56) | 5.89 (0.40) |

soil from a range of pastures shown earlier to be of contrasting performances with regard to their respective clover symbioses. Each partnership was grown for 12 weeks in the absence of inorganic nitrogen or subjected to a 4-week period of nitrate (5 mM $NO_3$) supply prior to an 8-week period of inorganic nitrogen starvation. Unexpectedly, the clover–WU95 partnership ranked poorly in comparison with most of the indigenous populations of rhizobia under both $NO_3$-rich and N-free conditions. Furthermore, certain soil-based inocula exhibited greater tolerance to $NO_3^-$ supply than others, suggesting that native rhizobial genotypes might be adapted to the concentrations of nitrate likely to be experienced in the pasture with which they were associated (Unkovich & Pate 1998).

7 The NA technique has also been used to evaluate $N_2$ fixation and %Ndfa within the context of the total nitrogen relations of a pasture and potential net transfer of nitrogen to a subsequent cereal crop. Studies by Unkovich et al. (1998) in subterranean clover-based pastures subjected to light or intensive grazing prior to cropping, found that a 3-year episode of low stocking promoted a mean above-ground pasture dry matter yield of 11.5 t ha$^{-1}$ yr$^{-1}$, compared with only 7.9 t ha$^{-1}$ yr$^{-1}$ under a similar period of intensive stocking. Lesser productivity under intensive grazing was mostly due to reduced grass growth, but total biomass nitrogen and $N_2$ fixed under intensive grazing (300 and 153 kg N ha$^{-1}$, respectively), proved to be as high as under light grazing (300 and 131 kg N ha$^{-1}$, respectively). Surprisingly, despite greater soil mineral nitrogen concentrations under high stocking, %Ndfa under such treatment (78%) was virtually identical to that under light grazing (84%). Carrying the study through to a subsequent cropping phase (Unkovich et al. 1998), plots previously subjected to intensive grazing generated considerably greater soil mineral nitrogen through the cropping season, leading to higher crop nitrogen yields and grain protein contents than in plots previously subjected to light grazing. The study indicated considerable scope for improving grain yield and quality by optimizing grazing regimes in seasons prior to cropping.

## Case studies on perennial native N₂-fixing associations in natural habitat

Our first application of the NA method to native Southwestern Australian legumes (Hansen & Pate 1987b) involved the same three understorey acacias (*A. pulchella*, *A. alata* and *A. extensa*) of jarrah forest as had been used previously for $C_2H_2$ reduction-based assays (see above). The assay system was first tested by raising sand-cultured plants on a range of constantly maintained concentrations of nitrate ($\delta^{15}N$ value of +5.5‰), with zero nitrate controls assessing $^{15}N$ natural abundances of fully symbiotic plants and using non-nodulated plants to assess possible isotopic discrimination during nitrate assimilation.

The results indicated very similar behaviour by the species (Fig. 8.4). Thus, non-nodulated plants produced dry matter carrying a $\delta^{15}N$ signature indistinguishable from that of the applied $NO_3$, indicating negligible discrimination during assimilation of this nitrogen source. Conversely, plants totally dependent on fixed nitrogen recorded $\delta^{15}N$ values close to atmospheric $N_2$, again suggesting essentially no discrimination. Finally, and as expected, nodulated plants of all species grown on different concentrations of nitrate achieved $\delta^{15}N$ values indicating predictable patterns of decreasing proportional dependence on fixation with increase in concentrations of nitrate supplied (Fig. 8.4).

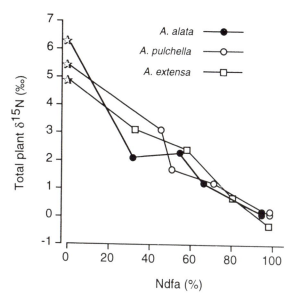

**Figure 8.4** Relationships between plant dependence on $N_2$ (%Ndfa) and $\delta^{15}N$ values of total plant nitrogen of nodulated seedlings of three acacias pot cultured on a range of constantly maintained levels of nitrate. Values marked by stars denote $\delta^{15}N$ values for non-nodulated nitrate-grown plants. Nitrate used throughout carried $\delta^{15}N$ signature of +5.5‰. Data from Hansen and Pate (1987b).

Armed with this encouraging result, a field study was conducted across sites representative of the complex lateritic soils of jarrah forest. $\delta^{15}N$ values of soil total nitrogen and soil mineral nitrogen ($NH_4^+ + NO_3^-$), and of shoot total nitrogen of acacias and woody reference species, were then assessed following appropriate sampling at each location. The data showed disappointingly wide ranges of variation in $\delta^{15}N$ values of soil samples (from $-2.1$ to $+5.4‰$ for total nitrogen and from $+3$ to $+14.2‰$ for soil mineral nitrogen) (Hansen & Pate 1987b). Equally disappointingly, reference plants bore little evidence of $\delta^{15}N$ values matching the discrimination values of their parent soils, nor did visibly well-nodulated legumes achieve $\delta^{15}N$ values closer to atmospheric $N_2$ than their partner reference species. Gross heterogeneity in $^{15}N$ discrimination of soil nitrogen pools at macro- and micro-scales had clearly precluded effective use of the NA technique. This contrasts with apparently successful use of $C_2H_2$ reduction assays on the same species at the same sites (see above).

Turning to the supposedly less complex system afforded by *Banksia* woodlands of the oligotrophic coastal sands near Perth, we examined the nitrogen relationships of over 30 common species at sites burnt from 0.3 to 23 years previously. Species were classified according to life form, root morphology and the presence or absence of nodules, cluster (proteoid) roots and various types of mycorrhizal symbioses (Stewart *et al.* 1993). Nitrate reductase (NR) activities of new leafy shoots in spring indicated substantially less assimilation in unburnt than recently burnt sites, a finding consistent with the high rates of release of $NO_3$ that occur after fire. Comparisons of mean NR rates and shoot $\delta^{15}N$ of non-$N_2$-fixing species across all sites indicated a strong correlation between the ability to utilize nitrate and the absence of mycorrhizas (Fig. 8.5; Pate *et al.* 1993). This applied particularly to herbaceous species whose xylem sap was exceptionally rich in nitrate and whose dry matter was well discriminated in favour of $^{15}N$. In one highly nitrophilous species, *Ptilotus polystachyus*, very similar $\delta^{15}N$ signals were evident for soil $NO_3$-N, xylem nitrogen and shoot total nitrogen, indicating minimal isotope discrimination during the utilization of nitrate.

The study sites examined by Pate *et al.* (1993) contained seven shrub legumes and a tree (*Allocasuarina fraseriana*) with the potential to engage in *Frankia*-based symbiosis. All bore apparently healthy symbiotic organs and showed negative or slightly positive $\delta^{15}N$ values in their shoot dry matter, compared with mostly positive values for non-fixing species (Fig. 8.5). $\delta^{15}N$ values for non-mycorrhizal herbaceous species were significantly higher (more positive) than for those that were mycorrhizal or potentially capable of $N_2$ fixing. When selecting woody mycorrhizal non-fixing species as the most appropriate references to the $N_2$ fixers, analysis of the $\delta^{15}N$ signals for these groups failed to even indicate the presence, let alone the extent, of $N_2$ fixation (Pate *et al.* 1993).

A further NA-based study related to *Macrozamia riedlei*, the cycad referred to earlier in connection with $C_2H_2$-based assessments of $N_2$ fixation. Nine study sites were investigated, two in jarrah forest south of Perth, five in sandplain heath north of Perth and two in *Banksia* woodlands of the same region. Mean $\delta^{15}N$ values of

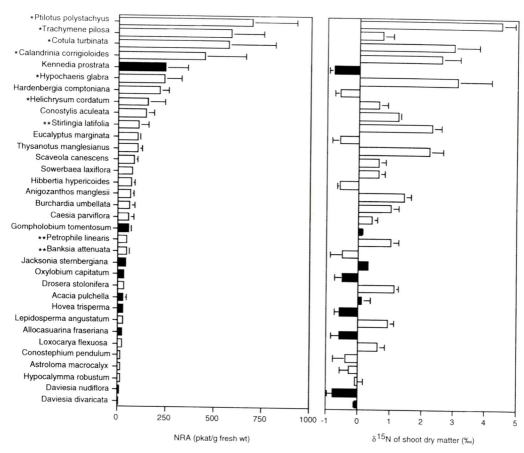

**Figure 8.5** Mean values of *in vivo* nitrate reductase activity (NRA) of young leafy shoots and corresponding δ15N values for total nitrogen of this shoot material for a range of common species sampled across a range of sites of different fire history in the bushland of Kings Park, Perth, Australia. All values are given as means with standard errors for each species across the eight sites investigated. Data for putative N₂ fixers are shown with shaded bars. Species marked with an asterisk are non-mycorrhizal herbaceous, those with two asterisks are non-mycorrhizal Proteacea bearing proteoid roots. All other species are variously mycorrhizal. Data from Pate *et al.* (1993).

non-fixing woody reference species (+1.3 and +0.4‰) at the two jarrah forest sites were not significantly different from the cycad samples (+0.8 and –0.3‰). The other seven sites showed reference plant species with δ15N values in the range +2.7 to +5.3‰, compared with a non-overlapping, much less positive range for the cycads (+0.9 to +2.2‰). The results tentatively indicated considerable activity in N₂ fixation, as might have been expected from the presence of coralloid roots on all specimens used in the study.

Further cases in which substantially different δ15N signals have been recorded between reference plants and putative N$_2$-fixing species relate to *Aeschynomene indica* and *Sesbania* sp. growing in monsoonal wetlands near Derby, Northwest Australia (J.S. Pate, unpublished). Reference plants, consisting of partly submerged herbaceous and woody species, showed uniformly high δ15N values (+5.8 to +8.1‰), possibly reflecting denitrification in the nitrogen-rich wetland. Corresponding legume material showed much lower δ15N values (ranging from +2 to +3.3‰), suggestive of appreciable symbiotic activity. Haemoglobin-pigmented nodules were abundant on both legume species.

When we turned our attention to *Acacia*-dominated mulga vegetation of the semi-arid regions of Australia (200–500 mm annual rainfall), a situation particularly well suited to the study of nitrogen cycling was encountered. Employing a combination of NR assays and xylem sap analyses on common species, it was concluded that nitrate was the principal form of nitrogen utilized by plants, especially in the case of herbaceous ephemerals flourishing after winter rains (Erskine *et al.* 1996). Artificial watering of soils during a temporary drought in spring increased the capacities of these ephemerals to utilize nitrate, implying that adequate availability of water was essential for the assimilation of nitrogen under such circumstances.

Investigations of mulga in both Queensland and Western Australia (Pate *et al.* 1998) showed that δ15N values of newly formed shoot nitrogen of all major flora lay within the somewhat high range of 7.5–15.5‰. Eight common species of *Acacia* at the sites carried a mean δ15N signal (9.10±0.6‰) indistinguishable from that of 37 non-fixing shrubs and trees (9.06±0.5‰) sampled across the study sites. With no evidence of nodulation on any of the acacias, the data indicated negligible symbiotic activity. By contrast, a cyanobacteria-containing lichen and worker members of termite colonies showed δ15N values (range 0.5–3‰) close to atmospheric N$_2$, indicating contributions of fixed nitrogen from these sources.

It was concluded that mulga vegetation is likely to be limited by phosphorus rather than nitrogen and that dominance of acacias across the region must therefore denote competitive advantages other than the potential to fix N$_2$. Interestingly, the ephemeral floras of mulga are dominated by nitrophilous species from the families Asteraceae, Poaceae, Amaranthaceae, Portulacaceae and Goodeniaceae (Pate 1983). By contrast, annual Fabaceae are absent or extremely poorly represented, possibly reflecting minimal advantage from nitrogen fixation in this high nitrogen environment.

The Western Australian mulga study region contained certain atypical populations of acacias and papilionoid legumes inhabiting heavily leached sand dune ridges bordering a lake. In stark contrast to the situation for general mulga, the legumes at this lake site were well nodulated and carried δ15N signals (2–3‰) noticeably lower than those of closely adjacent non-fixing taxa (6–11‰). Fixation was assumed to be appreciable in this unusual situation, indicating that the lack of fixation in surrounding general mulga reflected constraints attributable to high soil nitrogen status, rather than outright inability to engage in symbiosis.

A final example of our use of the NA technique concerns the nitrogen relation-ships of the woody root hemiparasite *Santalum acuminatum* and its hosts at three sites in coastal sandplain heathlands near Dongara, Western Australia (Tennakoon *et al.* 1997). *Acacia rostellifera* comprised the dominant host legume at one site, *A. pulchella* at another and *Allocasuarina campestris* at the third. All three species were well nodulated and haustorial attachment of *S. acuminatum* proved to be much more intense on these than on any other hosts (Fig. 8.6). $\delta^{15}N$ data for shoot dry matter of *S. acuminatum* and all hosts suggested that the three putative $N_2$ fixers were subsisting mostly on atmospheric $N_2$, and that *Santalum* was feeding indi-rectly on the same nitrogen source (Fig. 8.6). The ability to target haustorial initia-tion towards $N_2$-fixing species is likely to be especially important in the adaptation of *S. acuminatum* to nitrogen-deficient sites such as at Dongara.

## Contrasting nitrogen fixation inputs in natural bushland and adjacent agricultural systems: causes and consequences

Opportunities for study of this kind were provided in an area of relic native wood-land at Moora, 250 km north of Perth. The bush site under study contained trees of *Banksia prionotes*, *Eucalyptus todtiana* and *Xylomelum angustifolium*, and an understorey of predominantly myrtaceous and proteaceous shrubs. Cleared land adjacent to this bush has been committed to grazing of sheep and cattle or crop-ping of lupins (*Lupinus angustifolius*) and cereals since first cleared in the 1960s. The perennial fodder shrub legume tagasaste (*Chamaecytisus proliferus*) has been recently introduced into the region in an attempt to reverse dangerously rising water-tables.

Study of the nitrogen dynamics of the relic *Banksia* woodland at the site identi-fied potential $N_2$-fixing associations as *Acacia pulchella*, *Allocasuarina humilis*, *Gompholobium tomentosum*, *Mirbelia spinosa* and *Jacksonia* spp. These collectively comprised less than 5% of the standing biomass. Both the legumes and *Allocasuar-ina* were mostly non-nodulated and $\delta^{15}N$ analyses of their new shoot dry matter averaged $0.3 \pm 0.3‰$ *vs.* $0.8 \pm 0.4‰$ for appropriate reference species. These equiv-ocal findings in respect of symbiotic competence, combined with low proportions of biomass as potential $N_2$ fixers, indicated that $N_2$ fixation was currently insignifi-cant at this mature stage of a postfire sequence resulting from a burn at the site 35 years previously. Tight nitrogen cycling was also evident from the finding that groundwater at the bush site contained less than $1 \, mg \, NO_3\text{-}N \, l^{-1}$ and that seasonal peak concentrations for soil mineral nitrogen in the top 10 cm of soil were only $0.4 \, mg \, NO_3\text{-}N$ and $0.6 \, mg \, NH_4\text{-}N \, kg^{-1}$ dry soil.

Cleared land bordering the bushland site has regularly received dressings of superphosphate, and assuming that wheat–lupin crop rotations or legume-based annual pastures have comprised the normal agricultural practices, large annual returns of fixed nitrogen from legume residues would be expected. Mineralization of this nitrogen, combined possibly with moderate dressings of nitrogen to cereal and canola crops, would be likely to result in substantial rates of nitrate leaching

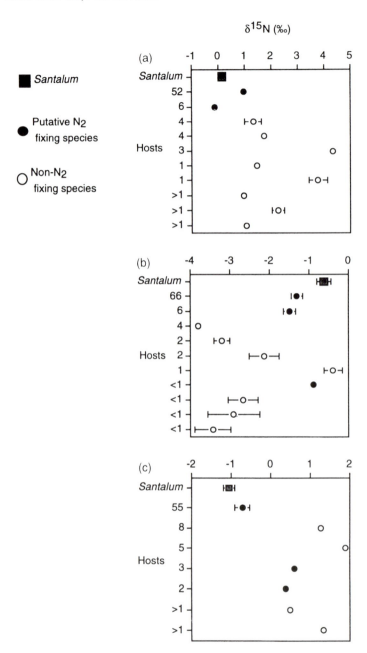

**Figure 8.6** Mean $\delta^{15}N$ signatures of new season's shoot dry matter of the root hemiparasite *Santalum acuminatum* and its principal hosts at three sites in coastal heath near Dongara, Western Australia. Numbers against each host refer to an 'availability index' in respect of parasitism by *Santalum*. This was computed from respective densities of the host species and intensities of haustorial occurrence on their roots. Putative $N_2$-fixing species are indicated by black circles, non-fixing hosts by open circles. Data from Tennakoon *et al.* (1997).

(Anderson *et al.* 1998). This conclusion is substantiated by finding of remarkably high concentrations of nitrate (6–19 mg/l) in groundwater under agricultural land of the region.

In a comparative study of nitrogen balance in a tagasaste–lupin alley system, conventional lupin plantings and a plantation of tagasaste, the lupin derived more nitrogen from soil than from $N_2$ fixation, and, after harvest of seed, showed a negative nitrogen balance in terms of return to the ecosystem in the season of study. A second season of study demonstrated the reverse, with large inputs of fixed nitrogen and an Ndfa value of 85%. Tagasaste fixed much more nitrogen than it obtained from soil or groundwater, and, after allowing for part removal of its biomass by grazing animals, would still generate net inputs of nitrogen to the system. It would thus be likely to exacerbate rather than ameliorate groundwater nitrate pollution.

## Concluding remarks

The case histories presented in this paper collectively show how difficult it is to make accurate measurements of $N_2$ fixation within ecosystems, particularly those involving mixed woody native vegetation. When applied to such situations both the $C_2H_2$ reduction and the NA techniques suffer serious disadvantages with respect to accuracy and reliability. However, the more extensively each method is used the more confidently one can pinpoint likely elements of success or failure in a particular set of circumstances.

As a general protocol for future research we would advocate combining detailed biological investigation of all potential $N_2$ fixers in a system to methodologically sound assessments of nitrogen flow from each of these players to pools of soil nitrogen and those of interacting biota. This should include the assessment of activities of free-living microbial agents involved in nitrogen cycling and corresponding fluxes through all system nitrogen pools. Only then will it be possible to assess the overall impact of symbiotic $N_2$ fixation on factors such as availability of water, fire, drought and cycling of potentially limiting nutrients other than nitrogen.

## References

Anderson, G., Fillery, I.R.P., Dunin, F.X., Dolling, P.J. & Asseng, S. (1998). Nitrogen and water flows under pasture-wheat and lupin-wheat rotations in deep sands in Western Australia. 2. Drainage and nitrate leaching. *Australian Journal of Agricultural Research*, 49, 345–362.

Armstrong, E.A., Heenan, D.P., Pate, J.S. & Unkovich, M.J. (1997). Nitrogen benefits of lupins, field pea and chickpea to wheat production in south-eastern Australia. *Aus-*

*tralian Journal of Agricultural Research*, **48**, 39–47.

Bergersen, F.J., Turner, G.L., Gault, R.R., Chase, D.L. & Brockwell, J. (1985). The natural abundance of [15]N in an irrigated soybean crop and its use for the calculation of nitrogen fixation. *Australian Journal of Agricultural Research*, **36**, 411–423.

Erskine, P.D., Stewart, G.R., Schmidt, S., Turnbull, M.H., Unkovich, M.J. & Pate, J.S. (1996). Water availability — a physiological constraint on nitrate utilization in plants of Australian semi-arid mulga woodlands. *Plant, Cell and Environment*, **19**, 1149–1159.

Halliday, J. & Pate, J.S. (1976). Symbiotic nitrogen fixation by coralloid roots of the cycad *Macrozamia riedlei*: physiological characteristics and ecological significance. *Australian Journal of Plant Physiology*, **3**, 349–358.

Handley, L.L. & Scrimgeour, C.M. (1997). Terrestrial plant ecology and [15]N natural abundance: the present limits to interpretation for uncultivated systems with original data from a Scottish old field. *Advances in Ecological Research*, **27**, 133–212.

Hansen, A.P. (1994). *Symbiotic N2 Fixation of Crop Legumes*. Margraf-Verlag, Weikersheim.

Hansen, A.P. & Pate, J.S. (1987a). Comparative growth and symbiotic performance of seedlings of *Acacia* spp. in defined pot culture or as natural understorey components of a eucalypt forest ecosystem in S.W. Australia. *Journal of Experimental Botany*, **38**, 13–25.

Hansen, A.P. & Pate, J.S. (1987b). Evaluation of the [15]N natural abundance method and xylem sap analysis for assessing $N_2$ fixation of understorey legumes in jarrah (*Eucalyptus marginata* Donn ex Sm.) forest in S.W. Australia. *Journal of Experimental Botany*, **38**, 1446–1458.

Hansen, A.P., Pate, J.S. & Atkins, C.A. (1987a). Relationships between acetylene reduction activity, hydrogen evolution and nitrogen fixation in nodules of *Acacia* spp.: experimental background to assaying fixation by acetylene reduction under field conditions. *Journal of Experimental Botany*, **38**, 1–12.

Hansen, A.P., Pate, J.S., Hansen, A. & Bell, D.T. (1987b). Nitrogen economy of post fire stands of shrub legumes in jarrah (*Eucalyptus marginata* Donn ex Sm) forest of SW Australia. *Journal of Experimental Botany*, **38**, 26–41.

Herridge, D.F., Bergersen, F.J. & Peoples, M.B. (1990). Measurement of nitrogen fixation by soybean in the field using the ureide and natural [15]N abundance methods. *Plant Physiology*, **93**, 708–716.

Högberg, P. (1997). [15]N natural abundance in soil–plant systems. *New Phytologist*, **137**, 179–203.

Ledgard, S.F. (1989). Nutrition, moisture and rhizobial strain influence isotopic fractionation during $N_2$ fixation in pasture legumes. *Soil Biology and Biochemistry*, **21**, 65–68.

Ledgard, S.F. & Peoples, M.B. (1988). Measurement of nitrogen fixation in the field. In *Advances in Nitrogen Cycling in Agricultural Ecosystems* (Ed. by J.R. Wilson), pp. 351–367. CAB International, Wallingford.

Ledgard, S.F., Freney, J.R. & Simpson, J.R. (1984). Variations in natural enrichment of [15]N in the profiles of some Australian pasture soils. *Australian Journal of Soil Research*, **22**, 155–164.

McNeill, A.M., Zhu, C. & Fillery, I.R.P. (1998). A new approach to quantifying the N benefit from pasture legumes to succeeding wheat. *Australian Journal of Agricultural Research*, **49**, 427–436.

Monk, D., Pate, J.S. & Loneragan, W.A. (1981). Biology of *Acacia pulchella* R. Br. with special reference to symbiotic nitrogen fixation. *Australian Journal of Botany*, **29**, 570–592.

Pate, J.S. (1976). Transport in symbiotic systems fixing nitrogen. In *Encyclopedia of Plant Physiology New Series*, Vol. 2, Part B (Ed. by U. Lüttge & M. Pitman), pp. 278–303. Springer-Verlag, Berlin.

Pate, J.S. (1983). Patterns of nitrogen metabolism in higher plants and their ecological significance. In *Nitrogen as an Ecological Factor* (Ed. by J.A. Lee, S. McNeill & I. Rorison), pp. 225–255. Blackwell Scientific Publications, Oxford.

Pate, J.S. & Dawson, T.E. (1999). Assessing the performance of woody plants in uptake and utilisation of carbon, water and nutrients: implications for designing agricultural mimic systems. In *Agriculture as a Mimic of Natural Ecosystems* (Ed. by E.C. Lefroy, R.J. Hobbs, M.H. O'Connor & J.S. Pate). Kluwer Academic, Amsterdam, in press.

Pate, J.S., Stewart, G.R. & Unkovich, M.J. (1993). [15]N natural abundance of plant and soil components of a *Banksia* woodland ecosystem in relation to nitrate utilization, life form, mycorrhizal status and $N_2$-fixing abilities of component species. *Plant, Cell and Environment*, **16**, 365–373.

Pate, J.S., Unkovich, M.J., Armstrong, E.L. & Sanford, P. (1994). Selection of reference plants for [15]N natural abundance assessment of $N_2$ fixation by crop and pasture legumes in southwest

Australia. *Australian Journal of Agricultural Research*, **45**, 133–147.

Pate, J.S., Unkovich, M.J., Erskine, P.D. & Stewart, G.R. (1998). Australian mulga ecosystems — $^{13}C$ and $^{15}N$ natural abundance of biota components and their ecophysiological significance. *Plant, Cell and Environment*, **21**, 1231–1242.

Peoples, M.B. & Herridge, D.F. (1990). Nitrogen fixation by legumes in tropical and subtropical agriculture. *Advances in Agronomy*, **44**, 155–223.

Peoples, M.B., Faizah, A.W., Rerkasem, B. & Herridge, D.F. (1989). *Methods for Evaluating Biological Nitrogen Fixation by Nodulated Legumes in the Field*. Australian Centre for International Agricultural Research, Canberra.

Russell, C.A. & Fillery, I.R.P. (1996). Estimates of lupin below-ground biomass N, dry matter and N turnover to wheat. *Australian Journal of Agricultural Research*, **47**, 1047–1059.

Sanford, P., Pate, J.S. & Unkovich, M.J. (1994). A survey of proportional dependence of subterranean clover and other pasture legumes on $N_2$ fixation in south-west Australia utilizing $^{15}N$ natural abundance. *Australian Journal of Agricultural Research*, **45**, 165–181.

Sanford, P., Pate, J.S., Unkovich, M.J. & Thompson, A.N. (1995). Nitrogen fixation in grazed and ungrazed subterranean clover pasture in south-west Australia assessed by the $^{15}N$ natural abundance technique. *Australian Journal of Agricultural Research*, **46**, 1427–1443.

Shearer, G. & Kohl, D.H. (1986). $N_2$ fixation in field settings: estimates based on natural $^{15}N$ abundance. *Australian Journal of Plant Physiology*, **13**, 699–756.

Stewart, G.R., Pate, J.S. & Unkovich, M.J. (1993). Characteristics of organic nitrogen assimilation of plants in fire-prone Mediterranean-type vegetation. *Plant, Cell and Environment*, **16**, 351–363.

Tennakoon, K.U., Pate, J.S. & Arthur, D. (1997). Ecophysiological aspects of the woody root hemiparasite *Santalum acuminatum* (R. Br.) A. DC and its common hosts in south western Australia. *Annals of Botany*, **80**, 245–256.

Unkovich, M.J. & Pate, J.S. (1998). Symbiotic effectiveness and tolerance to early season nitrate availability in indigenous populations of subterranean clover rhizobia from SW Australian pastures. *Soil Biology and Biochemistry*, **30**, 1435–1443.

Unkovich, M.J., Pate, J.S., Sanford, P. & Armstrong, E.L. (1994). Potential precision of the $\delta^{15}N$ natural abundance method in field estimates of nitrogen fixation by crop and pasture legumes in SW Australia. *Australian Journal of Agricultural Research*, **45**, 119–132.

Unkovich, M.J., Pate, J.S. & Sanford, P. (1997). Nitrogen fixation by annual legumes in Australian Mediterranean agriculture. *Australian Journal of Agricultural Research*, **48**, 267–293.

Unkovich, M.J., Sanford, P., Pate, J.S. & Hyder, M. (1998). Effects of grazing on plant and soil nitrogen relations of pasture-crop rotations. *Australian Journal of Agricultural Research*, **49**, 475–485.

Walker, B.A. & Pate, J.S. (1986). Morphological variation between seedling progenies of *Viminaria juncea* (Schrad. & Wendl.) Hoffmans. (Fabaceae) and its physiological significance. *Australian Journal of Plant Physiology*, **13**, 305–319.

Walker, B., Pate, J.S. & Kuo, J. (1983). Nitrogen fixation by nodulated roots of *Viminaria juncea* (Schrad. & Wendl.) Hoffmans. (Fabaceae) when submerged in water. *Australian Journal of Plant Physiology*, **10**, 409–421.

# Chapter 9

# Parasitic plants: physiological and ecological interactions with their hosts

*M.C. Press, J.D. Scholes and J.R. Watling*

## Introduction

Parasitic angiosperms are a taxonomically diverse group of organisms. Parasitism in angiosperms is considered to have evolved approximately 11 times and there are approximately 4000 species in total, accounting for just over 1% of all flowering plants. The higher level classification of some species is problematic for two reasons. First, modification or loss of morphological features during evolution has confounded phylogenetic analysis and, second, evolutionary convergence has resulted in similar features in very different groups. Nickrent *et al.* (1998) classify the plants into approximately 270 genera and 22 families (Fig. 9.1), using not only morphological characteristics but also molecular characteristics of both plastid and nuclear genomes.

Parasitic angiosperms rely, to varying degrees, on their host plant or plants for the acquisition of water and solutes, both inorganic and organic; movement of resources occurs through one or more haustoria, which serve to attach the parasite to the host. The position of the haustorium, either above or below ground, has led to the grouping of parasitic plants as either shoot or root parasites, respectively. Further, parasitic plants may be classified as either holo- or hemi-parasites, depending on the absence or presence of chlorophyll, respectively. The former evolved from the latter and are considered to be entirely heterotrophic. Hemiparasites, on the other hand, are capable of fixing a proportion of their carbon autotrophically and may also have a greater capacity to assimilate inorganic nutrients than achlorophyllous species.

Parasitic angiosperms differ markedly in host specificity. For example, the root holoparasite *Conopholis americana* is reported to infect only one species of oak, *Quercus borealis* (Kuijt 1969), while some parasitic angiosperms, such as root hemiparasites in the Scrophulariaceae, may infect large numbers of different species. Gibson and Watkinson (1989) collated data for 12 such hemiparasites which infected between four and 79 different host species from many diverse plant families, including species with different life histories and growth forms. It is

*Department of Animal and Plant Sciences, University of Sheffield, Sheffield, S10 2TN, UK. E-mail: M.C.Press@sheffield.ac.uk*

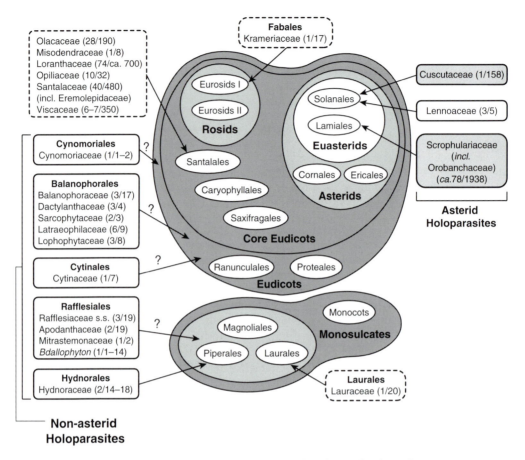

**Figure 9.1** Summary of the phylogeny of parasitic angiosperms based on molecular and morphological characteristics. Dashed borders = hemiparasites; solid borders = holoparasites; shaded boxes indicate both trophic modes. Arrows indicate strong evidence for placement of taxa and question marks indicate uncertain affinities. The number of genera and species is given in parentheses for each family. D. L. Nickrent (unpublished data).

becoming increasingly apparent that some species commonly infect multiple hosts (Gibson & Watkinson 1989; Pate *et al.* 1990a), a phenomenon which may have important implications for resource acquisition (see below). There is some evidence to suggest a latitudinal trend in specificity within the loranthaceous mistletoes, with these parasites showing low specificity in lowland tropical rain forests and high specificity at higher latitudes (Norton & Carpenter 1998). This trend, which is also apparent in some groups of animal parasites, may be driven by latitudinal trends in species diversity and abundance (Norton & Carpenter 1998).

Parasitic angiosperms are present in most ecosystems, from the equator (e.g. *Rafflesia* spp. in South-east Asian lowland dipterocarp rain forests) to the poles (e.g. *Pedicularis* spp. on Spitzbergen). Their abundance is usually low, except in certain nutrient-poor ecosystems, for example, subarctic heathlands and meadows, alpine meadows and South-west Australian coastal habitats. Species from four families can be important as agricultural weeds and, in decreasing order of importance, the families are the Scrophulariaceae (including the Oroban-chaceae), the Cuscutaceae, the Viscaceae and the Loranthaceae (Parker 1991). The most important parasitic weeds are in the genus *Striga* (the witchweeds), which primarily infect $C_4$ cereals in the African semi-arid tropics. Few detailed studies have been conducted of the distribution of parasitic angiosperms relative to their hosts, but it seems likely that parasitic plants are restricted to a narrow part of the range of their host or hosts (see e.g. Cochrane & Press 1997). An extreme example in the UK is the relationship between the distribution of the broomrape *Orobanche reticulata* and its thistle hosts (*Cirsium* and *Carduus* species). Although the hosts are widespread the parasite only occurs in Yorkshire, on magnesium limestone out-crops (Hughes & Headley 1996). Precisely which components of the environment differentially influence host and parasite is uncertain, but both abiotic and biotic factors have been implicated.

This chapter aims to examine the ways in which parasitic angiosperms interact with their hosts, both at a physiological and an ecological level. Physiological inter-actions are considered at two stages, first the establishment of the parasitic asso-ciation, specifically seed germination and haustorial initiation and second, the impact that parasites exert on host growth and metabolism. The role of parasitic angiosperms is then briefly explored at the community level. Our knowledge of interactions between angiosperm parasites and their hosts is heavily skewed towards a relatively small number of species in the orders Solanales, Lamiales and Santalales. This bias not only crudely reflects the number and abundance of species but also agricultural interests and experimental tractability.

## Establishment of the host–parasite interface

### Germination cues

In common with other symbiotic associations (*sensu* Douglas 1994) involving plants (both mutalistic and parasitic), critical developmental and physiological events are mediated by signal molecules (Lynn & Chang 1990; Boone *et al.* 1995; Estabrook & Yoder 1998). For parasitic associations between plants, the two most important cues are for seed germination and haustorial initiation. The control of seed germination for some root parasites (genera in the orders Lamiales, Bal-anophorales, Rafflesiales and Hydnorales) is unique in that it is triggered by chemical compounds present in the root exudate of host (and some non-host) species (Joel *et al.* 1994).

The best studied germination system is that between *Striga* spp. and $C_4$ grasses. The germination cues that operate in nature are thought to be sesquiterpene lactones, which differ slightly in structure among host species (Hauck *et al.* 1992; Sugimoto *et al.* 1998). The molecules are labile and active at very low concentrations in the soil (in some cases picomolar) and thus act as a chemical mechanism that regulates distance between parasite and host. Understanding the structure–function relations of the germination cues has been hindered by their low rates of production, laborious isolation procedures and complex stereochemistry, but their activity appears to be more dependent on the precise three-dimensional structure than the specific nature of functional groups on the molecule (Fischer *et al.* 1990; Sugimoto *et al.* 1998). Their mode of action within the seed is also poorly understood, but Logan and Stewart (1991) and Jackson and Parker (1991) propose a model whereby the host-derived signal molecules, in conjunction with other compounds, elicit ethylene production within the seed, thereby initiating biochemical changes that lead to germination.

### Haustorial initiation

Haustoria of parasitic angiosperms are diverse in size, structure and function (e.g. Fineran 1985; Heide-Jørgensen 1989, 1991; Pate *et al.* 1990b; Heide-Jørgensen & Kuijt 1995; Dörr 1997; Reiss & Bailey 1998). They also differ markedly from those of the more studied fungal biotrophs (Smith & Smith 1990). A chemical signal from the host is necessary for formation of the haustorium and, in some species, such chemicals may be released from host tissue through the action of oxidative enzymes secreted by the parasite (Chang & Lynn 1986). Hautorial initiation is achieved by a greater diversity of compounds than those that stimulate seed germination and active compounds can be classified into four classes: flavonoids, *p*-hydroxy acids, quinones and cytokinins (Lynn & Chang 1990). The first three classes are structurally related phenolics derived from phenylalanine and may operate through different mechanisms to the cytokinins (Estabrook & Yoder 1998).

In *Striga* it appears that a quinone, 2,6-dimethoxy-*p*-benzoquinone, is the haustorial initiating cue in nature. Higher concentrations (millimolar) are required than for the germination cue, and exposure for a minimum period of 6 hours is necessary for the completion of the developmental process (Smith *et al.* 1990). The next stage involves the penetration of host tissue; some species may penetrate host cells (e.g. the dodder *Cuscuta*, Dawson *et al.* 1994) while others move between them (e.g. the broomrape *Orobanche*), employing both mechanical pressure and pectolytic enzymes that change the properties of the host cell wall and middle lamella (Losner-Goshen *et al.* 1998).

### Haustorial structure and function

Close xylem connections are present in most haustoria, although the proportion of direct xylem-to-xylem contacts at the host–parasite interface with luminal continuity may be small (Pate *et al.* 1990b). By contrast, direct links with host phloem have not been reported in haustoria, although in some species fully differentiated sieve tubes have been observed, while in others specialist conductive cell types have been reported (Kuijt 1977).

The haustorium has been presumed largely to act only as a conduit for solute

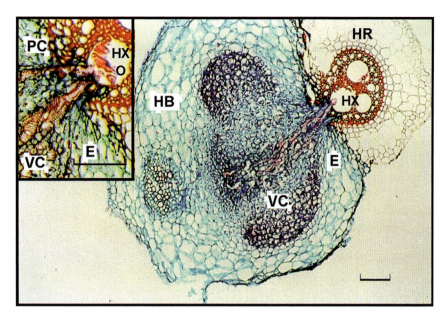

**Plate 9.1** A section through a haustorium of *Striga hermonthica* parasitizing a root of the C$_4$ grass *Eragrostis pilosa*. Xylem continuity (red stain) can be seen between the host and parasite. Inset shows parenchyma cells (PC) within the vascular core (VC). E = endophyte; HB = hyaline body; HR = host root; HX = host xylem; O = oscula (A. Leakey, J.R. Watling & M.C. Press, unpublished).

*[Facing p. 178]*

transfer, but there is mounting evidence that argues against this view. Again, taking *Striga* as an example, the haustorium consists of two regions, one external to the host (the hyaline body, HB in Plate 9.1, facing page 178) and the other within the host tissue (the endophyte, E in Plate 9.1), both of which are traversed by a vascular core. The endophyte consists of parenchyma cells and xylem vessels (Plate 9.1), the latter forming 'clustered intrusions' (or 'oscula') directly into the large host xylem vessels (Dörr 1997 and Plate 9.1, inset). Sieve elements, sieve plates and the presence of efficient symplastic channels for nutrient transfer have not been observed at the endophyte–host junction. However, transport bridges of parenchyma cells connecting the endophyte to host phloem have been found, although the parenchyma cells lack plasmodesmata (Dörr 1997).

Thus, although the initial movement of solutes is apoplastic, there is evidence to suggest rapid uptake of solutes into the parenchyma cells (Riopel & Timko 1995). In addition, the structure of the cells within the hyaline body is suggestive of high metabolic activity; they contain an abundance of rough endoplasmic reticulum, large nuclei and large numbers of dictyosomes (Visser *et al.* 1984; Mallaburn & Stewart 1987). Pate *et al.* (1991) suggested that the haustorium of the mistletoe *Amyema* influenced the composition of amino acid flux into the parasite, again indicative of an active role for this organ. The haustorium of *Santalum acuminatum* has been shown to play a major role in nitrogen assimilation (Tennakoon *et al.* 1997). Nitrate reductase activity was induced in the haustorium following nitrate feeding to host xylem and the haustorium also acted as a major site for the synthesis and export of proline.

## Carbon, nutrient and water acquisition: some generalizations

During the past decade great advances have been made in our understanding of the carbon, water and nutrient relations of certain parasitic angiosperms and these allow a number of generalizations to be made concerning resource acquisition by parasitic plants. However, our knowledge of the impacts of parasitic plants on host resource acquisition, metabolism and growth has progressed more slowly. Some of the key aspects of carbon, nutrient and water acquisition by parasitic angiosperms are summarized below, before examining the extent to which competition for these resources may account for the impact of these organisms on their hosts.

### Water and nutrients

The water relations of many root and shoot hemiparasites are very distinct from those of their hosts in two key respects: they have higher rates of transpiration and lower water potentials. Parasite transpiration rates commonly exceed those of their host by two- to fivefold or more, especially under conditions of low soil water availability (Press & Whittaker 1993). The stomatal behaviour of some hemiparasites is also distinct from that of their hosts, most noticeably in short-lived or annual species, where only partial nocturnal closure occurs (Press *et al.* 1988; Davidson *et al.* 1989).

A large number of studies have reported that water potentials are lower, of the order of $1–2$ MPa or more, in hemiparasites than in their hosts, probably resulting from a combination of high transpiration rates and low hydraulic conductivity across the haustorium (Ehleringer & Marshall 1995). In addition, both hemiparasites and holoparasites maintain osmotic potentials below those of their host, but the nature of the osmotica varies greatly between species, and includes amino acids, organic acids, polyols and other carbohydrates (Richter & Popp 1992; Press 1995; Tennakoon *et al.* 1997). In addition, hemiparasites often have higher concentrations of xylem-mobile cations than their hosts, although this may be a consequence of being unable to retranslocate these ions rather than an adaptation to maintain relatively low water potentials (Ehleringer & Marshall 1995). This characteristic is reflected in the high concentrations of nutrients in the leaves of hemiparasites, typically two- to 20-fold those in host leaves (Pate 1995).

The functional significance of these traits is to facilitate the acquisition of water and inorganic and organic solutes from the host. For hemiparasites, the greater the rate of transpiration, the greater the flux of inorganic and organic solutes from host xylem to the parasite. In both hemi- and holoparasites, the driving force is maintained by the accumulation of carbon- or nitrogen-containing compounds which act as osmoticants. Few studies have determined the source of water for root parasites and, like their shoot counterparts, they are largely considered to rely on their host for water rather than to abstract water directly from the soil.

## Carbon

An additional potential consequence of high transpiration rates in hemiparasites, and hence greater leaf conductances, is lower resistance to $CO_2$ diffusion into the leaves. This does not result in greater rates of photosynthesis; indeed the converse is the case. Rates of $CO_2$ fixation are typically towards the low end of the range for $C_3$ plants, although they differ markedly among species and are also a function of the nature and nutritional status of the host (Press 1995). This combination of relatively low rates of photosynthesis and relatively high rates of transpiration results in low water use efficiency.

The extent to which parasitic plants depend on heterotrophic sources of carbon differs markedly between species, ranging from complete dependency in holoparasites to complete autotrophy in unattached, facultative hemiparasites. In nature, however, such facultative species almost always form connections with a host, but they initially pass through a (sometimes lengthy) preparasitic phase (e.g. *Olax phyllanthi*). The degree of heterotrophy in parasites has been quantified through partial modelling (Graves *et al.* 1989; Jeschke *et al.* 1994a,b) and estimates based on the natural abundance of stable carbon isotopes (Press *et al.* 1987; Marshall *et al.* 1994; Tennakoon & Pate 1996a). The proportion of host-derived carbon in hemiparasites differs markedly between species, ranging from <10% to >60% (Press & Whittaker 1993), but it is also a function of the type and nutritional status of the host (Cechin & Press 1993; Tennakoon & Pate 1996a).

## Host quality

The performance of parasites that are not host specific differs markedly between hosts and may be controlled by the ability of the host to supply the parasite with nitrogen. The growth of some root hemiparasites is stimulated to a greater extent by legume than by non-legume hosts (Wilkins 1963; Gibson & Watkinson 1989), as is the growth of some mistletoe species (Schulze & Ehleringer 1984). Close association with legumes in the field, together with enhanced rates of growth and photosynthesis on legume hosts, have been reported in some hemiparasites (Press *et al.* 1993). It seems likely that the additional nitrogen allows the hemiparasites to synthesize more photosynthetic pigments and proteins, resulting in higher rates of autotrophic carbon gain (Seel *et al.* 1993). The importance of nitrogen as a resource has been emphasized by Schulze and Ehleringer (1984), who interpreted high transpiration rates in mistletoes as a nitrogen-acquiring strategy. They presented evidence to suggest that water use by the parasite was most profligate on hosts with a poor ability to supply the mistletoe with an enriched nitrogen source, and *vice versa*. However, evidence linking water-use efficiency to nitrogen acquisition has yet to be shown to be a general phenomenon in angiosperm parasites (Panvini & Eickmeier 1993; Seel *et al.* 1993).

## Interactions with multiple hosts

Although the exploitation of multiple hosts by some species is well documented, there is a paucity of studies that demonstrate the functional significance of the phenomenom. Intuitively, attachment to multiple hosts should provide access to a greater diversity of both nutritional resources and secondary metabolites. It is well known that parasitic angiosperms can accumulate high concentrations of unusual compounds synthesized by the host, for example, the accumulation of carbinolamide quinolizidine alkaloids in *Castilleja sulphurea* from the host plant *Lupinus argeneus* (Arslanian *et al.* 1990). This phenomenon may have important implications for multispecies interactions (see below).

Although Matthies (1996) failed to demonstrate benefits (measured as biomass accumulation) to the hemiparasite *Melampyrum arvense* from simultaneous attachments to different types of hosts (grasses and legumes) in a controlled environment study, such benefits may only become apparent during extreme conditions. An elegant field study of the South-west Australian hemiparasitic shrub, *Olax phyllanthi*, suggests that the ability to acquire water from different hosts is important for survivorship through the summer dry season (Pate *et al.* 1990a). Host species differ markedly in their rooting patterns, from shallow-rooted species with little lateral extension (e.g. *Acacia littorea*; *O. phyllanthi* is similarly shallow-rooted), to deep tap-rooted species with access to underground water during the summer (e.g. *Allocasuarina humilis*). During these periods, the water potentials of shallow-rooted hosts were up to 3 MPa more negative than those of the parasite, making acquisition of water from these species impossible. Conversely, the water potentials of the deep-rooted species were between 0.5 and

1.0 MPa above those of the parasite throughout the year and could thus supply the parasite with water.

## Impacts of parasitic plants on their hosts

The extent to which growth and reproductive output of host plants is compromised by parasitic angiosperms varies enormously and is not necessarily correlated with the degree of dependence of the parasite on its host. Furthermore, our understanding of the mechanisms by which parasites affect host vigour is incomplete and what we do know largely relates to parasites that are also agricultural weeds. Here, the impacts that parasitic plants have on hosts are discussed from two perspectives. First, interactions that are largely source–sink driven through competition for resources and second, those where non-source–sink processes exert an overriding effect on the host.

### Source–sink-dominated interactions

The impact of many parasitic angiosperms on host plants can be attributed largely to the redirection of host resources to the parasite and to the resulting disruption of source–sink relations within the host. The extent to which parasites compete with hosts for carbon and other nutrients will depend upon the degree of autotrophy of the parasite and the relative abilities of host and parasite sinks to attract resources, i.e. their sink strength. The competitive ability of the parasite sink is a function of the position of the haustoria with respect to host sinks, the efficiency of nutrient transfer across the haustorial interface and the relative developmental stages of host and parasite. The effect of the parasite sink on host growth and biomass is likely to be influenced by factors such as host genotype, the nutritional status of the host, the ability of host photosynthesis to meet the extra demand for carbon and prevailing environmental conditions.

Holoparasites, being completely dependent on their hosts for all resources, provide an excellent model system for examining the extent to which competition for resources determines the impact of the parasite on host growth, resource allocation and photosynthesis.

#### Orobanche–host interactions

While there are many holoparasitic plants, those from the genus *Orobanche* have received the most research attention, largely because of their importance as weeds of crops in Mediterranean regions. In a controlled environment study, tobacco infected with *Orobanche cernua* increased the net flux of carbon from shoots to roots by 77%, and 73% of this carbon was removed by the parasite almost entirely through the phloem (>99%) (Hibberd *et al.* 1999). Further, *O. cernua* relies heavily on host phloem for inorganic solutes and it exerts a large impact on the nitrogen relations of the plant; nitrate uptake is stimulated and amino acid content of xylem sap is lower.

Because *Orobanche* spp. are such efficient sinks, we can examine the extent to which host responses are driven by source–sink interactions in these associations. In the tobacco–*O. cernua* association, for example, infected plants achieved only 29% of the biomass of control plants over a 73-day period (Fig. 9.2a). If dry weights of host and parasite are combined, however, the biomass of the infected system does not differ significantly from that of uninfected tobacco plants (Fig. 9.2b), suggesting that the difference in biomass can be attributed solely to diversion of dry matter from host to parasite. This maintenance of productivity in the infected system, relative to uninfected tobacco, is achieved through a combination of physiological and morphological modifications in the host. First, the leaf area of infected

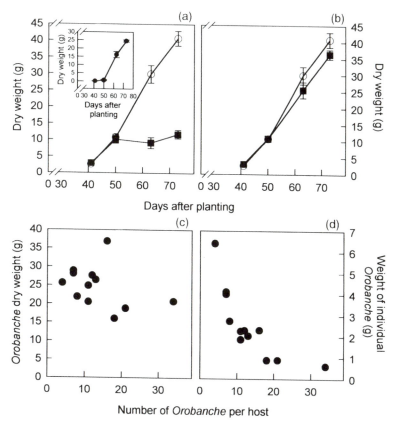

**Figure 9.2** (a) Dry weight (g) of uninfected tobacco (○) and tobacco infected with *Orobanche cernua* (■). Inset shows the dry weight (g) of *O. cernua* per host over the same period. (b) Dry weight (g) of uninfected tobacco plants (○) and the infected system (tobacco plus *O. cernua*) (■). All data are shown as means ±SE. (c) Dry weight of *O. cernua* (g per host plant) as affected by the number of *O. cernua* attached to the roots of the host. (d) Dry weight (g) of individual *O. cernua* spikelets as affected by the number of *O. cernua* attached to the roots of the host.

plants is similar to that of uninfected plants and second, leaf senescence is delayed in infected plants, resulting in a stimulation of canopy photosynthesis by approximately 20%, relative to uninfected plants (Hibberd *et al.* 1998a, 1999).

The extent to which infection with *O. cernua* can stimulate the relative productivity of tobacco is limited, however, as illustrated by the observation that the number of *O. cernua* attached to host roots has little effect on the total dry weight of parasite tissue supported (Fig. 9.2c). Thus, as the number of attachments increases, the size of each *O. cernua* spikelet decreases (Fig. 9.2d). These data suggest that the potential productivity of the host places an upper limit on biomass accumulation in the parasites, overriding parasite 'sink strength' as a determinant of host productivity (Hibberd *et al.* 1998a). This has been convincingly illustrated by using Rubisco antisense technology to modify, genetically, the productivity of tobacco plants and then infecting them with a single *O. cernua* parasite per host (J.M. Hibberd, W.P. Quick, J.D. Scholes & M.C. Press, unpublished data). By decreasing Rubisco content, rates of photosynthesis (Fig. 9.3a) and carbohydrate supply to the parasite are also lowered. In this modified system there is a linear relationship between system biomass and maximum photosynthetic rate (Fig. 9.3b) and also between parasite biomass and host photosynthetic rate (Fig. 9.3c). Thus, as the productivity of the host is decreased, both host and parasite biomass decrease proportionally, as can be seen from the relative sizes of the *O. cernua* spikelets growing on hosts with different photosynthetic capacities (Fig. 9.3d). System biomass (host plus parasite) remains the same as that of uninfected tobacco with a similar photosynthetic capacity. Thus, *O. cernua* is able to modify productivity of its tobacco host up to a limit set by the source capacity of the host. This results in the overall productivity of the infected system being comparable to that of uninfected tobacco.

Field-grown tobacco, tomato and faba beans infected with *O. ramosa* also have similar system biomass when compared with uninfected plants (ter Borg 1985). By contrast, productivity of tomato infected with *O. aegyptiaca* in a controlled environment study was sustained only when infection density was low or occurred early in the life cycle of the host. When parasite biomass increased beyond a critical level, system biomass was lower than that of uninfected control plants (Barker *et al.* 1996). Clearly, in this situation, the host plant could not compensate sufficiently (either by morphological or physiological means) to maintain the overall productivity of the system, i.e. parasite sink strength dominated the interaction. By contrast, productivity of *Trifolium repens* infected with *O. minor* is enhanced by growth at elevated $CO_2$ levels, but there is no difference in parasite biomass per host between plants grown at ambient or elevated $CO_2$ (Dale & Press 1998). Thus, in this association, sink strength of *T. repens* (and probably its *Rhizobium* symbionts) appears to be stronger than that of *O. minor*.

*Cuscuta–host interactions*
A second group of holoparasites that have attracted a significant amount of attention are the dodders. These are stem parasites belonging to the genus *Cuscuta*,

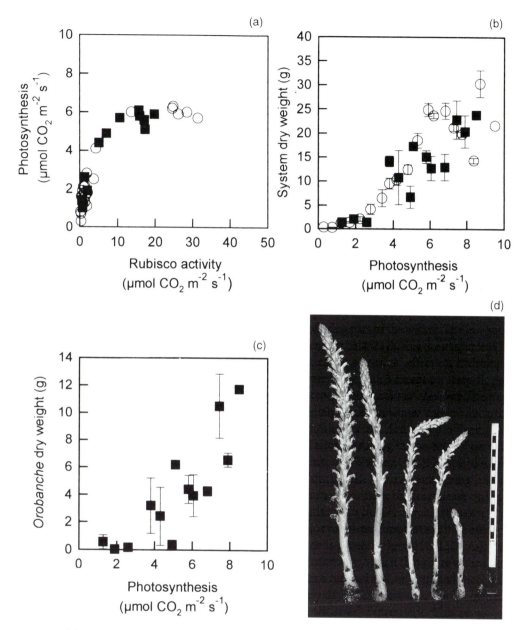

**Figure 9.3** (a) The impact of total Rubisco activity on the light-saturated rate of photosynthesis for uninfected tobacco plants (○) and plants infected with *Orobanche cernua* (■). The relationship between the light-saturated rate of photosynthesis and (b) total biomass of uninfected tobacco (○) and of the infected system (host plus *O. cernua*) (■) and (c) biomass of *O. cernua* per host. (d) *O. cernua* spikelets from tobacco plants with, from left to right, decreasing rates of photosynthesis.

among them *C. reflexa*, whose impact on host biomass, photosynthesis and source–sink relations has been studied extensively by Jeschke and co-workers (Jeschke *et al.* 1994a,b, 1997; Jeschke & Hilpert 1996, 1997). *C. reflexa* contains small amounts of chlorophyll (Hibberd *et al.* 1998b; Bungard *et al.* 1999), however, only 0.6% of its total carbon demand is supplied autotrophically (Jeschke *et al.* 1994b). *C. reflexa* parasitizes a number of species that also form associations with nitrogen-fixing bacteria of the genus *Rhizobium*. Studies with *Vicia faba* (Wolswinkel 1974), *Lupinus angustifolius* (Bäumel *et al.* 1993) and *L. albus* (Jeschke *et al.* 1994a,b) have shown that *C. reflexa* severely depresses host biomass accumulation and reproductive output. Indeed, sink demand can exceed current rates of resource acquisition by the host, thus necessitating mobilization of host reserves. For example, *C. reflexa* withdrew more than twice the amount of nitrogen being fixed by *L. albus* (Jeschke *et al.* 1994b). Infection also resulted in a sink-induced stimulation of photosynthesis in *L. albus*, but the increase was inadequate to compensate for losses of carbon to the parasite and the rhizobial symbionts.

The impact of *C. reflexa* on non-leguminous hosts appears to be more analogous to that of *O. cernua*–tobacco associations. When host productivity of *Ricinus communis* (Jeschke & Hilpert 1996, 1997) and *Coleus blumei* (Jeschke *et al.* 1997) is modified by nitrate supply, biomass of *C. reflexa* is restricted in parallel with that of the host, in a manner similar to that observed in the *O. cernua*–tobacco association. In addition, particularly at low nitrate supply, there is an increase in host nitrate uptake and host tissue N concentration, relative to uninfected plants. There is also a sink-stimulated increase in host photosynthesis, although this does not result from a delay in senescence as with *O. cernua*–tobacco, but rather from an increase in the photosynthetic rate of young leaves.

From the above examples a number of generalizations can be made about the ways in which holoparasitic angiosperms and their hosts interact. First, holoparasites are strong sinks for carbon and nitrogen and the removal of these resources exerts profound effects on host growth, allometry, reproduction and physiology. Second, biomass of both partners is controlled by a combination of sink strength and source activity, both of which can be influenced by genotype, environment and their interactions.

### Source–sink interactions involving hemiparasites

Holoparasite–host associations provide a relatively simple model for the study of parasite impacts on host growth. By contrast, it has been more difficult to resolve the processes underlying host associations with hemiparasites, where the parasite can make a significant contribution to its own carbon budget. The degree of autotrophy varies within and among species, according to developmental stage, host type and host nutritional status. For example, stable carbon isotope measurements have revealed that the proportion of autotrophic carbon, in tissue of *Striga hermonthica* infecting sorghum, increases from virtually none in non-photosynthetic, below-ground juveniles to around 60–70% in mature, emerged plants that are photosynthetically competent (Press *et al.* 1987). Furthermore, determining the extent to which some hemiparasites (e.g. mistletoes) impact on host carbon relations is confounded by the large size and indeterminate growth pattern of hosts. Such hosts may be able to respond to infection with compensatory growth, but the extent to which this can overcome loss of resources to parasites has not been quantified. Irrespective of whole-plant responses, mistletoes can cause severe, localized effects on individual branches of their woody hosts (Tennakoon & Pate 1996b).

### Interactions dominated by non-source–sink processes

Some parasites, as well as removing resources from their hosts, also appear to have other non-source–sink associated impacts. This is, perhaps, best illustrated by *Striga*-infected cereals, where parasite biomass is generally only a fraction of the observed difference between infected hosts and uninfected plants (Graves *et al.* 1989, 1990; Cechin & Press 1993). When field-grown sorghum is infected with *S. hermonthica* there is a linear relationship between host grain yield and parasite dry weight only when parasite biomass is very low ($\leq 1$ g *S. hermonthica* per plant) (Fig. 9.4). Thereafter, there is only a small decrease in grain yield for a large increase in parasite biomass. Thus, even allowing for potentially different rates of respiration between host and parasite tissue (Graves *et al.* 1992), the biomass not gained by infected plants cannot be accounted for by fluxes of resources from host to parasite.

*Host photosynthesis*
Carbon balance models for the association between *S. hermonthica* and sorghum indicate that only 20% of the predicted loss in host production over the lifetime of

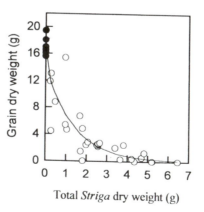

**Figure 9.4** The relationship between grain yield (g dry weight) in field-grown sorghum and total *Striga* dry weight (g) per host plant at the final harvest 100 days after planting. Plants were grown in the absence (●) or presence (○) of *S. hermonthica*. An exponential decay curve was fitted to the data. A.L. Gurney, M.C. Press & J.D. Scholes (unpublished data).

the association can be accounted for by direct loss of carbon to the parasite, with the major limitation to productivity being a parasite-induced lowering of host carbon fixation (Graves *et al.* 1989). This results from both lower investment in leaf area and lower rates of photosynthesis (Graves 1995). However, the mechanisms underlying these lower rates of photosynthesis remain elusive.

There is some evidence that lower stomatal conductances, rather than changes in photosynthetic metabolism, may be at least partially responsible. First, extractable activities of the major photosynthetic enzymes (Rubisco, phospho*enol*pyruvate carboxylase and NADP-malic enzyme), are unaltered in leaves of *Striga*-infected sorghum (Press & Cechin 1994; M. Adcock, J.D. Scholes and M.C. Press, unpublished data) and maize (Smith *et al.* 1995). In the latter study, however, an enhanced incorporation of $^{14}C$ into glycine and serine suggested that there was an increase in photorespiratory metabolism that may have contributed to the reduction in the rate of photosynthesis. Smith *et al.* (1995) suggested that this occurred as a result of 'leaky' bundle sheath cells, which tend to have thinner walls in infected plants. Second, for two cultivars of sorghum (CSH-1 and Ochuti), the relationship between the rate of photosynthesis and stomatal conductance is the same for both infected and uninfected plants, indicating that, in these cultivars, there are no biochemical limitations to photosynthesis in infected plants (Frost *et al.* 1997). Stomatal limitations arise, however, during photosynthetic induction (the increase in photosynthesis that occurs on a change from low to high light). Rates of photosynthesis and stomatal conductance increase more slowly in leaves of infected plants of both cultivars, however, final rates of photosynthesis and stomatal conductance are lower in the infected CSH-1 cultivar, relative to uninfected plants, but not in the infected Ochuti cultivar (Frost *et al.* 1997).

Changes in host photosynthesis, however, are not always correlated with changes in host biomass. The rate of steady-state photosynthesis of infected plants of the Ochuti cultivar is similar to that of control plants, but the effects of *Striga* on host biomass are as severe as those seen for the cultivar CSH-1. In the case of *Eragrostis pilosa*, a native savanna grass from east Africa, plants infected with *S. hermonthica* may accumulate only 50% of the biomass of their uninfected counterparts, despite no differences in light-saturated rates of photosynthesis (Watling & Press 1998). Similar responses in growth and photosynthesis have been recorded for *Poa alpina* infected with the facultative hemiparasite *Rhinanthus minor* (Seel & Press 1996). Furthermore, growth at elevated $CO_2$ results in higher rates of photosynthesis in both *Striga*-infected sorghum and *E. pilosa*, relative to infected plants grown at ambient $CO_2$. But in both cases elevated $CO_2$ does not alleviate the impact of the parasite on host biomass (Watling & Press 1997, 1998). Clearly, changes in leaf-level photosynthesis cannot always account for the changes in biomass accumulation brought about by infection with some hemiparasites, however, changes in canopy photosynthesis are probably a significant factor in these associations, as infected grasses invariably have significantly lower leaf areas than their uninfected counterparts (Graves *et al.* 1989; Seel & Press 1996; Frost *et al.* 1997; Watling & Press 1997).

*Changes in plant growth regulators*
As well as changes in host biomass, a number of hemiparasties also induce changes in host morphology, a good example being the distortions in branch architecture (witches' brooms) of hosts infected with some mistletoes such as *Arceuthobium*, *Viscum* and *Loranthus* (Stewart & Press 1990). Infection with *Striga* also induces changes in host biomass allocation and architecture; infected plants have higher root:shoot ratios and internode length is much shorter relative to uninfected controls. But again it is not clear what mechanisms underly these responses, although it has been suggested that changes in the balance of plant growth regulators may be responsible.

Changes in tissue or sap concentrations of abscisic acid (ABA) and cytokinins have been found for hosts infected by *Striga* (Drennan & El Hiweris 1979; Taylor *et al.* 1996; Frost *et al.* 1997), suggesting that perturbations in the balance of growth regulators may contribute to changes in host architecture. By contrast, the concentrations of four gibberellins ($GA_{53}$, $GA_{19}$, $GA_{20}$ and $GA_1$) were similar in control and *S. hermonthica*-infected sorghum plants, but the concentration of indole-3-acetic acid (IAA) was lower in infected plants (D. Frost, J. D. Scholes & M. C. Press, unpublished data). Certainly, changes in ABA concentration may contribute to the lower stomatal conductances found in some cultivars infected with *Striga*. In addition, greater concentrations of ABA in the xylem may reduce leaf expansion (e.g. Zhang & Davies 1990; Dodd & Davies 1994) and the size of individual leaves is typically reduced in sorghum plants infected with *S. hermonthica*. ABA can also enhance root:shoot ratio and reduce stem growth (Trewavas & Jones 1991). Clearly the effect of parasitic plants on the plant growth regulator status of their hosts merits further investigation. Finally, there have been attempts to explain the effects of some parasitic plants, notably *Striga* species, on host growth and physiology through the involvement of toxins (Musselman 1980; Ransom *et al.* 1996). However, to date, the movement of compounds from parasite to host has not been demonstrated nor have any putative toxins been identified.

In summary, the mechanisms by which parasitic angiosperms bring about changes in host growth, morphology, architecture and reproduction range from simple competition for resources to more subtle and elusive processes that interfere with host metabolism and, possibly, signalling systems. Changes in gene expression have been demonstrated for other symbiotic associations (e.g. plant–pathogen and plant–mycorrhizal associations); however, to date, little has been done to explore this possibility in plant–parasitic angiosperm associations. In an ecological context, a better understanding of the impacts of parasitic angiosperms on hosts in natural communities is also needed.

## Impacts of parasitic plants on community structure and function
There is a growing recognition of the impact that parasitic organisms can exert on community structure and function (Dobson & Crawley 1994) and understanding the complex nature of the underlying multi-species and multi-trophic interactions

is an important challenge for ecologists (see e.g. Gange & Brown 1997). In comparison with many other parasites, our knowledge of the impact of parasitic angiosperms on community processes is rudimentary, but some examples are illustrated below.

*Cuscuta salina* has been shown to exert marked impacts on the structure of a Californian salt marsh community (Pennings & Callaway 1996; Callaway & Pennings 1998), through two mechanisms. First, it has been shown to influence the location of the boundary between two neighbouring species, where one is a host (*Salicornia virginica*) and the other a non-host (*Arthrocnemon subterminale*). By depressing the growth of *S. virginica*, *C. salina* allowed *A. subterminale* to ingress further into the *Salicornia* zone. Second, infection by the parasite within the *Salicornia* zone stimulated both expansion and invasion of subordinate species, including *Limonium californicum*. This phenomenon occurred as a consequence of the creation of gaps by the parasite in the otherwise closed plant community. Thus, given the importance of gaps in controlling community dynamics in some ecosystems, parasitic angiosperms clearly have the potential to exert subtle but significant impacts on processes that control species abundances and hence ecosystem function.

Studies of the impact of *Rhinanthus* and related species on agricultural systems demonstrate that parasitic angiosperms can reduce productivity severely (ter Borg 1985). Davies *et al.* (1997) demonstrated that productivity of grassland swards was suppressed by between 8 and 73% in the presence of *Rhinanthus* species. Further, the degree of suppression was positively correlated with sward productivity. However, the impact of the parasites on species diversity indices are not always consistent. While Gibson and Watkinson (1992) report lower species diversity, an increase in diversity may be more typical (Matthies 1996; Davies *et al.* 1997), presumably through the alleviation of competitive relationships between host and non-host species, as in the case of the Californian salt marsh community. The precise impact of hemiparasites on species diversity may well depend on the position of preferred hosts in the dominance hierarchy of the vegetation type and the nutritional status of the community.

Both these examples illustrate that parasitic angiosperms may play an important role in determining species abundance and diversity in some ecosystems through their impact on competitive interactions. However, shifts in competitive balance may be controlled not only by the relative fitness of component species, but also through changes in soil microbial processes (Fig. 9.5). Bardgett *et al.* (1998) suggest that above-ground herbivory may influence soil microbial structure and function in two ways. The first mechanism through the effects of above-ground herbivory on rates of carbon allocation to root systems, root exudation or longer-term impacts on root system biomass and morphology. The second mechanism is through the effects of herbivores on litter quality, both with regard to nutrient content and the presence of secondary metabolites. Parasitic angiosperms may exert similar impacts on soil microbial processes in communities where they are at least locally abundant (Press 1998). For example, with root parasites infection commonly results in higher root to shoot quotients in infected plants, compared with their uninfected counterparts, with a stimulation of carbon flux below ground and higher concentrations of soluble carbohydrates in infected roots (Dale & Press 1998; Hibberd *et al.* 1999). This may lead to greater rates of carbon exudation in the rhizosphere. Further, the high nutrient content of leaves and litter of hemiparasites results in greater nutrient inputs to soil. Such comparatively nitrogen-rich litter in otherwise nutrient-poor environments is likely to result in both increased rates of decomposition and a more rapid release of nutrients from associated litter which is more resistant to breakdown (Press 1998).

Studies of tripartite interactions that involve parasitic angiosperms have lagged behind those of other symbioses, but what evidence there is suggests that interactions with a third organism may exert significant effects on angiosperm parasite–host interactions. For example, Davies and Graves (1998) showed that infection of *Lolium perenne* with arbuscular mycorrhizal fungi stimulates biomass accumulation and reproductive output of *Rhinanthus minor* plants supported by the grass. The outcome of these interactions appears to conform to the source–sink model described earlier, with the stimulation of parasite performance being attributed to additional carbon and nitrogen fluxes resulting from mycorrhizal infection. Thus, the presence of mycorrhizas may lead to a longer term increase in the abundance of the parasite, which in turn could influence species diversity and ecosystem productivity through the mechanisms described above.

Interactions at higher trophic levels may also be influenced by the presence of parasitic angiosperms. Host type may exert a strong influence on the nature of the herbivores feeding on parasitic angiosperms, both positively (through nutritional interactions) and negatively (through the transport of defence compounds). For example, both total nitrogen and secondary metabolite contents of *Castilleja wightii* were found to be strongly host dependent, which in turn impacted on the survivorship and fecundity of aphids feeding on the parasite (Marvier 1996). The consequences of such multi-trophic interactions are difficult to predict and may also be influenced by components of the abiotic environment, particularly nutrient, water and light availability. In the case of the association between *C. wightii*, *Lupinus arboreus* (the host) and *Nearctaphis kachena* (a specialist aphid that feeds on the genus *Castilleja*), aphid fecundity and survivorship were stimulated by the enriched nitrogen supply from the host (compared with that from non-nitrogen-fixing hosts) rather than being depressed by the presence of potentially toxic quinolizidine alkaloids (Marvier 1996).

These examples demonstrate that host traits can markedly influence interactions between parasites and herbivores, in a similar way to which host traits can influence interactions between herbivores and their predators and parasitoids (Marquis & Whelan 1996). Access to multiple hosts may further complicate interactions, as may the extent to which the host species themselves are subject to herbivory (Marvier 1996). Thus, in conclusion, as our understanding of the biology of parasitic angiosperms is scaled up from interactions with host organisms to interactions at the community level, the gaps in our knowledge become correspondingly wider, making generalizations about the physiological ecology of this interesting and important group of plants less secure. Future challenges include gaining a firmer understanding of the role of these organisms in both natural and agricultural communities—it may well turn out that although some species are certainly agricultural foes, others may be important ecological friends.

## Acknowledgements

We gratefully acknowledge the generosity of the following colleagues who have given us permission to use unpublished data: D.L. Nickrent, A.L. Gurney, J.M. Hibberd, W.P. Quick, A. Leakey and M. Adcock.

## References

Arslanian, R.L., Harris, G.H. & Stermitz, F.R. (1990). New quinolizidine alkaloids from *Lupinus argenteus* and its hosted root parasite *Castilleja sulphurea*. Stereochemistry and conformation of some naturally occurring cyclic carbinolamides. *Journal of Organic Chemistry*, **55**, 1204–1210.

Bardgett, R.D., Wardle, D.A. & Yeates, G.W. (1998). Linking above-ground herbivory and below-ground interactions: how plant responses to foliar herbivory influence soil organisms. *Soil Biology and Biochemistry*, **30**, 1867–1878.

Barker, E.R., Press, M.C., Scholes, J.D. & Quick, W.P. (1996). Interactions between the parasitic angiosperm *Orobanche aegyptiaca* and its tomato host: growth and biomass allocation. *New Phytologist*, **133**, 637–642.

Bäumel, P., Jeschke, W.D., Witte, L., Czygan, F.-C. & Proksch, P. (1993). Uptake and transport of quinolizidine alkaloids in *Cuscuta reflexa* parasitizing on *Lupinus angustifolius*. *Zeitschrift für Naturforschung*, **48c**, 436–443.

Boone, L.S., Fate, G., Chang, M. & Lynn, D.G. (1995). Seed germination. In *Parasitic Flowering Plants* (Ed. by M.C. Press & J.D. Graves), pp. 14–38. Chapman & Hall, London.

ter Borg, S.J. (1985). Population biology and habitat relations of some hemiparasitic Scrophulariaceae. In *The Population Structure of Vegetation* (Ed. by J. White), pp. 463–487. Dr W. Junk, Publishers, Dordrecht,.

Bungard, R.A., Ruban, A.V., Hibberd, J.M., Press, M.C., Horton, P. & Scholes, J.D. (1999). Unusual carotenoid composition and a new type of xanthophyll cycle in plants. *Proceedings of the National Academy of Sciences USA*, **96**, 1135–1139.

Callaway, R.M. & Pennings, S.C. (1998). Impact of a parasitic plant on the zonation of two salt marsh perennials. *Oecologia*, **114**, 100–105.

Cechin, I. & Press, M.C. (1993). Nitrogen relations of the sorghum–*Striga hermonthica* host–parasite association: growth and photosynthesis. *Plant, Cell and Environment*, **16**, 237–247.

Chang, M. & Lynn, D.G. (1986). The haustorium and the chemistry of host recognition in parasitic angiosperms. *Journal of Chemical Ecology*, **12**, 561–579.

Cochrane, V. & Press, M.C. (1997). Geographical distribution and aspects of the ecology of the hemiparasitic angiosperm *Striga asiatica* (L.) Kuntze: a herbarium study. *Journal of Tropical Ecology*, **13**, 371–380.

Dale, H. & Press, M.C. (1998). Elevated atmospheric $CO_2$ influences the interaction between the parasitic angiosperm *Orobanche minor* and its host *Trifolium repens*. *New Phytologist*, **140**, 65–73.

Davidson, N.J., True, K.C. & Pate, J.S. (1989). Water relations of the parasite : host relationship between the mistletoe *Amyema linophyllum* (Fenzl) Tieghem and *Casuarina obesa* Miq. *Oecologia*, **80**, 321–330.

Davies, D.M. & Graves, J.D. (1998). Interactions between arbuscular mycorrhizal fungi and the hemiparasitic angiosperm *Rhinanthus minor* during co-infection of a host. *New Phytologist*, **139**, 555–563.

Davies, D.M., Graves, J.D., Elias, C.O. & Williams, P.J. (1997). The impact of *Rhinanthus* spp. on sward productivity and composition: implications for the restoration of species-rich grasslands. *Biological Conservation*, **82**, 87–93.

Dawson, J.H., Musselman, L.J., Wolswinkel, P. & Dörr, I. (1994). Biology and control of *Cuscuta*. *Reviews of Weed Science*, **6**, 265–317.

Dobson, A. & Crawley, M. (1994). Pathogens and the structure of plant communities. *Trends in Ecology and Evolution*, **9**, 393–398.

Dodd, I.C. & Davies, W.J. (1994). Leaf growth responses to ABA are temperature dependent. *Journal of Experimental Botany*, **45**, 903–907.

Dörr, I. (1997). How *Striga* parasitizes its host: a TEM and SEM study. *Annals of Botany*, **79**, 463–472.

Douglas, A.E. (1994). *Symbiotic Interactions*. Oxford University Press, Oxford.

Drennan, D.S.H. & El Hiweris, S.O. (1979). Changes in growth regulating substances in *Sorghum vulgare* infected by *Striga hermonthica*. In *Proceedings of the Second Symposium of Parasitic Weeds* (Ed. by L.J. Musselman, A.D. Worsham & R.E. Eplee), pp. 144–155. North Carolina State University, Raleigh, USA.

Ehleringer, J.R. & Marshall, J.D. (1995). Water relations. In *Parasitic Flowering Plants* (Ed. by M.C. Press. & J.D. Graves), pp. 125–140. Chapman & Hall, London.

Estabrook, E.M. & Yoder, J.I. (1998). Plant–plant communications: rhizosphere signalling between parasitic angiosperms and their hosts. *Plant Physiology*, **116**, 1–7.

Fineran, B.A. (1985). Graniferous tracheary elements in haustoria of root parasitic angiosperms. *Botanical Review*, **51**, 389–441.

Fischer, N.H., Weidenhamer, J.D., Riopel, J.L., Quijano, L. & Menelaou, M.A. (1990). Stimulation of witchweed germination by sesquiterpene lactones: a structure–activity study. *Phytochemistry*, **29**, 2479–2483.

Frost, D.L., Gurney, A.L., Press, M.C. & Scholes, J.D. (1997). *Striga hermonthica* reduces photosynthesis in sorghum: the importance of stomatal limitations and a potential role for ABA? *Plant, Cell and Environment*, **20**, 483–492.

Gange, A.C. & Brown, V.K. (1997). *Multitrophic Interactions in Terrestrial Systems*. Blackwell Science, Oxford.

Gibson, C.C. & Watkinson, A.R. (1989). The host range and selectivity of a parasitic plant, *Rhinanthus minor* L. *Oecologia*, **78**, 401–406.

Gibson, C.C. & Watkinson, A.R. (1992). The role of the hemiparasitic annual *Rhinanthus minor* in determining grassland community structure. *Oecologia*, **89**, 62–68.

Graves, J.D. (1995). Host-plant responses to parasitism. In *Parasitic Plants* (Ed. by M.C. Press & J.D. Graves), pp. 206–225. Chapman & Hall, London.

Graves, J.D., Press, M.C. & Stewart, G.R. (1989). A carbon balance model of the sorghum–*Striga hermonthica* host–parasite association. *Plant, Cell and Environment*, **12**, 101–107.

Graves, J.D., Wylde, A., Press, M.C. & Stewart, G.R. (1990). Growth and carbon allocation in *Pennisetum typhoides* infected with the parasitic angiosperm *Striga hermonthica*. *Plant, Cell and Environment*, **13**, 367–373.

Graves, J.D., Press, M.C., Smith, S. & Stewart, G.R. (1992). The carbon canopy economy of the association between cowpea and the parasitic angiosperm *Striga gesnerioides*. *Plant, Cell and Environment*, **15**, 283–288.

Hauck, C., Müller, S. & Schildknecht, H. (1992). A germination stimulant for parasitic flowering plants from *Sorghum bicolor*, a genuine host. *Journal of Plant Physiology*, **139**, 474–478.

Heide-Jørgensen, H.S. (1989). Development and ultrastructure of the haustorium of *Viscum minimum*. I. The adhesive disk. *Canadian Journal of Botany*, **67**, 1161–1173.

Heide-Jørgensen, H.S. (1991). Anatomy and ultrastructure of the haustorium of *Cassytha pubescens* R. BR. I. The adhesive disk. *Botanical Gazette*, **152**, 321–334.

Heide-Jørgensen, H.S. & Kuijt, J. (1995). The haustorium of the root parasite *Triphysaria* (Scrophulariaceae), with special reference to xylem bridge ultrastructure. *American Journal of Botany*, **82**, 782–797.

Hibberd, J.M., Quick, W.P., Press, M.C. & Scholes, J.D. (1998a). Can source–sink relations explain responses of tobacco to infection by the root holoparasitic angiosperm *Orobanche cernua*? *Plant, Cell and Environment*, **21**, 333–340.

Hibberd, J.M., Bungard, R.A., Press, M.C., Jeschke, W.D., Scholes, J.D. & Quick, W.P. (1998b). Localisation of photosynthetic metabolism in the parasitic angiosperm *Cuscuta reflexa*. *Planta*, **205**, 506–513.

Hibberd, J.M., Quick, W.P., Press, M.C., Scholes, J.D. & Jeschke, W.D. (1999). Solute fluxes from tobacco to the parasitic angiosperm *Orobanche cernua* and the influence of infection on host carbon and nitrogen relations. *Plant, Cell and Environment*, **22**, 937–947.

Hughes, M. & Headley, A. (1996). The biology and ecology of the thistle broomrape, *Orobanche reticulata* Wallr. *Naturalist*, **121**, 3–10.

Jackson, M.B. & Parker, C. (1991). Induction of germination by a strigol analogue requires ethylene action in *Striga hermonthica* but not in *Striga forbsii*. *Journal of Plant Physiology*, **138**, 383–386.

Jeschke, W.D. & Hilpert, A. (1996). Does assimilate or nitrogen supply from the host limit growth of *Cuscuta*? Experiments with *Ricinus communis* and *Cuscuta reflexa*. In *Advances in Parasitic Weed Research* (Ed. by M.T. Moreno, J.I. Cubero, D. Berner, D. Joel, L.J. Musselman & C. Parker), pp. 373–383. Junta de Andalucía, Sevilla.

Jeschke, W.D. & Hilpert, A. (1997). Sink-stimulated photosynthesis and sink-dependent increase in nitrate uptake: nitrogen and carbon relations of the parasitic association *Cuscuta reflexa–Ricinus communis*. *Journal of Experimental Botany*, **20**, 47–56.

Jeschke, W.D., Räth, N., Bäumel, P., Czygan, F.-C. & Proksch, P. (1994a). Modelling the flow and partitioning of carbon and nitrogen in the holoparasite *Cuscuta reflexa* Roxb. and its host *Lupinus albus* L. I. Methods for estimating net flows. *Journal of Experimental Botany*, **45**, 791–800.

Jeschke, W.D., Bäumel, P., Räth, N., Czygan, F.-C. & Proksch, P. (1994b). Modelling the flow and partitioning of carbon and nitrogen in the holoparasite *Cuscuta reflexa* Roxb. and its host *Lupinus albus* L. II. Flows between host and parasite and within the parasitized host. *Journal of Experimental Botany*, **45**, 801–812.

Jeschke, W.D., Baig, A. & Hilpert, A. (1997). Sink-stimulated photosynthesis, increased transpiration and increased demand-dependent stimulation of nitrate uptake: nitrogen and carbon relations in the parasitic association

*Cuscuta reflexa–Coleus blumei*. *Journal of Experimental Botany*, **148**, 915–925.

Joel, D.M., Steffens, J.C. & Matthews, D.E. (1994). Germination of weedy root parasites. In *Seed Development and Germination* (Ed. by J. Kigel, M. Negbi, & G. Galili), pp. 567–597. Marcel Dekker, New York.

Kuijt, J. (1969). *The Biology of Parasitic Flowering Plants*. University of California Press, Berkeley.

Kuijt, J. (1977). Haustoria of phanerogamic parasites. *Annual Review of Phytopathology*, **17**, 91–118.

Logan, D.C. & Stewart, G.R. (1991). Role of ethylene in the germination of the hemiparasite *Striga hermonthica*. *Plant Physiology*, **97**, 1435–1438.

Losner-Goshen, D., Portnoy, V.H., Mayer, A.M. & Joel, D.M. (1998). Pectolytic activity by the haustorium of the parasitic plant *Orobanche* L. (Orobanchaceae) in host roots. *Annals of Botany*, **81**, 319–326.

Lynn, D.G. & Chang, M. (1990). Phenolic signals in cohabitation: implications for plant development. *Annual Review of Plant Physiology and Plant Molecular Biology*, **41**, 497–526.

Mallaburn, P.S. & Stewart, G.R. (1987). Haustorial function in *Striga*: comparative anatomy of *S. asiatica* (L.) Kuntze and *S. hermonthica* (Del.) Benth. (Scrophulariaceae). In *Parasitic Flowering Plants. Proceedings of the 4th International Symposium on Parasitic Flowering Plants* (Ed. by H.C. Weber & W. Forstreuter), pp. 523–536. Marburg, Phillips Universitat.

Marquis, R.J. & Whelan, C. (1996). Plant morphology and recruitment of the third trophic level: subtle and little recognised defences? *Oikos*, **75**, 330–334.

Marshall, J.D., Ehleringer, J.R., Schulze, E.-D. & Farquhar, G.D. (1994). Carbon isotope composition, gas exchange and heterotrophy in Australian mistletoes. *Functional Ecology*, **8**, 237–241.

Marvier, M.A. (1996). Parasitic plant–host interaction: plant performance and indirect effects on parasite-feeding herbivores. *Ecology*, **77**, 1398–1409.

Matthies, D. (1996). Interactions between the root hemiparasite *Melampyrum arvense* and mixtures of host plants: heterotrophic benefit and parasite-mediated competition. *Oikos*, **75**, 118–124.

Musselman, L.J. (1980). The biology of *Striga*,

Orobanche and other root-parasitic weeds. *Annual Review of Phytopathology*, **18**, 463–489.

Nickrent, D.L., Duff, R.J., Colwell, A.E. *et al.* (1998). Molecular phylogenetic and evolutionary studies of parasitic plants. In *Molecular Systematics of Plants II. DNA Sequencing* (Ed. by D.E. Soltis, P.S. Soltis & J.J. Doyle), pp. 211–241. Kluwer Academic, Boston.

Norton, D.A. & Carpenter, M.A. (1998). Mistletoes as parasites: host specificity and speciation. *Trends in Ecology and Evolution*, **13**, 101–105.

Panvini, A.D. & Eickmeier, W.G. (1993). Nutrient and water relations of the mistletoe *Phoradendron leucarpum* (Viscaceae): how tightly are they integrated? *American Journal of Botany*, **80**, 872–878.

Parker, C. (1991). Protection of crops against parasitic weeds. *Crop Protection*, **10**, 6–22.

Pate, J.S. (1995). Mineral relationships of parasitic plants and their hosts. In *Parasitic Flowering Plants* (Ed. by M.C. Press & J.D. Graves), pp. 80–102. Chapman & Hall, London.

Pate, J.S., Davidson, N.J., Kuo, J. & Milburn, J.A. (1990a). Water relations of the root hemiparasite *Olax phyllanthi* (Labill) R. Br. (Olacaceae) and its multiple hosts. *Oecologia*, **84**, 186–193.

Pate, J.S., Kuo, J. & Davidson, N.J. (1990b). Morphology and anatomy of the haustorium of the root hemiparasite *Olax phyllanthi* (Olacaceae), with special reference to the haustorial interface. *Annals of Botany*, **65**, 425–436.

Pate, J.S., True, K.C. & Raisins, E. (1991). Xylem transport and storage of amino acids by SW Australian mistletoes and their hosts. *Journal of Experimental Botany*, **42**, 441–451.

Pennings, S.C. & Callaway, R.M. (1996). Impact of a parasitic plant on the structure and dynamics of salt marsh vegetation. *Ecology*, **77**, 1410–1419.

Press, M.C. (1995). Carbon and nitrogen relations. In *Parasitic Plants* (Ed. by M.C. Press & J.D. Graves), pp. 103–124. Chapman & Hall, London.

Press, M.C. (1998). Dracula or Robin Hood? A functional role for root hemiparasites in nutrient poor ecosystems. *Oikos*, **82**, 609–611.

Press, M.C. & Cechin, I. (1994). Influence of nitrogen on *Striga hermonthica*–sorghum association. In *Biology and Management of Orobanche. Proceedings of the Third International Workshop on Orobanche and Related Striga Research* (Ed. by A.H. Pieterse, J.A.C. Verkleij & S.J. ter Borg),

pp. 301–311. Royal Tropical Institute, Amsterdam The Netherlands.

Press, M.C. & Whittaker, J.B. (1993). Exploitation of the xylem stream by parasitic organisms. *Philosophical Transactions of the Royal Society of London, B*, **341**, 101–111.

Press, M.C., Shah, N., Tuohy, J.M. & Stewart, G.R. (1987). Carbon isotope ratios demonstrate carbon flux from $C_4$ host to $C_3$ parasite. *Plant Physiology*, **85**, 1143–1145.

Press, M.C., Graves, J.D. & Stewart, G.R. (1988). Transpiration and carbon acquisition in root hemiparasitic angiosperms. *Journal of Experimental Botany*, **39**, 1009–1014.

Press, M.C., Parsons, A.N., Mackay, A.W., Vincent, C.A., Cochrane, V. & Seel, W.E. (1993). Gas exchange characteristics and nitrogen relations of two Mediterranean root hemiparasites: *Bartsia trixago* and *Parentucellia viscosa*. *Oecologia*, **95**, 145–151.

Ransom, J.K., Odhiambo, G.D., Eplee, R.E. & Diallo, A.O. (1996). Estimates from field studies of the phytotoxic effects of *Striga* spp. on maize. In *Advances in Parasitic Weed Research* (Ed. by M.T. Moreno, J.I. Cubero, D. Berner, D. Joel, L.J. Musselman & C. Parker), pp. 327–333. Junta de Andalucía, Sevilla.

Reiss, G.C. & Bailey, J.A. (1998). *Striga gesnerioides* parasitising cowpea: development of infection structures and mechanisms of penetration. *Annals of Botany*, **81**, 431–440.

Richter, A. & Popp, M. (1992). The physiological importance of accumulation of cyclitols in *Viscum album* (L.). *New Phytologist*, **121**, 431–438.

Riopel, J.L. & Timko, M.P. (1995). Haustorial initiation and differentiation. In *Parasitic Flowering Plants* (Ed. by M.C. Press & J.D. Graves), pp. 39–79. Chapman & Hall, London.

Schulze, E.-D. & Ehleringer, J.R. (1984). The effect of nitrogen supply on growth and water-use efficiency of xylem-tapping mistletoes. *Planta*, **162**, 268–275.

Seel, W.E. & Press, M.C. (1996). Effects of repeated parasitism by *Rhinanthus minor* on the growth and photosynthesis of perennial grass, *Poa alpina*. *New Phytologist*, **134**, 495–502.

Seel, W.E., Cooper, R.E. & Press, M.C. (1993). Growth, gas exchange and water use efficiency in the facultative hemiparasite *Rhinanthus minor* associated with hosts differing in foliar nitrogen concentration. *Physiologia Plantarum*, **89**, 64–70.

Smith, S.E. & Smith, F.A. (1990). Structure and function of the interfaces in biotrophic symbioses as they relate to nutrient transport. *New Phytologist*, **114**, 1–38.

Smith, C.E., Dudley, M.W. & Lynn, D.G. (1990). Vegetative/parasitic transition: control and plasticity in *Striga* development. *Plant Physiology*, **93**, 208–215.

Smith, L.H., Keys, A.J. & Evans, M.C.W. (1995). *Striga hermonthica* decreases photosynthesis in *Zea mays* through effects of leaf cell structure. *Journal of Experimental Botany*, **46**, 759–765.

Stewart, G.R. & Press, M.C. (1990). The physiology and biochemistry of parasitic angiosperms. *Annual Review of Plant Physiology and Plant Molecular Biology*, **41**, 127–151.

Sugimoto, Y., Wigchert, S.C.M., Thuring, J.W.J.F. & Zwanenburg, B. (1998). Synthesis of all eight stereoisomers of the germination stimulant sorghlactone. *Journal of Organic Chemistry*, **63**, 1259–1267.

Taylor, A., Martin, J. & Seel, W.E. (1996). Physiology of the parasitic association between maize and witchweed (*Striga hermonthica*): is ABA involved? *Journal of Experimental Botany*, **47**, 1057–1065.

Tennakoon, K.U. & Pate, J.S. (1996a). Heterotrophic gain of carbon from hosts by the xylem tapping root hemiparasite *Olax phyllanthi* (Olacaceae). *Oecologia*, **105**, 369–376.

Tennakoon, K.U. & Pate, J.S. (1996b). Effects of parasitism by a mistletoe on the structure and functioning of branches of its host. *Plant, Cell and Environment*, **19**, 517–528.

Tennakoon, K.U., Pate, J.S. & Stewart, G.R. (1997). Haustorium-related uptake and metabolism of host xylem solutes by the root hemiparasitic shrub *Santalum acuminatum* (R. Br.) A DC. (Santalaceae). *Annals of Botany*, **80**, 257–264.

Trewavas, A.J. & Jones, H.G. (1991). An assessment of the role of ABA in plant development. In *Abscicic Acid: Physiology and Biochemistry* (Ed. by W.J. Davies & H.G. Jones), pp. 169–188. Bios Scientific, Oxford.

Visser, J.H., Dörr, I. & Kollmann, R. (1984). The hyaline body of the root parasite *Alectra oroban-choides* Benth. (Scrophulariaceae) — its anatomy, ultrastructure and histochemistry. *Protoplasma*, **121**, 146–156.

Watling, J.R. & Press, M.C. (1997). How is the relationship between the $C_4$ cereal *Sorghum bicolor* and the root hemi-parasites *Striga hermonthica* and *Striga asiatica* affected by elevated $CO_2$? *Plant, Cell and Environment*, **20**, 1292–1300.

Watling, J.R. & Press, M.C. (1998). How does the $C_4$ grass *Eragrostis pilosa* respond to elevated carbon dioxide and infection with the parasitic angiosperm *Striga hermonthica*? *New Phytologist*, **140**, 667–675.

Wilkins, D.A. (1963). Plasticity and establishment of *Euphrasia*. *Annals of Botany*, **27**, 533–552.

Wolswinkel, P. (1974). Complete inhibition of setting and growth of fruits of *Vicia faba* L., resulting from the draining of the phloem system by *Cuscuta* species. *Acta Botanica Neerlandica*, **23**, 48–60.

Zhang, J. & Davies, W.J. (1990). Does ABA in the xylem control the rate of leaf growth in soil-dried maize and sunflower plants? *Journal of Experimental Botany*, **41**, 1125–1132.

# Chapter 10
# Herbivory

*M.J. Crawley*

## Introduction

Physiological ecologists sometimes regard herbivores and pathogens as inconvenient, because their activities are liable to muddy the otherwise clear waters of accurate physiological measurement. But real plants, living in real environments, are subject to all manner of insults from a range of plant-feeding taxa that impinge on every aspect of the plant's physiological functioning. Physiological responses of plants may be quite different when growing on their own in plant-pots in greenhouses than when they are exposed to the rigours of competition under field conditions. Thus, the impact of herbivory is also highly dependent on the environment in which the plant is situated.

The effects of herbivory on plant performance can be seen at a wide range of spatial scales. At the landscape scale, the identity of the dominant plant can be determined by selective grazing; for example, *Calluna vulgaris* is dominant on lightly grazed moorland but is replaced by *Nardus stricta* on adjacent heavily grazed sheep runs in the uplands of northern England. At the population scale, recruitment of palatable species may be prevented by herbivore feeding; rabbit grazing may preclude the regeneration of broom, *Cytisus scoparius*. At the scale of individuals, the growth of plants attacked by herbivores is often reduced relative to that of their neighbours. Thus, repeated selective browsing can keep saplings of *Quercus robur* at seedling size (<10 cm) in grasslands for several decades, preventing succession to woodland. Within plants, herbivores may show preferences for specific tissues, for example, young leaves may be preferred to old leaves, or some locations within tissues might be favoured, for example, gall-forming insects may be restricted to leaf veins of a certain diameter.

In order to attempt any generalizations about the impact of grazing on plant physiology, it is important, therefore, to distinguish different kinds of herbivores and different kinds of feeding damage. Perhaps the most fundamental distinction is between large, mobile, polyphagous herbivores (like deer, sheep or rabbits) that can have profound impacts at the level of the ecosystem and small, sedentary monophagous herbivores (like insects or fungal pathogens) that tend to have much more subtle ecosystem-level effects. It is true, of course, that insects and fungal

*Department of Biology, Imperial College of Science, Technology and Medicine, Silwood Park, Ascot, SL5 7PY, UK. E-mail: m.crawley@ic.ac.uk*

pathogens can sometimes have devastating outbreaks, but these tend to affect relatively few of the plant species within an ecosystem and to last for relatively short periods of time (Crawley 1997). Likewise, different kinds of feeding guild (e.g. chewers, miners, suckers, gall-formers) tend to have quite different effects on plant performance and the consumption of different plant tissues elicits different suites of plant responses. For example, the consumption of leaves (source tissues) tends to bring about an increase in the photosynthetic rate of surviving, non-grazed leaves, whereas the removal of young fruits (sink tissues) tends to cause a reduction in photosynthetic rate (Crawley 1983). Responses to herbivory may be both individualistic and also context specific. The composition of the community, spatial heterogeneity in resource supply rates, grazing history and the extent to which neighbouring plants are subject to herbivory can all impact on the response of a species to herbivory.

At least for perennial plants, it appears that there is a rather general overall pattern of response to herbivory. A number of studies point to the existence of a strict hierarchy for prioritizing resource allocation in grazed plants: top priority is given to survival, next to growth, and only low priority is given to reproductive investment. For example, work on the exclusion of insect herbivores from *Quercus robur* trees showed that low levels of herbivory reduced reproduction without any measurable effect on growth or survival (Crawley 1985). Again, there may be a rather general pattern to the relationship between productivity and herbivore impact (Oksanen *et al.* 1981). At extremely high productivity (e.g. bulky emergent macrophytes like *Phragmites* or *Typha* spp.), herbivory may have negligible effects because the plants have such high powers of compensation and regrowth. At extremely low productivity (e.g. in nutrient-poor deserts) plants may be so unpalatable, so long lived and so well protected that herbivore numbers are very low, and herbivore impacts per plant are relatively unimportant. At intermediate productivity, herbivores may have a range of effects (from devastating to negligible) depending upon the extent to which their numbers are depressed below food-limited equilibrium by the depredations of natural enemies (Crawley 1992).

## Herbivores and photosynthesis

Impacts of herbivores on photosynthesis may be crudely subdivided into two types, first direct removal of photosynthetic tissue and second, impacts on the rate or efficiency at which carbon dioxide is fixed, although the two processes are not necessarily mutually exclusive (Crawley 1997). Further, indirect consequences of leaf removal may arise through impacts on the microenvironment of the plant; for example, the removal of upper canopy leaves may increase irradiance for older leaves, lower in the canopy.

There are a wealth of data on the impacts of herbivores on rates of photosynthesis, stomatal conductance, and source–sink relations, showing a range of positive, negative and conditional responses. Thus, it is difficult to generalize on *the* response of such processes to herbivory. For example, phloem-feeding insects can create additional sinks for photosynthate and hence may increase rates of photosynthesis, but this is more likely to occur under sink-limited conditions. Damage to a leaf can lead to water loss, stomatal closure and hence reduced photosynthesis but may also lower the total water requirement of the plant in the medium term, a factor that could be of importance under conditions of soil moisture deficit (see e.g. Contouransel *et al.* 1996). Hormones injected by herbivores, or induced by herbivore feeding, may increase or decrease the photosynthetic rate, depending on, for example, the balance of co-occurring plant hormones, plant developmental stage and environmental conditions (see e.g. Regina & Carbonneau 1997).

The natural downward trend in net photosynthetic rate with leaf age (Crawley 1983) can be halted or even reversed by herbivory, in a process referred to a 'rejuvenation' (Dreccer *et al.* 1997). For example, high rates of herbivory by the beetle *Gastrophysa viridula* feeding on *Rumex obtusifolius* led to greater rates of photosynthesis and lower rates of dark respiration per unit leaf area, and caused increases in stomatal conductance (Pearson & Brooks 1996).

## Plant responses to defoliation

Removal of leaves by grazing herbivores has a number of immediate effects on plant ecophysiology, over and above impacts on photosynthetic rate, and these effects operate at a number of different scales. At the level of the leaf, it is important to distinguish between effects of defoliation on partially grazed and ungrazed leaves because, in general, damaged leaves perform less well following partial defoliation (Larson 1998) and undamaged leaves perform better. The precise location of feeding damage within a leaf can have important consequences. For example, a failure to consider leaf developmental patterns can result in gross overestimates of consumption by leaf-eating herbivores and severe underestimates of the effect of herbivory on leaf area display (Coleman & Leonard 1995). The way in which tissue is removed can also have a dramatic effect on photosynthetic capacity of the remaining tissue. Tip defoliation was found to enhance photosynthesis relative to undefoliated controls, whereas double edge or perforation defoliation depressed photosynthesis relative to controls (Morrison & Reekie 1995).

Some observations (Coleman *et al.* 1996) and theoretical models (Lehtila 1996) suggest that evenly distributed herbivory results in a smaller decrease in growth and reproduction than the same level of herbivory concentrated on only one part of the canopy. One possible reason for this is that plant parts are able to compensate for small amounts of local damage spread all over the plant, but not for concentrated damage of the same extent because of the resource distribution patterns in a plant. The model was proposed by Edwards and Wratten (1983), who predicted that wounding a plant would cause herbivores to take more meals of a smaller size and/or consume less foliage overall, to grow more slowly, and hence increase the period of vulnerability to parasitoids and predators. However, support for this hypothesis is equivocal (see Bergelson & Lawton 1988; Honkanen *et al.* 1994; Coleman *et al.* 1996, 1997).

As reported above, defoliation may reduce the rate of ontogenic decline in photosynthesis (Pearson & Brooks 1996) and shoot pruning can also slow down senescence (Delmolino *et al.* 1995). At the whole-plant level, defoliation is typically associated with reduced root growth, regrowth of new leaves, reduced reproductive output (Hendon & Briske 1997), increased death rates of plant parts and, presumably, reduced fitness (see Crawley 1997 for examples).

It is widely documented that differences between plant genotypes in important phenological events like the date of bud break can have profound implications for the average levels of herbivore damage they suffer (Varley & Gradwell 1968; Crawley & Akhteruzzaman 1988). It is less well known that defoliation itself can cause changes in budbreak phenology. For example, following total defoliation (and subsequent refoliation) of mountain birch (*Betula pubescens* ssp. *tortuosa*) by the spring-feeding geometrid, *Epirrita autumnata*, there was a significant delay of budbreak in the year after damage (Kaitaniemi *et al.* 1997).

## Responses to browsing

Browsing (the consumption of buds and twigs from woody perennial plants) has rather different effects to those of defoliation, and responses of woody plants to browsing by larger mammalian herbivores vary with both season and herbivore numbers. For example, shoot size increased after winter browsing by moose, but decreased after summer browsing; branching of annual shoots increased after winter browsing, but decreased after summer browsing; leaf size increased after winter browsing, but the response to summer browsing was variable (no response to increase); and winter browsing increased leaf concentrations of potassium, calcium and nitrogen, but decreased the concentration of protein-precipitating compounds, while early summer browsing decreased nitrogen concentration in leaves produced during the following summer (Danell *et al.* 1994). As a result of changes in leaf morphology and chemistry, the performance of invertebrate herbivores in one year is affected by the intensity of mammal browsing during the previous year. Further, responses may also be influenced by clipping intensity and other factors that determine plant productivity (see Danell *et al.* 1997), making generalizations difficult.

The ability to produce epicormic buds on old wood confers tremendous resilience to browsing, and many woody plants exhibit very low mortality even under extremely high, repeated herbivory (e.g. seedling-sized individuals of *Quercus robur* repeatedly browsed by rabbits may be 20 years old or more (Crawley & Long 1995)). There is great scope for future work on the interplay between herbivory and the relative abundance of auxin, cytokinins, gibberellins and abscisic acid in different tissues, especially in relation to bud-burst phenology and other aspects of plant regrowth.

## Responses to root pruning

Just as defoliation induces refoliation following shoot herbivory, so root pruning by herbivores brings about root regrowth. This is typically associated with an increase in the proportion of fixed carbon allocated to below-ground tissues (Pedersen & Hansen 1996) but it may also involve the mobilization of stored reserves. Severe root pruning causes wilting and shoot dieback, but the long-term impact of root herbivory is likely to depend upon whether or not plant growth is light limited. Many trenching experiments in forest communities elicit little or no response in tree growth, suggesting that in light-limited communities root pruning is of little consequence, presumably because competition for water and nutrients is relatively unimportant (but see Coomes & Grubb 1998). In communities where there is ample light, and plant growth is either water limited or nutrient limited, then selective root herbivory could have extremely severe consequences. The relative commonness of studies where simulated root herbivory is less damaging than simulated leaf herbivory (e.g. Houle & Simard 1996) might be taken as evidence that light competition is more prevalent than competition for water or nutrients.

In some cases where the impact of root-feeding and shoot-feeding herbivores has been compared, it has turned out that root feeding has the bigger impact on plant population dynamics. For example, the inconspicuous, root-feeding flea-beetle, *Longitarsus jacobaeae*, exerted a negative impact on the density of ragwort, *Senecio jacobaea*, because it caused a significant increase in rosette mortality (McEvoy *et al.* 1993). The more conspicuous cinnabar moth, *Tyria jacobaeae*, which had a devastating effect on shoot biomass and seed production in most years, had little effect on population density, partly because cinnabar feeding actually *reduced* plant death rate. The main cause of death of reproductive-sized ragwort plants was exhaustion of underground reserves following seed fill. Thus, when seed fill was prevented by cinnabar moth feeding, overwinter survival of the adult plants was enhanced.

There is increasing evidence that the incidence and impact of below-ground herbivory is strongly influenced by the presence of mycorrhizas (Brown & Gange 1990). The latter may mitigate the effects of herbivory at low larval densities and confer some degree of resistance to subterranean insect herbivores in some associations (Gange *et al.* 1994). Most of the available data suggest that severe above-ground herbivory reduces root colonization by vesicular-arbuscular and ectomycorrhizal fungi, but the reverse interaction has also been documented — mycorrhizal fungi deter herbivores and interact with fungal endophytes to influence herbivory (Gehring & Whitham 1994). Removal experiments demonstrated that scale insects (the pinyon needle scale, *Matsucoccus acalyptus*) negatively affected ectomycorrhizas of *Pinus edulis*, but there was no ectomycorrhizal effect on scale mortality when levels of ectomycorrhizal infection were experimentally enhanced (in contrast with studies showing that arbuscular mycorrhiza negatively affected herbivores). The resistance of *P. edulis* to scales mediated the asymmetrical interaction between fungal mutualists and scale herbivores; high-scale densities suppressed ectomycorrhizal colonization, but only on trees susceptible to scales (Gehring *et al.* 1997).

## Responses to phloem-feeding herbivores

Phloem sap is not a constant, uniform resource for sucking insects and within-plant, temporal variation in sap composition in nutrients (e.g. amino acids) and defensive compounds (e.g. sinigrin) is often high. Sectoriality of the plant's vascular system constrains the ability of sap-sucking insects to tap the entire resource base of a plant, but both the site and timing of attack mitigate the degree of limitation imposed by sectoriality. During peak periods of assimilation, photosynthate flow is mainly over short distances (between sources and sinks within the canopy), and thus sap-sucking insects have a small resource base to draw upon. By contrast, when sucking insects tap into vascular elements where the flow is from roots to leaves and *vice versa*, resource availability to the insect (and in turn, potential resource loss from the plant) are only limited by the resources present in those vascular elements (Marquis 1996).

In some cases, more damage is done to plant performance by plant diseases transmitted by phloem-feeders, than by direct removal of resources (see Crawley 1997). Nevertheless, there are important direct effects of sap feeding on plant performance and any processes which increase the rate of sap removal (e.g. attendance of aphids by ants) will tend to cause reduced seed production (Banks & Macaulay 1967).

## Responses to gall-formers

Galls are plant growths, induced by invertebrate herbivores, in which insect development occurs protected from enemies and desiccation by the structure of the gall and where the insect feeds via connection with the host phloem. Gall formation involves a modification of the developmental pattern of existing meristems in buds, or the induction of meristematic activity *de novo* in other tissues like leaves. In some cases (e.g. cynpid galls on *Quercus robur*), gall infestation is thought to have rather little impact on plant performance (Crawley & Long 1995), but in other cases, galls clearly represent a damaging sink, competing for limited resources suppressing plant sinks. For example, attack by the apical meristem galler *Antistrophus silphii* on rosinweed (*Silphium integrifolium*) reduced plant height, leaf area and inflorescence production. Under field conditions, rosinweed diverted biomass to stems, but produced no regrowth from axillary meristems, although rosinweed was much more tolerant of *Antistrophus* gall damage under competition-free conditions in a garden. Galls initially reduced plant height and leaf area, but axillary meristems grew profusely after gall formation, more than replacing leaf area initially lost to gall formation. Water- and nutrient-supplemented rosinweed were most tolerant of gall damage, experiencing little loss of total biomass or reproductive output. By contrast, field rosinweed failed to mount a tolerance-enhancing regrowth response because galls, resource availability and competition combined to constrain axillary meristem growth (Fay *et al.* 1996). This example serves to emphasize the importance of assessing herbivore impacts under competitive conditions.

In the case of *Tomoplagia rudolphi*, galling of its host plant *Vernonia polyanthes* induced a significant increase in the number of lateral shoots, but flower-head production was not affected. Seed viability, however, was lower on galled plants (Silva *et al.* 1996). Gall-formers can also have indirect effects on plant performance, as in the case of *Solidago altissima*, where breaking of underground clonal connections was enhanced by gall-makers (How *et al.* 1994).

A bud-galling midge *Rabdophaga* sp. attacks coyote willow (*Salix exigua*). Changes in bud populations for galled and ungalled shoots were followed over two seasons to measure the impact of galling on plant development. Galls were found in both apical and lateral buds and, in both instances, the gall arrested bud growth. Apical galls caused stunting of shoots and a significant reduction in the future growth and reproductive potential of galled shoots. Galled lateral buds were immediately lost from the bud population, but because lateral buds represent the beginnings of new bud lineages, bud losses were greater than just the loss of the attacked bud. For example, galled vegetative and reproductive buds represented losses from the next generation of 19–23 buds and 7–8 buds, respectively. *S. exigua* appeared to compensate for galling herbivory through the release of newly formed lateral buds close to galls within shoots. However, the majority of lateral shoots produced in response to galling had abscised by the following growing season (Declerck Floate & Price 1994).

## Compensation for herbivore damage

The rate at which plant performance varies as a function of herbivore feeding is a fundamental feature of any plant–herbivore interaction. In the simplest case, the rate of loss of plant performance is the same as the rate of herbivore feeding. This model would apply to a case where the herbivores took a fraction of plant production without having any influence on the subsequent rate of production. In real plant–herbivore interactions, it may be more reasonable to expect that the relationship between feeding and plant performance is non-linear. A concave relationship describes the case where low levels of herbivore feeding have disproportionately large effects on plant performance: this might be called the 'damage' case. By contrast, a convex relationship indicates that low levels of herbivore feeding have little or no impact on plant performance: this might be called 'compensation'. Last, and most intriguingly, it is possible that low levels of herbivore feeding might lead to increases in plant performance: this is 'overcompensation'.

The mechanisms by which plant compensation might arise are various: reduced shading of surviving leaf area; increased photosynthetic efficiency of surviving green leaf area; improved water and nutrient availability to surviving tissues;

release of downregulation following defoliation; leaf rejuvenation and delayed senescence; reallocation of photosynthate to new leaves; mobilization of stored reserves; reduced rate of floral abortion; and altered plant hormonal balance leading to the production of new shoots from dormant buds or the production of new epicormic buds. Given the constraints on these processes, it is more likely that compensation will be observed in circumstances where the resource supply rate is high, than where it is low. Most of the best examples of compensation are reported from field studies in which herbivore feeding is episodic rather than continuous, and the plant's growing season lasts longer than the herbivore's feeding period (see Crawley 1983, 1997; McNaughton 1983).

There has been considerable controversy about overcompensation. Most of the problems arising in the debate can be attributed to differences in the currencies used by different authors in measuring plant performance. Broadly, the protagonists can be divided into two camps: the agriculturalists and the evolutionists. The agriculturalists (e.g. McNaughton 1979; Dyer et al. 1991) use above-ground net production as their currency, while the evolutionists (e.g. Belsky 1986; Crawley 1987) use Darwinian fitness. There is overwhelming evidence that repeated defoliation of grasslands can lead to greater dry matter productivity; annual yield from a single biomass harvest taken at the end of the growing season is typically less than the total yield obtained from two or more cuts taken during the rapid growth phase (see DeAngelis & Huston 1993). Thus, the agriculturalists say, 'grazing is good for plants'. What they mean is that grazing has increased annual dry matter production. The waters are muddied by the fact that some of them also cite the fact that some plant species are found only in grazed grasslands and not in ungrazed communities as evidence that 'grazing is good for plants'. What they mean in this case is that grazing creates the ecological conditions in which certain grazing-tolerant species can flourish in competition with less tolerant (but otherwise more competitive) plants. The evolutionists do not dispute the increased production, but argue that overcompensation in yield was actually bought at the expense of a *decrease* in Darwinian fitness. The reason that grasses produce more dry matter under grazing is that flowering and seed production are prevented by repeated defoliation. Under mesic conditions, grazed fields are green at the end of the summer, while ungrazed fields are straw coloured from the grasses having gone to seed (Crawley 1997).

The evolutionists find little or no evidence in the literature for convincing cases where Darwinian fitness is increased by herbivory (but see Lennartsson et al. 1997). Most of the carefully controlled and well-replicated experiments show good evidence for compensation but no evidence at all for overcompensation (compare Bergelson & Crawley 1992 with Paige & Whitham 1987).

## Repair or discard: herbivores and abscission

The modular structure of vascular plants means that the option is open to respond to attack by herbivores by shedding the affected parts without disabling the remaining parts. This has the added benefit with small, sessile herbivores like insects or pathogens, that the herbivores are disposed of along with the leaf, possibly reducing subsequent infection. The alternative to premature abscission is to invest in repairing the damaged tissues. Circumstances favouring longevity of leaves, such as low resource supply or xeromorphy, will tend to favour repair over disposal, because the residual tissue is likely to be of relatively high value. Where leaves are relatively cheap, as in well-watered, high-nutrient environments, repair is unlikely to be cost effective, and we should expect to find premature abscission with efficient remobilization of resources prior to abscission as the most common response to damage.

There is little or no theoretical work on the repair/discard trade-off, but field evidence suggests that the balance of cost and benefit between abscission and repair is likely to be subtle. For instance, in England, the holly leaf miner *Phytomyza ilicis* (Diptera) has no significant impact in hastening leaf-fall of its evergreen host, *Ilex aquifolium*, but in the analogous system in eastern North America, attack by holly leaf miner causes premature abscission of the leaves of *Ilex opaca* (H.V. Cornell, personal communication).

## Competitor release: population-level compensation

Population-level compensation occurs when herbivore attack on one individual allows another individual to grow more rapidly. Broadly, the immediate neighbours of individuals attacked by herbivores are likely to benefit as a result of greater light availability or higher supply rates of water or nutrients. There are many examples of this from field experiments (Bach 1994), but if plant biomass is not high (e.g. because resources are scarce) then population-level compensation is less likely to be observed.

In a study of artificial herbivory on cotton plants, Sadras (1996) found the opposite of the expected pattern: plants with neighbours suffering herbivory did *less* well than unattacked plants. The important point was the kind of damage, and removal of vegetative buds had strikingly different effects to the removal of flowers or young fruits. Removal of vegetative buds did not reduce seed cotton production per unit ground area. In uniformly damaged crops, compensation was essentially the result of profuse branching after release of apical dominance and activation of axillary buds. As expected, in non-uniformly damaged crops in which every second plant was damaged, undamaged plants grown alongside damaged neighbours accumulated more root and shoot biomass and produced more seed cotton than undamaged plants in uniform crops (Sadras 1996). By contrast, removal of flowerbuds and young fruits caused a marked increase in vegetative growth of the damaged plants. This meant that undamaged plants with damaged neighbours were at a competitive disadvantage. Undamaged plants with damaged neighbours had up to 56% less mature fruit mass than their counterparts with undamaged neighbours (Sadras 1997).

## Nitrogen and interactions between herbivores

Most herbivores are seriously nitrogen limited, because their body composition has roughly 10 times the nitrogen content of the plant foods they eat; Southwood (1973) called this the 'nutritional hurdle'. Other things being equal, increases in plant nitrogen content are likely to benefit herbivores (White 1984; Price 1991; Kyto *et al.* 1996; Spiegel & Price 1996; Cobb *et al.* 1997). Nitrogen is intimately bound up in most of the major plant physiological processes, including photosynthesis, growth and seed production. Thus, there are many within-plant correlations involving nitrogen and plant performance. The best known is the correlation between photosynthetic rate and leaf nitrogen content (Field & Mooney 1986), but equally important for herbivores is the correlation between total leaf nitrogen and the abundance of nitrogen-based defensive chemicals (Lincoln *et al.* 1982).

Herbivores can have substantial direct effects on nitrogen availability in plant communities through the deposition of dung and urine, and the spatial heterogeneity of this input can have long-term consequences for grazing behaviour. Herbivores also have indirect, medium-term effects on plant nitrogen, because the presence of abundant regrowth foliage lowers the average leaf age of a sward and hence increases the average nitrogen concentration of the pasture. Thus, grazing history, both long term and short term, can have profound effects on nitrogen availability in above-ground biomass. The result is that the long-term and short-term effects of defoliation on leaf nitrogen content are often different. For example, Milchunas *et al.* (1995) found that current-year defoliation typically had positive effects, but long-term grazing had negative effects, on forage nitrogen concentrations and digestibilities in short-grass steppe. Vertebrate grazing can also influence the subsequent performance of invertebrate herbivores. Prior grazing of *Solidago missouriensis* affected leaf nutritional quality, growth and fecundity of *Trirhabda canadensis* (Coleoptera, Chrysomelidae). Larvae feeding on leaves from grazed plots accumulated biomass and nitrogen more slowly, used them less efficiently, and reached a lower final mass than did those receiving leaves from plots without prior herbivory. Following larval defoliation of host plants, however, adult beetles under field conditions may encounter regrowth foliage that has quite different nutritional characteristics. Thus, adults feeding in plots defoliated by larvae ate younger leaves than those feeding in undefoliated plots, and this had direct positive impacts upon the growth and egg production of adult *T. canadensis*. Adults converted these young leaves into biomass with greater efficiency and preferred them to older leaves from ungrazed plots in choice tests. On balance, however, the negative indirect effects of prior grazing upon initial adult mass negated any positive influence that these more nutritious leaves had upon growth, and made the overall impact of prior grazing upon egg production negative (Brown & Weis 1995).

Most plants support several to many different herbivores (often of different feeding guilds) at any given time, and species may interact in many ways. Direct effects include diverting resource flows, altering root and shoot architecture and injecting saliva. Indirect effects include transmitting plant diseases, inducing chemical defences, and modifying plant hormonal balances. These higher-order effects have been little studied, but it is clear that they can have important effects on the interacting herbivores as well as on overall plant performance (Salt *et al.* 1996; Masters & Brown 1997). Plant-mediated interactions between herbivores are likely to be asymmetric, but in principle they could be competitive or mutualistic. Feeding by herbivore A could reduce the resource supply to herbivore B (as when a leaf-feeder reduces the flow of carbohydrate to a root-feeder, or a mycorrhizal grazing collembolan reduces the phosphorus supply to a gall-former). Alternatively, feeding by B could increase the resource flow to A (as when a root-feeder induces water stress which leads to an increase in nitrogen mobilization, increasing the food quality of above-ground leaf-feeders or phloem-suckers (Gange *et al.* 1994; White 1984)).

## Impacts of elevated $CO_2$ and UV-B radiation

There has been a great deal of interest in recent years in the interacting impacts of climate change and air pollution on plant physiological ecology and this has included the impacts of elevated concentrations of $CO_2$ and increased fluxes of ultraviolet-B radiation (UV-B) on plant–herbivore associations (see e.g. Bezemer & Jones 1998; Chapter 12). For example, UV-B-induced changes in plant chemistry (carbon, nitrogen, water-soluble phenolics, free amino acids) may render plants more or less attractive to herbivores (Ballare *et al.* 1996; Pavia *et al.* 1997; Salt *et al.* 1998). Leaves exposed to above-ambient fluxes of UV-B commonly contain increased concentrations of phenolic compounds, which may influence herbivores. Studies with the noctuid moth *Autographa gamma* feeding on *Pisum sativum* grown at a range of plant-effective UV-B fluxes showed little deleterious effect on fifth instar larvae despite increases in total phenolics. However, tissue nitrogen also increased with increasing UV-B, and this was correlated with an increase in the efficiency with which larvae utilized their food, in larval growth rate, and with a reduction in the total amount of plant material consumed. In this case, therefore, the positive and negative effects of UV-B may have cancelled each other out (Hatcher & Paul 1994).

The impacts of elevated $CO_2$ on plants and the potential consequences for herbivores have been reviewed by Long (Chapter 13) and Bezemer and Jones (1998), respectively. Responses may be highly individualistic, as illustrated by the examples below. Second-instar larvae of the red-headed pine sawfly, *Neodiprion lecontei*, feeding on foliage of seedlings of the loblolly pine, *Pinus taeda*, grown in open-top field chambers, showed 21% higher relative consumption rates on plants grown under elevated $CO_2$. Insect consumption rate was negatively related to leaf nitrogen content and positively related to the starch : nitrogen ratio. Although consumption changed, the relative growth rates of larvae were not different among $CO_2$ treatments. Despite lower nitrogen consumption rates by larvae feeding on the plants grown in elevated $CO_2$, nitrogen accumulation rates were the same for all treatments, due to a significant increase in nitrogen utilization efficiency (Williams *et al.* 1994; Traw *et al.* 1996). A single clone of the aphid *Aulacorthum solani* was reared on *Vicia faba* and tansy (*Tanacetum vulgare*) under ambient and elevated $CO_2$. While insect performance was enhanced on both hosts at elevated $CO_2$, the mechanism was different for each plant species. On bean, the daily rate of production of nymphs was increased by 16% but there was no difference in the development time. On tansy, however, the development time was 10% shorter at elevated $CO_2$ but the rate of production of nymphs was not affected (Awmack *et al.* 1997). As with UV-B, it seems that generalizations are few and far between, but it is clear that many insects show a substantial ability to compensate for reductions in food quality (e.g. lower nitrogen concentrations under $CO_2$ enrichment) by increases in intake, or increases in nitrogen use efficiency, and that this compensation ability could ameliorate the effects of $CO_2$ enrichment (Kerslake *et al.* 1998).

It has generally been assumed that increasing atmospheric $CO_2$ concentrations will increase concentrations of plant carbon-based secondary or structural compounds and that these changes may have consequences for herbivory and plant litter decomposition (Peñuelas & Estiarte 1998). However, demonstrating clear general patterns of the effects of elevated $CO_2$ on plant secondary metabolism has proved difficult (Jones & Hartley 1998). Some studies show an increase in the concentration of phenolics and tannins in plants grown under elevated $CO_2$ (e.g. Ruth & Lindroth 1994), but many have failed to do so (e.g. Fajer *et al.* 1992). Thus, species differ widely in their responses to an elevated $CO_2$ atmosphere, particularly in the extent of nitrogen dilution within leaf tissue. Variations in both the nutritive value of tissues and the extent to which tissues are defended when grown under elevated concentrations of $CO_2$ are likely to determine the outcome of interactions under an elevated $CO_2$ atmosphere.

## Defence against herbivory

Plants display an astonishing range of chemical and morphological defences against herbivores (Hartley & Jones 1997) and the literature on plant defences is vast (Hay 1997; Karban & Baldwin 1997). Two broad strategies are observed: constitutive defence (where the plant invests in defensive chemicals, whether or not it is attacked by herbivores) and induced defence (where defensive chemicals are produced in response to feeding damage) (Adler & Karban 1994; Astrom & Lundberg 1994). The chemical defences may be dose dependent ('quantitative defence') (Ayres *et al.* 1997) or more potent toxins and feeding deterrents ('qualitative defence'). Here, we are concerned with the physiological trade-offs involved in plant defence, rather than the 'arms race' involved in the evolution of plant defence and herbivore counter-adaptation (see Jermy 1984; Thompson 1994).

The fundamental notion is that defence has a cost to the plant, the so-called 'costs of resistance' (Zangerl *et al.* 1997), thus plants must resolve a variety of trade-offs in deciding how to allocate their limited resources. Central to this resolution is what Pacala and Crawley (1992) call the 'palatability–competitive ability trade-off'. Plants that grow most quickly in the absence of herbivores are expected to be those that are most likely to be consumed by herbivores, because they have invested in growth rather than defence. Another important trade-off is that between resistance and tolerance; individuals that show high levels of resistance to a given kind of herbivore attack may show very low tolerance to this kind of damage if it is inflicted on them. For example, genotypes of *Ipomoea purpurea* that exhibit relatively high levels of resistance to insects that cause damage to apical meristems exhibit relatively low tolerance to this form of damage (Fineblum & Rausher 1995). Care must be taken when considering these issues, because trade-offs grounded in genetic constraints can differ fundamentally from those conceived in terms of limiting resources. Again, with resource-based trade-offs, there is often a failure to consider the critical issues of whether the resources involved are limiting, and whether this limitation is imposed by the assimilatory capacity of the organism or by a shortage in its environment (Mole 1994).

Some interesting recent work has returned to the question, first raised by Moran and Hamilton (1980), as to whether plants could defend themselves against herbivory by reducing their quality as food (e.g. by lowering their tissue nitrogen contents). The fundamental problem with this strategy is that the herbivore's most likely response would be to eat *more* plant material (to compensate for lower nutrient concentrations) not less (as a functional defence would require). Further, because tissue nitrogen content is positively correlated with photosynthetic rate, the plant would suffer a substantial cost, even when there were no herbivores present. Game-theoretic models demonstrate that the fact that herbivores may compensate for lowered nutrient quality does not, of itself, nullify the notion of low nutrient quality as a plant defence. It is clear, however, that compensatory feeding may restrict the conditions for the evolution of such defences (Augner 1995).

One of the major problems with induced defences is how to ensure that they work sufficiently rapidly such that the entire plant is not eaten before an effective defence has been mounted (Hay 1997; Karban *et al.* 1997). There is currently a great deal of interest in the biochemical mechanisms underlying postdamage signalling. The signals that activate many of the damage-induced responses in plants are endogenously produced in response to wounding. Jasmonates, for example, activate diverse wound-induced responses in plants, including induced nicotine production in *Nicotiana sylvestris*. The exogenous addition of methyl jasmonate (MJ) in small quantities to roots of hydroponically grown plants induces *de novo* nicotine synthesis and increases whole-plant nicotine concentrations, just as wounding does. The MJ-induced changes were proportional to the quantity of MJ applied. Moreover, the effects of MJ were additive to the effects of damage (Baldwin 1996). Octadecanoids are potent cyclic plant-signalling molecules and are involved in the regulation of a multitude of physiological processes such as senescence, herbivore and pathogen defence, mechanoperception and morphogenesis (Schaller & Weiler 1997). The activation of plant defensive genes in leaves of tomato plants in response to herbivore damage or mechanical wounding is mediated by a mobile 18-amino-acid polypeptide signal called systemin, derived from a larger, 200-amino-acid precursor called prosystemin, which is similar to polypeptide hormones and soluble growth factors in animals. Systemin activates a lipid-based signalling cascade, analogous to signalling systems found in animals (Bergey *et al.* 1996).

## Plant signals to predators and parasitoids

Natural enemies of herbivores are likely to be the friends of plants. Thus, any plant physiological response that increases the risk of predation to herbivores is likely to be favoured by natural selection. Tannins, toughness and low nutritional quality lengthen insect developmental times, making them more vulnerable to predators and parasitoids (Zareh *et al.* 1980). The widespread occurrence of these defences suggests that natural enemies are key participants in plant defences and may have influenced the evolution of these traits. To escape damage, leaves may expand rapidly, be flushed synchronously or, in seasonal climates, be produced during the drier periods when herbivores are less abundant. One strategy virtually limited to tropical forests is for plants to flush leaves but delay greening them until the leaves are mature (Coley & Barone 1996), although the functional significance of delayed greening is still the subject of some debate.

An alternative approach is for the plant to attempt to increase the numbers of natural enemies in attendance. Plants attacked by herbivores may release elevated levels of volatiles, which can serve as chemical signals that attract natural enemies to the damaged plant. Oral secretions from feeding herbivores may provide the initial chemical signal that triggers the release of plant volatiles, and these elicitors from oral secretions may allow the plant to differentiate herbivore feeding from mere mechanical wounding. Elicitors, in combination with mechanical wounding, trigger the release of compounds both locally and systemically. These volatiles, which may be a blend of constitutive and induced compounds, vary in their relative and absolute concentration over time (e.g. the signals are timed so that they are mainly released during the daytime, when natural enemies tend to forage (Turlings *et al.* 1995)). They serve as easily detectable and distinctive chemical cues for predators and parasitoids of the herbivores feeding on the plants (Pare & Tumlinson 1996, 1997). Studies of volatiles in corn and cotton plants indicate that the clarity of the volatile signals is high (they are unique for herbivore damage), they are produced in relatively large amounts and they are easily distinguishable from background odours. Specificity is limited when different herbivores feed on the same plant species, but is high as far as odours emitted by different plant species and genotypes are concerned. For example, cabbage plants respond to caterpillar (*Pieris brassicae*) herbivory by releasing a mixture of volatiles that makes them highly attractive to parasitic wasps (*Cotesia glomerata*) that attack the herbivores. Cabbage leaves that are artificially damaged and subsequently treated with gut regurgitant of *P. brassicae* caterpillars release a volatile blend similar to that of herbivore-damaged plants. Mattiacci *et al.* (1995) demonstrated the presence of β-glucosidase in *P. brassicae* regurgitant and showed that leaves treated with commercial β-glucosidase (from almonds) release a volatile blend similar to that of leaves treated with *P. brassicae* regurgitant. In a flight bioassay, leaves treated with almond β-glucosidase were highly attractive to the *C. glomerata* parasitoids, and the wasps did not discriminate between cabbage leaves treated with almond β-glucosidase and leaves treated with larval regurgitant.

## Conclusion

Virtually every aspect of plant metabolism may be influenced directly or indirectly by herbivore feeding and this can have profound effects on the performance of the species and the ways in which it interacts with both congenerics and other species. The problem, as ever, is that it is difficult or impossible to scale up from processes investigated at a small scale (e.g. cell or leaf) and make predictions at higher scales (e.g. whole plant, the plant population and beyond). The ubiquity of thresholds, non-linear relationships, interactions and compensation are not unique to plant–herbivore interactions, but they do make life difficult for those who wish to explain patterns that are manifest at large scales in terms of processes that operate at small scales. Thus, understanding the ways in which herbivores impact on plants and our ability to scale up such responses and understand their impacts at higher levels of organization are important challenges for physiological ecologists and modellers alike.

## References

Adler, F.R. & Karban, R. (1994). Defended fortresses or moving targets—another model of inducible defences inspired by military metaphors. *American Naturalist*, **144**, 813–832.

Astrom, M. & Lundberg, P. (1994). Plant defence and stochastic risk of herbivory. *Evolutionary Ecology*, **8**, 288–298.

Augner, M. (1995). Low nutritive quality as a plant defence—effects of herbivore-mediated interactions. *Evolutionary Ecology*, **9**, 605–616.

Awmack, C.S., Harrington, R. & Leather, S.R. (1997). Host plant effects on the performance of the aphid *Aulacorthum solani* (Kalt.) (Homoptera: Aphididae) at ambient and elevated $CO_2$. *Global Change Biology*, **3**, 545–549.

Ayres, M.P., Clausen, T.P., MacLean, S.F., Redman, A.M. & Reichardt, P.B. (1997). Diversity of structure and antiherbivore activity in condensed tannins. *Ecology*, **78**, 1696–1712.

Bach, C.E. (1994). Effects of a specialist herbivore (*Altica subplicata*) on *Salix cordata* and sand dune succession. *Ecological Monographs*, **64**, 423–445.

Baldwin, I.T. (1996). Methyl jasmonate-induced nicotine production in *Nicotiana attenuata*—inducing defences in the field without wounding. *Entomologia Experimentalis et Applicata*, **80**, 213–220.

Ballare, C.L., Scopel, A.L., Stapleton, A.E. & Yanovsky, M.J. (1996). Solar ultraviolet-B radiation affects seedling emergence, DNA integrity, plant morphology, growth-rate, and attractiveness to herbivore insects in *Datura ferox*. *Plant Physiology*, **112**, 161–170.

Banks, C.J. & Macaulay, E.D.M. (1967). Effects of *Aphis fabae* and its attendant ants and insect predators on yields of field beans (*Vicia faba* L.). *Annals of Applied Biology*, **60**, 445–453.

Belsky, A.J. (1986). Does herbivory benefit plants? A review of the evidence. *American Naturalist*, **127**, 870–892.

Bergelson, J.M. & Crawley, M.J. (1992). Herbivory and *Ipomopsis aggregata*—the disadvantages of being eaten. *American Naturalist*, **139**, 870–882.

Bergelson, J.M. & Lawton, J.H. (1988). Does foliage damage influence predation on insect herbivores of birch? *Ecology*, **69**, 434–445.

Bergey, D.R., Hoi, G.A. & Ryan, C.A. (1996). Polypeptide signaling for plant defensive genes exhibits analogies to defence signaling in animals. *Proceedings of the National Academy of Sciences USA*, **93**, 12053–12058.

Bezemer, T.M. & Jones, T.H. (1998). Plant–insect herbivore interactions in elevated atmospheric $CO_2$: quantitative analysis and guild effects. *Oikos*, **82**, 212–222.

Brown, V.K. & Gange, A.C. (1990). Insect herbivory below ground. *Advances in Ecological Research*, **20**, 1–58.

Brown, D.G. & Weis, A.E. (1995). Direct and indirect effects of prior grazing of goldenrod upon the performance of a leaf beetle. *Ecology*, **76**, 426–436.

Cobb, N.S., Mopper, S., Gehring, C.A., Caouette, M., Christensen, K.M. & Whitham, T.G. (1997). Increased moth herbivory associated with environmental stress of pinyon pine at local and regional levels. *Oecologia*, **109**, 389–397.

Coleman, J.S. & Leonard, A.S. (1995). Why it matters where on a leaf a folivore feeds. *Oecologia*, **101**, 324–328.

Coleman, R.A., Barker, A.M. & Fenner, M. (1996). Cabbage (*Brassica oleracea* var *capitata*) fails to show wound-induced defence against a specialist and a generalist herbivore. *Oecologia*, **108**, 105–112.

Coleman, R.A., Barker, A.M. & Fenner, M. (1997). A test of possible indirect mediation of wound-induced resistance in cabbage against *Pieris brassicae*. *Oikos*, **80**, 43–50.

Coley, P.D. & Barone, J.A. (1996). Herbivory and plant defences in tropical forests. *Annual Review of Ecology and Systematics*, **27**, 305–335.

Contouransel, D., Ilami, G., Ouarzane, A. & Louguet, P. (1996). Effect of water-stress on pyruvate, P-I dikinase and phosphoenol pyruvate-carboxylase activities in the leaves of 2 cultivars of sorghum (*Sorghum bicolor* L). *Journal of Agronomy and Crop Science–Zeitschrift Für Acker und Pflanzenbau*, **176**, 59–69.

Coomes, D.A. & Grubb, P.J. (1998). Responses of juvenile trees to above- and belowground competition in nutrient-starved Amazonian rain forest. *Ecology*, **79**, 768–782.

Crawley, M.J. (1983). *Herbivory: the Dynamics of Animal–Plant Interactions*. Blackwell Scientific Publications, Oxford.

Crawley, M.J. (1985). Reduction of oak fecundity by low-density herbivore populations. *Nature*, **314**, 163–164.

Crawley, M.J. (1987). Benevolent herbivores. *Trends in Ecology and Evolution*, **2**, 167–168.

Crawley, M.J. (1992). Population dynamics. In *Natural Enemies: The Population Biology of Predators, Parasites and Diseases* (Ed. by M.J. Crawley), pp. 40–89. Blackwell Scientific Publications, Oxford.

Crawley, M.J. (1997). Plant–herbivore dynamics. In *Plant Ecology* (Ed. by M.J. Crawley), pp. 401–474. Blackwell Science, Oxford.

Crawley, M.J. & Akhteruzzaman, M. (1988). Individual variation in the phenology of oak trees and its consequences for herbivorous insects. *Functional Ecology*, **2**, 409–415.

Crawley, M.J. & Long, C.R. (1995). Alternate bearing, predator satiation and seedling recruitment in *Quercus robur* L. *Journal of Ecology*, **83**, 683–696.

Danell, K., Bergstrom, R. & Iedenius, L. (1994). Effects of large mammalian browsers on architecture, biomass, and nutrients of woody plants. *Journal of Mammalogy*, **75**, 833–844.

Danell, K., Haukioja, E. & HussDanell, K. (1997). Morphological and chemical responses of mountain birch leaves and shoots to winter browsing along a gradient of plant productivity. *Ecoscience*, **4**, 296–303.

DeAngelis, D.L. & Huston, M.A. (1993). Further considerations on the debate over herbivore optimization theory. *Ecological Applications*, **3**, 30–31.

Declerck Floate, R. & Price, P.W. (1994). Impact of a bud-galling midge on bud populations of *Salix exigua*. *Oikos*, **70**, 253–260.

Delmolino, I.M.M., Martinezcarrasco, R., Perez, P., Hernandez, L., Morcuende, R. & Delapuente, L.S. (1995). Influence of nitrogen supply and sink strength on changes in leaf nitrogen compounds during senescence in 2 wheat cultivars. *Physiologia Plantarum*, **95**, 51–58.

Dreccer, M.F., Grashoff, C. & Rabbinge, R. (1997). Source–sink ratio in barley (*Hordeum vulgare* L) during grain filling: effects on senescence and grain protein concentration. *Field Crops Research*, **49**, 269–277.

Dyer, M.I., Turner, C.L. & Seastedt, T.R. (1991). Mowing and fertilization effects on productivity and spectral reflectance in *Bromus inermis* plots. *Ecological Applications*, **1**, 443–452.

Edwards, P.J. & Wratten, S.D. (1983). Wound-induced defences in plants and their consequences for patterns of insect grazing. *Oecologia*, **59**, 88–93.

Fajer, E.D., Bowers, M.D. & Bazzaz, F.A. (1992). The effects of nutrients and enriched $CO_2$ environments on production of carbon-based allelochemicals in *Plantago*: a test of the carbon/nutrient balance hypothesis. *American Naturalist*, **140**, 707–723.

Fay, P.A., Hartnett, D.C. & Knapp, A.K. (1996). Plant tolerance of gall-insect attack and gall-insect performance. *Ecology*, **77**, 521–534.

Field, C. & Mooney, H.A. (1986). The photosynthesis–nitrogen relationship in wild plants. In *On the Economy of Plant Form and Function* (Ed. by T.J. Givnish), pp. 25–55. Cambridge University Press, Cambridge.

Fineblum, W.L. & Rausher, M.D. (1995). Tradeoff between resistance and tolerance to herbivore damage in a morning glory. *Nature*, **377**, 517–520.

Gange, A.C., Brown, V.K. & Sinclair, G.S. (1994). Reduction of black vine weevil larval growth by vesicular-arbuscular mycorrhizal infection. *Entomologia Experimentalis et Applicata*, **70**, 115–119.

Gehring, C.A. & Whitham, T.G. (1994). Interactions between aboveground herbivores and the mycorrhizal mutualists of plants. *Trends in Ecology and Evolution*, **9**, 251–255.

Gehring, C.A., Cobb, N.S. & Whitman, T.G. (1997). Three-way interactions among ectomycorrhizal mutualists, scale insects, and resistant and susceptible pinyon pines. *American Naturalist*, **149**, 824–841.

Hartley, S.E. & Jones, C.G. (1997). Plant chemistry and herbivory, or why the world is green. In *Plant Ecology* (Ed. by M.J. Crawley), pp. 284–324. Blackwell Science, Oxford.

Hatcher, P.E. & Paul, N.D. (1994). The effect of elevated UV-B radiation on herbivory of pea by *Autographa gamma*. *Entomologia Experimentalis et Applicata*, **71**, 227–233.

Hay, M.E. (1997). The ecology and evolution of seaweed–herbivore interactions on coral reefs. *Coral Reefs*, **16**, S67–S76.

Hendon, B.C. & Briske, D.D. (1997). Demographic evaluation of a herbivory-sensitive perennial bunchgrass: does it possess an Achilles heel? *Oikos*, **80**, 8–17.

Honkanen, T., Haukioja, E. & Suomela, J. (1994). Effects of simulated defoliation and debudding on needle and shoot growth in Scots pine (*Pinus sylvestris*)—implications of plant source–sink relationships for plant-herbivore studies. *Functional Ecology*, **8**, 631–639.

Houle, G. & Simard, G. (1996). Additive effects of genotype, nutrient availability and type of tissue-damage on the compensatory response of *Salix planifolia* ssp. *planifolia* to simulated herbivory. *Oecologia*, **107**, 373–378.

How, S.T., Abrahamson, W.G. & Zivitz, M.J. (1994). Disintegration of clonal connections in *Solidago altissima* (Compositae). *Bulletin of the Torrey Botanical Club*, **121**, 338–344.

Jermy, T. (1984). Evolution of insect/host plant interactions. *American Naturalist*, **124**, 609–630.

Jones, C.G. & Hartley, S.E. (1998). Global change and plant phenolic concentrations: species level predictions using the protein competition model. In *Responses of Plant Metabolism to Air Pollution and Global Change* (Ed. by L.J. De Kok & I. Stulen), pp. 23–50. Backhuys Publishers, Leiden, The Netherlands.

Kaitaniemi, P., Ruohomaki, K. & Haukioja, E. (1997). Consequences of defoliation on phenological interaction between *Epirrita autumnata* and its host plant, Mountain Birch. *Functional Ecology*, **11**, 199–208.

Karban, R. & Baldwin, I.T. (1997). *Induced Responses to Herbivory*. University of Chicago Press, Chicago.

Karban, R., Agrawal, A.A. & Mangel, M. (1997). The benefits of induced defences against herbivores. *Ecology*, **78**, 1351–1355.

Kerslake, J.E., Woodin, S.J. & Hartley, S.E. (1998). Effects of carbon dioxide and nitrogen enrichment on a plant–insect interaction: the quality of *Calluna vulgaris* as a host for *Operophtera brumata*. *New Phytologist*, **140**, 43–53.

Kyto, M., Niemela, P. & Larsson, S. (1996). Insects on trees—population and individual response to fertilization. *Oikos*, **75**, 148–159.

Larson, K.C. (1998). The impact of two gall-forming arthropods on the photosynthetic rates of their hosts. *Oecologia*, **115**, 161–166.

Lehtila, K. (1996). Optimal distribution of herbivory and localized compensatory responses within a plant. *Vegetatio*, **127**, 99–109.

Lennartsson, T., Tuomi, J. & Nilsson, P. (1997). Evidence for an evolutionary history of overcompensation in the grassland biennial *Gentianella campestris* (Gentianaceae). *American Naturalist*, **149**, 1147–1155.

Lincoln, D.E., Newton, D.S., Ehrlich, P.R. & Williams, K.S. (1982). Coevolution of checkerspot butterfly *Euphydryas chalcedona* and its larval food plant *Diplacus aurantiacus*: larvae respond to protein and leaf resin. *Oecologia*, **52**, 216–223.

McEvoy, P.B., Rudd, N.T., Cox, C.S. & Huso, M. (1993). Disturbance, competition, and herbivory effects on ragwort *Senecio jacobaea* populations. *Ecological Monographs*, **63**, 55–75.

McNaughton, S.J. (1979). Grazing as an optimization process: grass–ungulate relationships in the Serengeti. *American Naturalist*, **113**, 691–703.

McNaughton, S.J. (1983). Compensatory plant growth as a response to herbivory. *Oikos*, **40**, 329–336.

Marquis, R.J. (1996). Plant architecture, sectoriality and plant tolerance to herbivores. *Vegetatio*, **127**, 85–97.

Masters, G.J. & Brown, V.K.B. (1997). Host–plant mediated interactions between spatially separated herbivores: effects on community structure. In *Multitrophic Interactions in Terrestrial Systems* (Ed. by A.C. Gange & V.K. Brown), pp. 217–237. Blackwell Science, Oxford.

Mattiacci, L., Dicke, M. & Posthumus, M.A. (1995). Beta-glucosidase — an elicitor of herbivore-induced plant odor that attracts host-searching parasitic wasps. *Proceedings of the National Academy of Sciences USA*, **92**, 2036–2040.

Milchunas, D.G., Varnamkhasti, A.S., Lauenroth, W.K. & Goetz, H. (1995). Forage quality in relation to long-term grazing history, current-year defoliation, and water-resource. *Oecologia*, **101**, 366–374.

Mole, S. (1994). Trade-offs and constraints in plant–herbivore defence theory — a life-history perspective. *Oikos*, **71**, 3–12.

Moran, N. & Hamilton, W.D. (1980). Low nutritive quality as defence against herbivores. *Journal of Theoretical Biology*, **86**, 247–254.

Morrison, K.D. & Reekie, E.G. (1995). Pattern of defoliation and its effect on photosynthetic capacity in *Oenothera biennis*. *Journal of Ecology*, **83**, 759–767.

Oksanen, L., Fretwell, S.D., Arruda, J. & Niemela, P. (1981). Exploitation ecosystems in gradients of primary productivity. *American Naturalist*, **118**, 240–261.

Pacala, S.W. & Crawley, M.J. (1992). Herbivores and plant diversity. *American Naturalist*, **140**, 243–260.

Paige, K.N. & Whitham, T.G. (1987). Overcompensation in response to mammalian herbivory: the advantage of being eaten. *American Naturalist*, **129**, 419–428.

Pare, P.W. & Tumlinson, J.H. (1996). Plant volatile signals in response to herbivore feeding. *Florida Entomologist*, **79**, 93–103.

Pare, P.W. & Tumlinson, J.H. (1997). *De novo* biosynthesis of volatiles induced by insect herbivory in cotton plants. *Plant Physiology*, **114**, 1161–1167.

Pavia, H., Cervin, G., Lindgren, A. & Aberg, P. (1997). Effects of UV-B radiation and simulated herbivory on phlorotannins in the brown alga *Ascophyllum nodosum*. *Marine Ecology-Progress Series*, **157**, 139–146.

Pearson, M. & Brooks, G.L. (1996). The effect of elevated $CO_2$ and grazing by *Gastrophysa viridula* on the physiology and regrowth of *Rumex obtusifolius*. *New Phytologist*, **133**, 605–616.

Pedersen, B.A. & Hansen, P. (1996). Source–sink relations in fruits 9. Effects of root treatments on growth and fruit-development in apple-trees. *Gartenbauwissenschaft*, **61**, 160–164.

Peñuelas, J. & Estiarte, M. (1998). Can elevated $CO_2$ affect secondary metabolism and ecosystem function? *Trends in Ecology and Evolution*, **13**, 20–24.

Price, P.W. (1991). The plant vigour hypothesis and herbivore attack. *Oikos*, **62**, 244–251.

Regina, M.D. & Carbonneau, A. (1997). Gas exchanges in *Vitis vinifera* under water stress regime 3. Abscisic acid and varietal behavior. *Pesquisa Agropecuaria Brasileira*, **32**, 579–584.

Ruth, S.K. & Lindroth, R.L. (1994). Effects of $CO_2$-mediated changes in paper birch and white pine chemistry on gypsy moth performance. *Oecologia*, **98**, 133–138.

Sadras, V.O. (1996). Population-level compensation after loss of vegetative buds — interactions among damaged and undamaged cotton neighbors. *Oecologia*, **106**, 417–423.

Sadras, V.O. (1997). Interference among cotton neighbours after differential reproductive damage. *Oecologia*, **109**, 427–432.

Salt, D.T., Fenwick, P. & Whittaker, J.B. (1996). Interspecific herbivore interactions in a high $CO_2$ environment — root and shoot aphids feeding on *Cardamine*. *Oikos*, **77**, 326–330.

Salt, D.T., Moody, S.A., Whittaker, J.B. & Paul, N.D. (1998). Effects of enhanced UVB on populations of the phloem feeding insect *Strophingia ericae* (Homoptera: Psylloidea) on heather (*Calluna vulgaris*). *Global Change Biology*, **4**, 91–96.

Schaller, F. & Weiler, E.W. (1997). Enzymes of octadecanoid biosynthesis in plants — 12-oxophytodienoate-10,11-reductase. *European Journal of Biochemistry*, **245**, 294–299.

Silva, I.M., Andrade, G.I., Fernandes, G.W. & Lemos, J.P. (1996). Parasitic relationships between a gall-forming insect *Tomoplagia rudolphi* (Diptera, Tephritidae) and its host-plant (*Vernonia polyanthes*, Asteraceae). *Annals of Botany*, **78**, 45–48.

Southwood, T.R.E. (1973). The insect–plant relationship — an evolutionary perspective. In *Insect Plant Relationships* (Ed. by H.F. van Emden), pp. 3–30. John Wiley, London.

Spiegel, L.H. & Price, P.W. (1996). Plant aging and the distribution of *Rhyacionia neomexicana* (Lepidoptera, Tortricidae). *Environmental Entomology*, **25**, 359–365.

Thompson, J.N. (1994). *The Coevolutionary Process*. University of Chicago Press, Chicago.

Traw, M.B., Lindroth, R.L. & Bazzaz, F.A. (1996). Decline in gypsy-moth (*Lymantria dispar*) performance in an elevated $CO_2$ atmosphere depends upon host-plant species. *Oecologia*, **108**, 113–120.

Turlings, T.C.J., Loughrin, J.H., McCall, P.J., Rose, U.S.R., Lewis, W.J. & Tumlinson, J.H. (1995). How caterpillar-damaged plants protect themselves by attracting parasitic wasps. *Proceedings of the National Academy of Sciences USA*, **92**, 4169–4174.

Varley, G.C. & Gradwell, G.R. (1968). Population models for the winter moth. In *Insect Abundance* (Ed. by T.R.E. Southwood), pp. 132–142. Blackwell Scientific Publications, Oxford.

White, T.C.R. (1984). The abundance of insect herbivores in relation to the availability of nitrogen in stressed food plants. *Oecologia*, **63**, 90–105.

Williams, R.S., Lincoln, D.E. & Thomas, R.B. (1994). Loblolly-pine grown under elevated $CO_2$ affects early instar pine sawfly performance. *Oecologia*, **98**, 64–71.

Zangerl, A.R., Arntz, A.M. & Berenbaum, M.R. (1997). Physiological price of an induced chemical defence: photosynthesis, respiration, biosynthesis, and growth. *Oecologia*, **109**, 433–441.

Zareh, N., Westoby, M. & Pimentel, D. (1980). Evolution in a laboratory host–parasitoid system and its effect on population kinetics. *Canadian Entomologist*, **112**, 1049–1060.

# Chapter 11

# SO$_2$ pollution: a bygone problem or a continuing hazard?

*T.A. Mansfield*

## Introduction

Until around 20 years ago, SO$_2$ was regarded as the most important phytotoxic air pollutant in western Europe, and also in many industrialized countries elsewhere. Since the 1970s the sharp decline in its contribution to urban pollution has been accompanied by much less research by plant scientists into its direct and indirect impacts, and the subject has not been comprehensively reviewed since the important volume edited by Winner *et al.* (1985), although there have been some authoritative accounts of specific aspects (e.g. Rennenberg & Polle 1994; Wellburn 1994; Rennenberg & Herschbach 1996).

Interest in the impacts of SO$_2$ has declined because of a dramatic fall in its contribution to air pollution in many developed countries. Figure 11.1 depicts the scale of changes that occurred in London over just 20 years from the mid 1960s. Over the preceding 30 years (1935–65) there had been little change in the annual mean concentrations. In 1952 there was a serious pollution episode that became known as the 'great London smog', to which around 4000 premature deaths were attributed, and this led to legislation to control smoke emissions in the UK. Commentators have, however, hesitated to give the legislation all the credit for the huge drop in urban SO$_2$ concentrations that quickly followed, because of co-occurring socioeconomic factors. The adoption of different types of fuel, especially gas to replace coal or heavy fuel oil, was also of great importance (Laxen & Thompson 1987).

Although emissions of SO$_2$ have fallen to a similar extent in most other industrialized countries over the past three decades, in much of the developing world the air quality is becoming steadily worse, with SO$_2$ playing a substantial part. In fact, the global rise in fossil fuel consumption is causing an annual increase of about 4% in total SO$_2$ emissions, causing serious local concerns in many countries where there is rapid economic development (Yunus *et al.* 1996). In Asia, five of 22 countries reviewed recently contributed 91% of the SO$_2$ emissions, and for the region as a whole coal burning accounted for 81% of all the SO$_2$ produced (Shrestha *et al.*

*Department of Biological Sciences, Institute of Environmental and Natural Sciences, Lancaster University, Lancaster, LA1 4YQ, UK. E-mail: t.mansfield@lancaster.ac.uk*

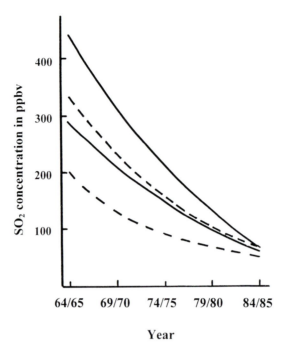

**Figure 11.1** Changes in highest daily SO$_2$ concentrations in London over a 20-year period. The curves were derived from published data, but are drawn to show only the major trends, and the paper by Laxen and Thompson (1987) should be consulted for more information on year-to-year variations. Since 1985 only minor further changes have occurred. The continuous lines apply to inner city areas, and the broken lines to outer city areas. In each case the upper curve represents concentrations in winter, and the lower line concentrations in summer. A similar fall in SO$_2$ concentrations has occurred in many cities in industrialized countries, but not in most developing countries.

1996). SO$_2$ is still regarded as the most important atmospheric sulphur compound deposited to vegetation (Rennenberg & Herschbach 1996), even though man-induced emissions of H$_2$S may be 10 times higher than those from natural sources (Wellburn 1988). In view of this, the decline in the amount of research on the phytotoxicity of SO$_2$, and on its ecological relevance, is short sighted. It has not, however, been the result of academic decisions by scientists, but has come about as research funds have been progressively withdrawn, or have been redirected to other areas judged to be of greater priority. In fact huge uncertainties still surround attempts to recommend 'critical levels' for the protection of crops and natural ecosystems (Rosenbaum *et al.* 1994).

In much of the research that has provided a basis for establishing critical levels for impacts of SO$_2$ on plants, fumigations have been with SO$_2$ alone, yet there is abundant experimental evidence that synergistic interactions with other pollutants are of great importance. It is also clear that interactions with biotic and abiotic factors need to be taken into account. On some of these issues, there is enough new information in publications over the past 10 years or so for several topics to be revisited or reassessed. Valuable new information became available after procedures for fumigating plants in the field with controlled doses of SO$_2$ were developed (McLeod & Baker 1988; Hendrey *et al.* 1992). Some long-term studies have been undertaken under realistic environmental conditions, and both primary and secondary impacts of the pollutant have been clarified.

The situation with nitrogenous air pollution is quite different from that concerning SO$_2$. Global emissions of oxidized nitrogen, principally as nitric oxide (NO) which is later converted to nitrogen dioxide (NO$_2$), have increased more than 10-fold this century, and the developed countries are those mainly responsible and are the most seriously affected (Fowler *et al.* 1998). Where socioeconomic factors have led to a decline in SO$_2$ pollution, they have generally been responsible for an increase in NO and NO$_2$, and these pollutants are indicators of affluence. In addition, ammonia production has increased alongside that of oxidized nitrogen, and developed countries with intensive agriculture are major contributors (Asman *et al.* 1998). Disturbance of the global nitrogen cycle is a subject that has been extensively covered in the literature in the past few years (Vitousek 1994; Galloway *et al.* 1995; Mansfield *et al.* 1998), and hence the main focus of this brief review will be on SO$_2$ pollution, with some reference to nitrogen compounds where they may interact with SO$_2$. This question of joint action is of considerable importance, but relatively neglected. In fact the co-occurrence of the two forms of pollution, which is the norm anyway, may lead to greater phytotoxicity than is revealed by studying them independently. Because the principal sources of SO$_2$ and NO$_2$ change with time of year, the SO$_2$:NO$_2$ ratio can differ markedly between winter and summer, but few, if any, of the long-term experiments have taken this into account. In a recent study of 30 urban sites in Poland, the SO$_2$:NO$_2$ ratio changed from approximately 3:2 in the winter to 1:3 in the summer (Krochmal & Kalina 1997).

## Dose–response characteristics

While there is little doubt that acute exposures to SO$_2$ pollution are toxic to most plants, experimental results with lower—and for most situations today, more realistic—concentrations are more difficult to interpret. Fowler *et al.* (1988) made use of a generalized dose–response curve that was first derived by Whitmore and Mansfield (1983) and supported, after analysis of several data sets, by Roberts (1984). This will be of value here because it can describe responses both to SO$_2$ alone and to SO$_2$ with NO$_2$ (Fig. 11.2). In the case of SO$_2$, lower concentrations, usually below 20 p.p.b., may often lead to growth stimulations, but the threshold between positive and negative effects of the pollutant cannot be regarded as fixed even within a single species. Much depends on environmental conditions, and because experimental procedures have varied considerably, data from different investigations often appear to be in conflict. Dose–response relationships also depend on the ways in which pollution episodes occur. In the field, concentrations

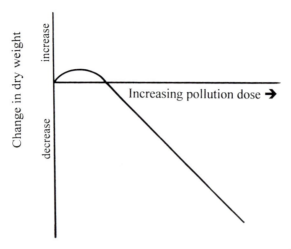

**Figure 11.2** The basic form of the dose–response curve that has been found when plants are exposed to increasing amounts of $SO_2$ or $SO_2 + NO_2$ pollution. In most experiments 'dose' has referred to $SO_2$ concentration in the atmosphere and not to uptake into the plants, and 'growth' has been determined as increases in above-ground dry mass. Because the pollutants often reduce the allocation of photosynthate to the roots, the 'positive' phase of the dose–response curve is usually diminished when the dry mass of roots together with shoots is determined.

of pollutants vary greatly over time, but under controlled conditions this variation has usually not been simulated. It is technically much more difficult to achieve than constant, or square wave, treatments, and accurate replication can become a problem. Garsed and Rutter (1984) conducted an elaborate factorial experiment over 2 years on *Pinus sylvestris*, in which the peak concentrations of $SO_2$ occurring in the field were simulated. They took advantage of the observation that the frequencies of the logarithms of daily mean concentrations approximate to a normal distribution. When this was used as a basis for 'realistic' simulations, they found that infrequent long peaks caused greater reductions in growth than frequent short peaks, even though the overall mean exposure remained the same. Experiments of this type are relevant to our understanding of the impacts of present-day $SO_2$ pollution, because irregular exposures to relatively high concentrations can still occur, even though the long-term means are much lower than previously. In the case of *Pinus contorta*, Legge et al. (1996) concluded that peak episodes of $SO_2$ are less important than the cumulative integral of exposures, i.e. concentration with respect to time. More comparative studies of this type are necessary for a range of species. It is often considered irrelevant to study the physiological and metabolic impacts of $SO_2$ at relatively high concentrations, but it is these events during peak exposures in the field which do in some cases make substantial contributions to overall effects in the long term. Mesanza et al. (1996) reported serious damage to several tree species in Spain, which was correlated with a single episode of $SO_2$ with a peak concentration of around 210 p.p.b.

## Metabolic basis of SO$_2$ tolerance

Although there can be no doubt that SO$_2$ which dissolves in water must form HSO$_3^-$ at the pH usually found in the apoplast, it is curious that HSO$_3^-$ has rarely been detected in apoplastic water. Rennenberg and Herschbach (1996) suggested that because the plasma membrane is relatively impermeable to both HSO$_3^-$ and SO$_3^{2-}$, the enzymatic conversion of SO$_3^{2-}$ to SO$_4^{2-}$ within the apoplast must be important not only as a detoxification mechanism, but also in enabling absorbed SO$_2$ to become involved in the normal sulphur metabolism of the plant. It may be this process, and the limitations within which it can operate, that determine whether there are positive responses of plants to low doses of SO$_2$. Superoxide and other radicals are generated during the aerobic oxidation of sulphite (Peiser & Yang 1985), which may be why SO$_2$ increases the activities of some antioxidant enzymes in the same manner as ozone pollution (Kubo *et al.* 1995; Okpodu *et al.* 1996; No *et al.* 1997). In the case of two wheat cultivars differing in sensitivity to SO$_2$, Ranieri *et al.* (1997) concluded that the difference in ability to maintain elevated levels of ascorbic acid was the determining factor.

Foliar sulphur content is dramatically increased during fumigations with SO$_2$ (Lorenzini *et al.* 1995), and in the field around industrial sources (Rautio *et al.* 1998). Importance has been ascribed to H$_2$S emissions in regulating the sulphur cycle in plant cells (Rennenberg 1991). Plants can take advantage of the SO$_2$ taken up for synthesizing the sulphur compounds needed for growth, and can maintain the cysteine pool at an appropriate level by emitting the excess sulphur as H$_2$S which is, in most circumstances, readily lost to the atmosphere (Rennenberg & Herschbach 1996). So long as SO$_2$ doses are not excessive, H$_2$S emissions appear to make a significant contribution to the SO$_2$ tolerance of some species (Filner *et al.* 1984; Takemoto *et al.* 1986; Kindermann *et al.* 1995).

The factors determining SO$_2$ tolerance are complex (Wilson & Bell 1990), but detoxification mechanisms are clearly of great importance, and we still do not completely understand the factors that affect their functioning. Jung and Winter (1992) worked with Caucasian fir (*Abies nordmanniana*) grown under nutrient-rich conditions, or with low magnesium, or low magnesium and low nitrogen supplies, and found that an SO$_2$ treatment had no effect on CO$_2$-saturated photo-synthetic O$_2$ evolution in the nutrient-rich plants, whereas there was a marked decline in the others. They suggested that the detoxification of SO$_2$ leading to amino acid formation depends on the capacity for protein synthesis, which is lower in nutrient-deficient plants. Unfortunately they applied a very high concentration of SO$_2$ (1 p.p.m. for 5 hours), and we do not know whether similar effects of nutrient deficiency would occur under more realistic doses of the pollutant.

This and other factors which may be important in determining the positive phase shown in Fig. 11.2 clearly need to be explored in detail if we are to define the critical levels of SO$_2$ for causing damage to plants and ecosystems. Because this phase of the dose–response curve is, at least in part, a reflection of the entry of SO$_2$ into the sulphur metabolism of the plant, we need to recognize that many of the experimental studies on which our present understanding of the toxicity of SO$_2$ is based have been performed under experimental conditions that favour metabolism, e.g. temperatures of 15–25°C in the laboratory. When plants have been exposed to $SO_2$ under real or simulated winter conditions, the picture can be very different.

## Seasonal differences in responses to $SO_2$

Kropff et al. (1990) found that for the same rate of uptake of $SO_2$ photosynthesis in Vicia faba was reduced much more at 8°C than at 18°C. They showed that this was the result of slower sulphite oxidation at the lower temperature, leading to the accumulation of more toxic S(IV) compounds in the tissues. They also drew attention to the reduced rate of sulphite oxidation in low irradiance (Rothermal & Alscher 1985). The low temperatures and irradiances in winter have been clearly shown to enhance injury by $SO_2$ and $SO_2$ plus $NO_2$ (Whitmore & Mansfield 1983; Mansfield & Jones 1985). Ashenden et al. (1996) compared the responses of 41 herbaceous species to $SO_2$ (continuous treatment with 100 p.p.b., then a series of peaks of 200 and 300 p.p.b., increasing in duration and frequency over time), in order to establish whether there was any correlation between response to the pollutant and C–S–R functional types as defined by Grime (1979). They concluded that slow-growing perennials (S types—stress tolerators) were more tolerant of $SO_2$ than were fast-growing perennials (C types—competitors) or fast-growing annuals (R types—ruderals). They conducted the fumigations over 83 days beginning on 1 September, under autumn/winter conditions, and it is necessary to ask whether the distinctions would have been so clear, i.e. would the C and R types have displayed greater $SO_2$ tolerance, if the study had been conducted in spring/summer?

## Effects of $SO_2$ in winter

Evidence pointing to the importance of wintertime effects of $SO_2$ and other pollutants has a long history, beginning with observations in the field that have been supported, more recently, by experimental studies. Table 11.1 outlines some of the contributions, mainly in older literature, based on field observations, and Table 11.2 summarizes some more recent experimental evidence which in general offers support for the earlier interpretations.

Keller (1978) was one of the first to supply experimental evidence of the importance of wintertime exposures to $SO_2$. He fumigated dormant young trees of Fagus sylvatica to 50, 100 or 200 p.p.b. $SO_2$ in winter, and found significantly greater sulphur concentrations in the leaves after they expanded in the spring. He also reported that the numbers of dead terminal buds increased linearly with the dose of $SO_2$. Further studies by Keller (1981) on Picea abies exposed to 25–225 p.p.b. $SO_2$ from October to April showed clearly that late frost injury was related to the $SO_2$ dose. The needle contents of ascorbic acid were reduced after the $SO_2$ treatments, and the decline in antioxidant capacity may have been responsible for reduced frost tolerance.

**Table 11.1** Evidence from observations in the field that pollution, particularly SO$_2$, is directly involved in causing damage to trees, in winter.

| Species | Observations | References |
|---|---|---|
| *Picea abies* | When SO$_2$ was absorbed in winter, injury appeared later, in the spring. Long-term exposures to mean SO$_2$ concentrations above about 7 p.p.b. appeared to increase frost sensitivity | Materna & Kohout (1963); Materna (1974, 1984) |
| *Picea abies* | Trees near a fertilizer factory were slow to develop cellular and tissue characteristics required for cold tolerance, e.g. increased solute content | Huttunen *et al.* (1981b) |
| *Picea abies* | Ultrastructural damage was found, particularly to chloroplasts, in trees exposed to industrial pollution in winter in Finland | Soikkeli & Tuovinen (1979) |
| *Picea abies* and other species | Where foliar damage has been reported in Central Europe it can often be attributed to SO$_2$ pollution, sometimes acting in combination with other stresses such as frost and insects | Kandler & Innes (1995) |
| *Pinus sylvestris* | Pines growing in areas affected by urban and/or industrial pollution (SO$_2$, NO$_x$ and HF were present) had visible lesions in spring. This was thought to be due to progressive injury to cell membranes in winter | Havas (1971); Havas & Huttunen (1972); Huttunen (1978, 1984) |
| *Pinus sylvestris* and *Picea abies* | Reduced water economy of trees in polluted areas. Evidence that erosion of cuticular wax may have contributed is inconsistent | Huttunen *et al.* (1981a), Cape (1983); Cape *et al.* (1995) |
| *Picea rubens* | The incidence of decline in red spruce is correlated with the distribution of air pollution. Cold tolerance may be reduced by 5°C or more in the polluted regions. This species may be specially sensitive to air pollution because it grows close to its limits of frost tolerance | Sheppard *et al.* (1989); Sheppard (1994); Johnson (1992) |

Frost damage has also been found in herbaceous plants after exposure to SO$_2$ pollution. Davison and Bailey (1982) grew *Lolium perenne* for 5 weeks with 87 p.p.b. SO$_2$, and afterwards kept the plants for 3 further weeks under reduced temperature to induce frost acclimation. They were then exposed to subzero temperatures, and the prefumigated plants showed greatly reduced frost resistance, i.e. frost hardening had been largely disabled by the pollution treatment. Baker *et al.* (1982) found comparable effects in the field with winter wheat (*Triticum aestivum*). Open-air fumigation with 110–120 p.p.b. SO$_2$ led to much greater frost injury when the temperature fell to −9°C in mid-January.

Although most of the experiments during real or simulated winters have involved single pollutants, some studies have included mixtures of SO$_2$ and NO$_2$. Freer-Smith and Mansfield (1987) exposed seedlings of Sitka spruce to SO$_2$, NO$_2$

**Table 11.2** Evidence from controlled experiments that $SO_2$ and/or nitrogenous pollutants may reduce frost tolerance.

| Species | Observations | References |
|---|---|---|
| *Fagus sylvatica* | Exposure of dormant trees to $SO_2$ in winter caused damage to terminal buds | Keller (1978) |
| *Picea abies* | 2 clones were fumigated during the 'dormant' season in an open-air fumigation system with 25, 75 and 225 p.p.b. $SO_2$. Frost injury to new shoots increased, even after exposure to the lowest concentration | Keller (1981) |
| *Lolium perenne* | Plants grown for 5 weeks in 87 p.p.b. $SO_2$ showed greatly reduced capacity for frost hardening | Davison & Bailey (1982) |
| *Triticum aestivum* | Winter wheat exposed to controlled doses of $SO_2$ in the field showed enhanced frost injury | Baker *et al.* (1982) |
| *Picea sitchensis* | Prefumigation with 30 p.p.b. $SO_2$ and 30 p.p.b. $SO_2$ + 30 p.p.b. $NO_2$ led to small increases in frost injury to needles when hardened plants were cooled to −5 or −10°C. Plants exposed to 45 p.p.b. $SO_2$ showed poor survival of lateral buds at these same temperatures | Freer-Smith & Mansfield (1987) |
| *Pinus sylvestris* | 3-year-old trees were exposed to $SO_2$ and/or $NH_3$ in various combinations. Frost hardiness at −10°C was affected by both pollutants individually, and there was synergism when both were combined | Dueck *et al.* (1990) |
| *Picea rubens* | 2-year-old seedlings were treated with mists containing $NH_4^+$, $SO_4^{2-}$ and $NO_3^-$ ions. Frost hardiness was reduced, and it was concluded that uptake of $NH_4^+$ and $SO_4^{2-}$ can disturb the hardening process, although acidity *per se* had no effect | Cape *et al.* (1991) |
| Moorland dominated by *Calluna vulgaris* | Additional N in the form of ammonium nitrate increased frost tolerance over the first 4–5 years of application, but winter injury appeared thereafter | Lee & Caporn (1998) |

and $SO_2$ with $NO_2$ in controlled environments during dormancy. 45 p.p.b. $SO_2$ applied for 12 weeks did not reduce dry weight in plants that were not subjected to freezing temperatures, but those receiving cold treatments did show changes in the form of new growth after dormancy. This was mainly attributed to the death of lateral buds on $SO_2$-treated plants subjected to −10°C. Exposure to 30 p.p.b. $SO_2$ and 30 p.p.b. $SO_2$ with 30 p.p.b. $NO_2$ led to consistent, though small, increases in frost injury to the needles of plants subsequently cooled to −5 and −10°C.

## Long-term exposures to realistic concentrations in winter

Wolfenden *et al.* (1991) worked with red spruce (*Picea rubens*) in controlled environments with simulated subzero winter conditions based on daily temperature cycles recorded at an elevated site in Virginia. Additions of 20 p.p.b. $SO_2$ with

20 p.p.b. $NO_2$ were made on a daily schedule amounting to a total of 40 hours per week, i.e. 23.8% of the time. This was a pollution dose comparable with that recorded in some locations where red spruce grows naturally in the southern Appalachians. After approximately 5 months of treatment, it was found that the fumigated trees had accumulated $SO_3^{2-}$ and $NO_2^-$ ions in their extracellular fluid, but not $SO_4^{2-}$ or $NO_3^-$. There were few after-effects of the pollution treatment on subsequent growth, but a change in the timing of bud burst was considered to be of possible importance. It was suggested that the small advance in first flushing of leader buds could be critically important in lowering a tree's ability to withstand late spring frosts. The data did not identify a specific association between the ionic accumulation and this response, although some pollution-induced changes in membrane lipids were indicated. There were increases in the saturation of fatty acids in the chloroplast membranes of polluted trees, which would be expected to reduce frost tolerance, but unfortunately the authors were unable to perform frost-hardiness tests on their pre-treated material.

This experiment by Wolfenden *et al.* (1991) is one of a small number of recent attempts to work with low, relevant concentrations of $SO_2$ and $NO_2$ pollution over long periods under simulated or real winter conditions. Strand (1993, 1995) reported a very important experiment with Scots pine (*Pinus sylvestris*) in northern Sweden. A 50-year-old stand was fumigated with 10–15 p.p.b. of $SO_2$ and $NO_2$ in two successive years from June to September, using an open-air fumigation system. The pollutants were released from vertical pipes in a 60-m-diameter circle for 12 hours daily, except in rainy weather. The values in an unfumigated plot never exceeded 2 p.p.b., and consequently this provided a control which would be un-attainable in most parts of Europe with higher background levels of the two pollutants. Needles were collected in January and measurements were made of $O_2$ evolution and chlorophyll *a* fluorescence in the laboratory, and it was found that both the $CO_2$-saturated rate of $O_2$ evolution and the photochemical efficiency of photosystem II had been adversely affected by the pollution treatment. Thus these very low concentrations of $SO_2$ and $NO_2$ had long-term effects on the photosynthetic capacity of pine needles in this northern environment. The sulphur:nitrogen ratio in the needles of the fumigated trees from the same experiment was found by Wingsle and Hällgren (1993) to be significantly increased. Sheppard (1994) found a correlation between increased foliar sulphur concentrations, or higher sulphur:nitrogen ratios, and decreases in frost hardiness in red spruce treated with acidic mists, and Strand (1995) thought that a greater sensitivity to frosts in early winter in the polluted needles of Scots pine may have explained his results.

### Recovery after winter injury

The importance of time of year in determining the impact of $SO_2$ plus $NO_2$ on growth was established by studies of *Poa pratensis* raised from seed in early autumn, and exposed to 62 p.p.b. $SO_2$, 62 p.p.b. $NO_2$, or 62 p.p.b. $SO_2$ with $NO_2$ for 11 months (Whitmore & Freer-Smith 1982; Whitmore & Mansfield 1983). More-than-additive effects of the two pollutants occurred during late winter, but not during summer when growth was more rapid. The changes in the responses to the individual pollutants, as well as to the mixture, were huge. In the case of $SO_2$ with $NO_2$, dry weight was reduced by over 80% in March, but by June the same plants had recovered and had almost caught up with the unpolluted controls.

## Impacts of co-occurring $SO_2$ and $NO_2$ pollution

Although the work described above suggests that responses to $SO_2$ with $NO_2$ might be determined by the differences in growth rate between winter and summer, there is also abundant evidence of damage resulting from synergism between the two pollutants under favourable conditions of growth. Tingey et al. (1971) exposed six crop species to $SO_2$ with $NO_2$ mixtures at relatively high concentrations of 50–500 p.p.b., but for short exposures of only 4 hours. Visible injury was assessed 2 days after treatment, and five of the six species showed traces of damage in 50 p.p.b. of each gas. The greatest foliar injury occurred at 150 p.p.b. $SO_2$ with 100 p.p.b. $NO_2$, or 100 p.p.b. $SO_2$ with 100 p.p.b. $NO_2$, depending on the species, and when the concentrations were increased further, injury was, surprisingly, reduced. Soybean (*Glycine max*), with 35% injury to the three most damaged leaves, was the most sensitive of the six species, but oats (*Avena sativa*) and radish (*Raphanus sativus*) displayed 25% injury assessed on the same basis. These strange results are relevant to present-day pollution levels because simultaneous exposures to $SO_2$ and $NO_2$ are the norm, and the concentrations that were found to cause maximum damage are experienced from time to time for short periods in some situations. The authors had no explanation for the reduced severity of injury at higher concentrations of the two pollutants, but subsequent studies on other species have suggested that the curious dose–response relationships could be related to effects of the pollutants on stomatal conductance.

At low concentrations of $SO_2$, or $SO_2$ with $NO_2$, increases in leaf conductance have often been found (Majernik & Mansfield 1971; Beckerson & Hofstra 1979; Black & Black 1979; Black & Unsworth 1980; Neighbour et al. 1988; Meng et al. 1994). The occurrence of foliar injury on plants in a natural community exposed to $SO_2$ from an erupting volcano was closely related to $SO_2$-induced changes in stomatal conductance (Winner & Mooney 1980). Black and Black (1979) found that stomatal conductance in *Vicia faba* increased by more than 20% in $SO_2$ concentrations as low as 17 p.p.b., and they showed that the enhanced opening was accompanied by damage to epidermal cells neighbouring the stomata. The wider stomatal openings were thought to be attributable to reduced mechanical resistance offered to the guard cells, which themselves were more tolerant of the pollutant. These findings were supported by observations by Neighbour et al. (1988) on two species of birch, *Betula pendula* and *B. pubescens*. Leaves excised from plants that had been exposed to mixtures of $SO_2$ with $NO_2$ at concentrations from 20 to 60 p.p.b. showed increased rates of water loss, and scanning electron microscopy of frozen hydrated specimens revealed that patches of stomata were surrounded by partially collapsed epidermal cells.

In the light of this later evidence, it can be suggested that the peculiar dose–response relationships reported by Tingey *et al.* (1971) might be explained as follows. At lower concentrations of mixtures of $SO_2$ and $NO_2$, selective damage to epidermal cells increases the mechanical advantage of guard cells so that stomata open more widely. This process occurs unevenly over the leaf surface because pollutant uptake is not uniform, being dependent on local differences in stomatal aperture, i.e. the 'patchiness' that is now well defined (Mansfield *et al.* 1990). At higher concentrations of the two pollutants, partial stomatal closure occurs as an early response, reducing pollutant uptake and avoiding the kinds of damage to epidermal cells found at lower concentrations, and also providing protection for tissues within the leaf. The stomata may close because of direct responses of the guard cells, or because photosynthesis in the mesophyll is quickly affected, causing a rise in intercellular $CO_2$ concentration which then determines the drop in stomatal aperture. Darrall (1989) provided a concise summary of the reported effects of $SO_2$, and of $SO_2$ with $NO_2$, on stomatal aperture and many of the data provide support for the above interpretation. In general, it appears that increases in stomatal conductance are restricted to low concentrations of pollutants, or to very short exposures to higher concentrations. Atkinson *et al.* (1991) found for spring barley (*Hordeum vulgare*) that $SO_2$ with $NO_2$ (24–35 p.p.b. of each gas) did not cause any change in stomatal aperture, but when the responsiveness of detached leaves to abscisic acid was tested the stomatal closure in the polluted leaves was considerably slower. This effect could have major implications for the drought tolerance of plants in polluted areas.

There is one recent piece of evidence that raises the possibility of a direct effect of $SO_2$ on the guard cells which could enhance stomatal opening. Veljovic-Jovanovic *et al.* (1993) found a stimulation of zeaxanthin formation in thylakoid membranes when *Pelargonium zonale* was fumigated for just 4 minutes with 4, 6 or 8 p.p.m. $SO_2$. The change in absorbance of the whole leaf, indicating the amount of zeaxanthin, was quantitatively related to the concentration of $SO_2$. Although the $SO_2$ concentrations were high the responses were seen almost instantaneously, suggesting that the effective dose of absorbed $SO_2$ was very small. Zeiger and Zhu (1998) have shown that the ability of stomata to respond to blue light (i.e. to the wavelengths primarily responsible for light-induced opening) is dependent on the zeaxanthin content of the guard cells. Is it possible that the acidification of the thylakoid lumen by $SO_2$, thought by Veljovic-Jovanovic *et al.* (1993) to be responsible for increases in zeaxanthin formation, makes guard cells more responsive to light?

## Implications for quantifying the phytotoxicity of SO₂

The responses to mixtures of $SO_2$ with other pollutants should be given high priority when the phytotoxicity of $SO_2$ under present-day conditions is discussed. $SO_2$ and $NO_2$ are a commonly encountered combination close to primary sources and, as seen above, there is abundant evidence that short-term exposures (a few hours only) may be highly damaging to some species. Much of the work hitherto judged to be relevant to the impact of $SO_2$ on crops and ecosystems has examined responses to low doses applied over long periods of time. Larcher (1980) outlined the factors limiting plant distribution, and wrote: 'The probability that a plant species will survive is a function of its ability to come unscathed through extreme weather conditions . . .'. It is clearly unsound that this elementary ecological principle should be disregarded when we establish the impacts of air pollutants on ecosystems. In those parts of the world where $SO_2$ concentrations have decreased and those of $NO_2$ have increased, the critical events are likely to be the co-occurrence of the two pollutants during infrequent short periods of elevated levels of $SO_2$. Much more needs to be understood about the basis of the synergistic interaction that sometimes leads to severe injury during exposure to both gases. Taylor and Bell (1992) found for *Plantago major* that selection for $SO_2$ tolerance does not confer tolerance to $SO_2$ with $NO_2$, or to $NO_2$ alone, clearly indicating that different underlying mechanisms are involved.

### How do mixtures of $SO_2$ and $NO_2$ damage plants?

In relation to the long-term effects of exposure to $SO_2$ with $NO_2$, the activities of the enzyme nitrite reductase appear to be important. Early studies with $^{15}NO_2$ showed clearly that nitrogen from $NO_2$ can be utilized via conversions to nitrite, nitrate, ammonia and subsequent assimilation into amino acids via the GOGAT pathway (Yoneyama & Sasakawa 1979). When $NO_2$ dissolves in water, $NO_2^-$ and $NO_3^-$ ions are formed and the removal of the toxic $NO_2^-$ must be a major factor in cellular tolerance of $NO_2$ pollution. Extracts from grasses that had been exposed to $SO_2$, $NO_2$ or $SO_2$ with $NO_2$ showed important differences in the activity of nitrite reductase (Wellburn et al. 1981; Wellburn 1982). When $NO_2$ was applied alone it induced significant increases in nitrite reductase activity that were both time and concentration dependent. $SO_2$ had no detectable effect on the normal levels of the enzyme, but when it was applied together with $NO_2$, no increase in nitrite reductase activity took place. This was the case both with high concentrations (250–500 p.p.b.) of the two pollutants over a few days, and with longer exposures to a lower concentration (68 p.p.b.).

This apparent elimination by $SO_2$ of the ability to regulate the levels of nitrite in cells is still regarded as one of the prime causes of the high toxicity of $SO_2/NO_2$ mixtures, and the comprehensive review by Wellburn (1982) of this and other possible metabolic impacts remains highly relevant.

More recent studies of the impacts of $SO_2$ and/or $NO_2$ have been reported by Petitte and Ormrod (1988, 1992). They treated non-tuberizing potatoes (*Solanum tuberosum*) with 110 p.p.b. of the two pollutants separately or in combination, and although these are relatively high concentrations, major effects appeared very quickly and consequently they are relevant to severe pollution episodes in the field. The $SO_2/NO_2$ mixture was much more damaging than either gas on its own. Leaf area was significantly reduced after 2 days, and root dry weight had fallen by 23% and 41% by the fourth and eighth days, respectively. At leaf level, $SO_2$ with $NO_2$ caused more negative osmotic potentials within the first day of treatment, accompanied by increased concentrations of reducing sugars. The authors concluded that an early response to SO$_2$ with NO$_2$ may be impairment of the partitioning of dry matter from the leaves to the roots. There was no significant inhibition of net photosynthesis in advance of the fall in root growth, and the reduction in shoot dry weight appeared later than that in the roots. Gould and Mansfield (1988) had found that similar concentrations of SO$_2$ with NO$_2$ (80–100 p.p.b.) applied to winter wheat selectively reduced root growth, while more [14]C-labelled photosynthate was retained in the leaves. The specific effect of SO$_2$ on translocation seems to be independent of tolerance, because even in species which are undamaged until SO$_2$ concentrations are very high, one of the clearest effects is reduced partitioning to the roots (Dodd & Doley 1998). The importance of the work of Petitte and Ormrod (1992) is the establishment of a time sequence of events, clearly indicating that photosynthesis, which in the past has often been regarded as the primary process damaged by air pollutants, is little affected when other impacts of SO$_2$ with NO$_2$ are substantial. An experiment by Steubing *et al.* (1989) lends support to the view that effects occurring as a result of short pollution episodes may have significant impacts on the functioning of plants in the field. They used moveable open-top chambers to fumigate a native herb layer in a beech forest in Germany. SO$_2$ (112 p.p.b.), SO$_2$ (112 p.p.b.) with NO$_2$ (52 p.p.b.), and SO$_2$ (112 p.p.b.) with NO$_2$ (52 p.p.b.) and O$_3$ (100 p.p.b.) were applied for just 4 hours per week. The treatments continued from March to October and many effects were detected, although there was much variation between species. Changes in carbohydrate metabolism were reported only for one species, *Allium ursinum*, and it was found that the SO$_2$ with NO$_2$ treatments (with or without O$_3$) caused substantial increases in the starch contents of the leaves, and the authors concluded that this was probably caused by interference with translocation of carbohydrates out of the leaves. Thus the phenomena recorded in more detail by Petitte and Ormrod (1992) may be of significance in the field where plants are exposed to episodic events involving SO$_2$ with NO$_2$.

Maurousset *et al.* (1992) simulated the subcellular effects of SO$_2$ by applying sodium sulphite to plasma membrane vesicles from leaves of *Vicia faba*. They found that although the proton-pumping ATPase of the vesicles was unaffected, there was strong inhibition of sucrose uptake, leading to the conclusion that sulphite derived from the uptake of SO$_2$ may inhibit transport of photoassimilates by its effect on the sucrose carrier of the plasma membrane. This may possibly have been involved in the disruption of translocation found by Petitte and Ormrod (1992), but there is currently no explanation of why SO$_2$ with NO$_2$ should be more inhibitory than SO$_2$ alone, and this is an important question worthy of further research. Alterations in patterns of translocation may be behind many secondary effects of these pollutants, such as inhibition of nitrogenase activity in nodules of legumes (Sandhu *et al.* 1992) and reduced colonization by vesicular-arbuscular (VA) mycorrhizal fungi (Clapperton & Reid 1992).

## Indirect effects of pollutants

Over the past 15 years it has become increasingly apparent that some of the more important impacts of air pollutants on plants may be the outcome of complex interactions involving other components of ecosystems, for example pests and diseases, and soil chemistry and microbiology.

The work of Hughes et al. (1981, 1982, 1983) provided the first sound experimental evidence that $SO_2$ pollution makes plants more susceptible to insect herbivores. Mexican bean beetles (Epilachna varivestis) were found to perform better, in terms of rate of growth, development and survival, on soybean plants exposed to $SO_2$. Many of the subsequent studies have been performed on sap-feeding insects, particularly aphids, because they are amenable to experimental manipulations. Dohmen et al. (1984) and Dohmen (1988) investigated the cause of higher infestations, in an area downwind of London, of Aphis fabae feeding on Vicia faba, and their experiments showed that the mean relative growth rate of the aphid was significantly increased after the plants had been prefumigated with $SO_2$ or $NO_2$. They also showed that the rate of biomass increase of the aphid was greater in ambient London air than in charcoal-filtered air. Dohmen (1988) extended the investigation to a different species of aphid, Macrosiphon rosae, on cultivated roses in Munich, and found that the ambient pollution, although less than that in London, was nevertheless sufficient to increase growth rate by an average of 20%. Thus aphids on both herbaceous and woody plants appear to respond in the same way, and it is interesting to note that Bolsinger and Flückiger (1984) had studied Aphis fabae on an alternative woody host, Viburnum opulus, and had found increased growth in urban polluted air.

Bolsinger and Flückiger (1989) worked with two plant species (Viburnum opulus and Phaseolus vulgaris) growing by roadsides and showing increased abudance of Aphis fabae, and discovered that the phloem sap contained increased concentrations of nearly all the amino acids that were detected. They reproduced the amino acid patterns in artificial diets, and found that those representing plants growing in ambient pollution produced larger larvae at birth, higher mean relative growth rates, and a shorter development time of the larvae. Although it would seem reasonable to conclude that the increased sulphur and nitrogen supplies came from the $SO_2$ and $NO_x$ in roadside air, caution is needed because $O_3$ pollution can also induce changes in nutritional quality, including increases in amino acid concentration (Bolsinger et al. 1991). This was tentatively attributed to the cellular damage caused by $O_3$, i.e. breakdown of structural proteins to soluble proteins and free amino acids. Whatever is the underlying cause of the change in nutritional quality under $SO_2$ pollution, it does seem to be a significant factor for a wide range of insects. Katzel and Moller (1993) found for the pest of pine needles, Bupalus piniarius, that $SO_2$-affected food offered nutritional benefits for the larvae.

Heightened performance of insects on plants in polluted air clearly implies that the insects are more tolerant of the pollution than the host plants. There are, however, strict limitations to the tolerance of the insects, as was clearly illustrated by the work of Warrington (1987) and Whittaker and Warrington (1990). The mean relative growth rate of pea aphids (*Acyrthosiphon pisum*) on *Pisum sativum* increased linearly with SO$_2$ concentration up to about 100 p.p.b., and thereafter the increase became smaller until the response became negative at about 300 p.p.b. (Fig. 11.3). When plants were fumigated for 4 days with SO$_2$ concentrations from 0 to 300 p.p.b., then transferred to clean air at which point the aphids were added, the mean relative growth rate of the aphids increased linearly over the whole range of concentrations. The difference between the two curves in Fig. 11.3 gives a clear indication of the sensitivity of the aphids to SO$_2$ *per se*. Kozlov *et al.* (1996) have also provided evidence of direct toxicity of SO$_2$ in the case of the leaf beetle *Melasoma lapponica*.

Whittaker and Warrington (1990) pointed out that as the response of the aphid to SO$_2$ concentration is linear to around 100 p.p.b. and the fitted curve goes through the origin, it is likely that any increase in SO$_2$ in the field may have an effect. This point is clearly relevant to the impacts of present-day SO$_2$ pollution, which has been substantially reduced in urban areas but not to the same degree in rural situations. Aminu-Kano *et al.* (1991) monitored populations of the grain aphid, *Sitobion avenae*, on winter wheat and barley, exposed to mean concentrations of 14–57 p.p.b. SO$_2$ in an open-air fumigation experiment. The quantitative relationship between SO$_2$ and aphid density was similar to that in Fig. 11.3, but the populations of the aphids' natural enemies did not increase with the SO$_2$ concentration, i.e. they did not apparently respond to the greater availability of prey.

Although it has been found that some fungal plant pathogens are sensitive to SO$_2$, the overall picture is much less clear than with insect pests. Mansfield *et al.* (1991) monitored the development of various fungal diseases during an open-air fumigation of winter barley (*Hordeum vulgare*) with SO$_2$ concentrations of 5–48 p.p.b. The only clear indications of SO$_2$-induced changes were increases in the

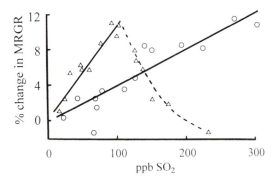

**Figure 11.3** Changes in mean relative growth rate (MRGR) of pea aphids feeding on peas exposed to different concentrations of SO$_2$. The upper curve and the triangles apply to plants containing aphids that were fumigated for 4 days. The lower curve and the circles apply to plants that were fumigated for 4 days without aphids present, after which aphids were added post-fumigation for another 4 days. Redrawn from Whittaker and Warrington (1990).

incidence and severity of powdery mildew, and decreases in leaf blotch. Similarly, the responses of phylloplane fungi were very variable, although pink yeasts were very sensitive to increased concentrations of $SO_2$ up to 47 p.p.b. (Magan & McLeod 1991).

Effects of the dry deposition of $SO_2$ on the soil microflora appear to be more clear cut than have been found on leaf-surface organisms. Steubing *et al.* (1989) noted that fumigation within a beech forest with 112 p.p.b. $SO_2$, or 112 p.p.b. $SO_2$ with 52 p.p.b. $NO_2$, for only 4 hours weekly led to decreases in numbers of soil bacteria and increases in soil fungi, and in consequence the bacteria:fungi ratio was reduced by 40–50%. The soil pH had not been affected by the fumigation. Wookey *et al.* (1991) found that a mean $SO_2$ concentration of 15 p.p.b. in an open-air fumigation reduced microbial respiration significantly in both pine and deciduous leaf litter. Pure cultures of two fungi likely to be important in cellulose decomposition, *Cladosporium cladosporoides* and *Coniothyrium olivaceum*, were shown to be sensitive to $SO_2$ fumigation in the laboratory. Four decomposer fungi were studied in more detail by Dursun *et al.* (1996b), and it was found that respiration and mycelial extension were inhibited by exposure to 40 p.p.b. $SO_2$. It was suggested by Wookey *et al.* (1991) that the effect of $SO_2$ in the field fumigation may have been the result of changes in the base status, particularly magnesium and calcium content, after oxidation of $SO_2$ to $SO_4^{2-}$ at or near the litter surface, and the subsequent leaching of base cations. Newsham *et al.* (1992) exposed leaf litters from deciduous woodlands to 10–30 p.p.b. $SO_2$ and found increased $SO_4^{2-}$ and protons in leachates, and decreased calcium and magnesium contents in the litters. A study of litter from a woodland heavily polluted with $SO_2$ showed that washout of $SO_4^{2-}$ could take place rapidly, i.e. the high-sulphate status was not very persistent. However, inhibitory effects of $SO_2$ solution products on decomposer fungi have also been clearly demonstrated, and when the fungi are growing on agar media they are also sensitive to realistic levels of $SO_2$ (Dursun *et al.* 1996a,b). Thus there appear to be direct effects as well those resulting from changes in base status within litter.

The observations from both the above studies suggest that the rates of deposition of $SO_2$ currently occurring in many parts of the world may be sufficient to interfere with decomposition processes in forest litter layers. Wookey *et al.* (1991) pointed out that in the long term there could be decreases in primary productivity, and changes in ecosystem structure and functioning.

The wider issues connected with the prolonged deposition of sulphur and nitrogen pollution on soil chemistry are beyond the scope of this brief review, and the paper by Thelin *et al.* (1998) should be consulted for recent information on this topic. It is important to recognize that $SO_2$-induced changes in soil chemistry are not now confined to northern latitudes. Deleterious effects of $SO_2$ deposition on soil chemistry, litter decomposition and populations of microarthropod decomposers, have been documented in Italy (Bressan & Paoletti 1997), and in one of very few reports from Africa, Onianwa and Babajide (1993) found that soil sulphate levels in Nigeria had increased with rising densities of vehicular traffic, but the effects on soil biota in that climate are not known. Cleaner air leads to increasing risks of sulphur deficiency in crops in some locations (McGrath & Zhao 1995), but this is easily remedied by the use of sulphur-containing fertilizers, and is not a problem to be compared with the damage that can result from excessive sulphur deposition.

## References

Aminu-Kano, M., McNeill, S. & Hails, R.S. (1991). Pollutant, plant and pest interactions: the grain aphid *Sitobion avenae* (F.). *Agriculture, Ecosystems and Environment*, **33**, 233–243.

Ashenden, T.W., Hunt, R., Bell, S.A. *et al.* (1996). Responses to SO$_2$ pollution in 41 British herbaceous species. *Functional Ecology*, **10**, 483–490.

Asman, W.A.H., Sutton, M.A. & Schjørring, J.K. (1998). Ammonia: emission, atmospheric transport and deposition. *New Phytologist*, **139**, 27–48.

Atkinson, C.J., Wookey, P.A. & Mansfield, T.A. (1991). Atmospheric pollution and the sensitivity of stomata on barley leaves to abscisic acid and carbon dioxide. *New Phytologist*, **117**, 535–541.

Baker, C.K., Unsworth, M.H. & Greenwood, P. (1982). Leaf injury on wheat plants exposed in the field in winter to SO$_2$. *Nature*, **299**, 149–151.

Beckerson, D.W. & Hofstra, G. (1979). Response of leaf diffusive resistance of radish, cucumber and soybean to O$_3$ and SO$_2$ singly and in combination. *Atmospheric Environment*, **13**, 1263–1268.

Black, C.R. & Black, V.J. (1979). The effects of low concentrations of sulphur dioxide on stomatal conductance and epidermal cell survival in field bean (*Vicia faba* L.). *Journal of Experimental Botany*, **30**, 291–298.

Black, V.J. & Unsworth, M. (1980). Stomatal responses to sulphur dioxide and vapour pressure deficit. *Journal of Experimental Botany*, **31**, 667–677.

Bolsinger, M. & Flückiger, W. (1984). Effect of air pollution at a motorway on the infestation of *Viburnum opulus* by *Aphis fabae*. *European Journal of Forest Pathology*, **14**, 256–260.

Bolsinger, M. & Flückiger, W. (1989). Ambient air pollution induced changes in amino acid pattern of phloem sap of host plants — relevance to aphid infestation. *Environmental Pollution*, **56**, 209–216.

Bolsinger, M., Lier, M.E., Lansky, D.M. & Hughes, P.R. (1991). Influence of ozone air pollution on plant–herbivore interactions. Part 1: Biochemical changes in ornamental milkweed (*Asclepias curassavica* L., Asclepiadaceae) induced by ozone. *Environmental Pollution*, **72**, 69–83.

Bressan, M. & Paoletti, M.G. (1997). Leaf litter decomposition and soil microarthropods affected by sulphur dioxide fallout. *Land Degradation and Development*, **8**, 189–199.

Cape, J.N. (1983). Contact angles of water droplets on needles of Scots pine (*Pinus sylvestris*) growing in polluted atmospheres. *New Phytologist*, **93**, 293–299.

Cape, J.N., Leith, I.D., Fowler, D., Murray, M.B. *et al.* (1991). Sulphate and ammonium in mist impair the frost hardening of red spruce seedlings. *New Phytologist*, **118**, 119–126.

Cape, J.N., Sheppard, L.J. & Binnie, J. (1995). Leaf surface properties of Norway spruce needles exposed to sulphur dioxide and ozone in an open-air fumigation system at Liphook. *Plant, Cell and Environment*, **18**, 285–289.

Clapperton, M.J. & Reid, D.M. (1992). Effects of low concentration sulphur dioxide fumigation and vesicular arbuscular mycorrrhizas on C-14 partitioning in *Phleum pratense* L. *New Phytologist*, **120**, 381–387.

Darrall, N.M. (1989). The effects of air pollutants on physiological processes in plants. *Plant, Cell and Environment*, **12**, 1–30.

Davison, A.W. & Bailey, I.F. (1982). SO$_2$ pollution reduces the freezing resistance of ryegrass. *Nature*, **297**, 400–402.

Dodd, I.C. & Doley, D. (1998). Growth responses of cucumber seedlings to sulphur dioxide fumigation in a tropical environment. *Environmental and Experimental Botany*, **39**, 41–47.

Dohmen, G.P. (1988). Indirect effects of air pollutants: changes in plant/parasite interactions. *Environmental Pollution*, **53**, 197–207.

Dohmen, G.P., McNeill, S. & Bell, J.N.B. (1984). Air pollution increases *Aphis fabae* pest potential. *Nature*, **307**, 52–53.

Dueck, Th.A., Dorel, F.G., Ter Horst, R. & van der Eerden, L.J. (1990). Effects of ammonia, ammonium sulphate and sulphur dioxide on the frost sensitivity of Scots pine (*Pinus sylvestris* L.). *Water, Air and Soil Pollution*, **54**, 35–49.

Dursun, S., Frankland, J.C., Boddy, L. & Ineson, P. (1996a). Sulphite and pH effects on $CO_2$ evolution by fungi growing on decomposing coniferous needles. *New Phytologist*, **134**, 155–166.

Dursun, S., Ineson, P., Boddy, L. & Frankland, J.C. (1996b). Sulphur dioxide effects on fungi growing on leaf litter and agar media. *New Phytologist*, **134**, 167–176.

Filner, P., Rennenberg, H., Sekiya, J. *et al.* (1984). Biosynthesis and emission of hydrogen sulfide by higher plants. In *Gaseous Air Pollutants and Plant Metabolism* (Ed. by M.J. Koziol & F.R. Whatley), pp. 219–312. Butterworth, London.

Fowler, D., Cape, J.N., Leith, I.D., Paterson, I.S., Kinnaird, J.W. & Nicholson, I.A. (1988). Effects of air filtration at small $SO_2$ and $NO_2$ concentrations on the yield of barley. *Environmental Pollution*, **53**, 135–149.

Fowler, D., Flechard, C., Skiba, U., Coyle, M. & Cape, J.N. (1998). The atmospheric budget of oxidized nitrogen and its role in ozone formation and deposition. *New Phytologist*, **139**, 11–23.

Freer-Smith, P.H. & Mansfield, T.A. (1987). The combined effects of low temperature and $SO_2 + NO_2$ pollution on the new season's growth and water relations of *Picea sitchensis*. *New Phytologist*, **106**, 237–250.

Galloway, J.N., Schlesinger, W.H., Levy, H., II, Michaels, A. & Schnoor, J.L. (1995). Nitrogen fixation: anthropogenic enhancement—environmental response. *Global Biogeochemical Cycles*, **9**, 235–252.

Garsed, S.G. & Rutter, A.J. (1984). The effects of fluctuating concentrations of sulphur dioxide on the growth of *Pinus sylvestris* L. & *Picea sitchensis* (Bong.) Carr. *New Phytologist*, **97**, 175–195.

Gould, R.P. & Mansfield, T.A. (1988). Effects of sulphur dioxide and nitrogen dioxide on growth and translocation in winter wheat. *Journal of Experimental Botany*, **39**, 389–399.

Grime, J.P. (1979). *Plant Strategies and Vegetation Processes*. John Wiley, Chichester.

Havas, P.J. (1971). Injury to pines growing in the vicinity of a chemical processing plant in northern Finland. *Acta Forestalia Fennica*, **121**, 1–21.

Havas, P.J. & Huttunen, S. (1972). The effects of air pollution on the radial growth of Scots pine (*Pinus sylvestris* L.). *Biological Conservation*, **4**, 361–368.

Hendrey, G.R., Lewin, K.F., Kolber, Z. & Evans, L.S. (1992). Controlled enrichment system for experimental fumigation of plants in the field with sulfur dioxide. *Journal of the Air and Waste Management Association*, **42**, 1324–1327.

Hughes, P.R., Potter, J.E. & Weinstein, L.H. (1981). Effects of air pollutants on plant–insect interactions: reaction of the Mexican bean beetle to $SO_2$-fumigated pinto beans. *Environmental Entomology*, **10**, 741–744.

Hughes, P.R., Potter, J.E. & Weinstein, L.H. (1982). Effects of air pollutants on plant–insect interactions: increased susceptibility of greenhouse-grown soybeans to the Mexican bean beetle after plant exposure to $SO_2$. *Environmental Entomology*, **11**, 173–176.

Hughes, P.R., Dickie, A.I. & Penton, M.A. (1983). Increased success of the Mexican bean beetle on field-grown soybeans exposed to sulfur dioxide. *Journal of Environmental Quality*, **12**, 565–568.

Huttunen, S. (1978). Effects of air pollution on provenances of Scots pine and Norway spruce in northern Finland. *Silva Fennica*, **12**, 1–16.

Huttunen, S. (1984). Interactions of disease and other stress factors with atmospheric pollution. In *Air Pollution and Plant Life* (Ed. by M. Treshow), pp. 321–355. John Wiley, Chichester.

Huttunen, S., Havas, P. & Laine, K. (1981a). Effects of air pollutants on wintertime water economy of the Scots pine (*Pinus sylvestris* L.). *Holarctic Ecology*, **4**, 94–101.

Huttunen, S., Karenlampi, L. & Kolari, K. (1981b). Changes in osmotic values and some related physiological variables in polluted coniferous needles. *Annales Botanici Fennici*, **18**, 63–71.

Johnson, A.H. (1992). The role of abiotic stresses in the decline of red spruce in high elevation forests of the Eastern United States. *Annual Review of Phytopathology*, **30**, 349–367.

Jung, I. & Winter, K. (1992). Mineral nutrient deficiency increases the sensitivity of photosynthesis to sulphur dioxide in needles of a coniferous tree, *Abies nordmanniana*. *Oecologia*, **90**, 70–73.

Kandler, O. & Innes, J.L. (1995). Air pollution and forest decline in Central Europe. *Environmental Pollution*, **90**, 171–180.

Katzel, R. & Moller, K. (1993). The influence of SO$_2$-stressed host plants on the development of *Bupalus piniarius* L. & *Dendrolimus pini* L. *Journal of Applied Entomology*, **116**, 50–61.

Keller, T. (1978). Wintertime atmospheric pollutants—do they affect the performance of deciduous trees in the ensuing growing season? *Environmental Pollution*, **16**, 243–247.

Keller, T. (1981). Folgen einer winterlichen SO$_2$—Belastung für die Fichte. *Gartenbauwissenschaft*, **46**, 170–181.

Kindermann, G., Hüve, K., Slovik, S., Lux, H. & Rennenberg, H. (1995). Emission of hydrogen sulphide by twigs of conifers—a comparison of Norway spruce (*Picea abies* (L.) Karst.), Scots pine (*Pinus sylvestris* L.) and blue spruce (*Picea pungens* Engelm.). *Plant and Soil*, **169**, 421–423.

Kozlov, M.V., Zvereva, E.L. & Selikhovkin, A.V. (1996). Decreased performance of *Melasoma lapponica* (Coleoptera, Chrysomelidae) fumigated by sulphur dioxide—direct toxicity versus host-plant quality. *Environmental Entomology*, **25**, 143–146.

Krochmal, D. & Kalina, A. (1997). Measurements of nitrogen dioxide and sulphur dioxide concentrations in urban and rural areas of Poland using a passive sampling method. *Environmental Pollution*, **96**, 401–407.

Kropff, M.J., Smeets, W.L.M., Meijer, E.M.J., van der Zalm, A.J.A. & Bakx, E.J. (1990). Effects of sulphur dioxide on leaf photosynthesis: the role of temperature and humidity. *Physiologia Plantarum*, **80**, 655–661.

Kubo, A., Saji, H., Tanaka, K. & Kondo, N. (1995). Expression of *Arabidopsis* cytosolic ascorbate peroxidase gene in response to ozone or sulfur dioxide. *Plant Molecular Biology*, **29**, 479–489.

Larcher, W. (1980). *Physiological Plant Ecology*, 2nd edn. Springer-Verlag, Berlin.

Laxen, D.P.H. & Thompson, M.A. (1987). Sulphur dioxide in Greater London. 1931–85. *Environmental Pollution*, **43**, 103–114.

Lee, J.A. & Caporn, S.J.M. (1998). Ecological effects of atmospheric reactive nitrogen deposition on semi-natural terrestrial ecosystems. *New Phytologist*, **139**, 127–134.

Legge, A.H., Nosal, M. & Krupa, S.V. (1996). Modeling the numerical relationships between chronic ambient sulfur dioxide exposures and tree growth. *Canadian Journal of Forest Research*, **26**, 689–695.

Lorenzini, G., Panicucci, A. & Nali, C. (1995). A gas-exchange study of the differential response of *Quercus* species to long-term fumigations with a gradient of sulfur dioxide. *Water Air and Soil Pollution*, **85**, 1257–1262.

McGrath, S.P. & Zhao, F.J. (1995). A risk assessment of sulphur deficiency in cereals using soil and atmospheric deposition data. *Soil Use and Management*, **11**, 110–114.

McLeod, A.R. & Baker, C.K. (1988). The use of open field systems to assess yield response to gaseous pollutants. In *Assessment of Crop Loss from Air Pollutants* (Ed. by W.W. Heck, O.C. Taylor & D.T. Tingey), pp. 181–210. Elsevier Applied Science, London.

Magan, N. & Mcleod, A.R. (1991). Effects of open-air fumigation with sulphur dioxide on the occurrence of phylloplanne fungi on winter barley. *Agriculture, Ecosystems and Environment*, **33**, 245–261.

Majernik, O. & Mansfield, T.A. (1971). Effects of SO$_2$ pollution on stomatal movements in *Vicia faba*. *Phytopathologische Zeitschrift*, **71**, 123–128.

Mansfield, P.J., Bell, J.N.B., McLeod, A.R. & Wheeler, B.E.J. (1991). Effects of sulphur dioxide on the development of fungal diseases of winter barley in an open-air fumigation system. *Agriculture, Ecosystems and Environment*, **33**, 215–232.

Mansfield, T.A. & Jones, T. (1985). Growth/environment interactions in SO$_2$ responses of grasses. In *Sulfur Dioxide and Vegetation: Physiology, Ecology, and Policy Issues* (Ed. by W.E. Winner, H.A. Mooney & R.A. Goldstein), pp. 332–345. Stanford University Press, Stanford, CA.

Mansfield, T.A., Hetherington, A.M. & Atkinson, C.J. (1990). Some current aspects of stomatal physiology. *Annual Review of Plant Physiology and Plant Molecular Biology*, **41**, 55–75.

Mansfield, T.A., Goulding, K.W.T. & Sheppard, L.J. (eds) (1998). *Disturbance of the Nitrogen Cycle*. Cambridge University Press, Cambridge.

Materna, J. (1974). Einfluss der SO$_2$—Immissionen auf Fichtenpflanzen in Wintermonaten. In *IXth Internationale Tagung Uber die Luftverunreini-*

*gung und Forstwirtschaft*, pp. 107–114. Vytiskl Tomos, Praha, Czechoslovakia.

Materna, J. (1984). Impact of atmospheric pollution on natural ecosystems. In *Air Pollution and Plant Life* (Ed. by M. Treshow), pp. 397–416. John Wiley, Chichester.

Materna, J. & Kohout, R. (1963). Die Absorption des Schwefeldioxids durch die Fichte. *Naturwissenschaften*, **50**, 407–408.

Maurousset, L., Lemoine, R., Gallet, O., Delrot, S. & Bonnemain, J.L. (1992). Sulfur dioxide inhibits the sucrose carrier of the plasma membrane. *Biochimica et Biophysica Acta*, **1105**, 230–236.

Meng, F.R., Cox, R.M. & Arp, P.A. (1994). Fumigating mature spruce branches with $SO_2$ — effects on net photosynthesis and stomatal conductance. *Canadian Journal of Forest Research*, **24**, 1464–1471.

Mesanza, J.M., Casado, H. & Encinas, D. (1996). Effects of a sulfur dioxide episode on trees in the surroundings of a refinery. *Journal of Environmental Science and Health, Part A*, **31**, 1025–1033.

Neighbour, E.A., Cottam, D.A. & Mansfield, T.A. (1988). Effects of sulphur dioxide and nitrogen dioxide on the control of water loss by birch (*Betula* spp.). *New Phytologist*, **108**, 149–157.

Newsham, K.K., Ineson, P., Boddy, I. & Frankland, J.C. (1992). Effects of dry-deposited sulphur dioxide on fungal decomposition of angiosperm leaf litter. 2. Chemical content of leaf litters. *New Phytologist*, **122**, 111–125.

No, E.G., Flagler, R.B., Swize, M.A., Cairney. J. & Newton. R.J. (1997). cDNAs induced by ozone from *Atriplex canescens* (saltbush) and their response to sulfur dioxide and water-deficit. *Physiologia Plantarum*, **100**, 137–146.

Okpodu, C.M., Alscher, R.G., Grabau, E.A. & Cramer, C.L. (1996). Physiological, biochemical and molecular effects of sulfur dioxide. *Journal of Plant Physiology*, **148**, 309–316.

Onianwa, P.C. & Babajide. A.O. (1993). Sulfate-sulfur levels of topsoils related to atmospheric sulfur dioxide pollution. *Environmental Monitoring and Assessment*, **25**, 141–148.

Peiser, G. & Yang, S.F. (1985). Biochemical and physiological effects of $SO_2$ on nonphotosynthetic processes. In *Sulfur Dioxide and Vegetation: Physiology, Ecology, and Policy Issues* (Ed. by W.E. Winner, H.A. Mooney & R.A. Goldstein),

pp. 148–161. Stanford University Press, Stanford, CA.

Petitte, J.M. & Ormrod, D.P. (1988). Effects of sulfur dioxide and nitrogen dioxide on shoot and root growth of Kennebec and Russet Burbank potato plants. *American Potato Journal*, **65**, 517–527.

Petitte, J.M. & Ormrod, D.P. (1992). Sulfur dioxide and nitrogen dioxide affect growth, gas exchange, and water relations of potato plants. *Journal of the American Society for Horticultural Science*, **117**, 146–153.

Ranieri, A., Castagna, A., Lorenzini, G. & Soldatini, G.F. (1997). Changes in thylakoid protein patterns and antioxidant levels in two wheat cultivars with different sensitivity to sulfur dioxide. *Environmental and Experimental Botany*, **37**, 125–135.

Rautio, P., Huttunen, S. & Lamppu, J. (1998). Seasonal foliar chemistry of northern Scots pines under sulphur and heavy metal pollution. *Chemosphere*, **37**, 271–287.

Rennenberg, H. (1991). The significance of higher plants in the emission of sulphur compounds from terrestrial ecosystems. In *Trace Gas Emissions by Plants* (Ed. by T.D. Sharkey, E.A. Holland & H.A. Mooney), pp. 217–260. Academic, San Diego.

Rennenberg, H. & Herschbach, C. (1996). Responses of plants to atmospheric sulphur. In *Plant Response to Air Pollution* (Ed. by M. Yunus & M. Iqbal), pp. 285–293. J. Wiley, Chichester.

Rennenberg, H. & Polle, A. (1994). Metabolic consequences of atmospheric sulphur influx into plants. In *Plant Response to the Gaseous Environment* (Ed. by A.R. Wellburn & R. Alscher), pp. 165–180. Chapman & Hall, London.

Roberts, T.M. (1984). Long-term effects of sulphur dioxide on crops: an analysis of dose–response relations. *Philosophical Transactions of the Royal Society, London B*, **305**, 299–316.

Rosenbaum, B.J., Strickland, T.C. & McDowell, M.K. (1994). Mapping critical levels of ozone, sulfur dioxide and nitrogen dioxide for crops, forests and natural vegetation in the United States. *Water, Air and Soil Pollution*, **74**, 307–319.

Rothermal, B. & Alscher, R. (1985). A light-enhanced metabolism of sulfite in cells of *Cucymis sativa* cotyledons. *Planta*, **53**, 105–110.

Sandhu, R., Li, Y. & Gupta, G. (1992). Sulfur dioxide

and carbon dioxide induced changes in soybean physiology. *Plant Science*, **83**, 31–34.

Sheppard, L.J. (1994). Causal mechanisms by which sulphate, nitrate and acidity influence frost hardiness in red spruce: review and hypothesis. *New Phytologist*, **127**, 69–82.

Sheppard, L.J., Smith, R.I. & Cannell, M.G.R. (1989). Frost hardiness of *Picea rubens* growing in spruce decline regions of the Appalachians. *Tree Physiology*, **5**, 23–37.

Shrestha, R.M., Bhattacharya, S.C. & Malla, S. (1996). Energy use and sulphur dioxide emissions in Asia. *Journal of Environmental Management*, **46**, 359–372.

Soikkeli, S. & Tuovinen, T. (1979). Damage in mesophyll ultrastructure of needles of Norway spruce in two industrial environments in central Finland. *Annales Botanici Fennici*, **16**, 50–64.

Steubing, L., Fangmeier, A., Both, R. & Frankenfeld, M. (1989). Effects of SO$_2$, NO$_2$, and O$_3$ on population development and morphological and physiological parameters of native herb layer species in a beech forest. *Environmental Pollution*, **58**, 281–302.

Strand, M. (1993). Photosynthetic activity of Scots pine (*Pinus sylvestris* L.) needles during winter is affected by exposure to SO$_2$ and NO$_2$ during summer. *New Phytologist*, **123**, 133–141.

Strand, M. (1995). Persistent effects of low concentrations of SO$_2$ and NO$_2$ on photosynthesis in Scots pine (*Pinus sylvestris*) needles. *Physiologia Plantarum*, **95**, 581–590.

Takemoto, B.K., Noble, R.D. & Harrington, H.M. (1986). Differential sensitivity of duckweeds (Lemnaceae) to sulphite. II. Thiol production and hydrogen sulphide emission as factors influencing phytotoxicity under low and high irradiance. *New Phytologist*, **103**, 541–548.

Taylor, H.J. & Bell, J.N.B. (1992). Tolerance to SO$_2$, NO$_2$ and their mixture in *Plantago major* L. populations. *Environmental Pollution*, **76**, 19–24.

Thelin, G., Rosengren-Brinck, U., Nihlgård, B. & Barkman, A. (1998). Trends in needle and soil chemistry of Norway spruce and Scots pine needles in South Sweden 1985–94. *Environmental Pollution*, **99**, 149–158.

Tingey, D.T., Reinart, R.A., Dunning, J.A. & Heck, W.W. (1971). Vegetation injury from the interaction of nitrogen dioxide and sulfur dioxide. *Phytopathology*, **61**, 1506–1511.

Veljovic-Jovanovic, S., Bilger, W. & Heber, U. (1993). Inhibition of photosynthesis, acidification and stimulation of zeaxanthin formation in leaves by sulfur dioxide and reversal of these effects. *Planta*, **191**, 365–376.

Vitousek, P.M. (1994). Beyond global warming: ecology and global change. *Ecology*, **75**, 1861–1876.

Warrington, S. (1987). Relationship between sulphur dioxide dose and growth of the pea aphid *Acyrthosiphon pisum* on peas. *Environmental Pollution*, **43**, 155–162.

Wellburn, A.R. (1982). Effects of SO$_2$ and NO$_2$ on metabolic function. In *Effects of Gaseous Air Pollution in Agriculture and Horticulture* (Ed. by M.H. Unsworth & D.P. Ormrod), pp. 169–187. Butterworth Scientific, London.

Wellburn, A.R. (1988). *Air Pollution and Acid Rain: The Biological Impact.* Longman, Harlow, Essex.

Wellburn, A.R. (1994). *Air Pollution and Climate Change: the Biological Impact.* Addison Wesley Longman, Harlow, Essex.

Wellburn, A.R., Higginson, C., Robinson, D. & Walmsley, C. (1981). Biochemical explanations of more than additive inhibitory effects of low atmospheric levels of sulphur dioxide plus nitrogen dioxide upon plants. *New Phytologist*, **88**, 223–237.

Whitmore, M.E. & Freer-Smith, P.H. (1982). Growth effects of SO$_2$ and/or NO$_2$ on woody plants and grasses observed during the spring and summer. *Nature*, **300**, 55–57.

Whitmore, M.E. & Mansfield, T.A. (1983). Effects of long-term exposure to SO$_2$ and NO$_2$ on *Poa pratensis* and other grasses. *Environmental Pollution (Series A)*, **31**, 217–235.

Whittaker, J.B. & Warrington, S. (1990). Effects of atmospheric pollutants on interactions between insects and their food plants. In *Pests, Pathogens and Plant Communities* (Ed. by J.J. Burdon & S.R. Leather), pp. 97–110. Blackwell Scientific Publications, Oxford.

Wilson, G.B. & Bell, J.N.B. (1990). Studies on the tolerance to sulphur dioxide of grass populations in polluted areas. VI. The genetic nature of tolerance in *Lolium perenne* L. *New Phytologist*, **116**, 313–317.

Wingsle, G. & Hällgren, J.-E. (1993). Influence of SO$_2$ and NO$_2$ exposure on glutathione, superoxide dismutase and glutathione reductase activities in Scots pine needles. *Journal of Experimental Botany*, **44**, 463–470.

Winner, W.E. & Mooney, H.A. (1980). Responses of Hawaiian plants to volcanic sulfur dioxide: stomatal behavior and foliar injury. *Science*, **210**, 789–791.

Winner, W.E., Mooney, H.A. & Goldstein, R.A., eds. (1985). *Sulfur Dioxide and Vegetation. Physiology, Ecology, and Policy Issues.* Stanford University Press, Stanford, CA.

Wolfenden, J., Pearson, M. & Francis, B.J. (1991). Effects of over-winter fumigation with sulphur and nitrogen dioxides on biochemical parameters and spring growth in red spruce (*Picea rubens* Sarg.). *Plant, Cell and Environment*, **14**, 35–45.

Wookey, P.A., Ineson, P. & Mansfield, T.A. (1991). Effects of atmospheric sulphur dioxide on microbial activity in decomposing forest litter. *Agriculture, Ecosystems and Environment*, **33**, 263–280.

Yoneyama, T. & Sasakawa, H. (1979). Transformation of atmospheric $NO_2$ absorbed in spinach leaves. *Plant and Cell Physiology*, **20**, 263–266.

Yunus, M., Singh, N. & Iqbal, M. (1996). Global status of air pollution: an overview. In *Plant Response to Air Pollution* (Ed. by M. Yunus & M. Iqbal), pp. 1–34. J. Wiley, Chichester.

Zeiger, E. & Zhu, J. (1998). Role of zeaxanthin in blue light photoreception and the modulation of light–$CO_2$ interactions in guard cells. *Journal of Experimental Botany*, **49**, 433–442.

# Chapter 12

# Terrestrial ecosystem responses to solar UV-B radiation mediated by vegetation, microbes and abiotic photochemistry

*M.M. Caldwell,[1] P.S. Searles,[1] S.D. Flint[1] and P.W. Barnes[2]*

## The nature of solar UV-B radiation

Stratospheric ozone depletion has elicited substantial interest in the significance of solar UV-B radiation for the biosphere, including biogeochemical processes. Solar UV-B increases that result from stratospheric ozone reduction must, however, be taken in context with the mosaic of other global environmental changes such as elevated atmospheric $CO_2$ concentrations and increasing temperatures. To set the stage, a brief perspective on solar UV-B radiation and how it varies geographically and seasonally will be given. The component of solar radiation that is affected by atmospheric ozone changes is largely within the UV-B waveband (280–320 nm) and constitutes less than 0.5% of the total solar energy reaching the earth's surface. It varies considerably in flux rate and wavelength composition at different geographical locations, and this variation is much greater than the corresponding change in total solar radiation (Caldwell *et al.* 1980; Madronich *et al.* 1995). This variation is caused by two primary factors: a latitudinal gradient of stratospheric ozone that decreases from polar to tropical regions, and differences in prevailing solar angles at different latitudes (Caldwell *et al.* 1980; Madronich 1993). Of course, for different habitats at the same latitude, solar radiation and UV-B can vary because of cloud cover and other local influences. There tends to be an inverse correlation between the normal solar UV-B levels and the relative increase of UV-B resulting from stratospheric ozone reduction (Madronich *et al.* 1995). The greatest relative increases in solar UV-B resulting from ozone reduction occur at high latitudes. Ozone reduction is moderate at temperate latitudes and scarcely detectable in the tropics. This inverse correlation also roughly applies to the timing of ozone reduction during the year at temperate and high latitudes. The greatest relative ozone depletions tend to occur in the winter and early spring. Thus, ecosystems historically accustomed to lower solar UV-B flux are those where the greatest relative increases are taking place.

[1] *The Ecology Center and Department of Rangeland Resources, Utah State University, Logan, UT 84322-5205, USA. E-mail: mmc@cc.usu.edu*
[2] *Department of Biology, Southwest Texas State University, San Marcos, TX 78666-4616, USA*

In efforts to understand the many implications of global climate change, considerable interest has been directed to retrospective analyses of how environmental factors such as temperature and $CO_2$ concentration have changed over historical and geological time scales (e.g. Berner 1997; Ehleringer *et al.* 1997). However, very little is known about how atmospheric ozone levels and therefore UV-B have fluctuated in the evolutionary history of organisms or, for that matter, even during this century. Apart from what is known about the ancient history of atmospheric oxygen on the earth (Margulis *et al.* 1976) and how this would be coupled with ozone levels (Caldwell 1979; Rozema *et al.* 1997), there are no clues to provide a history of ozone fluctuations prior to recent decades. The longest record of direct measurements from Arosa, Switzerland covers only about seven decades and indicates little trend apart from the last 20 years when anthropogenic activities are thought to have led to the decline of ozone (Fig. 12.1). Unlike other gases of the atmosphere such as the stable molecules of $CO_2$ and methane that can be recovered from glacial ice, ozone is both highly unstable and located principally in the stratosphere and, thus, would never be found in such a preserved, encapsulated form. Also, unlike temperature, whose history can be estimated by several proxies such as $^{18}O$, there are no known surrogates that can be used to glean a historical perspective on atmospheric ozone. A biological surrogate for ozone in the Antarctic was attempted in the form of phenolic compounds in herbarium specimens of mosses (Markham *et al.* 1990), but the record is roughly qualitative at best and of

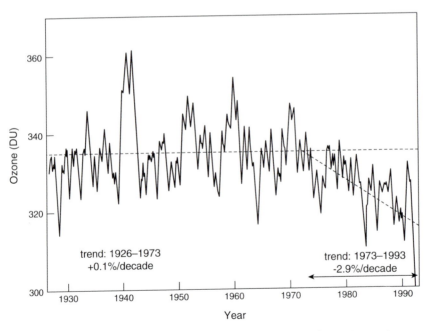

**Figure 12.1** Column ozone levels measured by the ground-based Dobson Meter at Arosa, Switzerland. Modified from McPeters and Hollandsworth (1996).

short duration. Björn *et al.* (1997) attempted a similar retrospective investigation of flavonoid compounds in herbarium specimens of the ericaceous subarctic shrub *Cassiope tetragona* collected in Sweden. Only the ratio of myricetin to quercetin was mildly suggestive of a trend through time. Thus, we have little perspective on past changes in ozone and solar UV-B.

## A thesis for terrestrial ecosystem responses to ozone reduction

There have been numerous reviews of UV-B effects on plant processes in the past decade (Caldwell & Flint 1994; Caldwell *et al.* 1995; Björn 1996; Jordan 1996; Rozema *et al.* 1997; earlier reviews cited therein). Thus, rather than adding another similar review, this chapter addresses terrestrial ecosystem responses that may be anticipated as a result of ozone reduction. The thesis of this essay is that rather than direct damaging effects, the more significant effects of solar UV-B in terrestrial ecosystems, both historically and in the future, are likely to involve indirect processes mediated by UV-B action on vegetation, microbes and abiotic photochemistry. Furthermore, many of these indirect responses may often involve species interactions, the direction of which will be difficult to predict. It is also contended that community primary plant production, *per se*, will probably not be greatly affected by further anticipated increases in solar UV-B. However, before addressing indirect responses, the direct damaging mode of UV action is first briefly considered.

## Damage, protection and repair

### Mechanisms of UV action and repair

Solar ultraviolet radiation can cause direct damage in organisms. Short wavelength solar UV contains actinic, potentially damaging radiation for biological systems. This can be in the form of nucleic acid damage, photooxidative effects and injury to specific targets in processes such as the photosystem II (PS II) reaction centre of photosynthesis. Direct photon-hit damage to DNA is most common in the form of two lesions, cyclobutane-type dimers of pyrimidine bases and (6–4) photoproducts (which are also dimers of pyrimidine bases). These two types of lesions differ from other DNA lesions in that most, but not all, organisms possess special enzymes (photolyases) that can effectively repair many of these lesions in a process known as photoreactivation if there is sufficient UV-A or visible light and temperatures are favourable (Britt 1996). Oxidative damage may also result from UV-B irradiation because active oxygen species can be produced (Foyer *et al.* 1994; Jordan 1996). Lipid peroxidation, chlorophyll and DNA damage are some of the common consequences of free radical production in plant tissues (Panagopoulos *et al.* 1990; Björn 1996). However, UV-B also stimulates antioxidant enzymes such as glutathione reductase, superoxide dismutase and ascorbate peroxidase (Jordan 1996). Photosynthetic damage can be seen in intact higher plants following UV-B exposure, but the exact mechanism of damage is still under investigation. Some lines of evidence point to a target in or near the reaction centre of PS II (Rozema *et al*. 1997), but other evidence suggests that PS II is unlikely to be the initial target and that other processes such as carboxylation and the enzyme ribulose bisphosphate carboxylase/oxygenase (Rubisco) may be affected long before PS II (Baker *et al*. 1997).

Most organisms that are normally exposed to sunlight have means of protection from solar UV-B. In large part, this amounts to shielding sensitive radiation targets by structural characteristics and pigments that screen out much of the most damaging radiation (Caldwell *et al*. 1983; Vogelmann 1993). Most studies show that the epidermis absorbs more than 90% of the incident UV-B for all lifeforms except for herbaceous dicotyledons, some of which can have UV-B epidermal transmittance up to 40% (Robberecht *et al*. 1980; Caldwell *et al*. 1983; Day *et al*. 1992; but see Day 1993). As discussed in greater detail later, phenolic pigments that absorb effectively in the UV waveband are an important constituent of this shielding. This UV shielding combined with DNA repair, antioxidants and polyamines that reduce membrane damage (Kramer *et al*. 1991) mitigate much, if not all, of the potential UV-B damage. Vivid evidence of the efficacy of this protection and repair comes from studies with mutant plants lacking the protective or repair mechanisms that exhibit pronounced sensitivity to UV-B damage (Reuber *et al*. 1996; Landry *et al*. 1997).

Although these various pathways of UV-B damage and protection are well supported by different lines of evidence as discussed above, there is also accumulating evidence that UV-B effects on plants are not caused by non-specific DNA damage. For example, genes encoding chloroplast proteins are specifically repressed by UV-B (Strid 1993; Jordan 1996). But, UV-B can also result in upregulation of other genes leading to increased UV protection. Genes encoding glutathione reductase and genes regulating different parts of the phenylpropanoid pathway that result in increased UV-absorbing phenolics can be specifically induced by UV-B (Strid 1993; Jordan *et al*. 1994).

## Little damage under field conditions

The degree of UV protection plants enjoy also usually reflects various environmental conditions during growth. For example, if plants have been cultured in open sunlight, they are generally much more resistant than when grown in growth-chamber or greenhouse conditions (Tevini & Teramura 1989; Teramura *et al*. 1990; Caldwell & Flint 1994; Fiscus & Booker 1995; Dai *et al*. 1997; Antonelli *et al*. 1998; Stephanou & Manetas 1998). Removal or reduction of the UV-B component of sunlight in filter exclusion experiments has led to small increases in height growth of several species (e.g. Tevini *et al*. 1990; Searles *et al*. 1995; Visser *et al*. 1997). Yet, such experiments do not reveal sizeable biomass changes, in most cases. Similarly, supplementing the normal solar radiation with UV-B from lamps can sometimes show height growth reductions (e.g. Teramura *et al*. 1990), but for the most part these are usually not large and there is often no detectable yield response (Miller *et al*. 1994; Dai *et al*. 1997).

Most of the research dealing with UV-B radiation influences on growth and yield in the field has been of short duration and most has involved annual plants or early growth stages of perennials (Caldwell & Flint 1994). One 3-year field study of UV-B impacts on a tree species (*Pinus taeda*) reported that detrimental UV-B effects might slowly accumulate through time (Sullivan & Teramura 1992). Johanson *et al.* (1995) also suggested some accumulation of morphological response to UV-B in an evergreen dwarf shrub heath species (*Vaccinium vitis-idaea*), but evidence for this was largely circumstantial. The possibility of a significant accumulation of deleterious effects of increased solar UV-B from year to year in perennial species certainly deserves further investigation. Analogous to the potential accumulation of UV-B effect in perennials, Musil (1996) reported that the effects of UV-B irradiation on allocation and growth in a sexually reproducing population of an annual desert plant appeared to accumulate as subsequent generations were exposed to elevated UV-B irradiation. Furthermore, after four generations of exposure to UV-B irradiation, effects persisted in the fifth generation that was not exposed to an elevated UV-B treatment (Musil *et al.* 1999).

Most of the reproductive parts of plants, such as pollen and ovules, are rather well shielded from solar UV-B radiation. For example, anther walls can absorb more than 98% of the incident UV-B radiation (Flint & Caldwell 1984). Only after transfer to the stigma might pollen be susceptible to solar UV-B radiation for a very brief time. Pollen germination in some cases (Flint & Caldwell 1984) and pollen tube growth of many species may be retarded (Torabinejad *et al.* 1998), but it is not clear that sufficient sunlight exposure would occur during the brief period of growth over the stigma surface to reduce successful fertilization, even under elevated solar UV-B.

In summary, while direct, detrimental effects of increased solar UV-B deriving from ozone reduction on terrestrial vegetation cannot be entirely dismissed, it is proposed that most of the significant ecosystem-level effects on terrestrial ecosystems are likely to involve other avenues of influence (Fig. 12.2). Likely pathways by which solar UV-B may influence terrestrial ecosystems include abiotic photochemical reactions in the lower atmosphere and in plant litter on the soil surface, various effects on microbes, and several indirect influences mediated by vegetation.

## UV-B effects mediated by vegetation

Two common classes of plant response to UV-B radiation are changes in the products of secondary metabolism, especially phenolic constituents, and morphological changes. These chemical and morphological changes are much more consistently documented in the literature dealing with plants and UV-B radiation than are indications of damage, especially under field conditions (Caldwell & Flint 1994). A frequently reported reaction of plants to UV-B is an increase of UV-B-absorbing phenolic compounds which may confer greater protection (Caldwell & Flint 1994; Reuber *et al.* 1996). Morphological alterations such as greater leaf thickness, shorter internode length and increased branching or tillering of graminoids

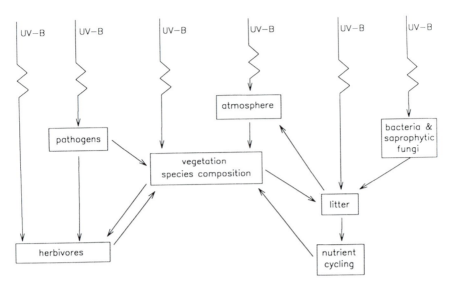

**Figure 12.2** Potential pathways of solar UV-B influence on terrestrial ecosystems.

are also common responses to UV-B (e.g. Barnes *et al.* 1990). There are several lines of evidence suggesting that these changes in morphology represent shifts in allocation rather than damage *per se* (Ballaré *et al.* 1991, 1995a,b; Ensminger 1993).

Changes in secondary chemistry and morphology appear to be mediated by a UV-B-absorbing photoreceptor, and action spectra (Fig. 12.3) suggest that this receptor, or receptors, is responsive only to the UV-B part of the spectrum (Wellmann 1983; Goto *et al.* 1993). It is an intriguing question why such a UV-B photoreceptor, or receptors, evolved in plants. Information gleaned by plants through this photoreceptor may well supplement that obtained by the major photoreceptors, phytochrome and the blue/UV-A receptor (also termed cryptochrome). Synthesis of UV-B-absorbing phenolics in response to elevated fluxes of solar UV-B would have obvious benefits for plants. Reduced hypocotyl extension by UV-B has been proposed as a mechanism to reduce the exposure of seedlings to sunlight at the soil surface while synthesis of UV-B-absorbing compounds is under-way (Ballaré *et al.* 1995b). The manifestations of the UV-B receptor(s) are likely mediated through gene up- or down-regulation as described earlier. While these are plausible advantages of a UV-B photoreceptor system for individual plants, there are also several other consequences of the resulting morphological and secondary chemistry changes at the community and ecosystem level.

### Competition

If morphological changes mediated by a UV-B photoreceptor are largely a matter of altered allocation, total growth or productivity of plants may not be changed. This appeared to be the case when monoculture stands were exposed to additional

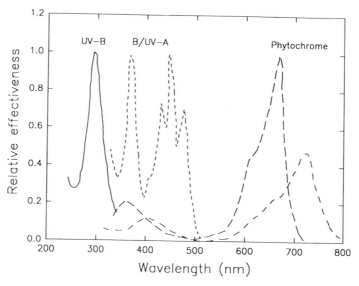

**Figure 12.3** Action spectra of three photoreceptor systems in higher plants that lead to photomorphogenetic changes and alterations of plant secondary chemistry. The phytochrome *in vitro* action spectra are from Butler *et al.* (1964), and the *in vivo* action spectra for the UV-A/blue light receptor from Hartmann (1967) and for the UV-B receptor from Beggs and Wellmann (1994).

UV-B in the field (Barnes *et al.* 1988, 1995). Yet, if plants are growing in mixed-species stands, even rather subtle morphological changes can alter the balance of competition among species (Bogenrieder & Klein 1982; Gold & Caldwell 1983). To date, the influence of enhanced UV-B on interspecific competition has been examined in 15 species pairs and, of these, six (40%) have shown detectable shifts in competitive balance (Table 12.1). However, the mechanism of the UV-B-induced shift in competitive balance has been well studied in only one species pair (Barnes *et al.* 1988, 1995). In this case, a 7-year study of competitive balance shifts in mixtures of wheat and a common weed, wild oat (Barnes *et al.* 1988, 1995), showed that the change in competition under supplemental UV-B could be attributed quantitatively to altered competition for visible sunlight to drive photosynthesis (Barnes *et al.* 1995).

An experimental survey of several annual species grown in dense monocultures in a greenhouse revealed that grasses were morphologically more responsive to UV-B than dicotyledons (Barnes *et al.* 1990). Based on these observations, mixtures with grasses might be more likely to experience UV-B-induced changes in competitive balance than mixtures of only dicotyledons. This hypothesis has yet to be tested, though findings from the studies to date are generally consistent with these ideas (i.e. five of the six cases showing alterations in competitive balance involved grasses; Table 12.1). Previous research has also shown that general UV-B

**Table 12.1** Summary of studies examining the influence of elevated UV-B on plant competition. When significant shifts in competitive balance were detected, the species that showed an increase in relative competitive status in response to UV-B treatment are indicated in **bold**.

| Competing species | Growth forms | Experimental conditions | Reference |
|---|---|---|---|
| Agricultural crops and associated weeds | | | |
| *Pisum sativum–Alyssum alyssoides* | Annual dicots | Field | 1 |
| ***Medicago sativa**–Amaranthus retroflexus* | Perennial/annual dicots | Field | 1 |
| *Medicago sativa–Brassica nigra* | Perennial/annual dicots | Field | 1 |
| *Allium cepa–Amaranthus retroflexus* | Perennial monocot/ annual dicot | Field | 1 |
| *Trifolium pratense–Setaria glauca* | Perennial dicot/ annual grass | Field | 1 |
| ***Triticum aestivum**–Aegilops cylindrica* | Annual grasses | Field | 2 |
| ***Triticum aestivum**–Avena fatua* | Annual grasses | Field | 2,3 |
| *Oryza sativa* var. Lemont–***Echinochloa crusgalli*** | Annual grasses | Greenhouse | 4 |
| *Oryza sativa* var. IR36–*Echinochloa crusgalli* | Annual grasses | Greenhouse | 4 |
| ***Phaseolus vulgaris**–Avena fatua* | Annual dicot/annual grass | Field | 5 |
| Species of disturbed areas | | | |
| *Bromus tectorum–Alyssum alyssoides* | Annual grass/annual dicot | Field | 1 |
| *Plantago patagonica–Lepidium perfoliatum* | Annual dicots | Field | 1 |
| *Setaria viridis–Echinochloa crusgalli* | Annual grasses | Greenhouse | 4 |
| Forage/grassland species | | | |
| ***Poa pratensis**–Geum macrophyllum* | Perennial grass/ perennial dicot | Field | 1 |
| *Sorghum halepense–Panicum virgatum* | Perennial grasses | Greenhouse | 4 |

References are: 1 = Fox and Caldwell (1978); 2 = Gold and Caldwell (1983); 3 = Barnes *et al.* (1988); 4 = J. Beck and P. Barnes, unpublished; 5 = R. Ryel, P. Barnes, S. Flint and M. Caldwell, unpublished data.

responsiveness often decreases under conditions of limited soil moisture (Sullivan & Teramura 1990) and nutrients (Murali & Teramura 1985; Hunt & McNeil 1998). Similarly, studies with wheat and wild oat indicated that UV-B-induced changes in competitive balance and morphology were more pronounced in years of high precipitation or with irrigation than in years of low precipitation and no supplemental water (Barnes *et al.* 1988). Thus, if light-competition-mediated change is indeed a general mechanism by which UV-B alters competitive balance, one might expect shifts in competitive balance under altered UV-B conditions to be most prevalent in moist, fertile habitats where stand density is high, plants are generally similar in stature, and light is a limiting resource for growth and productivity (e.g. a meadow).

In principle, allocation changes in root:shoot ratios (e.g. Ziska *et al.* 1993) and/or alterations of rhizosphere microorganisms (Klironomos & Allen 1995) caused by UV-B might also effect competitive balance shifts, but this has not been explicitly demonstrated. Greenhouse studies have, however, shown that UV-B can induce shifts in the competitive balance between wheat and wild oat when these species were grown in individual containers, thereby precluding below-ground interactions (Barnes *et al.* 1988).

Because of the inherent asymmetry of light competition (Schwinning & Weiner 1998), the rather subtle, but differential, effects of UV-B on plants have the potential to become greatly magnified over time. In a similar fashion, UV-B effects on shoot morphology may lead to altered size distributions of individual plants in dense monocultures if taller plants, which intercept more UV-B, show a greater inhibition in leaf or stem elongation than do subcanopy individuals. Variation in morphological responses to the ratio of red to far-red radiation has been shown to alter size inequality in *Arabidopsis thaliana* populations (Ballaré & Scopel 1997), but the influence of UV-B on size hierarchies has not been studied.

## Phenology and reproduction

UV-B radiation can alter both the timing of flowering (Caldwell 1968; Ziska *et al.* 1992; Staxén & Bornman 1994; Musil 1995) as well as the number of flowers in certain species (Musil 1995; Klaper *et al.* 1996; Saile-Mark & Tevini 1997) and such effects may be due to regulatory alterations in the plant rather than damage *per se*. Differences in timing of flowering may have consequences for the availability of pollinators. Allocation to reproductive effort may also be noticeably affected by additional UV-B. For example, even though vegetative growth characteristics and flowering phenology of the subarctic shrub *Vaccinium myrtillus* were not affected by supplemental UV-B in a field experiment, the number of fruits produced was increased by the additional UV-B (Gwynn-Jones *et al.* 1997). The Mediterranean shrub *Cistus creticus* also was induced to produce substantially more seeds per fruit and greater average seed mass when additional UV-B was given in a field experiment (Stephanou & Manetas 1998).

## Changes in plant secondary chemistry

Several potentially far-reaching effects of solar UV-B that might be mediated by vegetation involve changes in secondary plant chemistry and, specifically, the shikimic acid pathway. Approximately 20% of the carbon fixed by plants passes through this pathway (Herrmann 1995) and a complex array of products are produced including a variety of pigments, such as flavonoids and anthocyanins, phytoalexins such as pterocarpans and stilbenes, systemic signalling compounds such as salicylic acid, and lignin. The complex of secondary chemicals that are often regulated in part by UV-B and that might lead to ecological consequences fall primarily in the phenylpropanoid pathway. This pathway branches off the shikimic acid pathway where the key enzyme phenylalanine ammonia-lyase (PAL) forms cinnamate from phenylalanine (e.g. Dixon & Paiva 1995). The activity of PAL and several other enzymes of this pathway can be induced by different stress factors, including pathogen attack, low temperature, wounding, mineral nutrient deficiency and both visible and UV radiation. Sometimes greater activity of the branch point enzyme PAL can result in greater synthesis of compounds far downstream in the pathway, including lignin (Hahlbrock & Grisebach 1979; Campbell & Sederoff 1996). However, to generalize that UV-B induction of increased PAL activity will necessarily lead to an increase in all pathway products is not warranted. This pathway is regulated differently in various tissues and also varies among species (e.g. Lavola 1998).

Apart from PAL, several other enzymes of the phenylpropanoid pathway are stimulated by UV radiation, as well as by other stress factors, including 4-coumarate:CoA ligase (4CL), the last enzyme of the core phenylpropanoid pathway, chalcone synthetase (CHS), at the beginning of the branch leading to various flavonoid pathways, and resveratrol synthetase, an enzyme responsible for the synthesis of resveratrol (stilbene), a phytoalexin in peanut plants (Hahlbrock & Scheel 1989).

Given that several biotic and abiotic stress factors can elicit increased regulatory enzyme activity of the phenylpropanoid pathway, including visible and UV-A radiation, should one expect that increased solar UV-B in nature would have a noticeable effect on the accumulation of pathway products? One of the most commonly observed higher plant responses to elevated UV-B in both glasshouse and field experiments is an increase in UV-B-absorbing pigments, thought to be primarily flavonoids (Caldwell & Flint 1994). Thus, there does indeed appear to be a very specific stimulation of the phenylpropanoid pathway by UV-B. This is not necessarily expected, given that the UV-B component of sunlight is so small and plants exposed day after day to the full solar spectrum will still respond to a small supplement of UV-B (Flint et al. 1985; Barnes et al. 1987; Sullivan et al. 1994). The stimulation by UV-B of both PAL, at the initiation of the phenylpropanoid pathway, and particularly CHS, at the branch point leading to various flavonoid pathways, may be responsible for the relatively pronounced influence of UV-B on elevating flavonoids and related compounds in such experiments. Apart from flavonoids and related compounds, similar experiments in which plants were given different levels of UV-B have shown that more UV-B can result in greater levels of products from other branches of the phenylpropanoid pathway, e.g. furanocoumarins (McCloud & Berenbaum 1994) and tannins (Gehrke et al. 1995). Similarly, salicylic acid has been shown to be stimulated by UV-C (Yalpani et al. 1994) and might also be stimulated by UV-B. In the case of salicylic acid, it is thought not to have antimicrobial properties *per se*, but is part of a signalling process that leads to the development of systemic resistance and the production of pathogenesis-related proteins (Yalpani et al. 1994; Green & Fluhr 1995). However, an increase in all phenylpropanoid products should not necessarily be expected. Competition for substrate among different branches of the pathway may result in decreased products if regulatory enzymes for other branches are stimulated (Björn 1996), generalizations should be made with caution.

A variety of phenylpropanoid products can influence many interactions between vegetation, microbes and herbivores. Plant antimicrobial compounds induced by stress, commonly termed phytoalexins, include isoflavonoids, stilbenes, psoralens, coumarins and some flavonols (Dixon & Paiva 1995). Many of these compounds and related compounds such as tannins, lignin and suberin can play a role in defence against insect and vertebrate herbivores. Additionally, some flavonoids are important regulatory compounds, such as in *Rhizobium* infection of roots with resulting nitrogen-fixing activity (Peters *et al.* 1986; Stafford 1997). Thus, if increased solar UV-B elicits changes in these phenylpropanoid products, alterations in plant–herbivore relations, plant pathogen resistance and some plant mutualisms might be expected.

### Plant pathogenesis

Apart from the effects of UV-B on host chemistry, direct UV-B effects on pathogens also may be involved in the incidence and severity of plant diseases. Fungal pathogens of the genus *Septoria* can cause significant yield reduction in wheat and can be directly inhibited by UV-B (Paul 1997). Several other plant pathogens are also sensitive to UV-B *in vitro*, yet when these pathogens are in the host plants, the disease incidence can often be increased by elevated UV-B radiation (Paul 1997). This is most likely to be due to alterations in the host tissues as mentioned above. Furthermore, pathogens inhabiting the internal host tissues would enjoy the protection from UV radiation offered by UV-absorbing pigments and the structure of the host. Sporulation of fungal pathogens can be stimulated by UV radiation, but this is generally considered primarily a UV-A response (Manning & V-Tiedemann 1995) and UV-A is not affected by changes in atmospheric ozone concentrations.

The direction and degree of influence of solar UV on disease severity is not easily predicted. Twelve studies were summarized by Manning and V-Tiedemann (1995) and Paul (1997). Even though pathogens were shown *in vitro* to be UV sensitive in eight studies, disease severity was increased in many of these studies and, overall, in eight of the reports. Pathogens that were UV sensitive *in vitro* were less virulent in the host (two studies) or were unaffected in some of the results in eight studies. While changes in plant pathogen susceptibility resulting from elevated solar UV-B could have important ecosystem-level consequences, one cannot generalize about the directions of these changes at present.

### Insect herbivory

Most of the alterations of insect consumption of plants appear to be mediated primarily by changes of the host plants rather than direct UV-B effects on the insects. Many of these reports indicate that insect herbivory is reduced, but a couple of studies indicated the opposite response (Table 12.2). The reasons for the UV-induced changes in herbivory vary. Insect larvae (a cabbage looper, *Trichoplusia ni*) developed less well on a diet of rough lemon (*Citrus jambhiri*) foliage previously irradiated with a higher flux of UV-B (McCloud & Berenbaum 1994) and this was attributed to higher concentrations of furanocoumarins. This was one of the few studies in which the insect larvae were also directly exposed to UV-B during

**Table 12.2** Effects of UV radiation (primarily UV-B) on insect herbivory.

| Insect | Plant species | Insect response to UV-treated plant material | Possible mechanism of plant mediation | Reference |
|---|---|---|---|---|
| Bombyx mori (silkworm) | Morus alba (mulberry) | Less herbivory | Increased phytoalexins | Yasui & Yazawa (1995)[1] |
| Autographa gamma (moth larvae) | Pisum sativum (pea) | Less herbivory; but greater insect growth | Higher phenolic and nitrogen contents | Hatcher & Paul (1994) |
| Ostrinia nubilalis (European corn borer) | Zea mays (corn) | Less herbivory | Greater cell-wall-bound truxillic and truxinic acids | Bergvinson et al. (1994)[2] |
| Trichoplusia ni (cabbage looper) | Citrus jambhiri (rough lemon) | Decrease in survivorship and growth | Greater furanocoumarins | McCloud & Berenbaum (1994) |
| Coleoptera (leaf beetles) | Datura ferox (summer annual) | Less herbivory | Mechanism not known | Ballaré et al. (1996) |
| Acronicta, Nycteola, Orthosia, Ptilodon (caterpillars of night-flying moths) | Quercus robur (oak) | No UV-B effect[3] | Mechanism not known | Newsham et al. (1996) |
| Insects not identified | Vaccinium myrtillus, V. uliginosum, V. vitis-idaea (heathland shrubs) | More herbivory in V. myrtillus, less in V. uliginosum, no effect in V. vitis-idaea | Mechanism not known | Gwynn-Jones et al. (1997) |
| Lepidoptera: Noctuidae (moth larvae) | Gunnera magellanica (devil's strawberry) | Less herbivory | Slight increase in leaf nitrogen | Rousseaux et al. (1998) |
| Strophingia ericae (a psyllid) | Calluna vulgaris (heather) | Reduced insect populations | Reduced amino acid isoleucine | Salt et al. (1998) |
| Operophtera brumata (larvae of a winter moth) | Betula pendula (silver birch) | More herbivory | Leaf flavonoids increased, but flavonoids added to an artificial diet did not increase feeding | Lavola et al. (1998) |
| Caliothrips phaseoli (a thrip) | Glycine max (soybean) | Less herbivory; direct UV-B avoidance | Not known | Mazza et al. (1999) |

[1] UV-C response only was examined.
[2] UV-A and UV-B responses cannot be separated.
[3] More herbivory under UV-A lamps but no specific UV-B effect.

feeding and development. Both the greater furanocoumarins resulting from prior UV-B irradiation of the plants as well as direct UV-B irradiation of the larvae appeared to have negative effects on the larvae. Hatcher and Paul (1994) reported that if pea plants were previously irradiated with elevated UV-B, moth larvae (*Autographa gamma*) utilizing the pea plants grew much better, used their food more efficiently and also consumed less plant material. Although phenolic content was greater in the plants irradiated with UV-B, they ascribed the altered herbivore performance to increased nitrogen content of the host. From these few laboratory and greenhouse studies (Table 12.2), it would appear that both changes in host plant secondary chemistry (phenolics) and primary chemistry (nitrogen) can mediate the effects of elevated UV-B on plant–insect interactions.

Field experiments in which ambient solar UV-B radiation was modified by selective filters showed that ambient solar UV-B radiation can substantially reduce insect herbivory of agricultural and native plant foliage (Ballaré *et al.* 1996; Rousseaux *et al.* 1998; Mazza *et al.* 1999). Similarly, Bergvinson *et al.* (1994) showed that ambient solar UV (UV-B+UV-A) reduced leaf feeding damage by the European corn borer on maize plants. They attributed the reduced herbivore sensitivity of the UV-exposed plants to greater levels of specific cell-wall-bound phenolics (truxillic and truxinic acids that are photodimerized by UV), although it was not possible to separate the effects of UV-B from UV-A. Field studies involving supplementation of solar UV-B radiation with lamp systems also indicated a substantial reduction in populations of a phloem-feeding insect on a heathland plant (Salt *et al.* 1998). As mentioned above, most of the laboratory and field studies of elevated UV-B on insect herbivory have concentrated on influences mediated by changes in the plant host tissues. In the study reported by Mazza *et al.* (1999), changes in soybean host tissues led to decreased herbivory by thrips (*Caliothrips phaseoli*), but in addition, the thrips were shown to sense and avoid the solar UV-B in the field.

Apart from the phenylpropanoid products, several other effects of UV-B on plants including cuticular wax alterations (Steinmüller & Tevini 1985), changes in leaf morphology such as increased leaf thickness (Caldwell & Flint 1994), induction of plant pathogenesis-related proteins (Green & Fluhr 1995) and possibly singlet oxygen production (Berenbaum & Larson 1988) may also affect plant susceptibility to pathogens and herbivores.

## Direct UV-B effects on microbes

Microbes, such as the intestinal bacterium *E. coli*, have been the primary study object for much of what is known about the molecular mechanisms of UV action and repair systems. Less is known about how much terrestrial microbes exposed to sunlight may be affected by increased solar UV-B. Potential direct effects of solar UV-B on plant pathogens have been explored to some degree, as discussed earlier. Other pathogens and decomposer microbes can also be sensitive to solar UV-B. The half-lives of many animal and insect pathogens, e.g. viruses, bacteria, fungi and protozoa, in direct sunlight are often of the order of only a few hours (Zimmerman

1982; Ignoffo 1992). Because many of these pathogens affect the success of insect herbivores, there are obvious implications for vegetation. In the literature on biological control agents for insect pests, considerable attention has been directed to protecting disease agents against sunlight by UV-screening agents (e.g. Ignoffo & Batzer 1971; Shapiro 1984) or antioxidants (e.g. Ignoffo & Garcia 1994). With ozone reduction and consequent increases in the UV-B component of sunlight, insect pathogens might be decreased in effectiveness, but more UV-resistant strains would be likely to emerge.

Other free-living microbes also may be influenced by solar radiation. For example, the species composition of fungi that inhabit leaf surfaces have been shown to be altered by solar UV radiation (Newsham *et al.* 1997a). The decomposition of plant litter may be influenced by solar UV-B by several pathways (Fig. 12.2), including effects of the UV-B on microbes involved in litter decomposition as well as indirect effects mediated through changes in plant tissue chemistry, direct photochemical degradation of lignins and other litter components (Moorhead & Callaghan 1994; Gehrke *et al.* 1995; Zepp *et al.* 1995). Gehrke *et al.* (1995) used laboratory microcosms to show that the species composition of fungal decomposers could be changed by UV-B treatments. Newsham *et al.* (1997b) demonstrated similar changes in field experiments involving supplemental UV-B radiation. Of course, direct effects on decomposer microbes would be restricted to those directly exposed to sunlight.

Although direct UV effects on microbes certainly may play a role in several ecosystem-level phenomena, it is difficult to separate direct radiation effects on microbes from those mediated by UV-B influences on plant tissue chemistry. A compilation of fungi–plant interactions including pathogenic, endophytic, phylloplanic and decomposer fungi indicates some effects on the higher plants, or plant litter, but often more important effects on the fungal–plant interaction (Table 12.3). The direct UV effects on the fungi are not always readily distinguished.

## Effects mediated through abiotic photochemistry

Increased flux of solar UV-B may lead to accelerated photochemical reactions in the lower atmosphere, in plant litter and other organic debris, constituents in water and in various man-made compounds. Many of these abiotic photochemical reactions can exert indirect effects on ecosystems. The degradation of plant litter is, in part, non-biological when lignin and other molecules are broken down by solar radiation, with shorter wavelengths (e.g. the UV-B) being particularly effective (Moorhead & Callaghan 1994). The UV-B waveband has been shown to be more effective than longer wavelength radiation in stimulating the evolution of carbon monoxide (CO) from plant litter, but UV-A also contributes significantly to the photoproduction of CO (Tarr *et al.* 1995; Zepp *et al.* 1995). In otherwise unpolluted regions of tropical savannas, solar photoproduction of CO from plant litter might be a significant component of the total CO emissions and also a significant fraction of global CO emissions (Tarr *et al.* 1995).

**Table 12.3** Responses to UV-B radiation in fungi and plants. The studies were conducted under growth chamber, greenhouse or field conditions. Filtered UV lamps were used in all cases except Gunasekera *et al.* (1997) where sunlight was filtered.

| Fungal genus | Plant species | Fungal response to UV-B | Plant response to UV-B | Reference |
|---|---|---|---|---|
| *Puccinia* (pathogen) | *Triticum aestivum* (wheat) | Increased infection with results varying by wheat cultivar | Little, if any, reduction in dry weight and seed yield | Biggs and Webb (1986) |
| *Colletotrichum, Cladosporium* (pathogens) | *Cucumis sativus* (cucumber) | Increased infection in some cases (an indirect UV-B effect mediated through the plant) | Reduced plant height, leaf area, dry weight; increased leaf mass per area | Orth *et al.* (1990) |
| *Cercospora* (pathogen) | *Beta vulgaris* (sugar beet) | Not assessed | Reduced dry weight of leaf laminae and other plant parts in the presence of the pathogen | Panagopoulos *et al.* (1992) |
| *Pyricularia* (pathogen) | *Oryza sativa* (rice) | Greater lesions on plant in a few cases (an indirect UV-B effect mediated through the plant) | Reduced plant height, leaf area and dry weight in a few cases | Finckh *et al.* (1995) |
| *Septoria* (pathogen) | *Triticum aestivum* (wheat) | Fewer lesions on plant (a direct response to UV-B) or no effect, depending on time of year | Not assessed; if changes occurred, they did not affect the pathogen | Paul *et al.* (1997) |
| *Exobasidium* (pathogen) | *Camellia sinensis* (tea) | Reduced infection sites; no effect on sporulation | Not assessed | Gunasekera *et al.* (1997) |
| *Aureobasidium, Sporobolomyces* (phylloplane yeasts) | *Quercus robur* (oak) | Abundance on the upper, but not lower, leaf surface affected for some sampling dates, several other fungi not affected | Not assessed | Newsham *et al.* (1997a) |
| *Neotyphodium* (a leaf endophyte) | *Lolium perenne* (rye grass) | No effect | Reduced yield in the presence of the leaf endophyte | Newsham *et al.* (1998) |
| *Mucor, Truncatella, Penicillium* (saprotrophs) | *Vaccinium uliginosum* (a heathland shrub) | Reduced fungal colonization of decomposing leaves; change in fungal species composition | Altered leaf litter quality | Gehrke *et al.* (1995) |
| *Cladosporium, Acremonium,* and others (saprotrophs) | *Quercus robur* (oak) | Reduced fungal colonization of decomposing leaves; change in fungal species composition | Some transitory UV-B effect on mass loss of decomposing litter | Newsham *et al.* (1997b) |

As the stratosphere becomes more transparent to solar UV-B, this radiation penetrates more effectively into the troposphere where it can have effects on atmospheric photochemistry and, thereby, indirect effects on vegetation. For example, the increased UV-B flux can either increase or decrease the production of tropospheric ozone ($O_3$) and hydrogen peroxide ($H_2O_2$), both of which are quite detrimental to vegetation in sufficient concentrations. The direction of the changes in $O_3$ and $H_2O_2$ depend on the prevailing concentrations of nitrogen oxides ($NO_x$) (Tang & Madronich 1995).

Polycyclic aromatic hydrocarbons (PAHs) are widespread anthropogenically derived compounds resulting from refining and other industrial activities. Sunlight, and particularly the UV-B component, can photoconvert these compounds and the products can be much more toxic than the parent compounds, as demonstrated for anthracene, benzo(a)pyrene and phenanthrene with a duckweed, *Lemna gibba* (Huang *et al.* 1993).

## Conclusions

Although UV-B radiation is sufficiently actinic to cause several types of photochemical damage in plants, most of the field experimental work that has employed realistic supplemental UV-B or exclusion of ambient solar UV-B suggests that most plants will not suffer substantial damage or reduction in production as the ozone layer thins. Plant ultraviolet shielding and radiation damage repair systems appear to be adequate to prevent appreciable damage if effective UV-B increases by even some 30%—far more than predicted for most regions of the world. There are some data suggesting that UV-B effects might accumulate through several seasons in perennial foliage. Similarly, other UV-B effects appear to accumulate through subsequent generations of plants. Although UV-B damage should not be entirely dismissed, the most important ecosystem-level effects are likely to be indirect and the magnitude and even direction of these effects is not easily predicted in many cases. These indirect effects may be mediated through vegetation, direct effects on microbes and sometimes insects, or by abiotic photochemistry, i.e. altering the composition of the lower atmosphere, organic materials such as plant litter and man-made chemicals. Of the effects mediated through vegetation, the most important are likely to be the result of UV-B effects on regulatory processes resulting in changes of morphology and secondary chemistry, especially the phenylpropanoid pathway. Many of these effects, if not all, are likely to be the result of gene up- or down-regulation triggered by UV-B receptor(s). The competitive balance among plant species can be altered if the UV-B component of sunlight is augmented or reduced. Pathogens of plants and insects can be sensitive to UV-B damage. However, many of the indirect effects of UV-B on pathogen attack appear to primarily involve alterations of the host tissue chemistry and morphology. Similarly, insect herbivory on plants can be considerably altered, and often there is a reduction of insect consumption by UV-B. Decomposition processes can be influenced by UV-B radiation through several pathways, including effects on the decomposers, changes in plant litter quality and direct photochemical action on litter exposed to sunlight. Other direct photochemical effects on the atmosphere or man-made chemicals can elicit enhanced pollutant action on plants and animals.

### Acknowledgements

Many of the ideas in this chapter stemmed from research supported by the US Department of Agriculture (CSRS/NRICG 95–37100–1612, 95–35100–3166 and 98–35100–6107) and the US National Science Foundation (IBN 9524144) as part of the interagency 'Terrestrial Ecology and Global Change' Program (TECO).

### References

Antonelli, F., Bussotti, F., Grifoni, D., Grossoni, P., Mori, B., Tani, C. *et al.* (1998). Oak (*Quercus robur* L.) seedlings responses to a realistic increase in UV-B radiation under open space conditions. *Chemosphere*, **36**, 841–845.

Baker, N., Nogués, S. & Allen, D. (1997). Photosynthesis and photoinhibition. In *Plants and UV-B: Responses to Environmental Change* (Ed. by P. Lumsden), pp. 95–111. Cambridge University Press, Cambridge.

Ballaré, C.L., Barnes, P.W. & Kendrick, R.E. (1991). Photomorphogenic effects of UV-B radiation on hypocotyl elongation in wild type and stable-phytochrome-deficient mutant seedlings of cucumber. *Physiologia Plantarum*, **83**, 652–658.

Ballaré, C.L. & Scopel, A.L. (1997). Phytochrome signalling in plant canopies: testing its population-level implications with photoreceptor mutants of *Arabidopsis*. *Functional Ecology*, **11**, 441–450.

Ballaré, C.L., Barnes, P.W. & Flint, S.D. (1995a). Inhibition of hypocotyl elongation by ultraviolet-B radiation in de-etiolating tomato seedlings. I. The photoreceptor. *Physiologia Plantarum*, **93**, 584–592.

Ballaré, C.L., Barnes, P.W., Flint, S.D. & Price, S. (1995b). Inhibition of hypocotyl elongation by ultraviolet-B radiation in de-etiolating tomato seedlings. II. Time-course, comparison with flavonoid responses and adaptive significance. *Physiologia Plantarum*, **93**, 593–601.

Ballaré, C.L., Scopel, A.L., Stapleton, A.E. & Yanovsky, M.J. (1996). Solar ultraviolet-B radiation affects seeding emergence, DNA integrity, plant morphology, growth rate, and attractiveness to herbivore insects in *Datura ferox*. *Plant Physiology*, **112**, 161–170.

Barnes, P.W., Flint, S.D. & Caldwell, M.M. (1987). Photosynthesis damage and protective pigments in plants from a latitudinal arctic/alpine gradient exposed to supplemental UV-B radiation in the field. *Arctic and Alpine Research*, **19**, 21–27.

Barnes, P.W., Jordan, P.W., Gold, W.G., Flint, S.D. & Caldwell, M.M. (1988). Competition, morphology and canopy structure in wheat (*Triticum aestivum* L.) and wild oat (*Avena fatua* L.) exposed to enhanced ultraviolet-B radiation. *Functional Ecology*, **2**, 319–330.

Barnes, P.W., Flint, S.D. & Caldwell, M.M. (1990). Morphological responses of crop and weed species of different growth forms to ultraviolet-B radiation. *American Journal of Botany*, **77**, 1354–1360.

Barnes, P.W., Flint, S.D. & Caldwell, M.M. (1995). Early-season effects of supplemented solar UV-B radiation on seedling emergence, canopy structure, simulated stand photosynthesis and competition for light. *Global Change Biology*, **1**, 43–53.

Beggs, C. & Wellmann, E. (1994). Photocontrol of flavonoid biosynthesis. In *Photomorphogenesis in Plants* (Ed. by R. Kendrick & G. Kronenberg), pp. 733–751. Kluwer Academic Publishers, The Netherlands.

Berenbaum, M.R. & Larson, R.A. (1988). Flux of singlet oxygen from leaves of phototoxic plants. *Experientia*, **44**, 1030–1032.

Bergvinson, D., Arnason, J., Hamilton, R., Tachibana, S. & Towers, G. (1994). Putative role of photodimerized phenolic acids in maize resistance to *Ostrinia nubilalis* (Lepidoptera: Pyralidae). *Environmental Entomology*, **23**, 1516–1523.

Berner, R. (1997). The rise of plants and their effect on weathering and atmospheric $CO_2$. *Science*, **276**, 544–546.

Biggs, R.H. & Webb, P.G. (1986). Effects of enhanced ultraviolet-B radiation on yield, and disease incidence and severity for wheat under field conditions. In *Stratospheric Ozone Reduction, Solar Ultraviolet Radiation and Plant Life* (Ed. by R.C. Worrest & M.M. Caldwell), pp. 303–311. Springer-Verlag, Berlin.

Björn, L. (1996). Effects of ozone depletion and increased UV-B on terrestrial ecosystems. *International Journal of Environmental Studies*, **51**, 217–243.

Björn, L., Callaghan, T., Gehrke, C., Gunnarsson, T., Holmgren, B., Johanson, U. *et al.* (1997). Effects on subarctic vegetation of enhanced UV-B radiation. In *Plants and UV-B: Responses to Environmental Change* (Ed. by P. Lumsden), pp. 233–246. Cambridge University Press, Cambridge.

Bogenrieder, A. & Klein, R. (1982). Does solar UV influence the competitive relationship in higher plants? In *The Role of Solar Ultraviolet Radiation in Marine Ecosystems* (Ed. by J. Calkins), pp. 641–649. Plenum Press, New York.

Britt, A. (1996). DNA damage and repair in plants. *Annual Review of Plant Physiology and Plant Molecular Biology*, **47**, 75–100.

Butler, W.L., Hendricks, S.B. & Siegelman, H.W. (1964). Action spectra of phytochrome *in vitro*. *Photochemistry and Photobiology*, **3**, 521–528.

Caldwell, M.M. (1968). Solar ultraviolet radiation as an ecological factor for alpine plants. *Ecological Monographs*, **38**, 243–268.

Caldwell, M.M. (1979). Plant life and ultraviolet radiation: some perspective in the history of the earth's UV climate. *BioScience*, **29**, 520–525.

Caldwell, M.M. & Flint, S.D. (1994). Stratospheric ozone reduction, solar UV-B radiation and terrestrial ecosystems. *Climatic Change*, **28**, 375–394.

Caldwell, M.M., Robberecht, R. & Billings, W.D. (1980). A steep latitudinal gradient of solar ultraviolet-B radiation in the arctic-alpine life zone. *Ecology*, **61**, 600–611.

Caldwell, M.M., Robberecht, R. & Flint, S.D. (1983). Internal filters: prospects for UV-

acclimation in higher plants. *Physiologia Plantarum*, **58**, 445–450.

Caldwell, M.M., Teramura, A.H., Tevini, M., Bornman, J.F., Björn, L.O. & Kulandaivelu, G. (1995). Effects of increased solar ultraviolet radiation on terrestrial plants. *Ambio*, **24**, 166–173.

Campbell, M.M. & Sederoff, R.R. (1996). Variation in lignin content and composition. Mechanisms of control and implications for the genetic improvement of plants. *Plant Physiology*, **110**, 3–13.

Dai, Q., Peng, S., Chavez, A., Miranda, M., Vergara, B. & Olszyk, D. (1997). Supplemental ultraviolet-B radiation does not reduce growth or grain yield in rice. *Agronomy Journal*, **89**, 793–799.

Day, T.A. (1993). Relating UV-B radiation screening effectiveness of foliage to absorbing-compound concentration and anatomical characteristics in a diverse group of plants. *Oecologia*, **95**, 542–550.

Day, T.A., Vogelmann, T.C. & DeLucia, E.H. (1992). Are some plant life forms more effective than others in screening out ultraviolet-B radiation? *Oecologia*, **92**, 513–519.

Dixon, R.A. & Paiva, N.L. (1995). Stress-induced phenylpropanoid metabolism. *Plant Cell*, **7**, 1085–1097.

Ehleringer, J.R., Cerling, T.E. & Helliker, B.R. (1997). $C_4$ photosynthesis, atmospheric $CO_2$ and climate. *Oecologia*, **112**, 285–299.

Ensminger, P.A. (1993). Control of development in plants and fungi by far-UV radiation. *Physiologia Plantarum*, **8**, 501–508.

Finckh, M.R., Chavez, A.Q., Dai, Q. & Teng, P.S. (1995). Effects of enhanced UV-B radiation on the growth of rice and its susceptibility to rice blast under greenhouse conditions. *Agriculture, Ecosystems and Environment*, **52**, 223–233.

Fiscus, E.L. & Booker, F.L. (1995). Is increased UV-B a threat to crop photosynthesis and productivity? *Photosynthesis Research*, **43**, 81–92.

Flint, S.D. & Caldwell, M.M. (1984). Partial inhibition of *in vitro* pollen germination by simulated solar ultraviolet-B radiation. *Ecology*, **65**, 792–795.

Flint, S.D., Jordan, P.W. & Caldwell, M.M. (1985). Plant protective response to enhanced UV-B radiation under field conditions: leaf optical properties and photosynthesis. *Photochemistry and Photobiology*, **41**, 95–99.

Fox, F.M. & Caldwell, M.M. (1978). Competitive interaction in plant populations exposed to supplementary ultraviolet-B radiation. *Oecologia*, **36**, 173–190.

Foyer, C.H., Lelandais, M. & Kunert, K.J. (1994). Photooxidative stress in plants. *Physiologia Plantarum*, **92**, 696–717.

Gehrke, C., Johanson, U., Callaghan, T.V., Chadwick, D. & Robinson, C.H. (1995). The impact of enhanced ultraviolet-B radiation on litter quality and decomposition processes in *Vaccinium* leaves from the subarctic. *Oikos*, **72**, 213–222.

Gold, W.G. & Caldwell, M.M. (1983). The effects of ultraviolet-B radiation on plant competition in terrestrial ecosystems. *Physiologia Plantarum*, **58**, 435–444.

Goto, N., Yamamoto, K.T. & Watanabe, M. (1993). Action spectra for inhibition of hypocotyl growth of wild-type plants and of the hy2 long-hypocotyl mutant of *Arabidopsis thaliana* L. *Photochemistry and Photobiology*, **57**, 867–871.

Green, R. & Fluhr, R. (1995). UV-B-induced PR-1 accumulation is mediated by active oxygen species. *Plant Cell*, **7**, 203–212.

Gunasekera, T.S., Paul, N.D. & Ayres, P.G. (1997). The effects of ultraviolet-B (UV-B: 290–320 nm) radiation on blister blight disease of tea (*Camellia sinensis*). *Plant Pathology*, **46**, 179–185.

Gwynn-Jones, D., Lee, J.A. & Callaghan, T.V. (1997). Effects of enhanced UV-B radiation and elevated carbon dioxide concentrations on a sub-arctic forest heath ecosystem. *Plant Ecology*, **128**, 242–249.

Hahlbrock, K. & Grisebach, H. (1979). Enzymic controls in the biosynthesis of lignin and flavonoids. *Annual Review of Plant Physiology and Plant Molecular Biology*, **30**, 105–130.

Hahlbrock, K. & Scheel, D. (1989). Physiology and molecular biology of phenylpropanoid metabolism. *Annual Review of Plant Physiology and Plant Molecular Biology*, **40**, 347–369.

Hartmann, K. (1967). Ein Wirkungsspektrum der Photomorphogenese unter Hochenergiebedingungen und seine Interpretation auf der Basis des Phytochroms (Hypokotylwachstumshemmung bie *Lactuca sativa* L.). *Zeitschrift für Naturforschung*, **22b**, 1172–1175.

Hatcher, P.E. & Paul, N.D. (1994). The effect of elevated UV-B radiation on herbivory of pea by *Autographa gamma*. *Entomologia Experimentalis et Applicata*, **71**, 227–233.

Herrmann, K. (1995). The shikimate pathway: early steps in the biosynthesis of aromatic compounds. *Plant Cell*, **7**, 907–919.

Huang, X., Dixon, G. & Greenberg, B. (1993). Impacts of UV radiation and photomodification on the toxicity of PAHs to the higher plant *Lemna gibba* (duckweed). *Environmental Toxicology and Chemistry*, **12**, 1067–1077.

Hunt, J.E. & McNeil, D.L. (1998). Nitrogen status affects UV-B sensitivity of cucumber. *Australian Journal of Plant Physiology*, **25**, 79–86.

Ignoffo, C. (1992). Environmental factors affecting persistance of entomopathogens. *Florida Entomologist*, **75**, 516–525.

Ignoffo, C. & Batzer, O. (1971). Microencapsulation and ultraviolet protectants to increase sunlight stability of insect virus. *Journal of Economic Entomology*, **64**, 850–853.

Ignoffo, C. & Garcia, C. (1994). Antioxidant and oxidative enzyme effects on the inactivation of inclusion bodies of the *Heliothis baculovirus* by simulated sunlight-UV. *Environmental Entomology*, **23**, 1025–1029.

Johanson, U., Gehrke, C., Björn, O. & Callaghan, T.V. (1995). The effects of enhanced UV-B radiation on the growth of dwarf shrubs in a subarctic heathland. *Functional Ecology*, **9**, 713–719.

Jordan, B.R. (1996). The effects of ultraviolet-B radiation on plants: a molecular perspective. *Advances in Botanical Research*, **22**, 97–162.

Jordan, B.R., James, P.E., Strid, A. & Anthony, R.G. (1994). The effect of ultraviolet-B radiation on gene expression and pigment composition in etiolated and green pea leaf tissue: UV-B-induced changes are gene-specific and dependent upon the developmental stage. *Plant, Cell and Environment*, **17**, 45–54.

Klaper, R., Frankel, S. & Berenbaum, M.R. (1996). Anthocyanin content and UVB sensitivity in *Brassica rapa*. *Photochemistry and Photobiology*, **63**, 811–813.

Klironomos, J.N. & Allen, M.F. (1995). UV-B-mediated changes on below-ground communities associated with the roots of *Acer saccharum*. *Functional Ecology*, **9**, 923–930.

Kramer, G.F., Norman, H.A., Krizek, D.T. & Mirecki, R.M. (1991). Influence of UV-B radiation on polyamines, lipid peroxidation and membrane lipids in cucumber. *Phytochemistry*, **30**, 2101–2108.

Landry, L., Stapleton, A., Lim, J. *et al.* (1997). An *Arabidopsis* photolyase mutant is hypersensitive to ultraviolet-B radiation. *Proceedings of the National Academy of Sciences USA*, **94**, 328–332.

Lavola, A. (1998). Accumulation of flavonoids and related compounds in birch induced by UV-B irradiance. *Tree Physiology*, **18**, 53–58.

Lavola, A., Julkunen-Tiitto, R., Roininen, H. & Aphalo, P. (1998). Host-plant preference of an insect herbivore mediated by UV-B and CO₂ in relation to plant secondary metabolites. *Biochemical Systematics and Ecology*, **26**, 1–12.

McCloud, E.S. & Berenbaum, M.R. (1994). Stratospheric ozone depletion and plant–insect interactions: effects of UVB radiation on foliage quality of *Citrus jambhiri* for *Trichoplusia ni*. *Journal of Chemical Ecology*, **20**, 525–539.

McPeters, R. & Hollandsworth, S. (1996). Trends in global ozone as of 1995. *International Journal of Environmental Studies*, **51**, 165–182.

Madronich, S. (1993). The atmosphere and UV-B radiation at ground level. In *Environmental UV Photobiology* (Ed. by A.R. Young, L.O. Björn, J. Moan & W. Nultsch), pp. 1–39. Plenum Press, New York.

Madronich, S., McKenzie, R.L., Caldwell, M.M. & Björn, L.O. (1995). Changes in ultraviolet radiation reaching the earth's surface. *Ambio*, **24**, 143–152.

Manning, W.J. & V-Tiedemann, A. (1995). Climate change: potential effects of increased atmospheric carbon dioxide (CO₂), ozone (O₃), and ultraviolet-B (UV-B) radiation on plant diseases. *Environmental Pollution*, **88**, 219–245.

Margulis, L., Walker, J.C.G. & Rambler, M. (1976). Reassessment of roles of oxygen and ultraviolet light in Precambrian evolution. *Nature*, **264**, 620–624.

Markham, K.R., Franke, A., Given, D.R. & Brownsey, P. (1990). Historical antarctic ozone level trends from herbarium specimen flavonoids. *Bulletin de Liaison Du Groupe Polyphenols*, **15**, 230–235.

Mazza, C., Zavala, J., Scopel, A. & Ballaré, C. (1999). Perception of solar UVB radiation by photophagous insects: behavioral responses and ecosystem implications. *Proceedings of the National Academy of Sciences USA*, **96**, 980–985.

Miller, J.E., Booker, F.L., Fiscus, E.L., Heagle, A.S., Pursley, W.A., Vozzo, S.F. *et al.* (1994). Ultraviolet-B radiation and ozone effects on growth, yield, and photosynthesis of soybean. *Journal of Environmental Quality*, **23**, 83–91.

Moorhead, D.L. & Callaghan, T. (1994). Effects of increasing ultraviolet-B radiation on decomposition and soil organic matter dynamics: a synthesis and modelling study. *Biology and Fertility of Soils*, **18**, 19–26.

Murali, N.S. & Teramura, A.H. (1985). Effects of ultraviolet-B irradiance on soybean. VI. Influence of phosphorus nutrition on growth and flavonoid content. *Physiologia Plantarum*, **63**, 413–416.

Musil, C.F. (1995). Differential effects of elevated ultraviolet-B radiation on the photochemical and reproductive performances of dicotyledonous and monocotyledonous arid-environment ephemerals. *Plant, Cell and Environment*, **18**, 844–854.

Musil, C.F. (1996). Accumulated effect of elevated ultraviolet-B radiation over multiple generations of the arid-environment annual *Dimorphotheca sinuata* DC. (Asteraceae). *Plant, Cell and Environment*, **19**, 1017–1027.

Musil, C., Midgley, G. & Wand, S. (1999). Carry-over of enhanced ultraviolet-B exposure effects to successive generations of a desert annual: interaction with atmospheric CO₂ and nutrient supply. *Global Change Biology*, **5**, 311–329.

Newsham, K.K., McLeod, A.R., Greenslade, P.D. & Emmett, B.A. (1996). Appropriate controls in outdoor UV-B supplementation experiments. *Global Change Biology*, **2**, 319–324.

Newsham, K.K., Low, M.N.R., McLeod, A.R., Greenslade, P.D. & Emmett, B.A. (1997a). Ultraviolet-B radiation influences the abundance and distribution of phylloplane fungi on pedunculate oak (*Quercus robur*). *New Phytologist*, **136**, 287–297.

Newsham, K.K., McLeod, A.R., Roberts, J.D., Greenslade, P.D. & Emmett, B.A. (1997b). Direct effects of elevated UV-B radiation on the decomposition of *Quercus robur* leaf litter. *Oikos*, **79**, 592–602.

Newsham, K., Lewis, G., Greenslade, P. & McLeod, A. (1998). *Neotyphodium lolii*, a fungal leaf endophyte, reduces fertility of *Lolium perenne* exposed to elevated UV-B radiation. *Annals of Botany*, **81**, 397–403.

Orth, A.B., Teramura, A.H. & Sisler, H.D. (1990). Effects of ultraviolet-B radiation on fungal disease development in *Cucumis sativus*. *American Journal of Botany*, **77**, 1188–1192.

Panagopoulos, I., Bornman, J.F. & Björn, L.O. (1990). Effects of ultraviolet radiation and visible light on growth, fluorescence induction, ultra-weak luminescence and peroxidase activity in sugar beet plants. *Journal of Photochemistry and Photobiology B: Biology*, **8**, 73–87.

Panagopoulos, I., Bornman, J.F. & Björn, L.O. (1992). Response of sugar beet plants to ultraviolet-B (280–320 nm) radiation and Cercospora leaf spot disease. *Physiologia Plantarum*, **84**, 140–145.

Paul, N. (1997). Interactions between trophic levels. In *Plants and UV-B: Responses to Environmental Change* (Ed. by P. Lumsden), pp. 317–339. Cambridge University Press, Cambridge.

Paul, N., Rasanayagam, S., Moody, S., Hatcher, P. & Ayres, P. (1997). The role of interactions between trophic levels in determining the effects of UV-B on terrestrial ecosystems. *Plant Ecology*, **128**, 296–308.

Peters, N.K., Frost, J.W. & Long, S.R. (1986). A plant flavone, luteolin, induces expression of *Rhizobium meliloti* nodulation genes. *Science*, **233**, 977–980.

Reuber, S., Bornman, J.F. & Weissenböck, G. (1996). A flavonoid mutant of barley (*Hordeum vulgare* L.) exhibits increased sensitivity to UV-B radiation in the primary leaf. *Plant, Cell and Environment*, **19**, 593–601.

Robberecht, R., Caldwell, M.M. & Billings, W.D. (1980). Leaf ultraviolet optical properties along a latitudinal gradient in the arctic-alpine life zone. *Ecology*, **61**, 612–619.

Rousseaux, M., Ballaré, C., Scopel, A., Searles, P. & Caldwell, M. (1998). Solar ultraviolet-B radiation affects plant–insect interactions in a natural ecosystem of Tierra del Fuego (southern Argentina). *Oecologia*, **116**, 528–535.

Rozema, J., vandeStaaij, J., Björn, L.O. & Caldwell, M. (1997). UV-B as an environmental factor in plant life: stress and regulation. *Trends in Ecology and Evolution*, **12**, 22–28.

Saile-Mark, M. & Tevini, M. (1997). Effects of solar UV-B radiation on growth, flowering and yield of central and southern European bush bean cultivars (*Phaseolus vulgaris* L.). *Plant Ecology*, **128**, 114–125.

Salt, D., Moody, S., Whittaker, J. & Paul, N. (1998). Effects of enhanced UVB on populations of the phloem feeding insect *Strophingia ericae* (Homoptera: Psylloidea) on heather (*Calluna vulgaris*). *Global Change Biology*, **4**, 91–96.

Schwinning, S. & Weiner, J. (1998). Mechanisms determining the degree of size asymmetry in competition among plants. *Oecologia*, **113**, 447–455.

Searles, P.S., Caldwell, M.M. & Winter, K. (1995). The response of five tropical dicotyledon species to solar ultraviolet-B radiation. *American Journal of Botany*, **82**, 445–453.

Shapiro, M. (1984). Host tissues and metabolic products as ultraviolet screens for the gypsy moth (Lepidoptera: Lymantriidae) nucleopolyhedrosis virus. *Environmental Entomology*, **13**, 1131–1134.

Stafford, H.A. (1997). Roles of flavonoids in symbiotic and defense functions in legume roots. *Botanical Review*, **63**, 27–39.

Staxén, I. & Bornman, J.F. (1994). A morphological and cytological study of *Petunia hybrida* exposed to UV-B radiation. *Physiologia Plantarum*, **91**, 735–740.

Steinmüller, D. & Tevini, M. (1985). Action of ultraviolet radiation (UV-B) upon cuticular waxes in some crop plants. *Planta*, **164**, 557–564.

Stephanou, M. & Manetas, Y. (1998). Enhanced UV-B radiation increases the reproductive effort in the Mediterranean shrub *Cistus creticus* under field conditions. *Plant Ecology*, **134**, 91–96.

Strid, A. (1993). Alteration in expression of defence genes in *Pisum sativum* after exposure to supplementary ultraviolet-B radiation. *Plant and Cell Physiology*, **34**, 949–953.

Sullivan, J.H. & Teramura, A.H. (1990). Field study of the interaction between solar ultraviolet-B radiation and drought on photosynthesis and growth in soybean. *Plant Physiology*, **92**, 141–146.

Sullivan, J.H. & Teramura, A.H. (1992). The effects of ultraviolet-B radiation on loblolly pine. 2. Growth of field-grown seedlings. *Trees*, **6**, 115–120.

Sullivan, J.H., Teramura, A.H. & Dillenburg, L.R.

(1994). Growth and photosynthetic responses of field-grown sweetgum (*Liquidamber styraciflua*; Hamamelidaceae) seedlings to UV-B radiation. *American Journal of Botany*, **81**, 826–832.

Tang, X. & Madronich, S. (1995). Effects of increased solar ultraviolet radiation on tropospheric composition and air quality. *Ambio*, **24**, 188–190.

Tarr, M.A., Miller, W.L. & Zepp, R.G. (1995). Direct carbon monoxide photoproduction from plant matter. *Journal of Geophysical Research*, **100**, 11403–11413.

Teramura, A.H., Sullivan, J.H. & Lydon, J. (1990). Effects of UV-B radiation on soybean yield and seed quality: a 6-year field study. *Physiologia Plantarum*, **80**, 5–11.

Tevini, M. & Teramura, A.H. (1989). UV-B effects on terrestrial plants. *Photochemistry and Photobiology*, **50**, 479–487.

Tevini, M., Mark, U. & Saile, M. (1990). Plant experiments in growth chambers illuminated with natural sunlight. In *Environmental Research with Plants in Closed Chambers. Air Pollution Research Report 26* (Ed. by H.D. Payer, T. Pfirrman & P. Mathy), pp. 240–251. Commission of the European Communities, Belgium.

Torabinejad, J., Caldwell, M.M., Flint, S.D. & Durham, S. (1998). Susceptibility of pollen to UV-B radiation: an assay of 34 taxa. *American Journal of Botany*, **85**, 360–369.

Visser, A.J., Tosserams, M., Groen, M.W., Magendans, G.W.H. & Rozema, J. (1997). The combined effects of $CO_2$ concentration and solar UV-B radiation on faba bean grown in open-top chambers. *Plant, Cell and Environment*, **20**, 189–199.

Vogelmann, T.C. (1993). Plant tissue optics. *Annual Review of Plant Physiology and Plant Molecular Biology*, **44**, 231–251.

Wellmann, E. (1983). UV radiation in photomorphogenesis. In *Encyclopedia of Plant Physiology*, Vol. 16B (New Series). *Photomorphogensis* (Ed. by W. Shropshire, Jr & H. Mohr), pp. 745–756. Springer-Verlag, Berlin.

Yalpani, N., Enyedi, A.J., Leon, J. & Raskin, I. (1994). Ultraviolet light and ozone stimulate accumulation of salicylic acid, pathogenesis-related proteins and virus resistance in tobacco. *Planta*, **193**, 372–376.

Yasui, H. & Yazawa, M. (1995). Feeding response of the silkworm, *Bombyx mori*, to compounds induced by UV irradiation of mulberry leaves. *Bioscience, Biotechnology and Biochemistry*, **59**, 1326–1327.

Zepp, R.G., Callaghan, T.V. & Erickson, D.J. (1995). Effects of increased solar ultraviolet radiation on biogeochemical cycles. *Ambio*, **24**, 181–187.

Zimmermann, G. (1982). Effect of high temperatures and artificial sunlight on the viability of conidia of *Metarhizium anisopliae*. *Journal of Invertebrate Pathology*, **40**, 36–40.

Ziska, L.H., Teramura, A.H. & Sullivan, J.H. (1992). Physiological sensitivity of plants along an elevational gradient to UV-B radiation. *American Journal of Botany*, **79**, 863–871.

Ziska, L.H., Teramura, A.H., Sullivan, J.H. & McCoy, A. (1993). Influence of ultraviolet-B (UV-B) radiation on photosynthetic and growth characteristics in field-grown cassava (*Manihot esculentum* Crantz). *Plant, Cell and Environment*, **16**, 73–79.

# Chapter 13

# Understanding the impacts of rising $CO_2$: the contribution of environmental physiology

*S.P. Long*

## Introduction

Transpiration and photosynthesis, with the possible exception of respiration, are the two physiological processes by which plants sense directly, and respond to, the rising atmospheric partial pressure of $CO_2$ ($p_{CO_2}$). All other changes in plants growing in elevated $CO_2$ are, therefore, an indirect response to changes in these leaf-level processes (Drake *et al.* 1997). Understanding how photosynthesis and transpiration are affected by an increase in $p_{CO_2}$ is therefore fundamental to any sound prediction of future response of both natural and agricultural systems to atmospheric change. Predicting the feedback response of vegetation on atmospheric and climate change similarly depends on this understanding. Influential ecological discussions sometimes appear to have ignored physiological understanding of these primary effects, noting potential environmental and genetic restraints that could prevent plants utilizing the additional carbon and energy from increased photosynthesis. The first Intergovernmental Panel on Climate Change (IPCC) Scientific Assessment noted that while there were many uncertainties, it was unlikely that the short-term effects of elevated $p_{CO_2}$ on photosynthesis and transpiration had relevance in natural systems (Melillo *et al.* 1990). This chapter will show that physiological understanding of the responses of leaf transpiration and photosynthesis to long-term exposure to an elevation of $p_{CO_2}$ has profound relevance to understanding the feedback responses of terrestrial vegetation on the atmosphere–climate system. Both the theoretical arguments for and against this relevance will be evaluated, and compared with patterns emerging from field studies of $p_{CO_2}$ enrichment.

## Transpiration—from stomata to global warming

The aperture of stomata and, in turn, leaf transpiration ($E$), decrease with an increase in $p_{CO_2}$, with only a few exceptions. In intact leaves this response appears tightly linked to assimilation, so that at a given leaf–atmosphere vapour pressure

*Departments of Crop Sciences and Plant Biology, University of Illinois, 190 Edward R. Madigan Laboratory, 1201 W Gregory St, Urbana, IL 61801, USA. E-mail: stevel@life.uiuc.edu*

deficit (VPD) the ratio of intercellular $p_{CO_2}$ ($p_i$) to external $p_{CO_2}$ ($p_a$) remains constant with variation in light, assimilation and $p_a$. This has led to the development of phenomenological models of the response of stomatal conductance to $p_{CO_2}$ and humidity, in particular that of Ball *et al.* (1987). This model predicts that stomatal conductance ($g_s$) is directly proportional to the rate of leaf $CO_2$ uptake ($A$) and relative humidity (H), and inversely proportional to $p_{CO_2}$. Various modifications of the model have extended the range of conditions to which it may be applied, in particular use of the $CO_2$ compensation point of photosynthesis and of leaf–air water vapour pressure deficit (VPD) in place of H (Leuning 1995). An alternative approach has shown that the response of $g_s$ to elevated $p_{CO_2}$ and other environmental variables may be predicted by assuming that conductance is proportional to the extent to which photosynthetic capacity is realized (Jarvis & Davies 1998). This provides an alternative explanation of decreased $g_s$ under elevated $p_{CO_2}$, i.e. that the response results from increased realization of photosynthetic capacity at elevated $p_{CO_2}$ and not as a direct result of increased $p_{CO_2}$. A doubling of the $p_{CO_2}$ around a leaf, while increasing $A$, decreases $g_s$ and $E$. An acclimatory decrease in photosynthetic capacity causes a further decrease in conductance, as predicted by the models of both Ball *et al.* (1987) and Jarvis and Davies (1998). Because $g_s$ is mediated by changes in $A$, a decreased $g_s$ in plants with a reduced photosynthetic capacity is to be expected. Two quantitative syntheses, one of 23 and the other of 41 studies, of plants of varied form and origin, and grown under $p_{CO_2}$ elevated to 1.5–3 times current partial pressures, showed an average decrease in $g_s$ of 23% and 20%, respectively (Field *et al.* 1995; Drake *et al.* 1997). Most studies have reported $g_s$ at only one stage of growth. In wheat under free-air $CO_2$ enrichment (FACE) a persistent 29% decrease in $g_s$ was reported throughout the life of the crop, with elevation of $p_{CO_2}$ from 37 to 55 Pa (Garcia *et al.* 1998). Reduction of stomatal aperture and *gs* explains the reduction in leaf transpiration observed in plants grown in elevated $p_{CO_2}$. Does reduced *gs* in elevated $C_a$ limit photosynthesis in plants adapted to high $p_{CO_2}$?

### Stomata and photosynthesis

From the relationship of $A$ to $p_i$ of a leaf, the amount of assimilation lost because of a decrease in $p_{CO_2}$ as a result of the resistance to diffusion across the stomata may be estimated. The theoretical limitation imposed by stomata relative to a hypothetical leaf lacking any diffusive barrier from the leaf surface to the mesophyll can be quantified as the stomatal limitation to $CO_2$ assimilation (Farquhar & Sharkey 1982). Drake *et al.* (1997), summarizing studies of 26 species, noted by reference to $p_i$ and its relation to $A$ that under elevated $p_{CO_2}$ there was no evidence of any increase in the limitation to photosynthesis imposed by stomata. Thus, decreased transpiration is, at least on average, achieved without any additional limitation on carbon gain. Predicting change in $g_s$ with rising atmospheric $p_{CO_2}$ with phenomenological models such as that of Ball *et al.* (1987) requires that the response of stomata to $p_{CO_2}$, $A$ and H are the same at both current and elevated $p_{CO_2}$. Again, there is little evidence that stomata acclimate to elevated $p_{CO_2}$ independently of acclimation of photosynthesis (Eamus 1991; Long & Drake 1992; Sage 1994; Drake *et al.* 1997; Morison 1998).

## Reduced stomatal conductance and global warming

For closed canopies, the latent heat of evaporation used in evapotranspiration can remove half or more of the intercepted radiant energy. The observed and predicted decrease in $g_s$ and $E$ will result in a significant and predictable decrease in latent heat loss and a balancing increase in heat storage and sensible heat loss at the leaf level. Does this have relevance at the stand and landscape level?

The canopy and boundary layer conductances, which are independent of $p_{CO_2}$, will partially dampen this effect, their significance varying with windspeed and surface roughness. Where lower transpiration results in lower humidity over the vegetation, increase in VPD would increase transpiration for a given conductance and offset the effect of elevated $p_{CO_2}$. Conversely, the effect will be amplified if decreased atmospheric humidity induces decreased conductance. Increase in convection would alter mixing of air above the vegetation and transport of water vapour away from the surface. Therefore, many factors could negate or amplify the significance of the potential warming at the leaf level (Field *et al.* 1995). The IPCC 1990 Scientific Report noted that 'short-term measurements show that increased CO$_2$ reduces water-loss . . .', but went on to dismiss its significance by stating 'that for both physiological and meteorological reasons, high CO$_2$ might exert little or no effect on regional evapotranspiration . . .' (Melillo *et al.* 1990). The 'physiological' reasons mentioned in the report were that 'increases in leaf area and root extension observed in high-CO$_2$ plants tend to increase water use . . . counteracting the effect of lower transpiration per unit leaf area'. The second Scientific Report of the IPCC, while continuing to recognize the increased efficiency of water use at elevated $p_{CO_2}$, did not even mention decreased transpiration as a feedback in its consideration of how the responses of terrestrial vegetation to rising $p_{CO_2}$ may feedback on climate (Melillo *et al.* 1996). So is this effect as irrelevant as implied?

## Evidence that decreased stomatal conductance is linked to surface warming

Can increases in canopy surface temperature be detected within vegetation in the field, or do the potential feedbacks largely nullify the effect seen at the leaf level? Most field systems for elevating $p_{CO_2}$ around vegetation involve controlling the flow of air through the canopy atmosphere. This is achieved by total or, in the case of open-top chambers, partial enclosure. Thus, any additional heating of the canopy will be affected by the altered atmospheric coupling. Further, the practical limits on the size of enclosures, rarely more than 2 m diameter, would greatly diminish the ability to detect any local heating (McLeod & Long 1999). However, FACE systems with a minimal disturbance of daytime atmospheric coupling and a much greater scale, 20–30 m diameter, do provide an opportunity to test the hypothesis that elevated $p_{CO_2}$ leads to a daytime warming of a vegetated surface, relative to controls. During 1993 and 1994, wheat crops were grown at Maricopa,

Arizona, under $p_{CO_2}$ elevated to 55 Pa in four FACE rings of 23 m diameter, with four equivalent control rings within the same field (Pinter *et al.* 1996). Measurements of leaf and canopy evaporation, stomatal conductance and density, stem flow, soil moisture, water balance, energy exchange and canopy temperature, were made throughout the life of the crop. Elevated $p_{CO_2}$ decreased $g_s$ (Garcia *et al.* 1998), although stomatal density and frequency were not affected (Estiarte *et al.* 1994). These decreases at the leaf level corresponded to significant decreases in stem water flow, soil moisture use and canopy transpiration, even though the $p_{CO_2}$ treatment did not result in any change in leaf area index (Kimball *et al.* 1995; Pinter *et al.* 1996; Senock *et al.* 1996). Radiometry showed statistically significant 8% and 11% decreases in cumulative latent heat transfer over two consecutive growing seasons from these crops (Kimball *et al.* 1995; Pinter *et al.* 1996). This decrease in transpiration and latent heat transfer corresponded to a warmer canopy. Daytime canopy surface temperature, determined from false-colour thermal imaging of the FACE rings by aircraft remote sensing when canopy development was complete, showed a 1.2°C increase as a result of elevated $p_{CO_2}$ in fully irrigated subplots and 1.7°C in subplots subjected to a drought treatment. Over the life of the crop, infrared thermometry showed that the canopy surface temperature was increased by an average of 0.6°C (Kimball *et al.* 1995). If such increases occurred across continents then significant surface warming would occur simply because transpiration was suppressed. Further, this experiment may represent a minimum response. First, there was no acclimation of photosynthesis in the upper, exposed canopy leaves through most of the life of this well-fertilized crop (Garcia *et al.* 1998; Osborne *et al.* 1998). Second, evapotranspiration will be least responsive to a decrease in $g_s$ in aerodynamically smooth surfaces such as cereal crop canopies (Field *et al.* 1995). Conversely it could be argued that while the FACE rings were large enough to reveal significant local heating, they are an island within a much larger system functioning at the current atmospheric $p_{CO_2}$ and so avoid feedbacks that would operate at the landscape and regional level. To understand the significance of decreased leaf conductance at these larger scales requires the integration of physiological models of vegetation behaviour with atmospheric circulation models.

## Predicting the significance of stomatal response to global climate

The improved simple biosphere model (SiB2) incorporates a coupled photosynthesis-conductance submodel in its vegetation canopy model (Sellers *et al.* 1996a,b). This incorporates the stomatal response to $p_{CO_2}$ described by the model of Ball *et al.* (1987). By integrating SiB2 with satellite data of the type, density and greenness of vegetation, and then with a general global circulation model, Sellers *et al.* (1996c) were able to scale the effect of elevated $p_{CO_2}$ on stomata to global climate. This approach allowed simulation of the situation where all vegetation, rather than a localized patch, is affected by increased $p_{CO_2}$ and by the changes in humidity and surface temperature that would result. It also allowed analysis of the effect that vegetation would have on surface temperatures in the absence of any radiative warming resulting from increased absorbance of infrared radiation by greenhouse gases. Thus, their model was able to ask: in the absence of any direct radiative effect of doubled $pCO_2$ on climate, how much warming would result from decreased $g_s$ alone? The effect of decreased $g_s$ is termed 'physiological' warming, as distinct from the warming predicted from increased trapping of infrared radiation in the atmosphere by increased greenhouse gas partial pressures, termed 'radiative' warming. The simulation showed a 0.4°C increase in global air temperature averaged across the entire globe, including both the oceans and continents, as a result of physiological warming. On land, the predicted increase in average air temperature ranged from 0.6°C in the tropics to 1.1°C at 50.4–72.0°N. In the latter case this constituted one-third of the total warming of 3.3°C predicted when both radiative and physiological warming effects were combined. However, radiative and physiological effects were shown to combine in a non-additive way. The effect of adding in physiological effects to radiative produce a further increase of 0.9°C, compared to 0.6°C as a result of physiological effects in isolation in the tropics. Further, the increased surface heating resulting from physiological warming, which is temporally and spatially highly variable, is sufficient to influence the global circulation of air and heat. Generally, increases over current temperatures caused by decreased stomatal conductance will be greatest at the time of both seasonal and daily peaks of solar radiation (Sellers *et al.* 1996c). It is of particular significance that while maximum radiative warming resulting from increased greenhouse gas concentrations may require decades to develop, physiological warming is immediate. Thus, over the twenty-first century it could represent a major part of the warming that can be predicted. The finding indicates that incorporation of physiological response is vital to any meaningful prediction of global climate change (Sellers *et al.* 1997). Are there biological factors that could counteract the additional warming predicted by such general circulation models (GCMs) that incorporate this stomatal feedback response of vegetation?

## Limitations to the predicted warming caused by stomata

Melillo *et al.* (1990) suggested that decreased transpiration per unit leaf area could be counteracted by increased leaf area and root growth. If elevated $pCO_2$ suppresses regional transpiration so that soil moisture is increased, leaf growth and duration may be increased. This would be of particular significance in semi-arid areas lacking closed canopies. However, no significant increase in leaf area index was apparent in the average of a range of long-term growth studies (Drake *et al.* 1997). In the detailed studies of wheat crops within the FACE experiments, while initial increases in root growth (Wechsung *et al.* 1995) and leaf area (Pinter *et al.* 1996) were observed, these differences disappeared at canopy closure even though soil moisture use was decreased (Kimball *et al.* 1995). In other long-term studies very substantial decreases in transpiration were observed. For example, Field *et al.* (1997) found a 50% decrease in evapotranspiration during the long-term elevation of semiarid annual grassland. Two factors may be of particular significance to more precise prediction of physiological warming with rising $pCO_2$. These are interspecific variability in stomatal response and acclimatory decrease in photosynthetic capacity.

An analysis of stomatal responses to growth in elevated $p\text{CO}_2$ in trees has shown that while on average a doubling $p\text{CO}_2$ decreased $g_s$ by 23%, there is considerable variation between species and between growth conditions (Field *et al.* 1995). Of particular significance are coniferous species, which dominate the boreal forest biome. Laboratory observations have long suggested that at least some coniferous species are different, showing little or no response of stomatal aperture to large increases in $p\text{CO}_2$ (Saxe *et al.* 1998). Only one experiment has addressed the effect of elevated $p\text{CO}_2$ on mature coniferous forest in the open; the FACE experiment in mature *Pinus taeda* in the Duke Forest, North Carolina. Results from the first year have shown no direct effect of elevated $p\text{CO}_2$ on stomatal conductance (Ellsworth 1999). Even among flowering trees, although the majority of species appear to show similar responses in stomatal aperture to elevated $p\text{CO}_2$ to that observed in herbaceous plants, some species have been found to show little or no response (Saxe *et al.* 1998). Tropical forest trees represent a particular gap in knowledge. Improved characterization of stomatal responses to elevated $p\text{CO}_2$ in the dominant trees of different biomes will be crucial to predicting future climate, and improving the incorporation of vegetation feedback into GCMs. Equally, the range of conditions over which simple phenomenological models of stomatal response to elevated $p\text{CO}_2$ can operate must be extended, in particular to cope with drought. A major gap at present is the lack of understanding of the mechanism underlying the close relationship between assimilation, conductance and $p\text{CO}_2$. Understanding the underlying mechanism(s) will be fundamental to the development of more robust models. Despite a multitude of observations of photosynthetic acclimation, few studies have been designed to test the possibility of independent stomatal acclimation (Morison 1998). Again such information will be vital to accurate parameterization for interactive vegetation-GCMs.

Sellers *et al.* (1996c) showed that the degree of feedback of stomata on global climate was strongly dependent on assumptions about acclimation of photosynthetic capacity. Vegetation which had 'completely acclimated', i.e. showed complete loss of the initial stimulation of $A$ on transfer to elevated $p\text{CO}_2$, caused the greatest increase in global surface temperature. This is because phenomenologically, $g_s$ is proportional to $A$ so that decreases in $g_s$ and $E$ will be greatest when there is no increase in $A$ in elevated $p\text{CO}_2$. Acclimation of photosynthesis therefore has a double significance to global climate change—not only will it decrease the capacity of vegetation to absorb additional $CO_2$, it will also be a direct cause of global warming through its influence on $g_s$. Acclimation of photosynthesis is examined in the following section.

## Photosynthesis—does its short-term response to elevated $p\text{CO}_2$ predict long-term changes in production and distribution?

This section reviews expectations of plant responses to rising $p\text{CO}_2$ based upon an understanding of leaf photosynthesis. The evidence for and against the relevance of predictions based on photosynthetic mechanisms at the leaf level to the system level will be assessed, in part by analogy with $C_4$ photosynthesis. The role of acclimation is examined from a physiological perspective. Finally, the value of understanding photosynthetic mechanisms in predicting interactions of increase in $p_{CO_2}$ with other variables is illustrated by reference to the interactive effects of light and $p_{CO_2}$.

## Predicting the short-term responses of leaf photosynthesis to elevated $p_{CO_2}$

Photosynthesis is the primary process by which production in $C_3$ plants can respond to the change in atmospheric $p_{CO_2}$. It is the process that sets the potential limit on any increase in $CO_2$ uptake by plants and ecosystems with rising $p_{CO_2}$. All other changes in production, form and composition of plants must follow from this response. The only possible exception is respiration, which may be decreased independently by elevated $p_{CO_2}$ in some species (Drake *et al.* 1997). Photosynthesis can be regulated by $CO_2$ at a number of points within the chloroplast, including binding of Mn on the donor side of photosystem II (Klimov *et al.* 1995), the quinone binding site on the acceptor side of photosystem II (Govindjee 1993) and the activation of Rubisco (Portis 1995). However, these processes show a high affinity for $HCO_3^-$ or $CO_2$ and, unlike Rubisco, they are saturated at the current atmospheric $p_{CO_2}$ and cannot therefore respond to further increase in $p_{CO_2}$. Rubisco has a low affinity for $CO_2$ as the substrate for the carboxylation reaction and it is not saturated at the current atmospheric $p_{CO_2}$. In addition, Rubisco catalyses the oxygenation of ribulose-1,5-*bis*phosphate (RuBP), a reaction which is competitively inhibited by $CO_2$ (Bainbridge *et al.* 1995). Oxygenation of RuBP is the first step in the photosynthetic carbon oxidation or photorespiratory pathway (PCO), which decreases the net efficiency of photosynthesis by 20–50% (Zelitch 1973), by utilizing light energy and by releasing recently assimilated carbon as $CO_2$. $CO_2$ is a competitive inhibitor of the oxygenation reaction, such that a doubling of concentration at Rubisco will roughly halve the rate of oxygenation (Long 1991). This second effect on the PCO is of greater importance in an ecological context, because an increase in net photosynthesis will result regardless of whether photosynthesis is Rubisco or RuBP limited, and regardless of where metabolic control lies. The increase in uptake resulting from suppression of the PCO requires no additional light, water, or nitrogen, making the leaf more efficient in its net assimilation of $CO_2$ with respect to each of these resources.

Rubisco specificity ($\tau$) is the ratio of carboxylation to oxygenation activity when the concentrations of $CO_2$ and $O_2$ at Rubisco are equal. It determines directly the increase in efficiency of photosynthesis with rising $p_{CO_2}$. This value is therefore of fundamental importance in predicting the direct responses of plants to rising $p_{CO_2}$. $\tau$ has been suggested to vary little, from 88 to 110 across a range of $C_3$ higher plants, with an average of about 100 (Bainbridge *et al.* 1995). As temperature increases, $\tau$ declines sharply because both the solubility of $CO_2$ relative to $O_2$ and the affinity of Rubisco for $CO_2$ relative to $O_2$ decline (Jordan & Ogren 1984). The effect of this decline in $\tau$ with temperature is a progressive increase in the stimulation of photosynthesis by elevated $p_{CO_2}$ with an increase in temperature. The minimum stimulation of RuBP-limited photosynthesis by increasing $p_{CO_2}$ from 35 to 70 Pa rises from 4% at 10°C to 35% at 30°C (Drake et al. 1997). Because increasing $p_{CO_2}$ is predicted to increase leaf temperature, both directly by decreasing latent heat loss and indirectly through radiative forcing of the atmosphere, this interactive effect of $CO_2$ and temperature has profound importance to future photosynthesis. It also suggests a much greater stimulation of photosynthesis in hot *vs.* cold climates, in the absence of other limitations. In summary, rising $p_{CO_2}$ is therefore expected to increase net $CO_2$ assimilation, with the increase being more pronounced at higher temperatures. In addition, the net amount of $CO_2$ that may be fixed per unit of leaf nitrogen, water transpired or light absorbed is expected to rise. Thus, elevated $p_{CO_2}$ is expected to increase the efficiency with which plants can assimilate $CO_2$ relative to the resources of nitrogen, water and light.

Because the competing reactions of RuBP carboxylation and RuBP oxygenation dominate the response of photosynthesis to variation in $p_{CO_2}$, this can be mechanistically modelled. The model of Farquhar et al. (1980) was developed by combining the kinetic properties of Rubisco with a model of the light response to electron transport, to allow effective prediction of the light, temperature, $CO_2$ and $O_2$ responses of $C_3$ leaves. This model faithfully describes the immediate effects of elevation of $p_{CO_2}$ on the responses of A to temperature, nitrogen, water and light. This model has proved remarkably robust, presumably because of the conserved properties of photosynthesis and Rubisco across very different vegetation types and forms. It has been combined with well-defined canopy properties to allow scaling from the leaf to canopies and landscapes. It is commonly incorporated into both simple (Amthor 1995; Lloyd & Farquhar 1996) and complex canopy (Wang & Jarvis 1990; dePury & Farquhar 1997) models, and into models of atmosphere–biosphere bidirectional interaction (Field & Avissar 1998). The effective prediction of measured fluxes of $CO_2$ into large areas of forest is further evidence of the value of this approach (e.g. Lloyd et al. 1995). The model, and its $C_4$ counterpart (Collatz et al. 1992), have been incorporated in SVATs (Soil–Vegetation–Atmosphere Transfer models), such as SiB2, to construct a GCM with biosphere interactions (Randall et al. 1996). Is it, however, realistic to expect these short-term responses of leaf photosynthesis to elevated $CO_2$ to have relevance to the long term and to affect the future production and ecology of $C_3$ plants in an elevated $p_{CO_2}$ environment? Several points have been raised which can interfere or nullify predictions based on these simple leaf photosynthetic responses (Melillo et al. 1990; Oechel et al. 1994; Körner 1995; Melillo et al. 1996). These include acclimatory loss of photosynthetic capacity and limitation by other resources. In addition, a poor correlation between leaf photosynthesis and plant production might also seem a major barrier.

**Leaf photosynthesis is of relevance to productivity**
Surveys of crops and their ancestors have drawn attention to the fact that leaf photosynthesis can be poorly correlated or independent of productivity and yield (Gifford & Evans 1981). Such poor correlations should be no surprise, because net

carbon gain by a plant is a function of the size and arrangement of the canopy of leaves, whole-plant respiration, as well as the photosynthetic properties of the individual leaves (Chapter 3). The surveys used to establish this poor correlation are based largely on light-saturated rates measured at a single stage of crop development. The relationship of this measure to whole-plant photosynthesis is complex. First, 50% of whole-plant carbon gain may be from light-limited photosynthesis; here very different biochemical and biophysical properties determine photosynthetic rate (Long 1993). Second, increase in leaf area may often be achieved by decreased investment per unit leaf area, thus photosynthetic capacity per unit area is commonly lower in species with thinner leaves (Beadle & Long 1985). Contrary to the conclusions of Gifford and Evans (1981), Watanabe *et al.* (1994) showed a strong positive correlation between leaf photosynthetic rate and yield of Australian bread wheat cultivars. This difference might be explained by the fact that the latter study was limited to similar genotypes of a single species, where variability in leaf area and canopy arrangement would be small. Indeed an important contribution of elevated $CO_2$ research with crops and trees has been the demonstration of a tight linkage between increased leaf photosynthesis and increased production. Where the same $C_3$ genotype is grown in both current and elevated $p_{CO_2}$, an increase in leaf photosynthesis and a correlated increase in plant production, with little or no change in canopy size, is typically observed (Drake *et al.* 1997). Over a wide range of $C_3$ species, an approximate doubling of the current $p_{CO_2}$ produced no significant increase in leaf area, a 23–58% increase in leaf photosynthetic rate (Drake *et al.* 1997) and an average 35% increase in crop yield (Kimball 1983). In one of the few studies to follow leaf photosynthesis and production throughout the growth of a crop, a very close correlation between the increase in leaf photosynthesis and production occurs (Pinter *et al.* 1996; Garcia *et al.* 1998). It might be argued that this correlation between increased leaf photosynthesis and production under elevated $CO_2$ may be indirect. Evans (1997) noted that increased $p_{CO_2}$ improves plant water status and this could underlie the increases in production observed. Evidence that there is an independent increase due to increased leaf photosynthesis alone comes from two sources.

1 Large increases in production occurred under elevated $p_{CO_2}$ when wheat was irrigated in the field to the level required for maximum yield (Pinter *et al.* 1996) and when lowland rice was grown in a simulated paddy system in field chambers (Baker *et al.* 1990).

2 $C_4$ plants show similar reductions in stomatal aperture and dark respiration to $C_3$ plants when grown at elevated $p_{CO_2}$. In a mixed $C_3/C_4$ water-logged marsh community, elevated $p_{CO_2}$ caused a large and sustained increase in the net carbon gain of stands of the $C_3$, but not the $C_4$, species (Drake *et al.* 1995).

These findings suggest that while increased water use efficiency will contribute to production increase under elevated $p_{CO_2}$, when $C_3$ plants are grown with ample water, production increases still result and are attributable to increased leaf photosynthesis. In summary, the growth of crops at elevated $p_{CO_2}$ shows that suppression of photorespiration, leading to increased leaf net photosynthesis, corresponds to and explains the large observed increases in production of well-watered crops. In the absence of significant changes in plant canopy size, increased production may be expected to reflect increased leaf photosynthesis. Similar conclusions may be drawn from surveys of changes in leaf photosynthesis and production of trees grown in elevated $p_{CO_2}$ (Curtis & Wang 1998). However, this assumes that the increases in leaf photosynthesis observed in the short term will persist into the long term. While this is observed with well-watered and fertilized arable crops and with young trees, for which most data on $CO_2$ effects exist, the effect is less equivocal for natural and semi-natural vegetation (Körner 1995). Körner (1995) has highlighted the fact that, when plant growth responses are considered, there are plenty of exceptions to the simple predictions that can be made from mechanistic understanding of leaf photosynthesis. Others suggest that in natural environments, even in the absence of restrictions on root growth, increased photosynthetic carbon gain may be transient (Oechel et al. 1994; Hattenschwiler et al. 1997).

## Short-term increases in leaf photosynthesis and their significance to the long term

Can we have any expectation, then, that the instantaneous increases in the efficiency of use of nitrogen, water and light in the acquisition of carbon at the leaf level will have relevance to the ecology of wild plants in the long term? Three points should be considered here.

*Natural selection*

Within natural and semi-natural vegetation, there is considerable variation in the capacity of plants to respond to elevated $p_{CO_2}$. Although theories have been developed as to why one functional type should be more responsive than another, within any one type there is wide variability (Hunt et al. 1991, 1993; Stirling et al. 1997; Lüscher et al. 1998). This caused Körner (1995) to recommend that $CO_2$ manipulations should use actual mixed communities as the only sound basis for scaling. This, however, assumes that today's mixtures can represent those of tomorrow. Natural selection should surely favour the individuals that are able to utilize the potential increase in resource use efficiency provided by elevated $p_{CO_2}$. For example, individuals that are able to take advantage of a decreased requirement for nitrogen in the shoot by increasing root growth to capture more resources, can then use them to increase either their chance of survival during adverse seasons or their reproductive effort. By these means they may gain advantage. Although dependent on generation times relative to the time scale of change in $p_{CO_2}$, this suggests that species and individuals able to utilize the additional resource may be a better guide to the future than today's mixed communities. Few studies, so far, have had the scale and duration to provide indications of how natural or semi-natural populations may interact under elevated $p_{CO_2}$. Lüscher et al. (1998), in a 4-year study of pasture species mixtures under low input maintenance within a Free Air $CO_2$ Enrichment (FACE) experiment, has shown substantial differences between a range of species. The leguminous species, which have an additional sink for additional carbohydrate in their nitrogen-fixing nodules, showed a stronger response as a functional group than the grasses and non-leguminous dicots.

### $C_4$ analogy

Many processes downstream of leaf $CO_2$ fixation can influence the significance of any change at the leaf level to production and fitness of the individual (Chapter 2). Is it therefore reasonable to expect changed leaf photosynthetic properties to have any influence on the long-term ecology of $C_3$ plants under elevated $p_{CO_2}$? Analogy to $C_4$ plants provides evidence that this is a reasonable expectation. $C_4$ plants differ from $C_3$ only in the fact that their leaves concentrate $CO_2$ at the site of $C_3$ photosynthesis. They are essentially plants in which $C_3$ photosynthesis functions at internally elevated $p_{CO_2}$. As a result, $C_4$ species have the higher potential efficiencies of nitrogen, water and light use that we predict for $C_3$ species grown at elevated $p_{CO_2}$. However, because of energetic considerations these increased efficiencies will only be manifest in warm environments in the case of $C_4$ photosynthesis (Hatch 1987; Long 1999). Applying the reasoning that has been used to suggest that changes in leaf photosynthesis in $C_3$ plants have little relevance to production and distribution in the future, then the same should apply to $C_4$ species. Yet the broad ecology of $C_4$ species can be directly ascribed to the functioning of their $C_3$ photosynthesis at elevated $p_{CO_2}$. Their dominance of the herbaceous flora of the savanna and therophyte flora of hot deserts is attributed to their high water-use efficiency (Chapter 18). Their ability to outcompete $C_3$ species under conditions of low nitrogen availability in temperate grasslands is attributed to their higher nitrogen-use efficiency. Their record productivities in tropical habitats with a rich supply of water and nutrients is attributed to their higher photosynthetic efficiency (reviewed in Ehleringer & Monson 1993; Long 1999). $C_4$ species, like $C_3$, include a vast range of taxonomic groups, forms and functional types, from stress tolerators to competitors, and from shrubs to ephemerals. Despite this, broad patterns in the ecology and distribution of $C_4$ species can be attributed predominantly to their different leaf photosynthesis—explained by their ability to concentrate $CO_2$ at the site of Rubisco. $C_4$ plants provide an analogue of how the ecology of $C_3$ species may therefore be expected to change in the long term. That is, rising $p_{CO_2}$ should allow $C_3$ species to extend their potential range of habitats into sites with lower water and nitrogen availability than those they occupy at present. It should also allow them to be more productive on sites where these resources are limiting, in the absence of system-level feedbacks.

### Global $CO_2$-flux analysis

Flux analyses appear increasingly to suggest that a wide range of 'climax' terrestrial community types worldwide, from the tropics (e.g. Malhi et al. 1998) to the boreal zone (e.g. Chen et al. 1999), are currently net sinks for $CO_2$ from the atmosphere (Chapter 19). This could have many causes, but if this finding is a result of the stimulation of photosynthesis by the steady global increase in $p_{CO_2}$, then it would explain why an increase is observed simultaneously in communities from such disparate locations and contrasting climates.

## Acclimation and nitrogen

The ability of a plant or system to maintain elevated rates of photosynthesis will depend on downstream processes (Chapter 2). Acclimation of photosynthesis, in an ecological context, has been viewed as a negative response that may eliminate any long-term increase in $CO_2$ uptake by vegetation (e.g. Oechel et al. 1994). However, physiological consideration suggests a very different role for acclimation (Drake et al. 1997). If the plant is unable to use part or all of the additional carbon resulting from the initial stimulation of leaf $CO_2$ uptake, then decreased photosynthesis must follow. How widespread is complete limitation of response by an inability to use additional carbon and energy? Higher leaf carbohydrate concentrations, which are invariably found in $C_3$ plants grown in elevated $p_{CO_2}$ (Drake et al. 1997; Curtis & Wang 1998), have been suggested to indicate limitation in the ability of the plant to utilize additional carbon (Körner & Wurth 1996). However, it follows from the Münch hypothesis that if the capacity of the phloem remains the same, increased translocation can only be achieved with an increase in amounts of non-structural carbohydrate at the sites of phloem loading. There is abundant evidence that in the long term, photosynthesis acclimates to elevated $p_{CO_2}$, i.e. the photosynthetic properties of leaves developed at elevated $p_{CO_2}$ differ from those developed at the current $p_{CO_2}$ (Drake et al. 1997; Curtis & Wang 1998). The vast majority of studies report a decrease in $A$ of plants grown in elevated $p_{CO_2}$, relative to controls, when both are measured at the current ambient $p_{CO_2}$ (Drake et al. 1997; Curtis & Wang 1998). Two reasons for this acclimation are apparent. First, the plant may be unable to utilize all of the additional carbohydrate that photosynthesis in elevated $p_{CO_2}$ can provide, therefore a decrease in source activity has to result. Second, less Rubisco is required at elevated $p_{CO_2}$. Examples of a complete loss of the initial stimulation of photosynthesis by elevation of $p_{CO_2}$ are rare, with the important exception of arctic tundra (Oechel et al. 1994). Otherwise, quantitative analysis of field studies, where there is no restriction of rooting volume, show no clear evidence of any significant loss of the stimulation of photosynthesis by elevated $p_{CO_2}$ (Sage 1994; Drake et al. 1997; Curtis & Wang 1998). Yet acclimation occurs within the photosynthetic apparatus; how is this possible?

## Reconciling a sustained increase in photosynthesis with acclimation

Elevated $p_{CO_2}$ shifts control of photosynthesis away from Rubisco to processes limiting the rate of regeneration of RuBP. Thus, for leaves in which Rubisco was just sufficient to maintain the observed rate of photosynthesis at the current $p_{CO_2}$ Rubisco will be in considerable excess at 1.5 times and 2.0 times current $p_{CO_2}$ (Woodrow 1994). This explains why plants grown at elevated $p_{CO_2}$ may show a significantly lower $A$ than controls when measured at the current $p_{CO_2}$, and yet show a similar $A$ to controls when measured at the elevated $p_{CO_2}$ in which they were grown

(Drake *et al.* 1997). Acclimation in this context may represent a re-optimization of investment of resources away from a major leaf protein. Rubisco synthesis is inhibited at the level of gene expression by an increase in concentrations of soluble carbohydrates, providing a mechanism by which the plant can avoid sequestration of nitrogen and other resources into a protein for which there is a decreased requirement in elevated $p_{CO_2}$. Drake *et al.* (1997) showed an average reduction in the amount of Rubisco of 15% in eight studies that included 11 species. As a protein that can constitute 25% of leaf nitrogen, these reductions represent a major cause of the lower tissue nitrogen observed in foliage.

### Acclimation and increased efficiency of nitrogen use

The mechanism that may allow lower levels of Rubisco when nitrogen is in short supply is suggested by a study of *Lolium perenne* swards. In swards grown at elevated $p_{CO_2}$ using FACE, there was no acclimatory loss of Rubisco when nitrogen was in good supply. Plants grown at low nitrogen did, however, show a loss of Rubisco at elevated $p_{CO_2}$. This acclimation at low nitrogen was only apparent prior to cutting of the sward when the canopy was large. After partial defoliation to reduce the size of the source relative to the sink for carbohydrates, acclimation was eliminated. The findings suggest that conditions that would produce a source–sink imbalance trigger the loss of Rubisco. Despite acclimation, there was no decrease in the stimulation of leaf photosynthesis by elevated $p_{CO_2}$. Thus, acclimation rather than limiting the response of plants to elevated $p_{CO_2}$ can, from a physiological perspective, be shown to be a mechanism by which efficiency of nitrogen use can be increased (Drake *et al.* 1997; Rogers *et al.* 1998).

### Nitrogen and acclimation of photosynthesis to elevated $p_{CO_2}$

The above discussion suggests that the response of plants to elevated $p_{CO_2}$ should be as effective at low as at high nitrogen. However, several reviews have suggested that stimulation of carbon gain is decreased or removed by low nitrogen (e.g. Melillo *et al.* 1990; Körner 1995). Lloyd and Farquhar (1996) re-evaluated this suggestion that plants growing under nutrient-poor conditions will respond less to elevated $p_{CO_2}$ and could find no theoretical justification in physiology, nor any substantial experimental evidence in its support. Curtis and Wang (1998) highlighted the importance of assessing such questions, in surveys of the literature, by objective methods. In a meta-analysis of over 500 studies of effects of growth under elevated $p_{CO_2}$ on trees, they similarly found no evidence of an interaction of nutrient stress and $p_{CO_2}$ on photosynthesis. Thus, while a few examples of low nitrogen suppressing a $CO_2$ response may be found (e.g. Körner 1995), consideration of the statistically valid datasets as a whole provides no significant evidence in its favour.

### Production at low nitrogen and the complication of below-ground partitioning

Most assessments of the effects of elevated $p_{CO_2}$ on natural and semi-natural vegetation have concerned productivity, usually measured as gain of new above-ground biomass over a given time interval. Under nutrient-limiting conditions this may be a poor indicator of system carbon gain and potential fitness. Where growth is strongly limited by nutrients, selection may have favoured individuals that can utilize any increase in carbon and chemical energy in acquisition of nutrients. Thus, increased investment in roots, rhizodeposition or transfer to soil symbionts might be expected. Relatively few field studies have been of sufficient scale or design to allow quantification of the total below-ground inputs. Their omission could, however, distort the view of how elevated $p_{CO_2}$ affects plant fitness in low nutrient environments. The recent application of mass isotopes to this issue is beginning to indicate how significant such increases may be. When a $C_3$ species is planted on a soil which has developed under $C_4$ vegetation, the amount of carbon that the species adds to the soil can be determined from the shift in the [13]C signal. Nitschelm *et al.* (1997) investigated clover on such a soil and showed that an addition of about 200 g(C) m[-2] yr[-1] at ambient $p_{CO_2}$ increased significantly to over 300 g(C) m[-2] yr[-1] with $p_{CO_2}$ elevated to 60 Pa. Ineson *et al.* (1996) planted birch seedlings (*Betula pendula*) on a $C_4$ soil and showed, using [13]C, that after one growing season, elevated $p_{CO_2}$ had increased carbon in root matter by 69% and other soil carbon resulting from rhizodeposition or/and fine root turnover by 150%. Shoot biomass for the same plants did not increase in elevated $p_{CO_2}$. In summary, despite influential statements to the contrary, stimulation of carbon gain under elevated $p_{CO_2}$ may be as large under nitrogen-limiting conditions as under conditions of good nitrogen supply. This is consistent with the prediction that can be made both from consideration of the theoretical decrease in the amount of Rubisco required for carbon gain at elevated $p_{CO_2}$, and from analogy to $C_4$ photosynthesis.

### The interaction of light and elevated $p_{CO_2}$

Another expectation that may be developed from an understanding of photosynthesis at the leaf level is that stimulation of $CO_2$ uptake due to increased $p_{CO_2}$ will be most pronounced under conditions that are strongly light limiting. The following section evaluates this important implication for the ecology of plants in shade environments. Light is used here as an example of one of the many environmental variables where we can expect, by theoretical consideration of leaf photosynthesis, an interaction with rising $p_{CO_2}$.

   Photosynthesis is light limited for all leaves for part of the day, and for leaves and plants in the lower canopy or on the forest floor for most or all of the day. The initial slope of the response of photosynthesis to light defines the maximum quantum yield or photosynthetic light-use efficiency ($\phi$) of a leaf and determines the rate of $CO_2$ uptake under conditions where photosynthesis is strictly limited by light. At a given $p_{CO_2}$ and temperature, $\phi$ has been shown to be remarkably constant in $C_3$ terrestrial plants, regardless of their taxonomic and ecological origins (Long *et al.* 1993). This may reflect the constancy of the photosynthetic mechanism across $C_3$ species. Even under light-limited conditions, net photosynthesis is reduced by the PCO, which competes with photosynthesis for ATP and NADPH and releases $CO_2$. Inhibition of the PCO by elevated $p_{CO_2}$ will therefore increase $\phi$ and light-limited *A*. It follows that if dark respiration remains constant then the light compensation point of photosynthesis must also be lowered by elevated $p_{CO_2}$ (Long & Drake 1991). In contrast to light-saturated photosynthesis, no mechanism is known by which photosynthesis that is strictly light limited could acclimate to growth at elevated $p_{CO_2}$, thus this decrease in the light compensation point of photosynthesis may be expected to be persistent and universal (Drake et al. 1997). Analyses of light compensation points in two long-term elevated $p_{CO_2}$ studies in the field have shown this predicted decrease, with an approximate halving of the light compensation point at doubled atmospheric $p_{CO_2}$ (Long & Drake 1991; Osborne et al. 1997). Forest floor vegetation commonly exists close to the light compensation point of photosynthesis. Any decrease should therefore allow large increases in carbon gain at elevated $p_{CO_2}$, relative to current $p_{CO_2}$, and extend the potential range of species into more shaded habitats.

Is there evidence that these theoretical expectations, based on leaf photosynthetic properties and measurements of leaf photosynthesis, are reflected in plant production? Curtis and Wang (1998) in their meta-analysis of over 500 studies of tree growth under elevated $p_{CO_2}$ showed a highly significant interaction between $p_{CO_2}$ and light, where the stimulation was very much greater at low light. Thus, there is strong evidence that the interaction of elevated $p_{CO_2}$ and light, expected on the basis of leaf photosynthetic properties, is evident in plant growth.

## Conclusions

The pronounced effect on predicted global warming that results when vegetation feedback is included in General Circulation Models now shows the importance of the bidirectional interaction between vegetation and the atmosphere climate system. While much effort has been expended in perfecting physical circulation models, major gaps in the understanding of the biological system add very significant uncertainty to predictions of future climate change. The development of dynamic general vegetation models (DGVMs) to decrease this uncertainty has lagged behind (Neilson & Drapek 1998). The example given here, of stomatal influence of global climate warming, emphasizes the priority that should now be given to this aspect of GCM modelling. Uncertainties concern defining the extent of stomatal response to rising $p_{CO_2}$ in different vegetation types, and the degree of photosynthetic acclimation which will in turn have a major influence on future transpiration. Finally, the lack of mechanistic understanding of stomatal response to elevated $p_{CO_2}$ and photosynthesis is a barrier to the development of more robust models of stomatal response.

Photosynthesis is the primary process by which plant production can respond to rising $p_{CO_2}$, and the process that sets the upper limit on the rate of carbon removal into ecosystems. Mechanistic understanding of the response of leaf photosynthesis to rising $p_{CO_2}$, and of the basis for increased resource-use efficiency, allows broad predictions of the responses of production and fitness in an elevated $p_{CO_2}$ world. However, genetic constraints on growth and environmental limitations could modify or nullify any significance of the well-defined initial response of photosynthesis to elevated $p_{CO_2}$ to predictions of long-term responses. Indeed, it has been suggested that the primary effects of elevated $p_{CO_2}$ on physiology may have little or no relevance and that a new empirical approach should be developed to allow scaling of $p_{CO_2}$ effects on vegetation to the global scale. Here, it is shown, from an evolutionary perspective and by analogy to $C_4$ photosynthesis, that the potential increases in resource-use efficiency at the leaf level under elevated $p_{CO_2}$ may be expected to have a profound and predictable response on the future ecology of $C_3$ plants. It is also noted here that the significance of photosynthetic acclimation, the process previously suggested to nullify the initial increase in photosynthesis in the long term in natural communities, has been overemphasized by the design of early experiments. Recent long-term studies, where there is no artificial restriction of plant growth, show little evidence of any acclimatory loss of the stimulation of photosynthetic $CO_2$ uptake in the field. Analysis of acclimation from the viewpoint of a physiological mechanism suggests that its role may in part be a re-optimization of investment of resources, reflecting changed control of processes under elevated $p_{CO_2}$. From theory and experiment it is shown that rather than decreasing photosynthetic $CO_2$ uptake under elevated $p_{CO_2}$ the process can act to increase the efficiency of nitrogen use. Understanding of the effects of elevated $p_{CO_2}$ at the leaf level provides a basis for making predictions of interactions with other environmental variables, primarily nitrogen, temperature, water and light. This is key to the future development of Earth System Models which will include dynamically responding vegetation. Light is taken here as one example where an interactive effect of elevated $p_{CO_2}$ is predicted from a leaf-level mechanism. It is shown, by reference to an objective meta-analysis of production studies, that the predicted interaction is borne out in the pattern of plant production in relation to $p_{CO_2}$ and light. This analysis shows that the relatively simple and conserved predictions of plant response that may be made from an understanding of leaf photosynthesis, while modified by genotype and local conditions, provide an effective prediction of the wider picture and have relevance to the long term. Rather than abandoning this fundamental understanding for more empirical approaches, improved understanding of the bases of variation are needed for predicting how terrestrial vegetation will interact with global atmospheric change.

## Acknowledgement

I thank Dr P.A. Davey for his comments on the draft manuscript.

## References

Amthor, J.S. (1995). Terrestrial higher-plant response to increasing atmospheric [$CO_2$] in relation to the global carbon-cycle. *Global Change Biology*, **1**, 243–274.

Bainbridge, G., Madgwick, P., Parmar, S., Mitchell, R., Paul, M., Pitts, J. *et al.* (1995). Engineering Rubisco to change its catalytic properties. *Journal of Experimental Botany*, **46**, 1269–1276.

Baker, J.T., Allen, L.H. & Boote, K.J. (1990). Growth and yield responses of rice to carbon dioxide concentration. *Journal of Agricultural Science,* **115**, 313–320.

Ball, J.T., Woodrow, I.E. & Berry, J.A. (1987). A model predicting stomatal conductance and its contribution to the control of photosynthesis under different environmental conditions. In *Progress in Photosynthesis Research,* Vol. 4 (Ed. by J. Biggins), pp. 221–224. Nijhoff, Dordrecht.

Beadle, C.L. & Long, S.P. (1985). Photosynthesis — is it limiting to biomass production? *Biomass,* **8**, 119–168.

Chen, W.J., Black, T.A., Yang, P.C., Barr, A.G., Neumann, H.H., Nesic, Z. *et al.* (1999). Effects of climatic variability on the annual carbon sequestration by a boreal aspen forest. *Global Change Biology,* **5**, 41–53.

Collatz, G.J., Ribascarbo, M. & Berry, J.A. (1992). Coupled photosynthesis–stomatal conductance model for leaves of $C_4$ plants. *Australian Journal of Plant Physiology,* **19**, 519–538.

Curtis, P.S. & Wang, X.Z. (1998). A meta-analysis of elevated $CO_2$ effects on woody plant mass, form, and physiology. *Oecologia,* **113**, 299–313.

Drake, B.G., Peresta, G., Beugeling, E. & Matamala, R. (1995). Long-term elevated $CO_2$ exposure in a Chesapeake Bay wetland: ecosystem gas exchange, primary production, and tissue nitrogen. In *Carbon Dioxide and Terrestrial Ecosystems* (Ed. by G.W. Koch & H.A. Mooney), pp. 197–214. Academic Press, San Diego.

Drake, B.G., Gonzalez Meler, M.A. & Long, S.P. (1997). More efficient plants: a consequence of rising atmospheric $CO_2$? *Annual Review of Plant Physiology and Plant Molecular Biology,* **48**, 609–639.

Eamus, D. (1991). The interaction of rising $CO_2$ and temperatures with water use efficiency: commissioned review. *Plant Cell and Environment,* **14**, 843–852.

Ehleringer, J.R. & Monson, R.K. (1993). Evolutionary and ecological aspects of photosynthetic pathway variation. *Annual Review of Ecology and Systematics,* **24**, 411–439.

Ellsworth, D.S. (1999). $CO_2$ enrichment in a maturing pine forest: $CO_2$ exchange and water status in the canopy affected? *Plant, Cell and Environment,* **22**, 461–472.

Estiarte, M., Peñuelas, J., Kimball, B.A., Idso, S.B., Lamorte, R.L., Pinter, P.J. *et al.* (1994). Elevated $CO_2$ effects on stomatal density of wheat and sour orange trees. *Journal of Experimental Botany,* **45**, 1665–1668.

Evans, L.T. (1997). Adapting and improving crops: the endless task. *Philosophical Transactions of the Royal Society of London Series B Biological Sciences,* **352**, 901–906.

Farquhar, G.D. & Sharkey, T.D. (1982). Stomatal conductance and photosynthesis. *Annual Reviews of Plant Physiology,* **33**, 317–345.

Farquhar, G.D., von Caemmerer, S. & Berry, J.A. (1980). A biochemical model of photosynthetic $CO_2$ assimilation in leaves of $C_3$ species. *Planta,* **149**, 78–90.

Field, C.B. & Avissar, R. (1998). Bidirectional interactions between the biosphere and atmosphere — introduction. *Global Change Biology,* **4**, 459–460.

Field, C.B., Jackson, R.B. & Mooney, H.A. (1995). Stomatal responses to increased $CO_2$: implications from the plant to the global scale. *Plant Cell and Environment,* **18**, 1214–1225.

Field, C.B., Lund, C.P., Chiariello, N.R. & Mortimer, B.E. (1997). $CO_2$ effects on the water budget of grassland microcosm communities. *Global Change Biology,* **3**, 197–206.

Garcia, R.L., Long, S.P., Wall, G.W., Osborne, C.P., Kimball, B.A., Nie, G.Y. *et al.* (1998). Photosynthesis and conductance of spring wheat leaves: field response to continuous free-air atmospheric $CO_2$ enrichment. *Plant Cell and Environment,* **21**, 659–670.

Gifford, R.M. & Evans, L.T. (1981). Photosynthesis, carbon partitioning, and yield. *Annual Review of Plant Physiology and Plant Molecular Biology,* **32**, 485–509.

Govindjee (1993). Bicarbonate-reversible inhibition of plastoquinone reductase in photosystem-II. *Zeitschrift für Naturforschung C Journal of Biosciences,* **48**, 251–258.

Hatch, M.D. (1987). $C_4$ photosynthesis — a unique blend of modified biochemistry, anatomy and ultrastructure. *Biochimica et Biophysica Acta,* **895**, 81–106.

Hattenschwiler, S., Miglietta, F., Raschi, A. & Körner, C. (1997). Thirty years of *in situ* tree growth under elevated $CO_2$: a model for future forest responses? *Global Change Biology,* **3**, 463–471.

Hunt, R., Hand, D.W., Hannah, M.A. & Neal, A.M. (1991). Response to $CO_2$ enrichment in 27 herbaceous species. *Functional Ecology*, **5**, 410–421.

Hunt, R., Hand, D.W., Hannah, M.A. & Neal, A.M. (1993). Further responses to $CO_2$ enrichment in British herbaceous species. *Functional Ecology*, **7**, 661–668.

Ineson, P., Cotrufo, M.F., Bol, R., Harkness, D.D. & Blum, H. (1996). Quantification of soil carbon inputs under elevated $CO_2$: $C_3$ plants in a $C_4$ soil. *Plant and Soil*, **187**, 345–350.

Jarvis, A.J. & Davies, W.J. (1998). The coupled response of stomatal conductance to photosynthesis and transpiration. *Journal of Experimental Botany*, **49**, 399–406.

Jordan, D. & Ogren, W.L. (1984). The $CO_2/O_2$ specificity of ribulose-1,5-bisphosphate carboxylase/oxygenase dependence on ribulosebisphosphate concentration, pH and temperature. *Planta*, **161**, 308–313.

Kimball, B.A. (1983). Carbon-dioxide and agricultural yield—an assemblage and analysis of 430 prior observations. *Agronomy Journal*, **75**, 779–788.

Kimball, B.A., Pinter, P.J., Garcia, R.L., Lamorte, R.L., Wall, G.W., Hunsaker, D.J. *et al.* (1995). Productivity and water-use of wheat under free-air $CO_2$ enrichment. *Global Change Biology*, **1**, 429–442.

Klimov, V.V., Allakhverdiev, S.I., Feyziev, Y.M. & Baranov, S.V. (1995). Bicarbonate requirement for the donor side of photosystem-II. *FEBS Letters*, **363**, 251–255.

Körner, C. (1995). Towards a better experimental basis for upscaling plant-responses to elevated $CO_2$ and climate warming. *Plant Cell and Environment*, **18**, 1101–1110.

Körner, C. & Wurth, M. (1996). A simple method for testing leaf responses of tall tropical forest trees to elevated $CO_2$. *Oecologia*, **107**, 421–425.

Leuning, R. (1995). A critical-appraisal of a combined stomatal–photosynthesis model for C-3 plants. *Plant Cell and Environment*, **18**, 339–355.

Lloyd, J. & Farquhar, G.D. (1996). The $CO_2$ dependence of photosynthesis, plant-growth responses to elevated atmospheric $CO_2$ concentrations and their interaction with soil nutrient status.1. General principles and forest ecosystems. *Functional Ecology*, **10**, 4–32.

Lloyd, J., Grace, J., Miranda, A.C., Meier, P., Wong, S.C., Miranda, H.S. *et al.* (1995). A simple calibrated model of Amazon rainforest productivity based on leaf biochemical properties. *Plant, Cell and Environment*, **18**, 1129–1145.

Long, S.P. (1991). Modification of the response of photosynthetic productivity to rising temperature by atmospheric $CO_2$ concentrations—has its importance been underestimated? *Plant, Cell and Environment*, **14**, 729–739.

Long, S.P. (1993). The significance of light-limited photosynthesis to crop canopy carbon gain and productivity—a theoretical analysis. In *Photosynthesis—Photoreactions to Plant Productivity* (Ed. by Y.P. Abrol, P. Mohanty & Govindjee), pp. 547–560. Kluwer, Dordrecht.

Long, S.P. (1999). Environmental responses. In *The Biology of C4 Photosynthesis* (Ed. by R.F. Sage & R.K. Monson), pp. 209–243. Academic Press, San Diego.

Long, S.P. & Drake, B.G. (1991). Effect of the long-term elevation of $CO_2$ concentration in the field on the quantum yield of photosynthesis of the $C_3$ sedge, *Scirpus olneyi*. *Plant Physiology*, **96**, 221–226.

Long, S.P. & Drake, B.G. (1992). Photosynthetic $CO_2$ assimilation and rising atmospheric $CO_2$ concentrations. In *Crop Photosynthesis: Spacial and Temporal Determinants* (Ed. by N.R. Baker & H. Thomas), pp. 69–103. Elsevier Science, Amsterdam.

Long, S.P., Postl, W.F. & Bolharnordenkampf, H.R. (1993). Quantum yields for uptake of carbon-dioxide in $C_3$ vascular plants of contrasting habitats and taxonomic groupings. *Planta*, **189**, 226–234.

Luscher, A., Hendrey, G.R. & Nosberger, J. (1998). Long-term responsiveness to free air $CO_2$ enrichment of functional types, species and genotypes of plants from fertile permanent grassland. *Oecologia*, **113**, 37–45.

McLeod, A.R. & Long, S.P. (1999). Free-air carbon dioxide enrichment (face) in global change research: a review. *Advances in Ecological Research*, **28**, 1–56.

Malhi, Y., Nobre, A.D., Grace, J., Krujit, B., Pereira, M.G.P., Culf, A. *et al.* (1998) Carbon dioxide transfer over a Central Amazonian rain forest. *Journal of Geophysical Research—Atmospheres*, **103**, 31593–31612.

Melillo, J., Callaghan, T.V., Woodward, F.I., Salati, E. & Sinha, S.K. (1990). Effects on ecosystems. In

*Climate Change: The IPCC Scientific Assessment* (Ed. by J.T. Houghton, G.J. Jenkins & J.J. Ephraums), pp. 283–310. Cambridge University Press, Cambridge.

Melillo, J., Prentice, I.C., Farquhar, G.D., Schulze, E.-D. & Sala, O.E. (1996). Terrestrial biotic responses to environmental change and feedbacks on climate. In *Climate Change 1995: The Science of Climate Change* (Ed. by J.T. Houghton, L.G. Meira Filho, B.A. Callander, N. Harris, A. Kattenberg & K. Maskell), pp. 445–482. Cambridge University Press, Cambridge.

Morison, J.I.L. (1998) Stomatal response to increased CO$_2$ concentration. *Journal of Experimental Botany*, **49**, 443–452.

Neilson, R.P. & Drapek, R.J. (1998) Potentially complex biosphere responses to transient global warming. *Global Change Biology*, **5**, 505–522.

Nitschelm, J.J., Luscher, A., Hartwig, U.A. & van Kessel, C. (1997). Using stable isotopes to determine soil carbon input differences under ambient and elevated atmospheric CO$_2$ conditions. *Global Change Biology*, **3**, 411–417.

Oechel, W.C., Cowles, S., Grulke, N., Hastings, S.J., Lawrence, B., Prudhomme, T. *et al.* (1994). Transient nature of CO$_2$ fertilization in arctic tundra. *Nature*, **371**, 500–503.

Osborne, C.P., Drake, B.G., LaRoche, J. & Long, S.P. (1997). Does long-term elevation of CO$_2$ concentration increase photosynthesis in forest floor vegetation? Indiana strawberry in a Maryland forest. *Plant Physiology*, **114**, 337–344.

Osborne, C.P., LaRoche, J., Garcia, R.L., Kimball, B.A., Wall, G.W., Pinter, P.J. *et al.* (1998). Does leaf position within a canopy affect acclimation of photosynthesis to elevated CO$_2$? *Plant Physiology*, **117**, 1037–1045.

Pinter, P.J., Kimball, B.A., Garcia, R.L., Wall, G.W., Hunsaker, D.J. & LaMorte, R.L. (1996). Free-air CO$_2$ enrichment: responses of cotton and wheat crops. In *Carbon Dioxide and Terrestrial Ecosystems* (Ed. by G.W. Koch & H.A. Mooney), pp. 215–249. Academic Press, San Diego.

Portis, A.R. (1995). The regulation of Rubisco by Rubisco activase. *Journal of Experimental Botany*, **46**, 1285–1291.

dePury, D.G.G. & Farquhar, G.D. (1997). Simple scaling of photosynthesis from leaves to canopies without the errors of big-leaf models. *Plant, Cell and Environment*, **20**, 537–557.

Randall, D.A., Dazlich, D.A., Zhang, C., Denning, A.S., Sellers, P.J., Tucker, C.J. *et al.* (1996). A revised land surface parameterization (SiB2) for GCMs.3. The greening of the Colorado State University general circulation model. *Journal of Climate*, **9**, 738–763.

Rogers, A., Fischer, B.U., Bryant, J., Frehner, M., Blum, H., Raines, C.A. *et al.* (1998). Acclimation of photosynthesis to elevated CO$_2$ under low N nutrition is effected by the capacity for assimilate utilisation. Perennial ryegrass under free-air CO$_2$ enrichment (FACE). *Plant Physiology*, **118**, 683–692.

Sage, R.F. (1994). Acclimation of photosynthesis to increasing atmospheric CO$_2$—the gas-exchange perspective. *Photosynthesis Research*, **39**, 351–368.

Saxe, H., Ellsworth, D.S. & Heath, J. (1998). Tree and forest functioning in an enriched CO$_2$ atmosphere. *New Phytologist*, **139**, 395–436.

Sellers, P.J., Bounoua, L., Collatz, G.J., Randall, D.A., Dazlich, D.A., Los, S.O. *et al.* (1996a). Comparison of radiative and physiological effects of doubled atmospheric CO$_2$ on climate. *Science*, **271**, 1402–1406.

Sellers, P.J., Los, S.O., Tucker, C.J., Justice, C.O., Dazlich, D.A., Collatz, G.J. *et al.* (1996b). A revised land surface parameterization (SiB2) for atmospheric GCMs.2. The generation of global fields of terrestrial biophysical parameters from satellite data. *Journal of Climate*, **9**, 706–737.

Sellers, P.J., Randall, D.A., Collatz, G.J., Berry, J.A., Field, C.B., Dazlich, D.A. *et al.* (1996c). A revised land surface parameterization (SiB2) for atmospheric GCMs.1. Model formulation. *Journal of Climate*, **9**, 676–705.

Sellers, P.J., Dickinson, R.E., Randall, D.A., Betts, A.K., Hall, F.G., Berry, J.A. *et al.* (1997). Modeling the exchanges of energy, water, and carbon between continents and the atmosphere. *Science*, **275**, 502–509.

Senock, R.S., Ham, J.M., Loughin, T.M., Kimball, B.A., Hunsaker, D.J., Pinter, P.J. *et al.* (1996). Sap flow in wheat under free-air CO$_2$ enrichment. *Plant Cell and Environment*, **19**, 147–158.

Stirling, C.M., Davey, P.A., Williams, T.G. & Long, S.P. (1997). Acclimation of photosynthesis to elevated CO$_2$ and temperature in five British native species of contrasting functional type. *Global Change Biology*, **3**, 237–246.

Wang, Y.P. & Jarvis, P.G. (1990). Description and validation of an array model — Maestro. *Agricultural and Forest Meteorology*, **51**, 257–280.

Watanabe, N., Evans, J.R. & Chow, W.S. (1994). Changes in the photosynthetic properties of Australian wheat cultivars over the last century. *Australian Journal of Plant Physiology*, **21**, 169–183.

Wechsung, G., Wechsung, F., Wall, G.W., Adamsen, F.J., Kimball, B.A., Garcia, R.L. *et al.* (1995). Biomass and growth-rate of a spring wheat root-system grown in free air $CO_2$ enrichment (Face) and ample soil-moisture. *Journal of Biogeography*, **22**, 623–634.

Woodrow, I.E. (1994). Optimal acclimation of the $C_3$ photosynthetic system under enhanced $CO_2$. *Photosynthesis Research*, **39**, 401–412.

Zelitch, I. (1973). Plant productivity and the control of photorespiration. *Proceedings of the National Academy of Sciences USA*, **70**, 579–584.

# Chapter 14

# Genetic vs. environmental control of ecophysiological processes: some challenges for predicting community responses to global change

*F.A. Bazzaz and K.A. Stinson*

## Introduction

At the heart of many ecological investigations today is the search for understanding the responses of the Earth's organisms to a changing environment. Historically, investigators have pursued different aspects of ecological and evolutionary questions depending upon their specialties. Community and ecosystem ecologists have demonstrated how population dynamics and competition among plant species translate into larger-scale patterns such as succession, productivity and energy flux. Ecophysiologists have unravelled the details of how the processes of photosynthesis and nitrogen assimilation translate into allocation, and thereby influence larger-scale community dynamics. Population biologists have documented morphological, phenological and physiological variation *within* plant species since the early part of the century. Since then, much research has identified the adaptive significance and the genetic *vs.* environmental basis of within-species variation for a number of ecophysiological traits. Finally, molecular biologists have made great advances in understanding gene activity and identifying genes associated with the expression of specific traits.

A logical outcome of this broad understanding is that we have begun to make predictions that span the various disciplines in our field. For example, it is becoming commonplace to use model plants such as *Arabidopsis thaliana* to match specific genes to the expression of given traits and to compare them across populations and environmental conditions. In addition, agricultural studies provide ample evidence that crop productivity can be influenced by the choice of specially selected genotypes. In the natural world, however, there can be complications in the gene-for-trait model, which make it unrealistic to assume that the genetic pathways to phenotypic expression are direct or simply defined. This chapter reviews some of the recent progress made in elucidating the biochemical control of two

*Department of Organismic and Evolutionary Biology, Harvard University, Cambridge, MA 02138, USA.*
*E-mail: fbazzaz@oeb.harvard.edu*

critical components of plant growth: photosynthesis and nitrogen accumulation. Following this, the challenges that biologists face in linking this information to predictions about community-level responses of vegetation to global change is discussed.

Of central importance to this discussion is a clarification of what we, as biologists from any field, mean by the genetic and environmental control of phenotypic traits. It is axiomatic to say that the responses of plants to their environment are determined in part by their genetic make-up. There are numerous studies in which ecophysiological processes, among other traits, vary between plant species. For example, shorter-lived taxa often have higher rates of gas exchange (Bazzaz 1979, 1996), higher carbon isotope discrimination (Ehleringer & Cooper 1988) and higher activity and concentration of Rubisco (Wullschleger 1993). In addition to variation in the genetic code which sets one species apart from another, there is, of course, variation within species. Turesson (1922) first coined the term 'ecotypic variation' to describe variation among populations of a single plant species occupying different habitats. Clausen et al. (1940, 1948) conducted a classic study that demonstrated this phenomenon in a large number of species occupying different sites along a broad elevational gradient, using the reciprocal transplant approach. Experiments such as these have been especially useful in providing examples of differential selection among populations, such as the various genotypes of Trifolium repens associated with certain grass species (Turkington & Harper 1979). More recently, ecophysiologists interested in evolutionary differences between populations have found ecotypic variation in a number of physiological traits such as photosynthetic rate and water use efficiency (Kalisz & Teeri 1986; Dawson & Bliss 1989; Comstock & Ehleringer 1992; Gurevitch 1992). In addition, there is now substantial documentation of significant variation in ecophysiological traits within populations (e.g. Ehleringer & Cooper 1988; Geber & Dawson 1990, 1997; Schuster et al. 1992; Ehleringer 1993).

It may be tempting to dismiss the variation within a species as insignificant for predicting larger-scale responses. However, the studies cited above (and others) have provided evidence that variation within species can be quite large and that selection can indeed operate on ecophysiological traits at different scales. Yet the degree to which genetics vs. environmental conditions control responses often remains unclear. Theoretically, one could define the relationship between genetic make-up and any given response, if it is assumed that the genetic pathways to phenotypic expression are straightforward and linear. For many reasons, this may not always be an appropriate assumption. Many traits, and ecophysiological traits in particular, may be controlled by various factors over a series of steps that are difficult to track. For example, photosynthesis is controlled by a series of enzymes encoded by genes in both the nucleus and the chloroplast. Similarly, the molecular expression and structure of the class of multifunctional enzymes known as superoxide dismutases (SODs) is quite diverse; there are cytosolic, plastid and mitochondrial SODs that function broadly as oxygen scavengers. While SOD expression and function were originally thought to be highly conserved across taxa and across functions, it now appears that this diverse set of proteins arose on several, separate occasions.

In addition to genetic variation, most populations of a given species show at least some degree of variation based on environmental differences among individuals, so even if the same gene is responsible for a certain trait within a population of a given species, the expression of that gene may vary with the environment. Traditionally, quantitative geneticists have used analysis of variance techniques to determine the relative contribution of the genome *vs.* the environment within a species based on the population mean for a given observable trait. However, different genes can also produce the same phenotype in the same environment (Thoday 1955; Bradshaw 1965), plant characters can exhibit different degrees of plasticity (Schlichting & Levin 1984) and genotypes may display different reaction norm shapes across environments (Sultan & Bazzaz 1993). Furthermore, not all genes are turned on at a given moment during the life of a plant, and the environment itself often changes during the lifetime of an individual plant. There is increasing evidence that gene expression itself responds to environmental conditions at various points in time (see below). The response of genotypes to the average environmental conditions over time may be less meaningful, in terms of fitness and natural selection, than responses to the environment during key phases of development and/or gene expression.

At present, quantitative genetics can only measure the average contribution of the environment *vs.* the genetic code to the observed variation in a given trait. Exciting developments have recently made it possible to identify the molecular basis of a number of traits using quantitative trait loci (QTL) techniques. While this approach holds enormous promise for linking individual genes to traits, at present QTLs refer to sections on the chromosome and not necessarily to genes *per se.* The gene or genes involved in the expression of a given trait are yet to be identified. Finally, other mechanisms, such as methylation and demethylation of certain genes and RNA editing and frame shifts during transcription, may effectively change the active genes of an individual at different times. Thus, what may appear to be a 'genetically controlled trait' by standard quantitative genetics or molecular techniques, may depend as much upon stochastic environmental factors that affect gene expression itself over a number of steps, as on the genetic code of the organism. To understand better the potential phenotypic and genetic changes in the control of ecophysiological and other plant traits, further research that explores the pathways to phenotypic expression is required. Such research should include the study of genetic *vs.* environmental control across a diverse array of taxa, time scales and developmental stages.

A second critical point of clarification when predicting the community-level effects of a given ecophysiological response from a genetically based response of individuals is the degree to which that response is correlated with other traits. We know that plant traits do not usually evolve independently of each other, but that many are correlated to varying extents. Both genetic and phenotypic correlation between traits can offset the responses of genetically controlled characters to any number of environmental conditions (Via & Lande 1985; Via 1987). The independent evolution of correlated traits requires differential responses of genes to changing environments. Theoretically, it would be impossible for a given trait to evolve independently of another strongly correlated trait (particularly when a trait has strong effects on fitness) unless both change in the same direction with the environment. Furthermore, because the degree of correlation among traits may also change in different environments (Schlichting 1989) a single trait may demonstrate phenotypic plasticity under some conditions while remaining under genetic constraints in others. Research that examines correlations between traits and the potential outcomes for fitness will be critical for predicting genetically controlled responses to, as well as potential evolutionary changes under, future environmental conditions.

Despite these difficulties in assigning specific genes to plant traits, significant progress has been made in elucidating the molecular basis of ecophysiological processes for several species, especially crops. While this knowledge may hold enormous promise for understanding ecosystem function under changing environmental conditions, it also allows the examination of the genetic *vs.* environmental control of ecophysiological responses. In some cases there may be greater variation within species than among species. For example, one study mapped a number of unrelated herbaceous and woody species, along with eight clonal genotypes of the annual plant *Abutilon theophrasti*, according to their photosynthetic response to elevated $CO_2$. While some of the *Abutilon* genotypes were grouped together on the most parsimonious tree, others were less similar to their conspecific counterparts than they were to completely unrelated species (Fig. 14.1). In such cases it may be more appropriate to use genotypes, rather than species, to assign organisms to functional response groups (Catovsky 1998). Recent studies on the molecular basis of photosynthesis and nitrogen uptake provide a good illustration of the interactive roles played by environment, genetics and correlation among traits in current and future environments. Below, these and other examples are used to identify and discuss the current difficulties in scaling up directly from species-level to community-level response.

## Genetic control of ecophysiological processes and the molecular basis of photosynthesis

Elucidating the regulation of carbon gain is of great value in understanding the function of communities and ecosystems. We have begun to understand that photosynthetic gene expression is under biochemical control, and that the interactions between gene expression and photosynthetic metabolism are complex and self regulating.

It is now accepted that the control of the activity and amount of Rubisco is dependent on the rate of carbon fixation relative to the export of photosynthate from source leaves and its subsequent utilization in the sinks of the plant. This is especially apparent when plants are grown under elevated concentrations of $CO_2$.

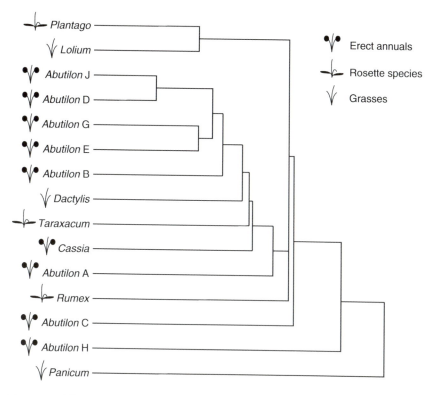

**Figure 14.1** The maximum parsimony tree for the photosynthetic response of several unrelated herbaceous and woody plant species to elevated $CO_2$. The variation observed among gentoypes within the single species, *Abutilon theophrasti*, suggests that species may be an inappropriate unit for assigning functional groups. Eight cloned genotypes of the annual plant were included in the study. The figure shows that *A. theophrasti* genotypes J, D, G, E and B were grouped together, whereas genotype A was closer to unrelated species than to conspecific genotypes. Genotypes C and H were grouped together in yet another section of the tree. From Thomas *et al.* (1999).

There is much evidence to suggest that the accumulation of soluble sugars that occurs in leaves of some species grown under elevated $CO_2$ (Delucia *et al.* 1985; Webber *et al.* 1994; Cheng *et al.* 1998) initiates a signal transduction pathway resulting in the repression of photosynthetic gene expression (Fig. 14.2). Carbon metabolites, particularly hexose sugars, have been shown to regulate the expression of photosynthetic genes in a number of species (Sheen 1990; Jang & Sheen 1994). Several additional studies have provided evidence for a conserved sugar-signalling pathway that uses hexokinases as a sugar sensor (Jang & Sheen 1994, 1997; Van Oosten *et al.* 1997; Smeekens 1998). Thus, the regulation of photosynthesis is twofold: the genetic code controls the transcription of Rubisco at the molecular level, but the leaf-level environment leads to feedback mechanisms that control the rate of carbon fixation.

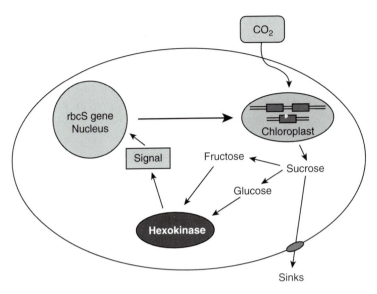

**Figure 14.2** Photosynthetic gene transcription via hexose sensing in higher plants.

Although these advances in our understanding of photosynthesis suggest that there is a genetic basis to differential rates of carbon uptake, environmental effects and the resulting feedbacks on carbon fixation may make it difficult to predict species-level responses to rising concentrations of $CO_2$. Because sugar-sensing mechanisms lead to a high degree of variation within a single individual, there may also be significant variation within, as well as among, taxa in the downregulation response of photosynthesis. Future research should address the degree of genetic and environmental control of photosynthesis both within and among species. In addition, it must be borne in mind that many steps occur between carbon gain and allocation to reproduction. As allocation itself plays a major role in carbon gain, photosynthetic responses may not necessarily be under selective pressure, especially if other allocation traits are more directly linked to fitness (Bazzaz *et al.* 1992). A further limitation to the research conducted to date is that many eco-physiological studies have concentrated on maximum photosynthetic rates measured on young, fully expanded, leaves under ideal light and moisture conditions. In addition to these measurements, future research aimed at understanding community-level responses to environmental perturbations needs to address the ability of whole plants to gain, allocate and deploy carbon.

## Correlations among ecophysiological traits

In addition to understanding the degree to which genetic variation in a single trait may result in different functional response groups, it is important to understand the extent to which such traits are likely to be under selective pressure in current and future environments. The literature reveals widespread correlation between ecophysiological, as well as other traits, within plant species (e.g. Gurevitch *et al.* 1986; Gurevitch 1988, 1992; Geber & Dawson 1990, 1997; Sandquist & Ehleringer 1997). However, it is not clear that these correlations are maintained in different environments. This suggests that selection based on any ecophysiological response may depend upon the strength and direction of selection on one or more other fitness-related traits. Understanding the correlations between ecophysiological and other traits is particularly important for predicting plant responses under future conditions and/or across habitats. For example, Thomas and Jasienski (1996) report that correlations between initial relative growth rate and plant height, biomass and leaf area at ambient $CO_2$ are not maintained under elevated $CO_2$ for *Abutilon theophrasti* (Fig. 14.3). This is just one of many examples suggesting that selection on ecophysiological characters under future conditions may be tightly linked to, and potentially constrained by, selection on other characters. Another example is that the ratio of two traits, such as the carbon:nitrogen ratio in plant tissue, may be held constant within a species, although the assimilation rates of the two elements will differ depending on environmental conditions (Bazzaz 1993). This makes it very difficult to predict differences in response within and among individuals using single trait measurements, suggesting that we do not always know that a given trait is under selection or that it will change at all with a changing environment.

Inherent links between two essential physiological processes may, in fact, constrain the evolutionary response of any single ecophysiological process. For example, photosynthesis and nitrogen assimilation are intimately linked. The assimilation of nitrogen into carbon skeletons has marked effects on plant productivity and biomass. Nitrogen deficiency can lead to lower rates of photosynthesis

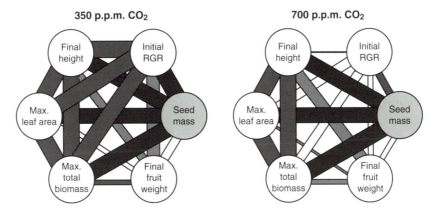

**Figure 14.3** Correlation structures for *Abutilon theophrasti* grown at ambient (350 p.p.m.) and high (700 p.p.m.) $CO_2$. Correlation of initial relative growth rate (RGR) with plant height, biomass and leaf area at ambient $CO_2$ is not maintained under elevated $CO_2$. From Thomas and Jasienski (1996).

(both chlorophyll concentration and Rubisco content are lower when compared with plants grown with an adequate supply of nitrogen) and to lower overall productivity (Lam *et al.* 1996). Again, with analogy to the regulation of photosynthesis described above, enzymes involved in nitrogen assimilation are subject to regulation at both a biochemical and molecular level. For example, in many species nitrate reductase acitivity is stimulated by the presence of nitrate. Recent studies have shown that genes involved in nitrogen assimilation are not constitutively expressed 'housekeeping genes' but are regulated by factors such as cell type, light and metabolites (Lam *et al.* 1996). Jang and Sheen (1994) discussed the possible cross-talk between light control of gene expression and regulation by sugars. Lam *et al.* (1996) noted that sucrose can mimic effects of light on the regulation of genes involved in nitrogen metabolism. Regulation of these genes by cellular carbon status again emphasizes the link between carbon and nitrogen metabolism in plants. Because each of these processes cannot proceed without the other, we cannot assume that selection acts on either process independently. Future evolutionary changes, and potential changes in community composition, are more likely to depend upon correlations between the two processes, their correlations with fitness, and the degree to which they remain coupled in altered environmental conditions such as elevated $CO_2$. A better understanding of trait correlations between photosynthesis and nitrogen uptake, and their potential effects on natural selection, is required in order to predict both evolutionary and community responses of plants to environmental changes.

## Using ecophysiological variation within and among species to predict community response

Recent progress in understanding the genetic control of ecophysiological processes should facilitate predictions concerning evolutionary responses of plants to a changing environment and provide the opportunity to identify response groups. As the above examples indicate, this requires the perspective not only to examine genetic variation among individuals, but also to recognize that there can be inherent levels of plasticity in the expression of that variation as well. For example, the fact that there is substantial genetic variation in the biochemical characteristics related to photosynthesis within populations of wild plant species (Wullschleger 1993), and among cultivars of agricultural species (Gimenez *et al.* 1992), must be examined in the light of the fact that photosynthesis exhibits a highly plastic response to feedback regulation.

Theory predicts that leaf biochemistry and stomatal conductance act together to control photosynthetic processes. Consistent with this prediction, Geber and Dawson (1997) report genetic variation for both biochemical and stomatal characteristics among family lines within a population of *Polygonum arenastrum*. For example, they found variation among families in the relationship between carbon assimilation and internal concentration of $CO_2$ in the leaves, suggesting that there are genetic differences in the ranges of response among families (Fig. 14.4).

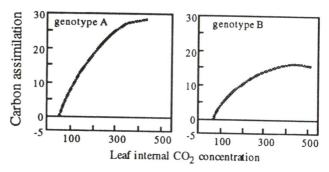

**Figure 14.4** Genetic variation among families for both biochemical and stomatal characteristics within a population of *Polygonum arenastrum*. The different relationship between photosynthesis and internal leaf $CO_2$ concentration for genotypes A and B suggests that there are genetic differences in the ranges of response among families. Adapted from Geber and Dawson (1997).

Whether this variation is sufficient for differential selection on gas-exchange characteristics is still in question, because the regulation of photosynthesis *via* feedback mechanisms indicates a high level of response to environmental conditions. If plants are using sugar-sensing mechanisms (as discussed earlier) to adjust their rate of photosynthesis to match environmental conditions throughout the day, and throughout the growing season, then gas-exchange characteristics should be in constant flux. It is possible therefore that there may be detectable, but biologically insignificant, genetic variation in any isolated physiological response. To determine the degree of selectively relevant variation in the ecophysiological responses of plants, future research should examine both the baseline genetic variation as well as variation in feedback mechanisms in response to environmental variables.

Scaling up from species to community responses is complicated by the fact that physiological responses do not remain constant. The discussion up to this point has centred on the variation in suites of life history and ecophysiological characters both within and among species, and the problem that the genetic control of ecophysiological traits may not be entirely conserved across taxa. In addition, even the current differences within and among species may not be conserved over time. For example, Bolker *et al.* (1995) showed that a number of species in a hardwood forest community had enhanced rates of photosynthesis during the first year of growth at elevated $CO_2$, leading to a greater increase in biomass at $700\,\mu mol\,mol^{-1}\,CO_2$ than at $350\,\mu mol\,mol^{-1}\,CO_2$ (Fig. 14.5). During the second year the enhancement of photosynthesis and growth were still evident when compared with that at the beginning of the experiment but had declined from their initial high values. The decline in growth (shown as plant mass ratio in Fig. 14.5) continued, at different rates for different species, during the third year.

By contrast, red oak (*Quercus rubra*) showed continued photosynthetic enhancement. Using these data in a simulation study with a mechanism-based,

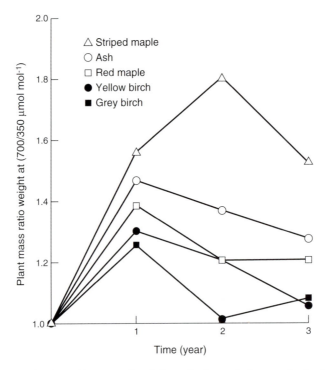

**Figure 14.5** Five hardwood forest species all showed enhanced photosynthesis when plants were grown at high $CO_2$ during the first year. Enhancement was still evident, but declined at different rates for different species in the second year. By the third year, the rate of downregulation tapered off or disappeared altogether.

spatially explicit model (SORTIE), Bolker *et al.* (1995) predicted that red oak would dominate the community after 150 years of growth. Because the single most responsive species became dominant, there was an increase in productivity. However, there was an overall decrease in total species diversity (Fig. 14.6). This is just one example of the difficulties involved in making general predictions about community-level responses to a changing environment based on ecophysiological responses of individuals within a species over short time scales. The immediate and longer-term effects of environmental conditions on the expression of ecophysiological traits need clarification before different species (or even subsets of species) can be assigned to ecophysiological functional groups. In addition, the relative importance of total productivity, species diversity and species composition should be considered when deciding which response variable(s) might exert the strongest effect on ecosystems in the future.

## Conclusions

Unprecedented rates of environmental change on a global scale underlie the importance of understanding biospheric responses to, for example, increases in

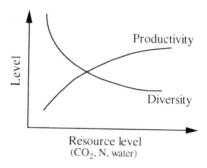

**Figure 14.6** Diversity declines with increasing productivity of a forest tree community. A simulation study with a mechanism-based, spatially explicit model (SORTIE) showed that after 150 years of growth, dominance of the single most responsive species led to an increase in total community productivity but a decrease in total species diversity.

$CO_2$ concentration. In this chapter cases have been pointed out in which a more detailed, evolutionary understanding could greatly enhance the accuracy of predictions about the responses of communities and ecosystems to global change. In particular, the control of ecophysiological processes is likely to be more complex than the simple scenario in which selection results in genetic, and hence phenotypic, change. Future research should examine the pathways to phenotypic expression.

For example, although there appears to be genetic variation in ecophysiological traits within and among populations of plant species, the regulation of genes controlling photosynthetic and nitrogen assimilation appears to be highly responsive to the biochemical environment within plant cells. Specifically, the regulation of photosynthetic genes is highly responsive to rising levels of $CO_2$. Selection on ecophysiological processes will therefore depend on both the degree of genetic variation among individuals, and the range of leaf-level ecophysiological response *within* individuals. In addition, ecophysiological processes are often correlated with other traits, and the strength of these correlations may also differ with respect to environmental conditions. As such, predictions regarding the responses of ecophysiological processes to selection in both current and future environments will require an understanding of both the genetic and environmental control of gene expression among species as well as within species. Lack of knowledge about the strength and direction of correlated traits also makes it difficult, at this point, to make general predictions about the consequences of selection on ecophysiological responses, even for given taxa. Furthermore, it is necessary to examine ecophysiological traits such as rates of carbon exchange and nitrogen assimilation within the context of other traits related to allocation, and gain an understanding of how correlations among traits may affect fitness in future environments. Finally, future research needs to consider both the immediate and longer-term effects of environmental change on ecophysiological responses. In particular, the interplay between species diversity, community composition and overall ecosystem productivity should be considered. In summary, it will be critical to identify the ways in which both the genetic code and environmental factors translate into functional responses, before we can base predictions of community response to rising $CO_2$ and other environmental changes on ecophysiological responses of individuals.

## References

Bazzaz, F.A. (1979). The physiological ecology of plant succession. *Annual Review of Ecological Systematics,* **10**, 351–371.

Bazzaz, F.A. (1993). Scaling in biological systems: population and community perspectives. In. *Scaling Physiological Processes: Leaf to Globe* (Ed. by J.R. Ehleringer & C. B. Field), pp. 233–254. Academic Press, San Diego.

Bazzaz, F.A. (1996). *Plants in a Changing Environment: Linking Physiological, Population and Community Ecology.* Cambridge University Press, Cambridge.

Bazzaz, F.A., Ackerly, D.D. & Woodward, F.I. (1992). $CO_2$ enrichment and dependence of reproduction on density in an annual plant and a simulation of its population dynamics. *Journal of Ecology,* **80**, 643–651.

Bolker, B.M., Pacala, S.W., Bazzaz, F.A., Canham, C.D. & Levin, S.A. (1995). Species diversity and ecosystem response to carbon dioxide fertilization: conclusions from a temperate forest model. *Global Change Biology,* **1**, 373–381.

Bradshaw, A.D. (1965). Evolutionary significance of phenotypic plasticity in plants. *Advances in Genetics,* **13**, 115–155.

Catovsky, S. (1998). Functional groups: clarifying our use of the term. *Bulletin of the Ecological Society of America,* **79**, 126–127.

Cheng, S.H., Moore, B.D. & Seeman, J.R. (1998). Effects of short and long term elevated $CO_2$ on the expression of ribulose-1,5-bisphosphate carboxylase/oxygenase genes and carbohydrate accumulation in leaves of *Arabidopsis thaliana* (L) Heynh. Plant Physiology, **116**, 715–723.

Clausen, J.D., Keck, D. & Hiesey, W.M. (1940). *Experimental Studies on the Nature of Species. I. The Effect of Varied Environments on Western North American Plants*, no. 520. Carnegie Institute of Washington, Washington.

Clausen, J., Keck, D. & Hiesey, W. (1948). *Experimental Studies on the Nature of Species. III. Environmental Responses of Climatic Races of* Achillea, no. 581. Carnegie Institute of Washington, Washington.

Comstock, J.P. & Ehleringer, J.R. (1992). Correlating genetic variation in carbon isotopic composition with complex climatic gradients. *Proceedings of the National Academy of Sciences USA,* **89**, 7747–7751.

Dawson, T.E. & Bliss, L.C. (1989). Intraspecific variation in the water relations of *Salix arctica*, an arctic-alpine dwarf willow. *Oecologia,* **79**, 322–331.

Delucia, E.H., Sasek, T.W. & Strain, B.R. (1985). Photosynthesis inhibition after long term exposure to elevated levels of atmospheric carbon dioxide. *Photosynthesis Research,* **7**, 175–184.

Ehleringer, J.R. (1993). Variation in carbon isotope discrimination in *Encelia farinosa*: implications for growth, competition, and drought survival. *Oecologia,* **95**, 340–346.

Ehleringer, J.R. & Cooper, T.A. (1988). Correlations between carbon isotope and microhabitat in desert plants. *Oecologia,* **76**, 562–566.

Geber, M.A. & Dawson, T.E. (1990). Genetic variation in and covariation between leaf gas exchange, morphology, and development in *Polygonum arenastrum*, an annual plant. *Oecologia,* **85**, 153–158.

Geber, M.A. & Dawson, T.E. (1997). Genetic variation in stomatal and biochemical limitations to photosynthesis in the annual plant, *Polygonum arenastrum. Oecologia,* **109**, 535–546.

Gimenez, C., Mitchell, V.J. & Lawlow, D.W. (1992). Regulation of photosynthetic rate of two sunflower hybrids under water stress. *Plant Physiology,* **98**, 516–524.

Gurevitch, J. (1988). Variation in leaf dissection and leaf energy budgets among populations of *Achillea* from an altitudinal gradient. *American Journal of Botany,* **75**, 1298–1306.

Gurevitch, J. (1992). Differences in photosynthetic

rate in populations of *Achillea lanulosa* from two altitudes. *Functional Ecology,* **6**, 568–574.

Gurevitch, J., Teeri, J.A. & Wood, A.M. (1986). Differentiation among populations of *Sedum wrightii* (Crassulaceae) in response to limited water availability: water relations, $CO_2$ assimilation, growth, and survivorship. *Oecologia,* **70**, 198–204.

Jang, J.C. & Sheen, J. (1994). Sugar sensing in higher plants. *Plant Cell,* **6**, 1665–1679.

Jang, J.C. & Sheen, J. (1997). Sugar sensing in higher plants. *Trends in Plant Science,* **2**, 208–214.

Kalisz, S. & Teeri, J. (1986). Population-level variation in photosynthetic metabolism and growth in *Sedum wrightii. Ecology,* **67**, 20–26.

Lam, H.M., Coschigano, K.T., Oliveira, I.C., Melo-Oliveira, R. & Coruzzi, G.M. (1996). The molecular genetics of nitrogen assimilation into amino acids in higher plants. *Annual Review of Plant Physiology and Plant Molecular Biology,* **47**, 569–593.

Sandquist, D.R. & Ehleringer, J.R. (1997). Intraspecific variation of leaf pubescence and drought response in *Encelia farinosa* associated with contrasting desert environments. *New Phytologist,* **135**, 635–644.

Schlichting, C.D. (1989). Phenotypic plasticity in Phlox. II. Plasticity of character correlations. *Oecologia,* **78**, 496–501.

Schlichting, C.D. & Levin, D.A. (1984). Phenotypic plasticity of annual *Phlox*: tests of some hypotheses. *American Journal of Botany,* **71**, 252–260.

Schuster, W.S.F., Sandquist, D.R. & Ehleringer, J.R. (1992). Heritability of carbon isotope discrimination in *Gutierrezia microcephala* (Asteraceae). *American Journal of Botany,* **79**, 216–221.

Sheen, J. (1990). Metabolic repression of transcription in higher plants. *Plant Cell,* **2**, 1027–1038.

Smeekens, S. (1998). Sugar regulation of gene expression in plants. *Current Opinion in Plant Biology,* **1**, 230–234.

Sultan, S.E. & Bazzaz, F.A. (1993). Phenotypic plasticity in *Polygonum persicaria*. I. Diversity and uniformity of genotypic reaction norms to light. *Evolution,* **47**, 1009–1031.

Thoday, J.M. (1955). Balance, heterozygosity and developmental stability. *Cold Spring Harbor Symposium. Quantitative Biology,* **20**, 318–326.

Thomas, S. & Jasienski, M. (1996). Genetic variability and the nature of microevolutionary responses to $CO_2$. In *Carbon Dioxide, Populations, and Communities* (Ed. by Ch. Korner & F.A. Bazzaz), pp. 51–81. Academic Press, New York.

Thomas, S., Jasienski, M. & Bazzaz, F.A. (1999). Early *vs.* asymptotic growth responses of herbaceous plants to elevated $CO_2$. *Ecology,* **80**, 1552–1567.

Turesson, G. (1922). The genotypical responses of the plant species to the habitat. *Heriditas,* **3**, 211–350.

Turkington, R. & Harper, J.L. (1979). The growth, distribution, and neighbour relationships of *Trifolium repens* in a permanent pasture. IV. Fine scale biotic differentiation. *Journal of Ecology,* **67**, 245–254.

Van Oosten, J.J.M., Gerbaud, A., Huijser, C., Dijkwel, P.P., Chua, N.H. & Smeekens, S.C.M. (1997). An *Arabidopsis* mutant showing reduced feedback inhibition of photosynthesis. *Plant Journal,* **125**, 1011–1020.

Via, S. (1987). Genetic constraints on the evolution of plasticity. In *Genetic Constraints on Adaptive Evolution* (Ed. by V. Loeschke), pp. 47–71. Springer-Verlag, Berlin.

Via, S. & Lande, R. (1985). Genotype–environment interaction and the evolution of phenotypic plasticity. *Evolution,* **39**, 505–523.

Webber, A.N., Nie, G.Y. & Long, S.P. (1994). Acclimation of photosynthetic proteins to rising atmospheric $CO_2$. *Photosynthesis Research,* **39**, 413–425.

Wullschleger, S.D. (1993). Biochemical limitations to carbon assimilation in $C_3$ plants. A retrospective analysis of the A/ci curves from 109 species. *Journal of Experimental Botany,* **44**, 907–920.

# Chapter 15
# Alpine plants: stressed or adapted?

*Ch. Körner*

## Introduction: Conceptional aspects of comparative ecology at high elevations

This chapter will first elaborate some conceptional aspects of functional and comparative ecology, that are key to an understanding of alpine plant life, but are relevant to other biomes as well. Second, a brief summary of current knowledge of plant functioning in the alpine life zone will be presented, followed by a comparison of seasonal and aseasonal alpine plant life in order to separate elevation from seasonality effects.

### Limitation, constraints, stress

Plant life in the alpine zone is commonly associated with terms like stress and limitation. This chapter will challenge these attributes and use alpine ecosystems as an illustration of widespread misconceptions in functional ecology. Flowers of *Ranunculus glacialis* frozen into an icy crust at 3000 m elevation, or those of *Soldanella pusilla* emerging right through spring snow, seem to symbolize the stressful existence of these and other alpine species. Similarly, solitary *Welwitschia mirabilis*, with its crinkly, distorted ball of old leaves in the Namib desert, would be rated as stressed by all common standards. By contrast, most people would consider a polar bear on an arctic ice floe an unstressed, 'happy' creature, perhaps, because a polar bear has a thick fur, is known for its strength and to be dangerous. These few lines illustrate the dilemma and how easily human perception can lead to biased judgements of what is stressed and what is not. If we were to 'improve' what might be considered stressful for cold climate organisms, for instance by warming the environment, they would suffer and may even die. How can conditions be considered limiting, if the long-term relief from the presumed constraint leads to extinction?

Watering the desert, fertilizing the tundra and warming the mountain tops would quickly eliminate many species which dominate these environments. Thus, we can argue that desert plants are not drought stressed, arctic tundra is not nutrient limited and alpine plants are not constrained by freezing. Ecologists have uncritically adopted the terminology from agro-horticulture, where it is obvious

*Institute of Botany, University of Basel, Schönbeinstrasse 6, CH-4056 Basel, Switzerland. E-mail: koerner@ubaclu.unibas.ch*

*what* is meant by limitation, namely the yield of a crop, commonly the biomass of useful components, per unit land area. For wild plants it is not mass production but successful reproduction that ultimately matters. In natural situations and in the long term, limitation exists only for the non-fit. If ecologists mean the limitation of mass production per unit land area irrespective of species present, they should be more explicit. It should be borne in mind that (in terms of its species structure) no natural late successional plant community is resource constrained, and alpine vegetation is no exception. As long as we deal with plant assemblages resulting from natural selection, all we can do is try to find out what contributes to species fitness, what explains persistence, and how success of species might change if resource supply were different. This is the first conceptual point of this chapter.

## Two different comparative approaches to identify adaptive traits

The second introductory remark also concerns a rather general problem that is of particular relevance to alpine ecology, namely the identification of adaptive properties. Obviously adaptive traits can only be identified in a comparative approach, which means studying plants (or animals) from different positions along a gradient of presumptive environmental impact. This can be done in two ways, each with its advantages and its shortcomings (Fig. 15.1).

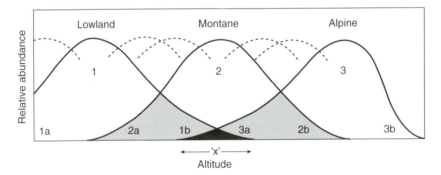

**Figure 15.1** Comparative approaches to an understanding of alpine plant life: abundance ranges of different plant species or groups of plant species (1, 2, 3) or ecotypes of species (1a,b; 2a,b; 3a,b) along an altitudinal gradient. Ranges of maximum abundance do not necessarily reflect physiological optima but are simply centres of distribution. Comparisons of plants with abundance peaks in the centre of environmental antipodes (e.g. 3 *vs.* 1) are more likely to exhibit strong, life-zone-specific attributes, but taxonomic and possibly functional diversity needs to be covered by comparing data for whole communities. By contrast, the comparative study of 1b *vs.* 1a, 2b *vs.* 2a and 3b *vs.* 3a has the advantage of taxonomic (and possibly functional) relatedness and the disadvantage of comparing distribution tails (ecotypes from marginal habitats) rather than plants from the centre of their range. Both approaches have been used in the past. For some questions, comparisons within overlapping zones, in particular the '*x*' zone, has much scientific potential (Woodward *et al.* 1986), but this has been explored to a lesser extent. Redrawn from Körner (1999).

Specialists, exclusively found at high altitudes (approach 1), will more likely reflect a high degree of 'adaptation' in their characteristics, hence they can be expected to behave in a more typically 'alpine' manner than plants that recently radiated from their lower-elevation centres of distribution to high-elevation outposts (Gjaerevoll 1990). However, a single specialist species with a narrow high-altitude range is still a weak indicator of life-zone-specific behaviour, because there is a large structural and functional diversity among plant species, even at the highest elevations (Körner 1991). It is the habitat- (altitude-) specific community of species and the relative frequency of traits among those species that bears the most solid message with respect to life-zone-specific adaptive responses (cf. Billings 1957). Provenances or ecotypes of single species from a wide altitudinal range (approach 2), extending far beyond the zone of their greatest abundance, have the advantage of closer taxonomic relatedness, but may be 'Jacks of all seasons', and hence are less likely to bear most of the characteristic features of the highest life zone of plants.

Thus, approach 1 aims at understanding habitat-specific adaptive traits at the suprataxon level (e.g. the community). In this case, variability illustrates long-term selection among species with most successful traits. The most abundant species could be considered to be the most successful species, those that have achieved the highest degree of ecotypic adaptation. This approach avoids marginal habitats for a given species. Approach 2 emphasizes stress physiology, i.e. it explores the limits and the variability within a given taxon along an environmental gradient. It includes central as well as peripheral parts of a taxon's distributional range.

The data referred to here are all type 1 approach data. Overcoming the risk associated with the use of different taxa at different elevations requires the investigation of many species, in an ideal case all species or at least all dominant species present in a given habitat. Traits to be compared are best presented in the form of frequency distribution diagrams rather than means (Körner 1991).

## Cause or consequence

The third lesson from many years of experimental work at high elevations again has wider relevance for functional ecology. Even the earliest attempts at understanding mechanisms of alpine plant life were driven by a misconception, which has its roots in a common prejudice: distinguishing between factors of primary and secondary significance. Without questioning this hierarchy, generations of biologists have been brought up with the belief that, because of its very elementary importance for life, photosynthesis is the one process that plants perform least efficiently, so that it would always remain the critical, most limiting process (the bottleneck of plant growth). This is very rarely the case. In most instances (deep shade excluded) photosynthesis is driven by the demand for assimilates (which in turn is triggered by nutrient and moisture availability) rather than being the primary controlling process itself. Alpine plant research contributed to this insight. It seems now well established that alpine plants do equally well or even better than lowland plants in terms of rates of $CO_2$ uptake per unit leaf area. As will be discussed below, insufficient carbon assimilation does not explain why alpine plants are so small and why biomass accumulation per unit land area is so low (Körner & Larcher 1988). These comments may also be true for many other environments.

## Time base for comparisons
The fourth and last conceptual remark relates to the use of adequate time scales for reference. To many people it may be surprising that alpine plant communities are, in fact, no less productive than low-altitude vegetation. This belief only developed because, once more, an agronomic perspective was adopted, i.e. determination of yield on an annual basis. In extratropical latitudes, alpine plants are under snow for 7–9 months of the year, thus making the 'year' a rather arbitrary reference. If alpine plant biomass production is related to the period of time during which plants receive solar radiation and can assimilate $CO_2$, alpine productivity (per unit of time during the growing period) differs little from lowland productivity (Fig. 15.2). In other words, what is commonly described as a stressful and limited world is as productive as any other part of the world, where sufficient moisture permits plant growth.

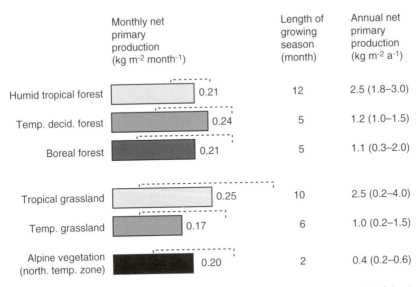

**Figure 15.2** Plant dry matter productivity (seasonal biomass production per unit of time) in the temperate alpine life zone compared with that of other biomes, when expressed per month of growing period. Note that alpine productivity is not lower, when extrapolated to a full year, albeit that the data base contains a lot of variability. From data compiled by Körner (1999).

## Functional traits in alpine plants—a brief review

The following is an attempt at a very brief summary of the current understanding of the functioning of plant species with a predominantly alpine distribution (see also Schröter 1908/1926; Pisek 1960; Billings & Mooney 1968; Bliss 1971; Körner & Larcher 1988; Friend & Woodward 1990; Körner 1999).

Without a doubt, alpine plant species are smaller than their lowland relatives. There is an order of magnitude difference between mean plant height and leaf size of herbaceous plants at 2600–3000 m elevation in the Alps compared with 600 m elevation (Körner *et al.* 1989; Fig. 15.3). No other plant physiological or morphological trait shows such a pronounced elevational differentiation. Small-sized alpine plants nest themselves in a microenvironment that allows metabolism to proceed at rates not much different from those of lowland plants. This adaptation is of critical importance in alpine environments.

### Direct thermal constraints

Those plants that passed the selective environmental 'sieve' at alpine altitudes are, of course, fit. Thus, for these species, low temperatures should not be considered to

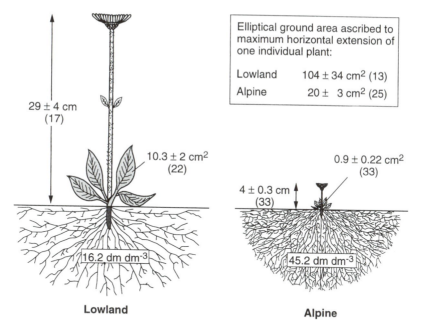

**Figure 15.3** Alpine plants and their leaves are significantly smaller than comparable plants and leaves found at low elevation (means for *c.* 25 species from each elevation). These size differences are largely genetic and represent the most important differentiation between lowland and alpine plants. Note that the rooting space occupied by alpine plant individuals is not proportionally smaller than in lowland plants, owing to greater fine root biomass fractions and greater specific root length. Drawn from data in Körner *et al.* (1989).

be extreme. Occasional tissue losses occur because of late spring or early autumn freezing events in the temperate zone and random freezing events in the tropical alpine zone may cause partial tissue losses, similar to that which may occur at low elevation, but such damage is rarely fatal to the whole plant (Sakai & Larcher 1987; Körner 1999). The critical period for freezing damage is the growing season, not the dormant period. An important protective feature against fatal freezing damage during the growing season is a morphological one: burial of apical meristems several centimetres below the soil surface. This is one explanation for the small size of many alpine plant species. By contrast, trees at the treeline have their shoot meristems fully exposed to the ambient climate. Dwarf shrubs, occupying intermediate elevational belts, have intermediate heights of bud position and intermediate atmospheric coupling. In winter, buds of dwarf shrubs are commonly under the snow.

As with the dangers of freezing damage, it is also an inappropriate assumption that meristem and leaf temperatures in alpine plants are generally lower than those in lowland plants. This is only true during the night and overcast weather, but not during sunshine hours, when most photosynthetic production occurs. A large literature provides evidence of radiative warming in low, prostrate plant canopies at high elevation (see reviews cited above). As a result, thermal optima for light-saturated rates of photosynthesis in most alpine forbs are not much lower than those for lowland forbs (Körner & Diemer 1987). Processes in the top-soil, such as nutrient cycling, also benefit from radiative heating. As a rule, alpine grassland soils (and buried shoot meristems) are warmer (not colder) than soils below closed forest near the treeline (Körner 1998; Fig. 15.4).

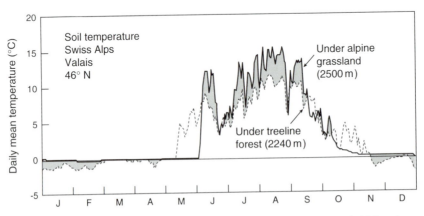

**Figure 15.4** Soil temperatures (−10 cm) below alpine grassland are commonly higher than those measured under treeline forests. This is a consequence of enhanced soil heat flux and canopy warming in low-stature alpine vegetation, and points at an obvious disadvantage of trees at higher elevations (cf. Körner 1998).

## Carbon, nutrient and water relations

Photosynthetic capacity at ambient partial pressures of $CO_2$ is similar for alpine and lowland plants, but is higher for alpine than for lowland plants when measured at equal partial pressures of $CO_2$ (Körner & Diemer 1987). Carbon isotope studies suggest that this higher specific carbon binding capacity in alpine plants is a global phenomenon (Körner *et al.* 1991). Integrated over a full season, carbon gain per unit leaf area does not greatly differ among forbs across a 2000-m elevational range (Diemer & Körner 1996). One explanation for the high photosynthetic efficiency of alpine plants is comparatively thick leaves, on average one additional palisade layer, compared with lowland forbs (Körner *et al.* 1989; see below).

Leaf nitrogen concentration measured on either a leaf area or leaf dry weight basis is higher in alpine plants, worldwide, as long as similar morphotypes are compared, reflecting greater investment in photosynthetic machinery (Körner 1989). In an ecological sense, alpine plants are not nutrient limited (see above), but fertilizer addition often strongly stimulates growth, with fast-responding species overgrowing, and thus eliminating, others (e.g. Bowman *et al.* 1993; Schäppi & Körner 1997). In the latter study it was shown that the addition of nitrogen at $40\,kg\,ha^{-1}\,a^{-1}$ to an alpine grassland (a deposition rate common in central lowland Europe) caused seasonal biomass accumulation to double, whereas doubling atmospheric $CO_2$ concentration for four seasons had no effect (Körner *et al.* 1997).

Respiratory losses in alpine plants are equal to, or smaller than, those in comparable low-elevation plants. This is largely caused by lower temperatures during the night and during overcast periods. The $Q_{10}$ does not differ between alpine and lowland plants and is 2.3 for non-acclimated and 1.7 for acclimated rates (Körner *et al.* 1997). Although irrelevant in an ecological context, experiments have demonstrated that alpine plants tend to respire more than lowland plants when (briefly) exposed to the same, higher temperatures. However, because such elevated temperatures are not encountered in the field, true respiratory losses are certainly no greater than those for comparable lowland species. Thus, the leaf carbon balance of alpine forbs is not at a disadvantage (see above). Nevertheless, our understanding of alpine plant respiration is poor, given the significance of respiratory metabolism. For instance, it is not known what causes plant acclimation to changing temperatures, what the time constants of such acclimatory processes are, and why some plant species from extremely high elevations seem to have very little potential for acclimation to higher temperatures and exhibit extreme values for $Q_{10}$ of 3–4 (Larigauderie & Körner 1995).

Water is not commonly a limiting resource, but top-soil desiccation may periodically impair the nutrient cycle. This indirect effect of moisture shortage is far more important than any direct effects, because deep soil moisture rarely becomes depleted. What is known of plant water relations in adult alpine plants suggests that the supply of this resource is adequate, even in semi-arid mountains, such as the Pamirs and parts of the Andes. Alpine plants have high maximum leaf conductances, little or no diurnal stomatal closure, and high leaf water potentials (Körner 1999). One explanation for this observation is low ground cover (small leaf area index). Seedling establishment may, however, be difficult at times because of top-soil dryness, hence the often observed dominance of clonal growth.

## Carbon investments

Alpine plants in the temperate zone are rich in mobile carbon compounds throughout the year; some even accumulate large amounts of lipids that are not recovered during leaf senescence (Tschager *et al.* 1982; Körner 1999). Total non-structural carbohydrate (TNC) concentrations in the great majority of alpine species examined during mid-season in the central Alps account for between 15 and 20% of leaf dry matter (Fig. 15.5; Körner 1999). Many alpine plants produce more assimilates than can be invested in structural growth, and the shedding of lipid-rich structures is just one example. Little is known about turnover, internal allocation and export through roots. The phenomenon of mobile carbon 'over-flow' in cold-grown plants does not seem to be restricted to alpine plants, but is widely observed (Farrar 1988).

Dry matter allocation, for example the leaf mass fraction (*LMF*, green leaf mass

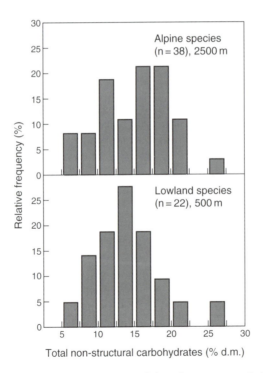

**Figure 15.5** Leaves of alpine plants are rich in mobile carbon compounds. The data shown here are for the central Alps. Total non-structural carbohydrates (TNC) represent starch and sugars (glucose, fructose, sucrose). Fructans do occur in some alpine species (e.g. some Asteraceae), but the absolute amounts are not high. Most of TNC is starch, except for legumes, in which sugars tend to dominate TNC at high elevation. Data from Körner (1999).

per total plant mass) is similar, but the stem fraction is smaller and the fine root fraction is larger in alpine compared with lowland forbs (Körner & Renhardt 1987). Specific leaf area (*SLA*, leaf area per unit leaf mass) varies by a factor of 2–3 within a single alpine or lowland plant community (similar to *LMF*), but means tend to be slightly lower at high elevation (Körner *et al.* 1989). Given the large differences observed in *SLA* in co-ocurring species, it seems impossible that this variable could be a decisive trait for determining plant carbon balance at high elevation (as was suggested by Atkin & Lambers 1998). Examination of large, randomly selected sets of species from contrasting elevations is required to achieve an understanding of the functional significance of this highly variable trait. Because it is highly unlikely that alpine plants are carbon limited, investigating partial processes of the carbon budget is unlikely to contribute much to an understanding of the specific features of alpine plant life. There is no experimental evidence supporting any causal link between the small size of alpine plants and carbon shortage. The enormous structural variability of the highly diverse alpine flora may be seen as an obvious support of this statement. Also the ineffectiveness of $CO_2$ enrichment (see above) is consistent with this supposition. Diemer (1996) reported that partial removal of leaves of *Ranunculus glacialis* at 2600–3100 m elevation in the Alps by snow vole had no effect on survival, vigour or flowering in the subsequent year, an observation that would be hard to explain if leaf carbon assimilation were critical.

**Tissue formation and development**

No obvious differences in the structural development of leaves, cell division in particular, are apparent between high- and low-elevation species (see Körner 1999). Mitotic indices are similar in comparable plants from different elevations or across a range of mountain plants from polar to equatorial latitudes (mitotic cells are always between 6 and 8% of all meristematic cells). Further, significant temperature-related shifts in the abundance of mitotic phases are not apparent. The temperature dependence of the duration of the cell cycle in primordial leaves is similar to that for dark respiration (Fig. 15.6).

In addition to these still rudimentary experimental data, the concept of cell division limiting growth seems implausible. If there were a critical slowing of the cell cycle as a result of low temperatures, plants could easily mitigate this by only sightly enhancing the size of the cycling meristematic cell cohort, because of the exponential consequences on the total number of embryonic cells produced (Fig. 15.7). Growth-constraining low temperatures do not appear to limit cell division to the extent that meristems become short of embryonic cells (there is no division bottleneck). Hence, if anywhere, it is most likely to be cell differentiation and cell wall formation, in particular, where critical physiological growth constraints come into play. The investment in differentiating cells is the process with the highest energetic and substrate demand. As illustrated in Fig. 15.7, cell differentiation needs to be seen as controlling cell division and not *vice versa*.

Related to leaf tissue formation discussed above is embryogenesis in developing seeds. The processes involved have only recently been explored, and it seems that a

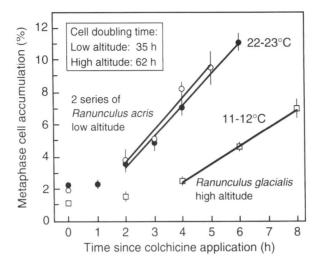

**Figure 15.6** Cell doubling time in primordial leaves of the alpine *Ranunculus glacialis* as compared with *R. acris* from low elevation. Cycling cells were forced to accumulate in the metaphase by adding colchicine (cell doubling time is calculated from the rate of metaphase accumulation). Temperatures during this treatment are natural local temperatures which differed by 11K. The 'Q$_{10}$' of cell doubling time between the two species is approximately 1.8. From Körner (1999).

variety of developmental typologies linked to flowering phenology exist (Wagner & Mitterhofer 1998 and references therein). The most time-consuming part of seed maturation is the formation of reserve tissues, whereas embryo establishment occurs rapidly. Low temperatures during histogenesis delay seed and fruit development, whereas seed filling and seed maturation seem to be largely independent of temperature.

It has long been known that alpine plants have little difficulty in producing viable seed, even at very high altitudes (Braun 1913), and pollination (largely by flies at highest elevations) seems assured in most cases. If annual seed crops are lost, it is most commonly through early freezing events, which hit premature seeds. Dormant seed banks, clonal growth and high longevity of plants mitigate against such losses (Körner 1999).

## Comparison of seasonal and aseasonal alpine plant life

Neither the carbon, nutrient nor water relations of alpine plants seems to provide adequate explanation for their small size and low biomass accretion per unit land area. Why then are alpine plants small and stunted? Meristematic processes may still be involved, but this is a largely unexplored field of alpine research. From the above it is doubtful that any specific physiological bottlenecks will be discovered. Perhaps much of the search for critical constraints was driven by the preoccupation

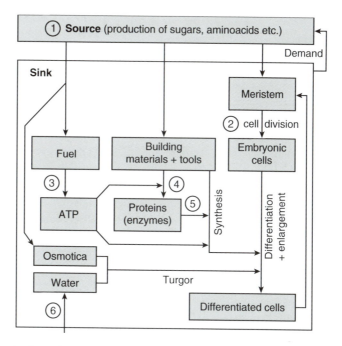

**Figure 15.7** A schematic presentation of the major processes involved in tissue growth. *Numbers* indicate where low-temperature effects could play a role and thus limit the overall process, given that all partial processes are interlinked. For the sake of clarity, fueling and substrate provision to cell division itself (2) were omitted in this scheme, because the quantities required are very small compared with those for subsequent supplies for growth and differentiation of the emergent embryonic cell populations. From Körner (1999).

that there must be a mechanistic, process-related explanation. What if there is none? What if the only way to survive in the alpine climate is to be small, to control phenology carefully, and thereby reduce early and end-of-season risks of tissue loss by freezing, and on top of that, ensure the production of viable seed or clonal propagules? The seasonal alpine plant life cycle and morphogenetic determinants (e.g. winter buds at or below the ground) seem to play such a key role that other components of plant growth, those classically explored in ecophysiology, may be subordinate.

Current knowledge points to incredibly good physiological adjustments for the optimal use of increasingly shorter growing periods at temperate and higher latitude alpine elevations. The adjustment seems so perfect, that the amount of plant mass produced during the short season is no different from major lowland biomes, including the tropics (Fig. 15.2), if expressed on a daily or monthly basis. Is this true for the tropical alpine zone as well?

### Comparison of pulsed and continuous alpine productivity

Because of the year-round replacement of old by new leaves, it is nearly impossible to obtain reliable productivity measurements for tropical-alpine vegetation. Standing biomass is not a useful substitute, because one needs to know, at least, rates of leaf turnover. The mean leaf duration of 16 herbaceous species between 4000 and 4600 m altitude in Ecuador was found to be $193 \pm 19$ days, compared with 80–90 days for comparable forbs in the Alps (Diemer 1998). Hence, leaf life-spans may be more than twice as long in the tropical zone than in the temperate zone, which means leaf biomass may be replaced roughly twice a year. Hofstede *et al.* (1995) reported even longer leaf life-spans, 300 days for leaves of tussock grasses in the Colombian paramos, and from these data an annual above-ground productivity of *c.* $200 \, g \, m^{-2}$ was estimated (the turnover rate of the below-ground biomass, $1–2 \, kg \, m^{-2}$, was unknown). Values for tropical-alpine tussock grasslands in Venezuela (Smith & Klinger 1985) and Papua New Guinea (Hnatiuk 1978) range from 150 to $630 \, g \, m^{-2} \, a^{-1}$, with mean rates of approximately $400 \, g \, m^{-2} \, a^{-1}$ (equivalent to $30 \, g \, m^{-2} \, month^{-1}$). These rates may be underestimates, because of unaccounted leaf losses between census dates. Thus, it seems likely that productivity may be between 500 and $600 \, g \, m^{-2} \, a^{-1}$, clearly far below the rates that would be obtained for temperate zone mountains by extrapolation from monthly productivity to a full year (Fig. 15.2). In summary, on a monthly basis, tropical alpine vegetation seems to be, at most, half as productive as temperate alpine vegetation.

Besides differences in day length, the discrepancy between tropical and temperate alpine productivity may also result from the consequences of continuous (non-seasonal) vs. pulsed (seasonal) growth. While the dormant season in seasonal alpine climates is useless for C-aquisition, it is not useless for mineralization (alpine soils often don't freeze under snow; Fig. 15.4) and is important with respect to atmospheric nutrient deposition. Snowpack in the Alps and Rocky Mountains contains between 5 and $15 \, kg \, N \, ha^{-1}$, released all at once at snow melt. Together with an annual nitrogen release by mineralization of similar magnitude (see Körner 1999), massive pulsed growth is permitted at the beginning of the growing period. This growth, though bound to the snow-free period, strongly profits from nutrients (nitrogen in particular) accumulated over the long dormant season. Hence, with respect to nutrient availability, the calculations for Fig. 15.2 are somewhat optimistic. If there were no dormant season, the same amount of atmospheric nitrogen deposition would have to serve a continuous period of 12 months' growth.

This discussion illustrates that assessing plant growth determinants at high elevations in the temperate zone is confounded by seasonality phenomena, not related to elevation *a priori*. Much more work is needed in the aseasonal tropics, where no such confounding effects exist, and where comparisons of plant functioning along elevational gradients reflect purely altitudinal phenomena (reduced temperature and pressure; cf. Körner *et al.* 1991), but in the absence of biologically significant precipitation gradients.

## Future research

The above summary of functional traits of alpine plants highlights a number of research areas, where knowledge is very poor. While influences of biotic interactions commonly exceed those of physical impacts at low elevation, the reverse becomes increasingly true as altitude increases. Hence, the alpine life zone, and its upper parts in particular, represents an ideal environment to study climatic influences on wild plants. Physiological plant ecology of alpine plants has, and will continue to, contribute much to our very basic understanding of plant functioning. Some examples have been provided in this chapter.

Because of plant specialization, poor soil conditions, short seasons and physical disturbances, alpine ecosystems are very susceptible to all sorts of environmental forcings, including global change effects (e.g. nitrogen deposition, warmer climate, altered snow cover duration). Understanding the impacts of such changes is a second and timely motive for experimental ecology at high elevation. A most promising field is 'space for time' studies in the global warming context, as will be specified below.

Of the many interesting and important fields of research, six topics have been selected which, I feel, deserve particular attention in the future. Listing such topics does not indicate discrimination against others; they merely reflect a personal impression from a recent review of the field (Körner 1999), and many other, highly relevant questions could have been added. Thus, priorities for research should include: (i) explanation of the temperature dependency of tissue formation and cell wall synthesis; (ii) examination of the mechanisms and time constants of thermal adjustments of mitochondrial respiration; (iii) assessment of the amount and fate of 'excess' carbon assimilates, lipids in particular; (iv) elucidation of plant growth and plant dry matter production in non-seasonal tropical alpine environments (transect studies); (v) examination of temperature responses of plants and ecosystems along natural thermal gradients (natural models of potential climate warming); and (vi) examination of the long-term influence of enhanced atmospheric nitrogen deposition on alpine plant community dynamics.

Motives for the first four topics were given in the text above. The two global-change-related topics (v and vi) seem important for two reasons: first, soluble nitrogen deposition is a most influential agent on alpine plant growth, and long-term species replacement through altered differential vigour could substantially change alpine vegetation and its stability. Natural nutrient availability gradients combined with very low dose nutrient addition experiments, run over longer periods, could add a lot to the understanding of the consequences of the ongoing global loading of the atmosphere with soluble nitrogen compounds. At the same time such experiments would contribute to the understanding of alpine nutrient relations in general. Second, climatic warming would be of overwhelming significance for plant life in the cold. However, manipulative warming experiments have very limited predictive value. Whatever warming treatment is applied, it represents a pulsed event with a pulsed answer, and thus may not reflect persistent long-term

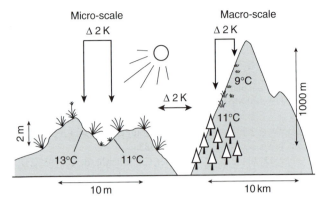

**Figure 15.8** Space for time 'experiments' in global change research. Plants and soils from thermally contrasting micro- and macro-habitats represent long-term responses to climate similar to the ones to be expected under global warming scenarios. In alpine terrain such thermal contrasts can be found over very short distances. Because nothing is stable in the alpine life zone, these relief gradients will always reflect dynamic life conditions, but still, the 'treatment' does not occur in a single step, hence permitting biological adjustments to longer-term means, impossible to simulate experimentally. In addition, there are almost no limitations in terms of spatial replication.

responses. I am convinced that natural thermal gradients within a given common environmental matrix (overall climate, geology, flora) are the most powerful tool to tackle this issue (Fig. 15.8), an approach little utilized to date.

With its mosaics of life conditions, its steep gradients of environmental conditions over very short distances, the alpine environment offers 'experiments by nature' much more powerful than many artificial experiments. The advantage of such existing experiments in the field is their long duration and the coupling of plant roots to a natural soil and rhizosphere.

## References

Atkin, O. & Lambers, H. (1998). Slow-growing alpine and fast-growing lowland species: a case study of factors associated with variation in growth rate among herbaceous higher plants under natural and collected conditions. In *Inherent Variation in Plant Growth* (Ed. by H. Lambers, H. Poorter & M.M.I. Van Vuuren), pp. 259–288. Backhuys, Leiden.

Billings, W.D. (1957). Physiological ecology. *Annual Reviews in Plant Physiology*, **8**, 375–391.

Billings, W.D. & Mooney, H.A. (1968). The ecology of arctic and alpine plants. *Biological Reviews*, **43**, 481–529.

Bliss, L.C. (1971). Arctic and alpine plant life cycles. *Annual Reviews in Ecology and Systematics*, **2**, 405–438.

Bowman, W.D., Theodose, T.A., Schardt, J.C. & Conant, R.T. (1993). Constraints of nutrient availability on primary production in two alpine tundra communities. *Ecology*, **74**, 2085–2097.

Braun, J. (1913). Die Vegetationsverhältnisse der Schneestufe. *Neue Denkschriften der Schweizerischen Naturforschenden Gesellschaft*, **18**, 1–347.

Diemer, M. (1996). The incidence of herbivory in high-elevation populations of *Ranunculus glacialis*: a re-evaluation of stress-tolerance in alpine environments. *Oikos*, **75**, 486–492.

Diemer, M. (1998). Life span and dynamics of leaves of herbaceous perennials in high-elevation environments: 'news from the elephant's leg'. *Functional Ecology*, **12**, 413–425.

Diemer, M. & Körner, Ch. (1996). Lifetime leaf carbon balances of herbaceous perennial plants from low and high altitudes in the central Alps. *Functional Ecology*, **10**, 33–43.

Farrar, J.F. (1988). Temperature and the partitioning and translocation of carbon. *Symposium of the Society of Experimental Biology*, **42**, 203–235.

Friend, A.D. & Woodward, F.I. (1990). Evolutionary and ecophysiological responses of mountain plants to the growing season environment. *Advances in Ecological Research*, **20**, 59–124.

Gjaerevoll, O. (1990). *Alpine Plants*. The Royal Norwegian Society of Sciences and Tapir Publishers, Trondheim.

Hnatiuk, R.J. (1978). The growth of tussock grasses on an equatorial high mountain and on two sub-antarctic islands. In *Geoecological Relations Between the Southern Temperate Zone and the Tropical Mountains*, Vol. 9. *Erdwissenschaftliche Forschung* (Ed. by C. Troll & W. Lauer), pp. 159–190. Steiner, Wiesbaden.

Hofstede, R.G.M., Chilito, E.J. & Sandovals, E.M. (1995). Vegetative structure, microclimate, and leaf growth of a paramo tussock grass species, in undisturbed, burned and grazed conditions. *Vegetatio*, **119**, 53–65.

Körner, Ch. (1989). The nutritional status of plants from high altitudes. A worldwide comparison. *Oecologia*, **81**, 379–391.

Körner, Ch. (1991). Some often overlooked plant characteristics as determinants of plant growth: a reconsideration. *Functional Ecology*, **5**, 162–173.

Körner, Ch. (1998). A re-assessment of high elevation treeline positions and their explanation. *Oecologia*, **115**, 445–459.

Körner, Ch. (1999). *Alpine Plant Life*. Springer, Berlin.

Körner, Ch. & Diemer, M. (1987). *In situ* photosynthetic responses to light, temperature and carbon dioxide in herbaceous plants from low and high altitude. *Functional Ecology*, **1**, 179–194.

Körner, Ch., Diemer, M., Schäppi, B., Niklaus, P. & Arnone, J. (1997). The responses of alpine grassland to four seasons of $CO_2$ enrichment: a synthesis. *Acta Oecologica*, **18**, 165–175.

Körner, Ch., Farquhar, G.D. & Wong, S.C. (1991).

Carbon isotope discrimination by plants follows latitudinal and altitudinal trends. *Oecologia*, **88**, 30–40.

Körner, Ch. & Larcher, W. (1988). Plant life in cold climates. *Symposium of the Society of Experimental Biology*, **42**, 25–57.

Körner, Ch., Neumayer, M., Pelaez Menendez-Riedl, S. & Smeets-Scheel, A. (1989). Functional morphology of mountain plants. *Flora*, **182**, 353–383.

Körner, Ch. & Renhardt, U. (1987). Dry matter partitioning and root length/leaf area ratios in herbaceous perennial plants with diverse altitudinal distribution. *Oecologia*, **74**, 411–418.

Larigauderie, A. & Körner, Ch. (1995). Acclimation of leaf dark respiration to temperate in alpine and lowland plant species. *Annals of Botany*, **76**, 245–252.

Pisek, A. (1960). Pflanzen der Arktis und des Hochgebirges. In *Handbuch der Pflanzenphysiologie*, Vol. 5 (Ed. by W. Ruhland), pp. 377–413. Springer, Berlin.

Sakai, A. & Larcher, W. (1987). *Frost Survival of Plants. Responses and Adaptation to Freezing Stress. Ecological Studies*, Vol. 62. Springer, Berlin.

Schröter, C. (1908/1926). *Das Pflanzenleben der Alpen. Eine Schilderung der Hochgebirgsflora*. Raustein, Zürich.

Schäppi, B. & Körner, Ch. (1997). *In situ* effects of elevated $CO_2$ on the carbon and nitrogen status of alpine plants. *Functional Ecology*, **11**, 290–299.

Smith, J.M.B. & Klinger, L.F. (1985). Aboveground: belowground phytomass ratios in Venezuelan paramo vegetation and their significance. *Arctic and Alpine Research*, **17**, 189–198.

Tschager, A., Hilscher, H., Franz, S., Kull, U. & Larcher, W. (1982). Jahreszeitliche Dynamik der Fettspeicherung von *Loiseleuria procumbens* und anderen Ericaceen der alpinen Zwergstrauchheide. *Acta Oecologica*, **3**, 119–134.

Wagner, J. & Mitterhofer, E. (1998). Phenology, seed development, and reproductive success of an alpine population of *Gentianella germanica* in climatically varying years. *Botanica Acta*, **111**, 159–166.

Woodward, F.I., Körner, Ch. & Crabtree, R.C. (1986). The dynamics of leaf extension in plants with diverse altitudinal ranges. I. Field observations on temperature responses at one altitude. *Oecologia*, **70**, 222–226.

# Chapter 16

# Arctic plants: adaptations and environmental change

*J.A. Lee*

## Introduction

### The physical environment

A simple definition of the Arctic is that region lying to the north of the latitudinal treeline. Much of the Arctic is in fact a land-locked ocean, but the land surface, even near sea level, provides one of the most extreme environments for plant growth. The Arctic is characterized by a continuous photoperiod during at least some of the summer months and continuous darkness during part of the winter. The low solar angle during the summer and the negative energy balance during the winter lead to permanently frozen soils. The permafrost exerts a marked effect on hydrology, leading in summer to the prevalence of waterlogged soils except on steep slopes, where water moves laterally through the active (unfrozen) surface layer. The permafrost results in cold soils that are completely frozen in winter and in which the active layer may extend from just a few centimetres to *c.* 1 m in depth during the summer. Thus, biological activity in the soil is low, resulting in slow rates of decomposition and nutrient cycling (see e.g. Robinson & Wookey 1997). Repeated freezing and thawing in the surface soil layers during the autumn and spring lead to marked disturbance in the rooting environment, causing the frost heave of both seedlings and mature plants. In some soils this results in considerable stone sorting, polygon formation and mud boils with major problems for plant establishment or clonal spread. Precipitation in arctic regions is generally low (100–150 mm), and falls mainly as snow. A feature of arctic environments, which they share with alpine regions, is that wind and topography lead to a marked redistribution of precipitation, with exposed ridges being made largely devoid of snow cover and snowbeds accumulating often to great thickness in sheltered depressions. This results in winter in the direct exposure of plants on the ridges to the extremes of low temperature (perhaps as low as −60°C) and desiccating winds frequently carrying abrading ice crystals. It also has a major influence on the length of the growing season for plants colonizing snowbeds. When the sun appears in spring, the radiant energy is

*Sheffield Centre for Arctic Ecology, Department of Animal and Plant Sciences, University of Sheffield, Sheffield, S10 2TN, UK. E-mail: j.a.lee@sheffield.ac.uk*

either reflected from the snow surface (high albedo, *c*. 0.95) or is absorbed and used to melt the lying snow. Plants are usually thought of as being largely physiologically inactive under a snow cover (but see below) and thus snow-lie can seriously reduce the already short growing season. The growing season in some high arctic plant communities may be as little as *c*. 6 weeks and in deep snowbeds the snow-free period may be too short in most years throughout the arctic to support the growth of vascular plants.

**Plant adaptations**

The adaptations of tundra plants to arctic environments have been well documented over many years (see e.g. Savile 1972). Plants have to be frost tolerant at all times, including during the summer growth period. The vegetation, in which cryptogams make a more important contribution to primary production than in temperate regions, is largely composed of long-lived perennial plants of low stature. There are very few annual species. Temperature, either directly on the plant or indirectly through effects on soil processes, appears to have a major effect on species distribution. For example, Rannie (1986) showed a high correlation ($r=0.97$) between vascular plant species number and mean July temperature for 35 local floras in the Canadian Arctic and three such floras in northwest Greenland (Fig. 16.1). However, low stature allows tundra plants to capitalize on the warm climate near the ground, a fact reflected in their relatively high temperature optima for photosynthesis (15–25°C for tundra dwarf shrubs compared with 20–25°C for temperate deciduous trees (e.g. Larcher 1995; cf. Chapter 15)).

Morphological features, such as the cushion habit, help to enhance shoot temperatures and also aid in maintaining their full hydration. Enhancement of tissue temperatures may also occur through dark pigmentation caused by high concentrations of anthocyanins. A number of reproductive adaptations have also been

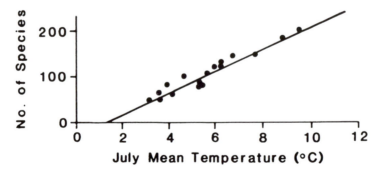

**Figure 16.1** Relationship between number of vascular plant species and mean daily temperature of July (the warmest month). Floristic data are from 35 local floras in the Canadian Arctic and three in northwest Greenland. Temperature data are from either weather station records, or long-term field records, or interpolation from a map of July mean temperature. From Billings (1992) after Rannie (1986).

identified in tundra vascular plants, including pre-formed flower buds to maximize the time for flowering and seed set in the short arctic summer. Heliotropic flowers apparently encourage pollination by attracting basking insects and also by raising the temperature of the gynaecium, accelerating fertilization and seed development events. A further reproductive feature is the abundance of species that show pseudovivipary (Lee & Harmer 1980). Although this is a form of vegetative reproduction, the bulbils or plantlets produced instead of seeds allow ready dispersal from the parent plant of a relatively large (compared with seed) propagule and this aids establishment. In many tundra communities in most years seedling establishment is very rare. Apart from frost tolerance during the growing season, there are few distinctive physiological features of arctic tundra plants (cf. a similar situation in alpine plants, Chapter 15). However, another feature, possibly related to frost tolerance, which requires further investigation, is the apparent ability of many tundra plants to survive ice encasement and potential anoxia during winter (Crawford 1997). Tundra plants show $C_3$ photosynthesis, tolerate soils that are often extremely deficient in nitrogen and also phosphorus, and are able to absorb nutrients and metabolize them at temperatures close to 0°C (Chapin & Bloom 1976). They efficiently withdraw nitrogen and phosphorus from senescing tissues (up to 81% of leaf nitrogen and 88% of leaf phosphorus in *Petasites frigidus*) and redistribute the elements to new growth (following storage) as a means of reducing their reliance on the soil nutrient supply (Chapin & Shaver 1989; Jonasson 1989). Polar desert and semi-desert vascular species are also adapted to withstand physiological drought, which may be particularly important in winter and early spring when the soils remain frozen and snow cover may be thin or absent. Carbon gain by cryptogams in these ecosystems may also be severely limited by moisture availability (Longton 1988).

## Global change

The difficulty in establishing clearly distinctive physiological adaptations in arctic plants has resulted in less emphasis on autecological investigations in recent years, and more attention to ecosystem processes. Research on arctic ecosystems has been encouraged by the predictions from General Circulation Models of the effects of increases in greenhouse gases (Maxwell 1992). These models predict a surface warming for the Arctic of 2–2.4 times greater than the annual average for the globe in winter and in summer 0.5–0.7 times, as the result of a doubling of present atmospheric $CO_2$ concentrations. These models also predict enhanced winter precipitation at high latitudes, which may have profound effects on the radiation balance and ecosystem processes. These predictions have stimulated a wide range of perturbation studies in the Arctic in an attempt to understand how tundra ecosystems might respond to future global change, and to provide possible early indicators or markers of such change. It is these investigations which have provided most of the research on arctic plants in the last decade.

## Perturbation studies

### Treatments

Perturbation studies of arctic ecosystems have largely concentrated on studying the responses of the vegetation and soil processes over one or more years to increasing air temperatures during the growing season. Most attempts at raising air temperature have involved the use of passive open-top devices (Marion *et al.* 1997), which also affect other environmental variables, notably wind speed. A few experiments have examined the effects of elevated $CO_2$ (Tissue & Oechel 1987; Gwynn-Jones *et al.* 1997), and colleagues have used water and/or fertilizer additions to simulate increases in summer precipitation and putative effects of soil warming on decomposition (Shaver & Chapin 1986; Wookey *et al.* 1995). Other experiments have involved the manipulation of snow cover (Walsh *et al.* 1997), the use of heating cables to increase soil temperatures (R. Crabtree, personal communication) and fluorescent tubes to enhance UV-B radiation (Gwynn-Jones *et al.* 1997). At most high arctic sites, the remoteness and absence of power has severely restricted the scope of experimentation, hence the emphasis on passive methods of field manipulation.

The advantage of passive chamber studies is that they are cheap to employ. Typically, they have involved the use either of translucent plexiglas transmitting *c.* 90% of visible wavelengths, or a translucent fibre glass or polythene (see Marion *et al.* 1997 for details), and a variety of different designs have been employed from open rectangular corners to almost enclosed tents (Fig. 16.2). The degree of warming depends on the chamber design and the nature of the site, and is largely due to daytime warming. Figure 16.3 shows the diurnal pattern of warming inside and outside a plastic tent at Abisko, subarctic Sweden. This shows marked air warming inside the tent after noon, with some cooling in the early morning. In a comparison of the use of chambers at four arctic sites, Marion *et al.* (1997) showed that the mean daily air temperature increase ranged from 1.2 to 1.8°C. They concluded that the more closed the system, the higher the temperature enhancement, but the greater the unwanted effects of chamber overheating, and altered light, moisture and windspeed.

### Responses

Despite the potential problems, these passive devices have been used widely, notably in ITEX, the International Tundra Experiment (Henry & Molau 1997). This experiment aims to monitor growth, reproduction and phenology of major circumpolar vascular plant species in response to climate variations and environmental manipulation, using standard protocols, at over 20 arctic sites. All species investigated showed short-term responses (1–3 years) to the small temperature amelioration produced by the open-top chambers, but the responses were individualistic and no general pattern for functional types or phenology class was observed (Henry & Molau 1997).

**Figure 16.2** Passive warming devices: (a) ITEX Corner; (b) cone chamber; (c) hexagon chamber; (d) plastic tent. From Marion *et al.* (1997).

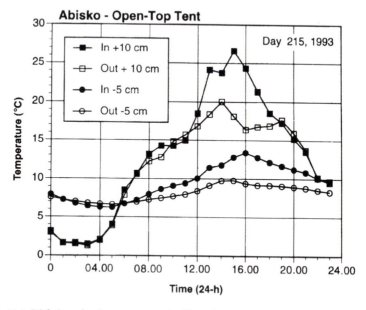

**Figure 16.3** Diel air and soil temperatures inside and outside a plastic tent at Abisko, Sweden. From Marion *et al.* (1997).

Although even high arctic ecosystems show responses to environmental perturbations within 1 year of treatments (Wookey *et al.* 1995), it may be unwise to predict longer-term responses from short-term studies. Chapin *et al.* (1995) manipulated light, temperature and nutrients in tussock tundra on the north slope of Alaska. They showed that short-term responses (3 years) in general were poor predictors of longer-term changes (9 years) in community composition. The nutrient addition treatment increased the biomass and production of deciduous shrubs, notably of *Betula nana*, but reduced the growth of cryptogams, graminoids and evergreen shrubs. Temperature enhancement increased shrub production, but reduced the production of cryptogams. The differential responses of species to these perturbation treatments buffered changes in ecosystems' characteristics such as biomass and productivity. The major effect of temperature in the longer term was to increase nutrient availability through changes in nitrogen mineralization in the soil. To some extent the changes in the perturbation treatments were mirrored in the longer-term changes in the control plots. The study coincided with the warmest decade on record for the region. During the study there was a large decline in the biomass of *Eriophorum vaginatum* in the control plots, which was consistent with the response of the species in the perturbation experiments. This study is the longest running arctic climate change experiment, and highlights the dangers of predictions based on short-term studies and the need for long-term investigations. Several other studies are now more than 6 years old (Gwynn-Jones *et al.* 1997; Press *et al.* 1998; Robinson *et al.* 1998), and should aid prediction of the effects of global change on arctic ecosystems. These are necessary because extrapolation from one arctic ecosystem to the whole arctic is unlikely to be wise. For example, and in contrast to the Alaskan tussock tundra, Press *et al.* (1998) showed that the most dramatic response in a Swedish subarctic dwarf shrub heath to nutrient addition was an increase in abundance of the grass *Calamagrostis lapponica* by a factor of more than 18 compared with plots not subject to nutrient addition.

In general, perturbation studies show much greater responses of arctic vegetation to nutrients than to other factors. The major limiting nutrient is nitrogen supply (Henry *et al.* 1986; Defoliart *et al.* 1988; Lee 1998), pointing to the importance of temperature effects acting indirectly on plants through soil mineralization processes. Nitrogen limitation provides one possible explanation of the rapid (within 3 weeks) homeostatic adjustment of the photosynthesis of *Eriophorum vaginatum* when grown *in situ* in elevated $CO_2$ (Oechel *et al.* 1994). However, long-term studies on a representative range of arctic communities are required to understand further the importance of $CO_2$ enrichment to tundra ecosystems. Of all perturbation studies to date, probably only the study involving temperature and nutrient treatment on the Alaskan tussock tundra reported by Chapin and colleagues (1995) is sufficiently large scale and over an adequate time period to demonstrate fully the potential importance of global change phenomena. There is also the need for much more long-term experimentation on the effects of winter warming in line with predictions from General Circulation Models. Despite the criticism that all perturbation studies are artificial and represent only a pulsed response to a pulsed event (Chapter 15), there is still much of ecological interest that has been and can be learnt from them.

## Recent advances in the ecophysiology of arctic plants

### Nitrogen nutrition

The marked effects of temperature on nitrogen mineralization in arctic soils (Nadelhoffer *et al.* 1991) point to the importance of nitrogen acquisition in the ecology of tundra plants. Annual rates of nitrogen mineralization in cold arctic soils are extremely low and result in the low availability of ammonium and especially nitrate, despite often large pools of organic nitrogen. Although annual rates may be of little relevance because of the very short growing season, and the fact that there may be substantial short-term pulses of mineralized nitrogen at snow melt as the ground thaws, there is little doubt that nitrogen supply limits the growth of plants in many arctic soils.

Nitrification in particular may be extremely limited in wet tundra soils, but evidence is accumulating that many tundra plants utilize both nitrate and ammonium. Atkin and Cummins (1994) showed that *Oxyria digyna* (from fertile habitats) and *Dryas integrifolia* (from nutrient-poor habitats) were both capable of utilizing nitrate despite typically growing on ammonium-dominated soils in the Canadian High Arctic. These workers also showed (Atkin *et al.* 1993) high concentrations of nitrate (up to 20% of total nitrogen in roots and 27% in shoots) of seven common species from habitats in Truelove Lowland, Devon Island, suggesting that these species acquire a substantial proportion of their nitrogen as nitrate even when the apparent availability of nitrate is low. Atkin (1996) also suggests that many high arctic plants show low rates of nitrate reductase activity even after $NO_3^-$ fertilization because of their inherently low growth rates, a factor that may, in some species, contribute to high tissue nitrate concentrations. High tissue nitrate concentrations suggest that this ion may play a significant role as an osmoticant, and is probably largely confined to vacuoles. However, further data are required from a wider range of species and geographical positions before the importance of the phenomenon can be assessed.

Despite the presence of high tissue nitrate concentrations, much recent interest has centred on the potential importance of short circuiting the nitrogen mineralization process by tundra plants tapping the organic nitrogen pool. This method of nitrogen acquisition has been well established in mycorrhizal plants (Chapter 7). Many tundra species are mycorrhizal and have the potential to utilize organic nitrogen compounds for growth via the fungal symbiont, but the quantitative importance of this is only now beginning to be recognized. Michelsen *et al.* (1996) found more negative [15]N natural abundance signatures in the leaves of ericoid mycorrhizal plants than those that were non-mycorrhizal or that were infected with arbuscular mycorrhizal fungi. Ectomycorrhizal plants had an intermediate signature. These data suggest that these different signatures result from the differential use of soil nitrogen pools, to at least some extent. However, the importance of the use of one major organic nitrogen pool, the free amino acids, is most easily evaluated in non-mycorrhizal species, such as some members of the Cyperaceae which are important components of many tundra communities. Chapin *et al.* (1993) showed that *Eriophorum vaginatum* seedlings grew better in hydroponic culture on a mixture of amino acids than on nitrate and ammonium, in contrast to seedlings of *Hordeum vulgare*. These workers also predicted from a knowledge of the amino acid and ammonium concentrations in the soil solutions at Toolik Lake, Alaska and laboratory [14]C uptake studies, that the field uptake of glycine by the sedge was greater than ammonium (Table 16.1). They concluded that *Eriophorum* plants probably absorbed the amino acids directly rather than as ammonium following bacterial deamination in the growth solution. Schimel and Chapin (1996) demonstrated, using $(^{15}NH_4)_2SO_4$ and $^{15}N^{13}C$ glycine or aspartate, that *Eriophorum vaginatum* and *Carex aquatilis* growing *in situ* in tundra soil competed well for glycine- and aspartate-N relative to $NH_4^+$. At high concentrations ($25\,\mu g\,N\,g\,soil^{-1}$) *Eriophorum* took up amino nitrogen more rapidly than $NH_4^+$. Given the seasonal average concentrations of free amino acids in the soil solutions, the capacity of *Eriophorum* roots to absorb at least some of these amino acids, and the importance of this species in the vegetation, it is probable that amino acid uptake plays a major role in the nitrogen economy of tussock tundra. Kielland (1994), in a study of 10 tundra species, showed that the percentage contribution of glycine to the estimated field uptake rate was greater than ammonium (53.6% as against 40.8%). There was no clear distinction between mycorrhizal and non-mycorrhizal plants. He concluded that inorganic nitrogen is inadequate as a measure of plant available nitrogen, and that differences between species in their ability to absorb amino acids may provide a basis for niche differentiation among tundra species. He also suggested that by short circuiting mineralization, plants may accelerate nitrogen turnover.

Further evidence for the importance of amino acids in the nutrition of tundra plants comes from studies of *Kobresia myosuroides* in the Rocky Mountains. Raab *et al.* (1996) showed that non-mycorrhizal plants of this sedge took up the amino acid glycine from nutrient solutions at greater rates than nitrate (Fig. 16.4) and grew better on glycine than on ammonium nitrate. Remarkably, *Kobresia* roots

**Table 16.1** Seasonal average concentrations of ammonium and free amino acids in tussock tundra at Toolik Lake, Alaska and kinetics of uptake. From Chapin *et al.* (1993).

| N form | Field concentration ($\mu M$) | $V_{max}$ ($\mu mol\,g^{-1}\,h^{-1}$) | $K_m$ ($\mu M$) | Predicted field uptake rate ($\mu mol\,g^{-1}\,h^{-1}$) |
|---|---|---|---|---|
| Ammonium | 10.7 | 13.7 | 242 | 0.58 |
| Glycine | 8.4 | 2.0 | 12 | 0.82 |
| Aspartate | 7.9 | 4.1 | 293 | 0.11 |
| Glutamate | 5.2 | 0.6 | 118 | 0.03 |
| Other amino acids | 84.1 | – | – | – |

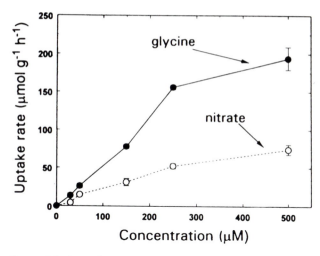

**Figure 16.4** Rates of glycine and nitrate uptake by *Kobresia myosuroides* plants from solutions containing different glycine and nitrate concentrations. Values are means ± SE, *n* = 5. From Raab *et al.* (1996).

were unable to take up ammonium ions, although under field conditions, where mycorrhizal infection may occur, ammonium utilization cannot be excluded. Raab *et al.* were also able to show, using differences in $\delta^{13}C$ signatures of $CO_2$ and glycine, that a significant amount of the glycine was taken up as intact molecules, contributing 16% to the total carbon assimilation over the 4-month growth period. This raises the prospect that amino acid uptake may make an appreciable contribution to the carbon economy of tundra plant roots as well. Näsholm *et al.* (1998) injected double universally labelled glycine (U-$^{13}C_2$, $^{15}N$ glycine) into the mor layer of a coniferous forest soil supporting *Picea abies* and two common sub-arctic plants, *Vaccinium myrtillus* and *Deschampsia flexuosa*. Figure 16.5 shows the relationship between excess $^{13}C$ and excess $^{15}N$ in the soluble nitrogen fraction of roots after 6 hours. The ratios of $^{13}C:^{15}N$ indicate that at least 91% of the nitrogen from the absorbed glycine was taken up as intact glycine by *V. myrtillus* and 64% by *D. flexuosa*. There can be little doubt that glycine absorption is an important method of nitrogen acquisition in some tundra plants at least. Raab *et al.* (1996) also demonstrated high rates of uptake of (L)-proline and glutamate by *Kobresia*, and given that glycine may represent only a small percentage of the soil amino pool in some tundra ecosystems (Table 16.1), there is the potential for other amino acids to be important nitrogen sources for plant growth as well.

Kielland (1997) also demonstrated that the cryptogams *Sphagnum rubellum* and *Cetraria richardsonii* had higher absorption capacities for glycine than for methylamine, an ammonium analogue. The importance of amino acid absorption in the nitrogen economy of cryptogams remains to be assessed fully, but given that dissolved organic nitrogen in spring run off may be twice that of inorganic nitrogen (Whalen & Cornwell 1985), this should be further researched as a matter of priority.

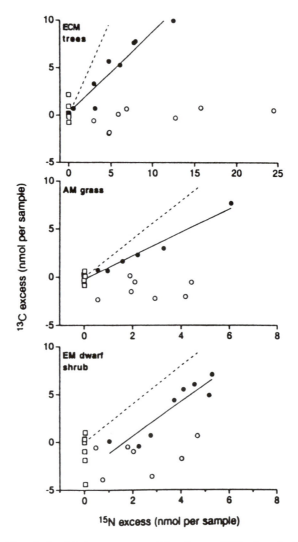

**Figure 16.5** The relationship between excess $^{13}C$ and excess $^{15}N$ in the soluble nitrogen fraction in plant roots from experimental plots 6 hours after injecting the mor layer with water (control plots, □) or solutions containing $[U^{-13}C_2{}^{15}N]$glycine (●) or $^{15}NH_4{}^+$ (○). The $^{13}C:{}^{15}N$ ratio of the glycine (2:1) injected (------) and the regressions relating to glycine-treated plots (———) are shown. AM = grass, *Deschampsia flexuosa* (slope = 1.24, $r^2 = 0.97$); ECM = ectomycorrhizal trees (slope = 0.85, $r^2 = 0.89$); EM = ericoid mycorrhiza, *Vaccinium myrtillus* (slope = 1.89, $r^2 = 0.84$); $n = 6$–8. From Näsholm *et al.* (1998).

## Nutrition and winter injury

In the High Arctic sparsely vegetated polar desert and polar semi-desert cover 93% of the land area (Bliss & Matveyeva 1992). Tundra with high vegetation cover including polar heath is very limited in extent, for example covering only 6% of the Canadian Arctic archipelago, but may be of considerable ecological importance. This is particularly true of coastal regions influenced by great numbers of nesting seabirds. Here, areas below bird-cliffs have markedly luxuriant vegetation, in stark contrast to surrounding plant communities (Summerhayes & Elton 1923). The vegetation of the bird-enriched sites differs not only in productivity from the surrounding tundra, but also in species composition (Scholander 1934). Thus *Dryas octopetala*, a dominant mat-forming dwarf shrub of high arctic fellfield vegetation, is largely absent from rock crevices and exposed talus below bird-cliffs. Similarly, the shrub *Cassiope tetragona*, which is typically found in more mesic conditions of greater snow-lie than *Dryas*, is also less abundant below bird-cliffs. By contrast, nitrophilous species, such as *Cochlearia officinalis*, are largely confined to sites of organic enrichment.

Henry *et al.* (1986) showed that single fertilizer additions of 5 and 25 g m$^{-2}$ N as NH$_4$NO$_3$ and 5 and 25 g m$^{-2}$ NPK to high arctic vegetation resulted, 2 years later, in reduced biomass of *Dryas integrifolia* and *Cassiope tetragona* in some habitats, but not in others. Fertilizer addition to tundra generally favours graminoids at the expense particularly of evergreen shrubs (Fox 1992), but experiments involving single annual doses of fertilizer addition can be difficult to interpret because of the potential for direct injury from the high doses (and therefore high concentrations) of salts. Robinson *et al.* (1998) studied the impacts of simulated climate change over five summer seasons (1991–95) on a high arctic polar semi-desert at Ny Ålesund, Svalbard. Polythene tents were used to increase temperature, summer precipitation was increased by *c.* 50% through watering six or seven times during each of the five seasons, and fertilizer was supplied in solution four times during each of the five seasons to give a total annual increase of 5, 5 and 6.3 g m$^{-2}$ of nitrogen, phosphorus and potassium, respectively.

During the early years of the experiment fertilizer addition increased the cover of *Dryas octopetala*, the dominant plant, and there was no indication of any direct toxic effects or leaf scorch as the result of the treatments. In addition, leaves of *Dryas* in the fertilizer treatments remained green into late August (Table 16.2) when plants in the control and other treatments showed marked chlorosis. In the spring of 1994, winter injury of *Dryas* plants was observed which resulted in a marked reduction in live cover of the plants in 1994 (Table 16.2). The autumn and early winter of 1993 was exceptionally mild with high precipitation when compared with the average for 1961–90, 230 mm precipitation in November 1993 compared with the 19 years' average of 69 mm, and a mean temperature of –2.8°C in November 1993 compared with the average of –7.0°C. The likely explanation of the *Dryas* injury is that arctic plants, like temperate ones, show cyclicity in cold hardiness despite always having some frost resistance during the summer. Thus, the fertilizer treatments delayed hardening and thereby increased the potential for early

**Table 16.2** Mean (±SE, *n* = 5) reduction in cover of living *Dryas octopetala* between July 1993 and July 1994 at a polar semidesert site (arcsin transformed) and a subjective chlorosis score in early autumn 1993 (28 August). The chlorosis score is a mean (*n* = 6) of (1) most leaves chlorotic, (2) some leaves green, but many showing some chlorosis, and (3) most leaves green, little or no yellowing. For further details see Robinson *et al.* (1998).

| Treatment | % Reduction in *Dryas* cover | Chlorosis score |
|-----------|------------------------------|-----------------|
| –T – W – F | 6.90 ± 1.42 | 1.00 |
| –T – W + F | 14.63 ± 3.29 | 2.50 |
| –T + W – F | 7.97 ± 2.85 | 1.00 |
| –T + W + F | 22.84 ± 5.63 | 2.33 |
| +T – W – F | 3.8 ± 2.40 | 1.36 |
| +T – W + F | 17.77 ± 5.60 | 2.40 |
| +T + W – F | 3.77 ± 2.07 | 1.42 |
| +T + W + F | 18.66 ± 3.81 | 2.50 |

Treatments as follows: T = temperature (tents), W = increased summer precipitation, F = fertilizer.

winter frost injury. Overall this effect caused a reduction in live cover of *Dryas* and an increase in standing dead cover (by up to 22%) between 1991 and 1995. By contrast, seedlings of nitrophilous species established on bare ground only in the fertilized plots were not eliminated by winter injury.

This example points to the importance of interactions between nutrient supply and climatic factors in determining the distribution of high arctic vegetation. Other species, e.g. *Saxifraga oppositifolia* (Robinson *et al.* 1998) and *Cassiope tetragona* (I.J. Alexander & S.J. Woodin, personal communication), were similarly damaged by winter injury in fertilizer treatments at Ny Ålesund in 1993, and there is clearly a need to understand the physiological basis of the interaction between frost tolerance and nutritional status through experimentation. If this effect was primarily caused by increased nitrogen supply, then the potential for increased reactive atmospheric nitrogen deposition to modify arctic ecosystems is great (Woodin 1997).

## Physiological activity under snow

Although it is frequently assumed that the growing season for tundra plants is defined by the snow-free period, leaf senescence in high arctic plants often occurs well before prolonged snow-lie is imminent (perhaps because of nutrient limitation) and during the continuous photoperiod. Similarly, important events must occur before and during snow melt in the spring, particularly in polar desert habitats where snow melt is a major source of water and nutrient supply. These may be particularly important for cryptogams that are only physiologically active when moist and many of which depend on the atmospheric supply of solutes for nutrition (Woodin 1997).

Figure 16.6 shows the effects of temperature and photon flux density on photosynthesis in the lichen *Cetraria nivalis* (Kappen *et al.* 1995). The lichen is capable of net photosynthesis even at –5°C, and at low photon flux densities

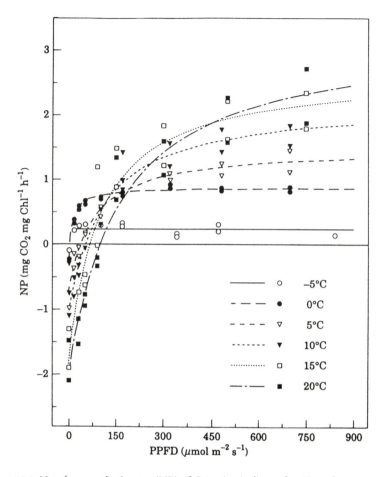

**Figure 16.6** Net photosynthetic rates (NP) of *Cetraria nivalis* as a function of photosynthetic photon flux densities (PPFD) at different temperatures. From Kappen *et al.* (1995).

($<150\,\mu mol\,m^{-2}\,s^{-1}$) appreciable photosynthesis occurs at 0°C (*c.* 30% of the maximum observed rate at any temperature). Temperatures close to 0°C can occur over several months beneath a snow layer, during which appreciable irradiance may penetrate at least thin snow cover (*c.* 10 cm) particularly in the spring (e.g. Kappen *et al.* 1995). Kappen was able to calculate that at a site in subarctic Sweden (near Abisko) the lichen *C. nivalis* accumulated over 200 mmol $CO_2$ kg dry wt$^{-1}$ over a 6-day period under melting snow.

The importance of lichen physiological activity under snow was also demonstrated by Crittenden (1998). He showed that 92 and 87% of the $NO_3^-$ and $NH_4^+$, respectively, in melting snow in the Antarctic was absorbed by the lichen *Usnea sphacelata*. Solutes in melting snow are not released uniformly, and commonly a pulse of high concentration occurs early in the snow melt event (Crittenden 1998).

For this to be sequestered by cryptogams (as it appears to be in the case of $NO_3^-$ and $NH_4^+$), these plants have to be able to absorb the ions efficiently even under low photon flux densities and at temperatures close to 0°C. We still have an imperfect understanding of how this is achieved. Under summer conditions in subarctic Sweden, there was a close correlation between $NO_3^-$ deposition events and the induction of nitrate reductase activity in *Sphagnum fuscum* (Woodin *et al.* 1985). When grown under laboratory conditions, *Sphagnum* species showed more rapid induction of nitrate reductase by nitrate at higher temperatures, but were capable of appreciable measurable enzyme activity at the lowest temperature used (5°C) (Woodin & Lee 1987). Woolgrove and Woodin (1996) demonstrated that the snowbed bryophyte *Kiaeria starkei* was able to absorb more than 90% of the nitrate in snow melt, and also showed measurable nitrate reductase activity at 2°C. Probably the key events in nutrient acquisition in tundra plants, whether under snow-pack or following snow melt, occur at temperatures around 0°C. These events have received relatively little attention from ecophysiologists. There has also been little attention paid to the integration of carbon and nitrogen metabolism in cryptogams under snow, which may be an important adaptation to growth in the polar regions.

## Scaling up

Arctic tundra covers *c.* 5% of the world's terrestrial surface, but contains *c.* 14% of the world's total soil organic carbon (Post *et al.* 1982). The fate of this organic carbon pool is of considerable current interest. It has the potential to provide a major feedback to climate change as soils warm, perhaps turning several tundra ecosystems from carbon sinks into carbon sources (Oechel *et al.* 1993). The results from perturbation studies so far are equivocal, but there is clearly the need to be able to scale up observations made on small experimental plots to the landscape scale.

Until recently, much of the detailed work on the ecology and ecophysiology of tundra plants has been conducted in tussock tundra on the North Slope of Alaska (Chapin *et al.* 1995), but clearly there are problems of using data from this ecosystem for scaling up to the whole arctic because it represents only a small proportion even of northern Alaskan terrestrial ecosystems (*c.* 20%). Walker *et al.* (1998) have examined energy and trace gas fluxes at sites on either side of a soil pH boundary in Alaska. They characterized moist non-acidic tundra and moist acidic tundra at two sites 7 km apart on either side of the boundary. The vegetation at the former is the *Dryadio integrifoliae–Caricetum bigelowii* community and at the latter is the *Sphagno–Eriophoretum* (tussock tundra). All the soil horizons of the former have a pH > 6.5 and are highly cryoturbated, whereas the surface soil of the latter has a pH of 4.0 at the surface increasing to 6.5 in the frozen C horizons. The moist acidic tundra had twice the gross photosynthesis and three times the respiration, over the measurement period than the moist non-acidic tundra, and was a much greater carbon sink (55.2 *vs.* 27.6 g C m$^{-2}$ season$^{-1}$). By contrast, methane efflux from tussock tundra was over six times higher than from non-acidic tundra. These workers point out that extrapolations based solely on numbers from the more intensively studied tussock tundra result in large errors. They conclude that for northern Alaska alone this would overestimate gross photosynthesis by at least 35%, respiration by 140%, net $CO_2$ uptake by at least 15% and methane flux by 140%. It is therefore important that further studies are made of energy balance and gas fluxes of the major arctic ecosystems, so that appropriate information is available to parameterize global models of atmospheric dynamics and climate (McFadden *et al.* 1998).

Although experiments show fairly rapid responses of arctic vegetation to perturbation treatments, the responses are not always easy to predict, particularly from short-term (1–3 years) experimentation. Physiological studies of individual tundra species or functional types have not so far proved very useful in predicting ecosystem responses to global change. Although McFadden and Chapin (1998) conclude that the conversion of tussock tundra to shrub tundra as the result of climatic warming may have little effect on water or energy balance, this may be difficult to predict from a knowledge of the physiology of *Eriophorum vaginatum* and *Betula nana* (the putative dominant shrub) alone. There would seem to be no substitute for long-term empirical experimentation combined with landscape-scale observations of energy, water and trace gas fluxes in predicting the responses of tundra ecosystems to climate change and their potentially major global importance.

## Acknowledgements

The author is grateful to Dr C.H. Robinson for permission to use her data on the effects of fertilizer treatment on *Dryas octapetala* and to Dr A.N. Parsons, Dr P.A. Wookey and Mrs B. Lee for assistance in the field.

## References

Atkin, O.K. (1996). Reassessing the nitrogen relations of arctic plants: a mini-review. *Plant, Cell and Environment*, **19**, 695–704.

Atkin, O.K. & Cummins, W.R. (1994). The effect of nitrogen source on growth, nitrogen economy and respiration of two high arctic plant species differing in relative growth rate. *Functional Ecology*, **8**, 389–399.

Atkin, O.K., Villar, R. & Cummins, W.R. (1993). The ability of several high arctic plant species to utilize nitrate nitrogen under field conditions. *Oecologia*, **96**, 239–245.

Billings, W.D. (1992). Phytogeographic and evolutionary potential of the Arctic flora and vegetation in a changing climate. In *Arctic Ecosystems in a Changing Climate: An Ecophysiological Perspective* (Ed. by F.S. Chapin III, R.L. Jefferies, J.F. Reynolds, G.R. Shaver & J. Svoboda), pp. 91–108. Academic Press, San Diego.

Bliss, L.C. & Matveyeva, N.V. (1992). Circumpolar arctic vegetation. In *Arctic Ecosystems in a Changing Climate: An Ecophysiological Perspective* (Ed. by F.S. Chapin III, R.L. Jefferies, J.F. Reynolds, G.R. Shaver & J.Svoboda), pp. 39–89. Academic Press, San Diego.

Chapin, F.S., III, & Bloom, A. (1976). Phosphate absorption. Adaptation of tundra graminoids to a low temperature, low phosphorus environment. *Oikos*, **26**, 111–121.

Chapin, F.S., III & Shaver, G.R. (1989). Differences

in growth and nutrient use among arctic plant growth forms. *Functional Ecology*, **3**, 73–80.

Chapin, F.S., III, Moilanen, L. & Kielland, K. (1993). Preferential usage of organic nitrogen for growth by a non-mycorrhizal arctic sedge. *Nature*, **361**, 150–153.

Chapin, F.S., III, Shaver, G.R., Giblin, A.E., Nadelhoffer, K.J. & Laundre, J.A. (1995). Response of arctic tundra to experimental and observed changes in climate. *Ecology*, **76**, 694–711.

Crawford, R.M.M. (1997). Habitat fragility as an aid to long-term survival in arctic vegetation. In *Ecology of Arctic Environments* (Ed. by S.J. Woodin & M. Marquiss), pp. 113–136. Blackwell Science, Oxford.

Crittenden, P.D. (1998). Nutrient exchange in an Antarctic macrolichen during summer snowfall-snow melt events. *New Phytologist*, **139**, 697–707.

Defoliart, L.S., Griffith, M., Chapin, F.S., III & Jonasson, S. (1988). Seasonal patterns of photosynthesis and nutrient storage in *Eriophorum vaginatum* L., an arctic sedge. *Functional Ecology*, **2**, 185–194.

Fox, J.F. (1992). Responses of diversity and growth-form dominance to fertility in Alaskan tundra fellfield communities. *Arctic and Alpine Research*, **24**, 233–237.

Gwynn-Jones, D., Lee, J.A. & Callagham, T.V. (1997). Effects of enhanced UV-B radiation and elevated carbon dioxide concentrations on a sub-arctic forest heath ecosystem. *Plant Ecology*, **128**, 242–249.

Henry, G.H.R., Freedman, B. & Svoboda, J. (1986). Effects of fertilization on three tundra plant communities of a polar desert oasis. *Canadian Journal of Botany*, **64**, 2502–2507.

Henry, G.R.H. & Molau, U. (1997). Tundra plants and climate change: the International Tundra Experiment (ITEX). *Global Change Biology*, **3** (Suppl. 1), 1–9.

Jonasson, S. (1989). Implications of leaf longevity, leaf nutrient re-absorption and translocation for the resource economy of five evergreen plant species. *Oikos*, **56**, 121–131.

Kappen, L., Sommerkorn, M. & Schroeter, B. (1995). Carbon acquisition and water relations in lichens in polar regions — potentials and limitations. *Lichenologist*, **27**, 531–545.

Kielland, K. (1994). Amino acid absorption by arctic plants: implications for plant nutrition and nitrogen cycling. *Ecology*, **75**, 2373–2383.

Kielland, K. (1997). Role of free amino acids in the nitrogen economy of arctic cryptogams. *Ecoscience*, **4**, 75–79.

Larcher, W. (1995). *Physiological Plant Ecology*, 3rd edn. Springer, Berlin.

Lee, J.A. (1998). Unintentional experiments with terrestrial ecosystems: ecological effects of sulphur and nitrogen pollutants. *Journal of Ecology*, **86**, 1–12.

Lee, J.A. & Harmer, R. (1980). Vivipary, a reproductive strategy in response to environmental stress? *Oikos*, **34**, 254–265.

Longton, R.E. (1988). *Biology of Polar Bryophytes and Lichens*. Cambridge University Press, Cambridge.

McFadden, J.P., Hollinger, D.Y. & Chapin, F.S., III (1998). Subgrid-scale variability in the surface energy balance of arctic tundra. *Journal of Geophysical Research*, **103 D22**, 28947–28961.

Marion, G.M., Henry, G.H.R., Freckman, D.W., Johnstone, J., Jones, G., Jones, M.H. *et al.* (1997). Open top designs for manipulating field temperature in high-latitude ecosystems. *Global Change Biology*, **3** (Suppl. 1), 20–32.

Maxwell, B. (1992). Arctic climate: potential for change under global warming. In *Arctic Ecosystems in a Changing Climate: An Ecophysiological Perspective* (Ed. by F.S. Chapin III, R.L. Jefferies, J.F. Reynolds, G.R. Shaver & J. Svoboda), pp. 11–34. Academic Press, San Diego.

Michelsen, A., Schmidt, I.K., Jonasson, S., Quarmby, C. & Sleep, D. (1996). Leaf [15]N abundance of subarctic plants provides field evidence that ericoid, ectomycorrhizal and non and arbuscular mycorrhizal species access different sources of soil nitrogen. *Oecologia*, **105**, 53–63.

Nadelhoffer, K.J., Giblin, A.E., Shaver, G.R. & Laundre, J.L. (1991). Effects of temperature and substrate quality on element mineralization in six arctic soils. *Ecology*, **72**, 242–253.

Näsholm, T., Ekblad, A., Nordin, A., Giesler, R., Högberg, M. & Högberg, P. (1998). Boreal forest plants take up organic nitrogen. *Nature*, **392**, 914–916.

Oechel, W.C., Hastings, S.J., Vourlitis, G.L., Jenkins, M., Riechers, G. & Grulke, N. (1993). Recent change of arctic tundra ecosystems from a net carbon dioxide sink to a source. *Nature*, **361**, 520–523.

Oechel, W.C., Cowles, S., Grulke, N., Hastings, S.J., Lawrence, B., Prudhomme, T. *et al.* (1994). Transient nature of $CO_2$ fertilization in arctic tundra. *Nature*, **311**, 500–503.

Post, W.M., Emanuel, W.R., Zinke, P.J. & Stangenberger, A.G. (1982). Soil carbon pools and world life zones. *Nature*, **298**, 156–159.

Press, M.C., Potter, J.A., Burke, M.J.W., Callaghan, T.V. & Lee, J.A. (1998). Responses of a sub-arctic dwarf shrub heath community to simulated environmental change. *Journal of Ecology*, **86**, 315–327.

Raab, T.K., Lipson, D.A. & Monson, R.K. (1996). Non-mycorrhizal uptake of amino acids by roots of the alpine sedge *Kobresia myosuroides*: implications for the alpine nitrogen cycle. *Oecologia*, **108**, 488–494.

Rannie, W.F. (1986). Summer air temperature and number of vascular species in arctic Canada. *Arctic*, **39**, 133–137.

Robinson, C.H. & Wookey, P.A. (1997). Microbial ecology, decomposition and nutrient cycling. In *Ecology of Arctic Environments* (Ed. by S.J. Woodin & M. Marquiss), pp. 41–68. Blackwell Science, Oxford.

Robinson, C.H., Wookey, P.A., Lee, J.A., Callaghan, T.V. & Press, M.C. (1998). Plant community responses to simulated environmental change at a high arctic polar semi-desert, Svalbard (79°N). *Ecology*, **79**, 856–866.

Savile, D.B.O. (1972). *Arctic Adaptations in Plants*. Monograph 6. Canada Department of Agriculture Information, Ottawa.

Schimel, J.P. & Chapin, F.S., III (1996). Tundra plant uptake of amino acid and $NH_4^+$ nitrogen *in situ*: plants compete well for amino acid N. *Ecology*, **77**, 2142–2147.

Scholander, P.F. (1934). *Skrifter Om Svalbard Og Ishavet*. Det Kongelige Department for Handel, Sjøfart, Industri, Handverk og Fiskeri. I Kommisjon hos Jacob Dybwad, Oslo.

Shaver, G.R. & Chapin, F.S., III (1986). Effect of fertilizer on production and biomass of tussock tundra Alaska, USA. *Arctic and Alpine Research*, **18**, 261–268.

Summerhayes, V.S. & Elton, C.S. (1923). Contributions to the ecology of Spitzbergen and Bear Island. *Journal of Ecology*, **11**, 214–286.

Tissue, D.T. & Oechel, W.C. (1998). Response of *Eriophorum vaginatum* to elevated $CO_2$ and temperature in the Alaskan tussock tundra. *Ecology*, **68**, 401–410.

Walker, D.A., Auerbach, N.A., Bockheim, J.G., Chapin, F.S., III, Eugster, W., King, J.Y. *et al.* (1998). Energy and trace-gas fluxes across a soil pH boundary in the Arctic. *Nature*, **394**, 469–472.

Walsh, N.E., McCabe, T.R., Welker, J.M. & Parsons, A.N. (1997). Experimental manipulations of snow-depth: effects on nutrient content of caribou forage. *Global Change Biology*, **3** (Suppl. 1), 158–164.

Whalen, S.C. & Cornwell, J.C. (1985). Nitrogen, phosphorus and organic carbon cycling in an arctic lake. *Canadian Journal of Fisheries and Aquatic Sciences*, **42**, 797–808.

Woodin, S.J. (1997). Effects of acidic deposition on arctic vegetation. In *Ecology of Arctic Environments* (Ed. by S.J. Woodin & M. Marquiss), pp. 219–237. Blackwell Science, Oxford.

Woodin, S.J. & Lee, J.A. (1987). The effects of nitrate, ammonium and temperature on nitrate reductase activity in *Sphagnum* species. *New Phytologist*, **105**, 103–115.

Woodin, S.J., Press, M.C. & Lee, J.A. (1985). Nitrate reductase activity in *Sphagnum fuscum* in relation to atmospheric nitrate deposition. *New Phytologist*, **99**, 381–388.

Wookey, P.A., Robinson, C.H., Parsons, A.N., Welker, J.M., Press, M.C., Callaghan, T.V. *et al.* (1995). Environmental constraints on the growth, photosynthesis and reproductive development of *Dryas octopetala* at a high arctic polar semi-desert, Svalbard. *Oecologia*, **102**, 478–489.

Woolgrove, C.E. & Woodin, S.J. (1996). Ecophysiology of a snowbed bryophyte, *Kiaeria starkei* during snowmelt and uptake of nitrate from polluted meltwater. *Canadian Journal of Botany*, **74**, 1095–1103.

# Chapter 17

# Ecophysiology of mangroves: challenges in linking physiological processes with patterns in forest structure

*M.C. Ball[1] and M.A. Sobrado[1,2]*

## Introduction

Mangroves are woody trees and shrubs that dominate saline, tidal wetlands along tropical and subtropical coastlines. One of the most striking features of mangrove forests is the tendency for species to become distributed differentially in a banded zonation pattern orientated roughly parallel to shore (Duke *et al.* 1998). These zones may be either monospecific or strongly dominated by only one or two species and are well correlated with the frequency and duration of tidal immersion. Many biological processes contribute to the development of zonation patterns in mangrove forest (Smith 1992). These processes include dispersal (Rabinowitz 1978), herbivory (Smith 1987b) and competition (Clarke & Hannon 1970), all of which can vary in relative importance along tidal gradients (Clarke & Myerscough 1993). Much less attention has been given to understanding ecophysiological processes, although these too play critical roles in the development of forest structure, because the physiological attributes of species form the framework for their interactions with other organisms and the environment.

In a recent review, which should be consulted for a detailed discussion and references, Ball (1996) considered how interspecific differences in salt tolerance might relate to mangrove forest structure along tidally maintained salinity gradients. In summary, responses of mangroves to salinity are as varied as the environments in which they are found, with interspecific differences in salt tolerance reflecting a physiological continuum from moderately salt tolerant glycophytes to highly salt tolerant and apparently obligate halophytes. Like most halophytes, mangroves typically show maximum growth under relatively low salinity conditions, but they differ in the range of salinities over which high growth rates are sustained. There is a tendency for the more salt tolerant species to grow more slowly under optimal conditions in comparison with less tolerant species (Fig. 17.1). Slower growth in

[1] *Ecosystem Dynamics Group, Research School of Biological Sciences, Institute of Advanced Studies, The Australian National University, Box 475, Canberra, ACT 2601 Australia. E-mail: mball@rsbs.anu.edu.au*
[2] *Laboratorio de Biologia Ambiental de Plantas, Departamento Biologia de Organismos, Universidad Simon Bolivar, Apartado 89.000, Caracas 1080 A, Venezuela*

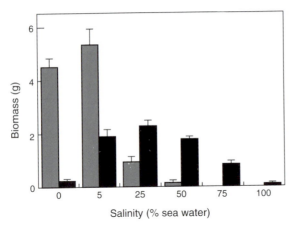

**Figure 17.1** Biomass accumulation in *Sonneratia lanceolata* (grey) and *S. alba* (black) in response to salinity. Values are means ± SE, $n = 5$. Redrawn from Ball and Pidsley (1995).

the more salt tolerant species is associated with two key attributes. First, the carbon cost of water uptake increases with increasing salinity and is greater in the more salt tolerant species. This is reflected in the greater allocation of biomass to roots. Second, mangroves typically have low transpiration rates and high water-use efficiencies. These water-use characteristics become more conservative with increases in salinity and in those species which demonstrate greater salt tolerance. Maintenance of high water-use efficiency may be a consequence of both high carbon costs of water uptake by the roots and restrictions in the rates of water usage by the leaves to maintain favourable water and salt relations. However, conservative water use requires changes in the display of foliage: the angle of leaf inclination as well as the shape and size of leaves (there is little change in reflective properties). These changes maintain favourable leaf temperatures with minimal evaporative cooling, resulting in decreased light interception with increasing salinity. These two attributes, the high carbon cost of water gain by the roots and conservative water use by the leaves, may contribute to the enhancement of salt tolerance but at the expense of the growth rate, such that species tolerant of broad ranges of salinity tend to grow more slowly than less tolerant species even under optimal salinities for growth. Thus, species from an available pool could become distributed differentially along a salinity gradient because of differences in tolerance limits and because of the ways in which physiological attributes, associated with differences in salt tolerance, might affect competitive interactions for resources along the gradient (Ball 1996).

Salinity gradients, however, do not exist in isolation from other edaphic and climatic factors. This chapter considers some of the challenges in understanding how mangroves respond to salinity in combination with other variables, and how such responses might influence the structure of mangrove forests.

## Interactive effects of salinity and aridity

The size, productivity and diversity of mangrove forests decline with increasing salinity and aridity (Saenger *et al.* 1977; Cintrón *et al.* 1978; Smith & Duke 1987), which affect the availability of water at the roots and the evaporative demand at the leaves. Mangroves growing in seasonally arid environments show adjustments in the components of water relations. This plays an important role in maintaining favourable carbon and water balances as soils become drier and more saline and as the evaporative demand increases with progression from the wet to the dry season (Rada *et al.* 1989; Smith *et al.* 1989; Medina & Francisco 1997; Suárez *et al.* 1998). For example, Suárez *et al.* (1998) compared water relations in leaves of *Avicennia germinans* growing in sites with low and high soil salinity. During a wet season, soil water salinity ranged from 5 to 15 parts per thousand (p.p.t.) at the low-salinity site and from 30 to 55 p.p.t. at the high-salinity site. The leaves maintained high water contents over a wide range of water potentials, with maintenance of turgor at the high-salinity site being achieved largely through accumulation of ions for osmotic adjustment (Fig. 17.2a). During the dry season, the soil water salinity at the low-salinity site increased to 50 p.p.t. Leaves maintained turgor during the drought by lowering their osmotic potential to values below those encountered at an equivalent salinity during the wet season, coupled with a slight increase in cell wall elasticity (Fig. 17.2b). Similar measurements were made on leaves of *Rhizophora mangle* growing near the sea, where salinities varied little between wet and dry seasons (Rada *et al.* 1989). This species showed less reduction in osmotic potential during the dry season, consistent with smaller changes in the salinity of interstitial water;

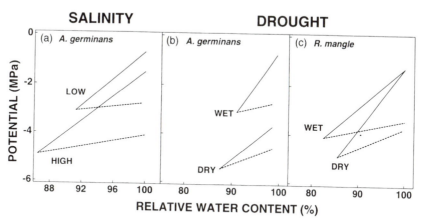

**Figure 17.2** Changes in leaf water relations in response to salinity and drought in mangroves grown under natural field conditions. (a) Leaves of *Avicennia germinans* grown in low- and high-salinity habitats. Redrawn from Suárez *et al.* (1998). (b) Leaves of *A. germinans* grown in a low-salinity habitat during wet and dry seasons (M.A. Sobrado, unpublished). (c) Leaves of *Rhizophora mangle* grown in a shoreline habitat subject to regular tidal inundation by seawater during wet and dry seasons. Redrawn from Rada *et al.* (1989). Lines indicate water potential (solid) and osmotic potential (dashed).

however, the cell walls were slightly more rigid during the dry season (Fig. 17.2c). Elastic properties of cell walls are an important component of plant water relations because they affect the change in volume for a given change in pressure. With a decrease in relative water content, leaves with more elastic cell walls can maintain higher turgor pressures, whereas those with more rigid cell walls can generate greater water potential differences between leaves and soil. Thus, there are interspecific differences in the way in which turgor is maintained with variation in salinity and evaporative demand (Rada *et al.* 1989; Suarez *et al.* 1998).

Water use in relation to carbon gain also varies between species and populations of species distributed along gradients in salinity and aridity. Stomatal conductance declines in response to increasing salinity and decreasing humidity, under both field (Clough & Sim 1989; Youssef & Saenger 1998; Sobrado 1999) and laboratory conditions (Ball & Farquhar 1984; Ball 1988; Ball *et al.* 1997). At lower humidities, both assimilation rate and stomatal conductance usually decline in unison; however, the decline in stomatal conductance is usually insufficient to exert large impacts on the transpiration rate, although rates of water loss are less than they would be in the absence of any stomatal closure. Consequently, water-use efficiencies, whether determined from instantaneous gas-exchange measurements on individual leaves (Ball & Farquhar 1984; Clough & Sim 1989; Youssef & Saenger 1998; Sobrado 1999) or based on gas-exchange characteristics averaged over time (Ball 1988; Ball *et al.* 1997), are lower under dry than under humid conditions. The more salt tolerant species maintain higher rates of transpiration with greater water-use efficiencies at higher salinities and lower humidities than less salt tolerant species (Ball 1988; Ball *et al.* 1997). These responses are consistent with species distributions along both salinity and aridity gradients (Ball *et al.* 1997). These observations are consistent with those of Youssef and Saenger (1998), who studied two populations of *Aegiceras corniculatum*. Plants growing in the seasonally dry tropics of northern Australia maintained higher rates of transpiration at higher water-use efficiencies than those growing in the humid wet tropics, when the two groups were subjected to similar dry microclimatic conditions. Mechanistic bases of adaptation to drought between and within species growing in saline environments are poorly understood, but may well involve changes in hydraulic architecture that reduce the occurrence of catastrophic embolism (Sperry *et al.* 1988).

## Interactive effects of salinity and high irradiance

Mangroves generally have low photosynthetic rates that become light saturated at quantum flux densities ranging from 30 to 50% of incident sunlight (Ball & Critchley 1982; Björkman *et al.* 1988; Cheeseman *et al.* 1991; Jin-Eong *et al.* 1995; Farnsworth & Ellison 1996). When exposed to high light, dissipation of excess excitation energy is important to prevent photodamage occurring (Chapter 1). Exposure to excess irradiance is reduced by changes in leaf orientation and physiological properties of the leaves, effecting a compromise between the requirements for irradiance and the maintenance of favourable leaf temperatures with minimal evaporative cooling (Ball *et al.* 1988). However, total avoidance of excess irradiance is not possible, and while the radiation load on a leaf can be greatly reduced, leaves still intercept more light than can be used in photosynthesis, particularly when high salinities reduce the capacity for photosynthetic carbon assimilation. Under such conditions, fully exposed sun leaves of mangroves become photoinhibited in response to high irradiance, as shown by a depression in the quantum yield of $O_2$ evolution and quenching of chlorophyll *a* fluorescence (Björkman *et al.* 1988). Similarly, other studies have also found quantum yield to be lower in naturally displayed sun leaves than shade leaves of mangrove canopies (Cheeseman *et al.* 1991; Lovelock & Clough 1992). This depression in quantum yield of mangrove leaves in response to high irradiance was found to be correlated with the concentration of zeaxanthin per unit leaf area (Lovelock & Clough 1992), consistent with reversible photoprotective dissipation of excess excitation energy through the xanthophyll cycle (Demmig-Adams *et al.* 1989). Similarly, the fluorescence ratio Fv/Fm, a measure of the photochemical efficiency of photosystem II, declines during midday, but returns to maximal values overnight, indicating no chronic photoinhibition in response to either drought (Sobrado 1999) or salinity (Sobrado & Ball 1999; Sobrado 1999) stress under field conditions. To date, there is no evidence for direct effects of salinity on primary photochemistry, except when salinity interacts with other stresses (Larcher *et al.* 1990).

Under benign conditions, several studies have shown good agreement between rates of electron transport calculated from fluorescence characteristics and rates of photosynthetic $CO_2$ assimilation (Genty *et al.* 1989). However, Lovelock and Winter (1996) found that rates of electron transport can exceed demands for photosynthesis, implying that photorespiration and other electron sinks might also play major roles in the dissipation of excess excitation energy. Photorespiratory consumption of electrons may increase when environmental stresses that induce changes in leaf anatomy, such as thickening of cell walls, restrict $CO_2$ diffusion toward the chloroplast (Evans & von Caemmerer 1996). This causes $CO_2$ concentrations at carboxylation sites to be substantially lower than in intercellular air spaces, thereby favouring photorespiration. Under saturating irradiance, changes in the $CO_2$ concentration within the chloroplast alter relative rates of $CO_2$ assimilation and photorespiration, while the rates of electron transport required to sustain both processes remain constant (Evans & von Caemmerer 1996). For example, Sobrado and Ball (1999) found no change in rates of electron transport in leaves of *Avicennia marina* in the field, despite a 43% decrease in assimilation rates from 7.6 to 4.3 $\mu$mol $CO_2$ m$^{-2}$ s$^{-1}$ in leaves from seawater and hypersaline sites, respectively. The observed rates of electron transport were clearly in excess of the requirements for photosynthesis and would have been sufficient to support midday assimilation rates of 10–11 $\mu$mol $CO_2$ m$^{-2}$ s$^{-1}$ in both sites. Using the methods of Evans and von Caemmerer (1996), Sobrado and Ball (1999) calculated the $CO_2$ concentration within the chloroplast ($C_c$) to be 144 $\mu$mol mol$^{-1}$ when the intercellular $CO_2$ concentration ($C_i$) was 245 $\mu$mol mol$^{-1}$ in leaves of seawater-grown plants, and $C_c$ to be 78 when $C_i$ was 212 $\mu$mol mol$^{-1}$ in leaves of hypersaline-grown plants. These differences between chloroplastic and intercellular $CO_2$ concentrations would correspond to a decrease in cell wall conductance from 75 mmol m$^{-2}$ s$^{-1}$ at the seawater site to 40 mmol m$^{-2}$ s$^{-1}$ at the hypersaline site. Thus rates of photorespiration might be much higher than expected and play an increasingly important role in photochemical dissipation of absorbed light energy with increasing salinity stress. This would account for similarities in both rates of electron transport and xanthophyll-dependent thermal dissipation of excess radiant energy in leaves from the two sites, despite major differences in assimilation rates (Sobrado & Ball 1999; Chapter 1).

Other studies suggest that the Mehler reaction and associated processes could account for substantial rates of $O_2$-dependent electron transport. Cheeseman *et al.* (1997) found rates of electron transport, estimated from fluorescence measurements in sunlit leaves of the mangrove *Rhizophora stylosa*, to be three times higher than the calculated rate of electron consumption through photosynthesis and photorespiration, assuming no significant restriction to diffusion of $CO_2$ to the chloroplast. High activities of superoxide dismutase and other enzymes involved in $H_2O_2$ metabolism were measured in the leaves of *R. stylosa*, implying that $O_2$ may be an important alternative sink for electrons under field conditions (Cheeseman *et al.* 1997). Other compounds might also contribute to the removal of $H_2O_2$. These include phenolics (Cheeseman *et al.* 1997) which occur in high concentrations in mangrove leaves grown under naturally high irradiance (Lovelock *et al.* 1992), and low-molecular-weight carbohydrates capable of scavenging oxygen radicals (Tarczynski *et al.* 1993; Shen *et al.* 1997). Indeed, concentrations of mannitol, proline and other low-molecular-weight compounds typically increase in mangrove leaves in response to increasing salinity (Popp 1984; Richter *et al.* 1990), and their potential roles in protection from photo-oxidative damage warrant further research.

In summary, the requirements for the dissipation of excess energy increase with salinity-induced decreases in photosynthetic activity (Björkman *et al.* 1988). Changes in leaf angle with increasing salinity reduce the interception of excess irradiance (Ball *et al.* 1988), and most of the excess absorbed excitation energy is dissipated via the xanthophyll pigments (Björkman *et al.* 1988; Lovelock & Clough 1992; Cheeseman *et al.* 1997; Sobrado & Ball 1999). However, the relative importance of photorespiration (Osmond 1981) and alternative electron sinks (Cheeseman *et al.* 1997) in photochemical dissipation of excess absorbed light energy, as increasing salinity progressively restricts light use, remains to be established. Interspecific differences in the capacity of leaves to maintain relatively high photosynthetic rates and to provide protection from excessive irradiance, with spatial and temporal variation in environmental factors that affect photosynthesis, may influence the relative performance of species along salinity gradients.

## Interactive effects of salinity and low irradiance

The establishment of mangrove seedlings appears to require high levels of irradiance. Demographic studies have shown that densities of naturally occurring seedlings are greater beneath porous canopies than dense canopies (Ball 1980) and in gaps rather than in closed forests (Sukardjo 1987). Further, juvenile trees are clumped in gaps or former light-gap areas (Jimenez 1988). Experimental studies have found that seedling growth and survivorship are lower in the dense shade of closed canopies than in gaps (Smith 1987a). One problem in the interpretation of these observations is the extent to which other factors, such as predation (Smith 1987a; Farnsworth & Ellison 1993) or environmental stresses (McKee 1995b), for example salinity and waterlogging, could affect apparent responses to irradiance under field conditions.

Understorey vegetation in mangrove forests generally occurs in areas of low salinity (Chapman 1975; Lugo 1986). The basis for this distribution is unknown, but physiological attributes associated with increasing salt tolerance (Ball 1996) might preclude the occurrence of understorey species in highly saline environments (Ball 1998). With increases in salinity, the carbon allocation to roots increases at the expense of leaf area (Ball 1988; Soto 1988) and specific leaf area decreases (Camilleri & Ribi 1983; Ball *et al.* 1988). Together, the processes described above will act to lower both carbon fixation and the proportion of assimilated carbon that is available for growth (e.g. Farnsworth & Ellison 1996). In turn, seedlings will have a higher light compensation point for growth (Givnish 1988), which could preclude establishment and survival in the understorey.

## Interactive effects of salinity and nutrients

The role of nutrients in determining the structure and function of mangrove forests along tidally maintained gradients in edaphic conditions is, perhaps, the most neglected area of mangrove research. There have been some attempts to develop generalities about nutritional limitations in mangrove forests based on the effects of flooding regimes on the organic content and redox status of sediments. In a series of field experiments, Boto and Wellington (1983, 1984) showed that mangrove growth was limited by phosphorus in aerobic, infrequently flooded mineral soils at high tidal elevations and by nitrogen in anaerobic, frequently flooded organic sediments at low tidal elevations. Their results were consistent with predictions based on chemical transformations of nutrients in aerobic and anaerobic marine sediments (Valiela 1984). However, more recent research shows that it is not possible to generalize on the basis of tidal flooding alone. In an elegant study, Feller (1995) identified phosphorus deficiency in a stunted stand of *Rhizophora mangle* growing in calcareous sediments in a basin subject to persistent flooding. Under these conditions, phosphorus may bind to carbonates, thereby becoming unavailable for uptake by plants. By contrast, growth in a nearby frequently flooded, well-developed mangrove forest underlain with peat was limited by nitrogen, as expected (Feller 1995). Extensive field-based studies by Twilley and colleagues (see review by Twilley 1995) have also highlighted limitations to growth and productivity by phosphorus deficiency in some mangrove systems and demonstrated the importance of externally derived nutrients to the nitrogen and phosphorus budgets of mangrove forests.

There may be interspecific differences in nutrient requirements for growth and in the physiological capacity to satisfy these requirements among species that characteristically occupy different positions along gradients of tidal inundation. Indeed, three sympatric mangrove species (*Avicennia germinans*, *Laguncularia racemosa* and *Rhizophora mangle*) showed marked differences in growth responses to high and low nitrogen levels in low salinity under controlled environment conditions (McKee 1995a). However, little is known of species responses to differences in the form and availability of nutrients under field conditions and how such differences might relate to species distributions.

Finally, interactive effects of nutrition and salinity on plant growth have received little attention. This is surprising given the well-established relationships between nitrogen nutrition and photosynthetic capacity (Evans 1989) and corresponding nitrogen requirements for photoprotective dissipation of excess light energy via xanthophyll pigments (Verhoeven *et al.* 1996). Indeed, nitrogen nutrition plays an important role in alleviating light stress when environmental factors, such as water, limit growth and photosynthesis (Ferrar & Osmond 1986). It follows that nitrogen nutrition could also be important in reducing salinity stress, particularly given the role of nitrogen-containing compounds in osmotic adjustment (Popp *et al.* 1984) and other protective functions. The diversion of nitrogen to protective functions may increase apparent requirements for nitrogen in growth and photosynthesis. Such differences in the relative distribution of nitrogen among physiological processes may partly explain observations of decreases in photosynthesis with increasing salinity and drought, despite constancy in the nitrogen concentration in leaves of mangroves growing under natural conditions in Venezuela (Sobrado 1999). Thus, salinity-dependent nutrient deficiencies could occur if adverse salinities, or edaphic conditions associated with waterlogging, were to interfere with nutrient uptake (Lin & Sternberg 1992), or if greater nutrient availability were required to support a given growth rate under adverse salinities. Clearly, there is a need to understand how interactions between resource use and stress tolerance relate to the structure and function of mangrove forests along environmental gradients.

## Impact of elevated $CO_2$ on responses of coastal wetland vegetation to tidally maintained gradients of salinity and waterlogging

Predicted changes in the global environment may exert marked impacts on mangrove forests, specifically through changes in sea level, climate and atmospheric $CO_2$ concentration. Estimated increases in sea level over the next century (largely arising from the thermal expansion of sea water) range from 16 to 55 cm (Warrick *et al.* 1995). As sea level rises, changes in patterns of tidal inundation and associated changes in geomorphological and climatic processes will alter the coastal landscape. Depending on the extent to which sedimentation keeps pace with rising sea levels, some habitats may disappear while others expand landwards with the advance of tidal floodwaters onto low-lying coastal lands. It is extremely difficult to predict the form and distribution of these future habitats, partly because of the uncertainty associated with predicting future sea levels. However, less uncertainty is associated with atmospheric $CO_2$ concentrations. The atmospheric $CO_2$ concentration is rising and the preindustrial concentration is expected to double by the middle of the twenty-first century, exerting major impacts on both carbon gain and water-use characteristics (Chapter 13). Such changes may be sufficient to alter the structure and function of coastal wetlands (see e.g. Drake 1992).

The prediction of community-level responses to elevated concentrations of $CO_2$ along salinity gradients suffers from a paucity of experimental data and is complicated by the presence of other environmental gradients which may also influence responses to elevated atmospheric $CO_2$. To date, the only field studies that address responses of halophytic vegetation to elevated $CO_2$ are those conducted on three communities in a temperate salt marsh, using open-top chambers (Drake 1992): monospecific stands of the $C_3$ sedge, *Scirpus olneyi*, the $C_4$ grass, *Spartina patens*, and a mixed community of these two species together with another $C_4$ grass, *Distichlis spicata*. Elevation of $CO_2$ for 4 years reduced water loss, increased water potential and delayed senescence in all three species, but only stimulated growth in the $C_3$ species, together with increasing the number of shoots, roots and rhizomes (Curtis *et al.* 1989a). These responses of *S. olneyi* were associated with a higher quantum yield (Long & Drake 1991) and photosynthetic capacity (Ziska *et al.* 1990; Arp & Drake 1991), lower rates of dark respiration, and lower concentrations of nitrogen in all tissues (Curtis *et al.* 1989b). At the ecosystem level, nitrogen fixation and carbon accumulation were stimulated (Curtis *et al.* 1990). In the mixed community, the biomass of the $C_3$ species increased by over 100% while the biomass of the $C_4$ species declined (Arp *et al.* 1993). These results were obtained in a salt marsh where edaphic conditions were conducive to vigorous growth. Extrapolating from these results to a complex salt marsh system or mangrove forest, encompassing a broader range of edaphic conditions, remains a challenge.

Projected rises in sea level may greatly affect mangrove vegetation because there are marked interspecific differences in the tolerance of mangroves to flooding (McKee 1993). Mangroves possess specialized root structures that promote aeration of roots under otherwise anaerobic conditions (Tomlinson 1986). In general, the roots are shallow, and possess numerous lenticels and extensive aerenchyma (Curran 1985; Youssef & Saenger 1996). These structures allow effective aeration (Curran 1985; Thibodeau & Nickerson 1986; McKee & Mendelssohn 1987; McKee 1993), provided that the roots are in direct contact with air at least during low tides. Such exposure is essential because aeration of the root systems depends on the diffusion of oxygen down partial pressure gradients from the atmosphere to sites of respiration in the roots, and oxygen diffuses 10 000 times more rapidly through air than water (Scholander *et al.* 1955). Nevertheless, oxygen concentrations in roots can drop to inhibitory levels during prolonged tidal submergence (Hovenden *et al.* 1995). Thus, increases in the duration and frequency of tidal flooding could profoundly influence the functioning of existing trees and the establishment of seedlings along changing gradients of tidal inundation (Ellison & Farnsworth 1996a).

There is little evidence that elevated $CO_2$ concentrations will alleviate the adverse effects of waterlogging on mangroves. For example, rates of water uptake and carbon gain are lower in waterlogged mangroves than in those grown under well-drained conditions (Naidoo 1985). Similarly, maximum rates of photosynthetic carbon gain and stomatal conductance were reduced when seedlings of *Rhizophora stylosa* were grown under flooding regimes that simulated a 16-cm increase in tidal height (Ellison & Farnsworth 1996b). If such reductions in carbon gain were a result of stomatal closure, then elevated concentrations of $CO_2$ might overcome stomatal limitations to $CO_2$ diffusion into leaves. However, a doubling of the $CO_2$ concentration enhanced the growth of salt marsh species when the rooting medium was aerated but not when it was anaerobic, as would occur with waterlogging (Rozema *et al.* 1991).

Many studies have predicted that elevated concentrations of $CO_2$ will enhance the growth of plants subject to salinity stress (see e.g. Idso & Idso 1994). However, a diversity of responses to elevated $CO_2$ has been reported, ranging from growth stimulation to growth depression, and varying markedly between species and growth conditions (Rozema *et al.* 1990, 1991; Drake 1992; Lenssen *et al.* 1993; Farnsworth *et al.* 1996; Ball *et al.* 1997).

With increases in salinity above optimal levels, stomatal conductance typically declines with a concomitant reduction in transpiration rates (Ball & Farquhar 1984; Clough & Sim 1989; Lin & Sternberg 1992). This may reflect increasing limitation in the rate at which water can be supplied to the shoots, as the water potential in the soil declines with increasing salinity. However, the restriction of water loss by partial stomatal closure would also restrict the diffusion of $CO_2$ into the leaf. This, in turn, can lead to low intercellular $CO_2$ concentrations with correspondingly low $CO_2$ assimilation rates, particularly in $C_3$ species. Relationships between water vapour efflux and $CO_2$ influx are such that lower stomatal conductance results in the enhancement of water-use efficiency, but at the expense of carbon gain (Cowan & Farquhar 1977). This trade-off between water use and carbon gain in $C_3$ species can be greatly ameliorated under elevated concentrations of $CO_2$, because higher intercellular concentrations of $CO_2$ can be maintained at lower stomatal conductances than under ambient $CO_2$ concentrations. In this way, a leaf could operate with higher water-use efficiency under elevated $CO_2$ without a penalty in terms of carbon gain. Such changes in water-use characteristics could have far-reaching consequences for plant function.

The results of growth analyses show interspecific differences in the extent to which changes in leaf area ratio and/or net assimilation rate contribute to changes in relative growth rates with increasing salinity. However, increase in net assimilation rate accounted for most of the growth enhancement under elevated $CO_2$ under both field (Drake 1992) and laboratory conditions (Rozema *et al.* 1991; Farnsworth *et al.* 1996; Ball *et al.* 1997). In these studies, growth enhancement occurred under moderate salinity stress and was associated with greater rates of photosynthesis and higher water-use efficiencies. Presumably, elevated $CO_2$ improved plant water status and alleviated stomatal limitations to $CO_2$ diffusion into leaves.

However, under more stressful salinities, elevated $CO_2$ had little or no beneficial effect on growth, despite improved water status and increased water-use efficiencies (Rozema *et al.* 1991; Farnsworth *et al.* 1996; Ball *et al.* 1997). Presumably, growth under these highly saline conditions was limited by effects such as ion toxicity or induction of deficiencies in nutrients such as $K^+$, factors which are unlikely to be affected by elevated $CO_2$ (Ball & Munns 1992). It thus appears that elevated concentrations of $CO_2$ stimulate the growth of plants under mild salinity stress but do not overcome limitations imposed on growth by severe salinity stress.

Plants growing in coastal wetlands of arid or seasonally dry areas may benefit from rising $CO_2$ concentrations. With a doubling of atmospheric $CO_2$, assimilation rates in $C_3$ species may operate close to their $CO_2$ saturation point. Thus, they may be less affected by decreases in stomatal conductance in response to large gradients in water vapour between leaves and air, resulting from low atmospheric humidity under arid conditions (Morison & Gifford 1983). In one study (Ball *et al.* 1997), growth responses of the mangroves *Rhizophora apiculata* and *R. stylosa* to salinity and humidity were consistent with their natural distributions along salinity and aridity gradients. Both species grew best in low salinity, but differed in response to humidity. *R. apiculata* had twice the relative growth rate of *R. stylosa* under high humidity, but relative growth rates of both species were similar under low humidity. However, *R. apiculata* gained a growth advantage under low humidity when $CO_2$ concentration was doubled. By contrast, growth responses to elevated $CO_2$ were minimal when salinity was high. Thus, elevated $CO_2$ enhanced growth more when carbon gain was limited by evaporative demand at the leaves than when it was limited by salinity at the roots (Ball *et al.* 1997). Such responses could alter competitive hierarchies among species whose distributions along salinity gradients are affected by aridity.

The extent to which changes in mangrove forest composition have already occurred in response to increases in $CO_2$ concentration is unknown. However, climate change may also result in greater levels of disturbance. Mangroves habitats can be highly dynamic and disturbance may effect changes in forest structure, both through geomorphological processes and catastrophic climatic events. For example, erosion and accretion of sediments, as well as uplift and subsidence of the land surface, can occur over time scales significantly less than the life-span of the trees (Thom 1967). Such topographical changes can cause the physicochemical environment of adult trees to be very different from that in which they established as seedlings, thereby leading to changes in vegetation structure and composition (Thom 1967). Such change can be quite rapid. For example, in South Florida major changes in hydrological characteristics of coastal wetlands have occurred since the early 1900s because of human activities and subsidence of the land. This has resulted in the expansion of mangrove forests into areas previously dominated by freshwater marsh (Ball 1980). Cyclonic storms occur with sufficient frequency that major destruction of mangrove forests can occur over a span of 50–100 years, depending on the locality. While destructive, these events provide opportunities for the regeneration of mangroves (Lugo & Snedaker 1974). Even without the projected increase in storm frequency with global climate change, the present frequency of such disturbance events would tend to accelerate rates of vegetation change in response to changing $CO_2$ concentrations by providing opportunities for relative increase in species favoured by elevated $CO_2$.

In summary, results of field and laboratory studies show that responses of halophytes to elevated concentrations of $CO_2$ may be sufficient to induce substantial changes in coastal wetland vegetation along natural salinity gradients. The greatest stimulation of growth can be expected under relatively low to moderate salinity regimes in which the species already grow well. By contrast, there is no evidence to suggest that elevated $CO_2$ will increase the range of salinities under which a species can grow or alleviate adverse effects of waterlogging. Halophytes are therefore unlikely to expand into areas of extreme edaphic conditions. Whatever growth enhancement may occur at salinities near the limits of tolerance of a species, it is unlikely to have a significant effect on ecological patterns. Thus, a more likely result of rising $CO_2$ concentration is preferential enhancement of growth under conditions already supporting the most productive coastal wetlands, which may further increase disparities in community structure and productivity along natural salinity gradients (Ball *et al.* 1997).

## References

Arp, W.J. & Drake, B.G. (1991). Increased photosynthetic capacity of *Scirpus olneyi* after 4 years of exposure to elevated $CO_2$. *Plant, Cell and Environment*, **14**, 1003–1006.

Arp, W.J., Drake, B.G., Pockman, W.T., Curtis, P.S. & Whigham, D.F. (1993). Interactions between $C_3$ and $C_4$ salt marsh plant species during four years of exposure to elevated atmospheric $CO_2$. *Vegetatio*, **104/105**, 133–143.

Ball, M.C. (1980). Patterns of secondary succession in a mangrove forest of southern Florida. *Oecologia*, **44**, 226–234.

Ball, M.C. (1988). Salinity tolerance in the mangroves, *Aegiceras corniculatum* and *Avicennia marina*. I. Water use in relation to growth, carbon partitioning and salt balance. *Australian Journal of Plant Physiology*, **15**, 447–464.

Ball, M.C. (1996). Comparative ecophysiology of mangrove forest and tropical lowland moist rainforest. In *Tropical Forest Plant Ecophysiology* (Ed. by S.S. Mulkey, R.L. Chazdon &

A.P. Smith), pp. 461–496. Chapman & Hall, New York.

Ball, M.C. (1998). Mangrove species richness in relation to salinity and waterlogging: a case study along the Adelaide River floodplain, northern Australia. *Global Ecology and Biogeography Letters*, **7**, 73–82.

Ball, M.C. & Critchley, C. (1982). Photosynthetic responses to irradiance by the grey mangrove, *Avicennia marina*, grown under different light regimes. *Plant Physiology*, **70**, 1101–1106.

Ball, M.C. & Farquhar, G.D. (1984). Photosynthetic and stomatal responses of two mangrove species, *Aegiceras corniculatum* and *Avicennia marina*, to long term salinity and humidity conditions. *Plant Physiology*, **74**, 1–6.

Ball, M.C. & Munns, R. (1992). Plant responses to salinity under elevated atmospheric concentrations of $CO_2$. *Australian Journal of Botany*, **40**, 515–526.

Ball, M.C. & Pidsley, S.M. (1995). Growth

responses to salinity in relation to distribution of two mangrove species, *Sonneratia alba* and *S. lanceolata*, in northern Australia. *Functional Ecology*, **9**, 77–85.

Ball, M.C., Cowan, I.R. & Farquhar, G.D. (1988). Maintenance of leaf temperature and the optimisation of carbon gain in relation to water loss in a tropical mangrove forest. *Australian Journal of Plant Physiology*, **15**, 263–276.

Ball, M.C., Cochrane, M.J. & Rawson, H.M. (1997). Growth and water use of the mangroves *Rhizophora apiculata* and *R. stylosa* in response to salinity and humidity under ambient and elevated concentrations of atmospheric $CO_2$. *Plant, Cell and Environment*, **20**, 1158–1166.

Björkman, O., Demmig, B. & Andrews, T.J. (1988). Mangrove photosynthesis: response to high irradiance stress. *Australian Journal of Plant Physiology*, **15**, 43–61.

Boto, K.G. & Wellington, J.T. (1983). Phosphorus and nitrogen nutritional status of a northern Australian mangrove forest. *Marine Ecology Progress Series*, **11**, 63–69.

Boto, K.G. & Wellington, J.T. (1984). Soil characteristics and nutrient status in a northern Australian mangrove forest. *Estuaries*, **7**, 61–69.

Camilleri, J.C. & Ribi, G. (1983). Leaf thickness of mangroves (*Rhizophora mangle*) growing in different salinities. *Biotropica*, **15**, 139–141.

Chapman, V.J. (1975). *Mangrove Vegetation*. J. Cramer, Vaduz.

Cheeseman, J.M., Clough, B.F., Carter, D.R., Lovelock, C.E., Eong, O.J. & Sim, R.G. (1991). The analysis of photosynthetic performance in leaves under field conditions: a case study using *Bruguiera* mangroves. *Photosynthesis Research*, **29**, 11–22.

Cheeseman, J.M., Herendeen, L.B., Cheeseman, A.T. & Clough, B.F. (1997). Photosynthesis and photoprotection in mangroves under field conditions. *Plant, Cell and Environment*, **20**, 579–588.

Cintrón, G., Lugo, A.E., Pool, D.J. & Morris, G. (1978). Mangroves of arid environments in Puerto Rico and adjacent islands. *Biotropica*, **10**, 110–121.

Clarke, L.D. & Hannon, N.J. (1970). The mangrove swamp and salt marsh communities of the Sydney district. III. Plant growth in relation to salinity and waterlogging. *Journal of Ecology*, **58**, 351–369.

Clarke, P.J. & Myerscough, P.J. (1993). The intertidal distribution of the grey mangrove (*Avicennia marina*) in southeastern Australia: the effects of physical conditions, interspecific competition, and predation on propagule establishment and survival. *Australian Journal of Ecology*, **18**, 307–315.

Clough, B.F. & Sim, R.G. (1989). Changes in gas exchange characteristics and water use efficiency of mangroves in response to salinity and vapour pressure deficit. *Oecologia*, **79**, 38–44.

Cowan, I.R. & Farquhar, G.D. (1977). Stomatal function in relation to leaf metabolism and environment. In *Integration of Activity in the Higher Plant* (Ed. by D.H. Jennings), pp. 471–505. Cambridge University Press, Cambridge.

Curran, M. (1985). Gas movements in the roots of *Avicennia marina* (Forsk.) Vierh. *Australian Journal of Plant Physiology*, **12**, 97–108.

Curtis, P.S., Drake, B.G., Leadly, P.W., Arp, W.J. & Whigham, D.F. (1989a). Growth and scenescence in plant communities exposed to elevated $CO_2$ concentrations on an estuarine marsh. *Oecologia*, **78**, 20–26.

Curtis, P.S., Drake, B.G. & Whigham, D.F. (1989b). Nitrogen and carbon dynamics in $C_3$ and $C_4$ estuarine marsh plants grown under elevated $CO_2$ *in situ*. *Oecologia*, **78**, 297–301.

Curtis, P.S., Balduman, L.M., Drake, B.G. & Whigham, D.F. (1990). Elevated atmospheric $CO_2$ effects on belowground processes in $C_3$ and $C_4$ estuarine marsh communities. *Ecology*, **71**, 2001–2006.

Demmig-Adams, B., Winter, K., Krüger, A. & Czygan, F.C. (1989). Zeaxanthin and the induction and relaxation kinetics of the dissipation of excess excitation energy in leaves in 2% $O_2$, 0% $CO_2$. *Plant Physiology*, **90**, 887–893.

Drake, B.G. (1992). A field study of the effects of elevated $CO_2$ on ecosystem processes in a Chesapeake Bay wetland. *Australian Journal of Botany*, **40**, 579–595.

Duke, N.C., Ball, M.C. & Ellison, J.C. (1998). Factors influencing biodiversity and distributional gradients in mangroves. *Global Ecology and Biogeography Letters*, **7**, 27–47.

Ellison, A.M. & Farnsworth, E.J. (1996a). Spatial and temporal variability in growth of *Rhizophora mangle* saplings on coral cays: links with variation in insolation, herbivory, and local sedimentation rate. *Journal of Ecology*, **84**, 717–731.

Ellison, A.M. & Farnsworth, E.J. (1996b). Simulated sea level alters anatomy, physiology, growth, and reproduction of red mangrove (*Rhizophora mangle* L.). *Oecologia*, **112**, 435–446.

Evans, J.R. (1989). Photosynthesis and nitrogen relationships in leaves of $C_3$ plants. *Oecologia*, **78**, 9–19.

Evans, J.R. & von Caemmerer, S. (1996). Carbon dioxide diffusion inside leaves. *Plant Physiology*, **110**, 339–346.

Farnsworth, E.J. & Ellison, A.M. (1993). Dynamics of herbivory in Belizean mangal. *Journal of Tropical Ecology*, **9**, 435–453.

Farnsworth, E.J. & Ellison, A.M. (1996). Sun–shade adaptability of the red mangrove, *Rhizophora mangle* (Rhizophoraceae): changes through ontogeny at several levels of biological organization. *American Journal of Botany*, **83**, 1131–1143.

Farnsworth, E.J., Ellison, A.M. & Gong, W.K. (1996). Elevated $CO_2$ alters anatomy, physiology, growth, and reproduction of red mangrove (*Rhizophora mangle* L.). *Oecologia*, **108**, 599–609.

Feller, I. (1995). Effects of nutrient enrichment on growth and herbivory of dwarf red mangrove (*Rhizophora mangle*). *Ecological Monographs*, **65**, 477–505.

Ferrar, P.J. & Osmond, C.B. (1986). Nitrogen supply as a factor influencing photoinhibition and photosynthetic acclimation after transfer of shade grown *Solanum dulcamara* to bright light. *Planta*, **168**, 563–570.

Genty, B., Briantais, J.M. & Baker, N.R. (1989). The relationship between the quantum yield of photosynthetic electron transport and quenching of chlorophyll fluorescence. *Biochimica et Biophysica Acta*, **990**, 87–92.

Givnish, T.J. (1988). Adaptation to sun and shade: a whole plant perspective. *Australian Journal of Plant Physiology*, **15**, 63–92.

Hovenden, M.J., Curran, M., Cole, M.A., Goulter, P.F.E., Skelton, N.J. & Allaway, W.G. (1995). Ventilation and respiration in roots of one-year-old-seedlings of the grey mangrove *Avicennia marina* (Forsk.) Vierh. *Hydrobiologia*, **295**, 23–29.

Idso, K.E. & Idso, S.B. (1994). Plant responses to atmospheric $CO_2$ enrichment in the face of environmental constraints: a review of the past 10 years' research. *Agricultural and Forest Meteorology*, **69**, 153–203.

Jimenez, J.A. (1988). The dynamics of *Rhizophora racemosa* Meyer, forests on the Pacific coast of Costa Rica. *Brenesia*, **30**, 1–12.

Jin-Eong, O., Khoon, G.W. & Clough, B.F. (1995). Structure and productivity of a 20-year-old stand of *Rhizophora apiculata* Bl. mangrove forest. *Journal of Biogeography*, **22**, 417–424.

Larcher, W., Wagner, J. & Thammathaworn, A. (1990). Effects of superimposed temperature stress on *in vivo* chlorophyll fluorescence of *Vigna unguiculata* under saline stress. *Journal of Plant Physiology*, **136**, 92–102.

Lenssen, G.M., Lamers, J., Stroetenga, M. & Rozema, J. (1993). Interactive effects of atmospheric $CO_2$ enrichment, salinity and flooding on growth of $C_3$ (*Elymus athericus*) and $C_4$ (*Spartina anglica*) salt marsh species. *Vegetatio*, **104/105**, 379–388.

Lin, G. & da Sternberg, L.S.L. (1992). Effects of growth form, salinity, nutrient, and sulphide on photosynthesis, carbon isotope discrimination and growth of red mangrove (*Rhizophora mangle* L.). *Australian Journal of Plant Physiology*, **19**, 509–517.

Long, S.P. & Drake, B.G. (1991). Effect of the long-term elevation of $CO_2$ concentration in the field on the quantum yield of photosynthesis of the $C_3$ sedge, *Scirpus olneyi*. *Plant Physiology*, **96**, 221–226.

Lovelock, C.E. & Clough, B.F. (1992). Influence of solar radiation and leaf angle on xanthophyll concentrations in mangroves. *Oecologia*, **91**, 518–525.

Lovelock, C.E. & Winter, K. (1996). Oxygen-dependent electron transport and protection from photoinhibition in leaves of tropical tree species. *Planta*, **198**, 580–587.

Lovelock, C.E., Clough, B.F. & Woodrow, I.E. (1992). Distribution and accumulation of ultra-violet-radiation-absorbing compounds in leaves of tropical mangroves. *Planta*, **188**, 143–154.

Lugo, A.E. (1986). Mangrove understorey: an expensive luxury? *Journal of Tropical Ecology*, **2**, 287–288.

Lugo, A.E. & Snedaker, S.C. (1974). The ecology of mangroves. *Annual Review of Ecology and Systematics*, **5**, 39–64.

McKee, K.L. (1993). Soil physicochemical patterns and mangrove species distribution—reciprocal effects? *Journal of Ecology*, **81**, 477–487.

McKee, K.L. (1995a). Interspecific variation in growth, biomass partitioning, and defensive characteristics of neotropical mangrove seedlings: response to light and nutrient availability. *American Journal of Botany*, **82**, 299–307.

McKee, K.L. (1995b). Seedling recruitment patterns in a Belizean mangrove forest: effects of establishment ability and physico-chemical factors. *Oecologia*, **101**, 448–460.

McKee, K.L. & Mendelssohn, I.A. (1987). Root metabolism in the black mangrove (*Avicennia germinans* (L.) L): response to hypoxia. *Environmental and Experimental Botany*, **27**, 147–156.

Medina, E. & Francisco, M. (1997). Osmolality and $\delta^{13}C$ of leaf tissue of mangrove species from environments of contrasting rainfall and salinity. *Estuarine, Coastal and Shelf Science*, **45**, 337–344.

Morison, J.I.L. & Gifford, R.M. (1983). Stomatal sensitivity to carbon dioxide and humidity. A comparison of two $C_3$ and two $C_4$ grass species. *Plant Physiology*, **71**, 789–796.

Naidoo, G. (1985). Effects of waterlogging and salinity on plant water relations and on the accumulation of solutes in three mangrove species. *Aquatic Botany*, **22**, 133–143.

Osmond, C.B. (1981). Photorespiration and photoinhibition: some implications for the energetics of photosynthesis. *Biochimica et Biophysica Acta*, **639**, 77–89.

Popp, M. (1984). Chemical composition of Australian mangroves. II. Low molecular weight carbohydrates. *Zietschrift Pflanzenphysiology*, **113**, 411–421.

Popp, M., Larher, F. & Weigel, P. (1984). Chemical composition of Australian mangroves. III. Free amino acids, total methylated onium compounds and total nitrogen. *Zietschrift Pflanzenphysiology*, **114**, 15–25.

Rabinowitz, D. (1978). Early growth of mangrove seedlings in Panama, and an hypothesis concerning the relationship of dispersal and zonation. *Journal of Biogeography*, **5**, 113–133.

Rada, F., Goldstein, G., Orozco, A., Montilla, M., Zabala, O. & Azocar, A. (1989). Osmotic and turgor relations of three mangrove ecosystem species. *Australian Journal of Plant Physiology*, **16**, 477–486.

Richter, A., Thonke, B. & Popp, M. (1990). 1D-1-*O* methyl-*muco*-inositol in *Viscum album* and members of the Rhizophoraceae. *Phytochemistry*, **29**, 1785–1786.

Rozema, J., Lenssen, G.M., Broekman, R.A. & Arp, W.P. (1990). Effects of atmospheric carbon dioxide enrichment on salt-marsh plants. In *Expected Effects of Climatic Change on Marine Coastal Ecosystems* (Ed. by J.J. Beukema), pp. 49–54. Kluwer Publishers, Amsterdam.

Rozema, J., Dorel, F., Janissen, R., Lenssen, G., Broekman, R., Arp, W. *et al.* (1991). Effect of elevated atmospheric $CO_2$ on growth, photosynthesis and water relations of salt marsh grass species. *Aquatic Botany*, **39**, 45–55.

Saenger, P., Specht, M.M., Specht, R.L. & Chapman, V.J. (1977). Mangal and coastal salt-marsh communities in Australasia. In *Wet Coastal Ecosystems* (Ed. by V.J. Chapman), pp. 293–345. Elsevier, Amsterdam.

Scholander, P.F., van Dam, L. & Scholander, S.I. (1955). Gas exchange in the roots of mangroves. *American Journal of Botany*, **42**, 92–98.

Shen, B., Jensen, R.G. & Bohnert, H.J. (1997). Increased resistance to oxidative stress in transgenic plants by targeting mannitol biosynthesis to chloroplasts. *Plant Physiology*, **113**, 1177–1183.

Smith, T.J. III (1987a). Effects of light and intertidal position on seedling survival and growth in tropical tidal forests. *Journal of Experimental Marine Biology and Ecology*, **110**, 133–146.

Smith, T.J., III (1987b). Seed predation in relation to tree dominance and distribution in mangrove forests. *Ecology*, **68**, 266–273.

Smith, T.J., III (1992). Forest structure. In *Tropical Mangrove Ecosystems* (Ed. by A.I. Robertson & D.M. Alongi), pp. 101–136. Coastal and Estuarine Studies No. 41, American Geophysical Union, Washington, DC.

Smith, T.J., III & Duke, N.C. (1987). Physical determinants of inter-estuary variation in mangrove species richness around the tropical coastline of Australia. *Journal of Biogeography*, **14**, 9–19.

Smith, J.A.C., Popp, M., Luttge, U., Cram, W.J., Diaz, M., Griffiths, H. *et al.* (1989). Ecophysiology of xerophytic and halophytic vegetation of a coastal alluvial plain in northern Venezuela. VI. Water relations and gas exchange of mangroves. *New Phytologist*, **111**, 293–307.

Sobrado, M.A. (1999). Drought effects on photosynthesis of the mangrove, *Avicennia germinans*, under contrasting salinities. *Trees*, in press.

Sobrado, M.A. & Ball, M.C. (1999). Light use in relation to carbon gain in the mangrove *Avicennia marina* under hypersaline conditions. *Australian Journal of Plant Physiology*, **26**, 245–251.

Soto, R. (1988). Geometry, biomass allocation and leaf life-span of *Avicennia germinans* (L.) L. (Avicenniaceae) along a salinity gradient in Salinas, Puntarenas, Costa Rica. *Revista de Biologia Tropical*, **36**, 309–323.

Sperry, J.S., Tyree, M.T. & Donnelly, J.R. (1988). Vulnerability of xylem to embolism in a mangrove vs an inland species of Rhizophoraceae. *Physiologia Plantarum*, **74**, 276–283.

Suárez, N., Sobrado, M.A. & Medina, E. (1998). Salinity effects on the leaf water relations components and ion accumulation patterns in *Avicennia germinans* (L.) seedlings. *Oecologia*, **114**, 299–304.

Sukardjo, S. (1987). Natural regeneration status of commercial mangrove species (*Rhizophora apiculata* and *Bruguiera gymnorrhiza*) in the mangrove forest of Tanjung Bungin, Banyuasin District, South Sumatra. *Forest Ecology and Management*, **20**, 233–252.

Tarczynski, M.C., Jensen, R.G. & Bohnert, H.J. (1993). Stress protection in transgenic tobacco producing a putative osmoprotectant, mannitol. *Science*, **259**, 508–510.

Thibodeau, F.R. & Nickerson, N.H. (1986). Differential oxidation of mangrove substrate by *Avicennia germinans* and *Rhizophora mangle*. *American Journal of Botany*, **73**, 512–516.

Thom, B.G. (1967). Mangrove ecology and deltaic geomorphology, Tabasco, Mexico. *Journal of Ecology*, **55**, 301–343.

Tomlinson, P.B. (1986). *The Botany of Mangroves*, pp. 62–115. Cambridge University Press, Cambridge.

Twilley, R.R. (1995). Properties of mangrove ecosystems in relation to the energy signature of coastal environments. In *Maximum Power* (Ed. by C.A.S. Hall), pp. 43–62. University of Colorado, Niwot.

Valiela, I. (1984). *Marine Ecological Processes*. Springer-Verlag, New York.

Verhoeven, A.S., Demmig-Adams, B. & Adams, W.W., III (1996). Enhanced employment of the xanthophyll cycle and thermal energy dissipation in spinach exposed to high light and N stress. *Plant Physiology*, **113**, 817–824.

Warrick, R.A., Le Provost, C., Meier, M.F., Oerlemans, J. & Woodworth, P.L. (1995). Changes in sea level. In *Climate Change 1995* (Ed. by J.T. Houghton, L.G. Meira Filho, B.A. Callander, N. Harris, A. Kattenberg & K. Maskell), pp. 363–405. Cambridge University Press, Cambridge.

Youssef, T. & Saenger, P. (1996). Anatomical adaptive strategies to flooding and rhizosphere oxidation in mangrove seedlings. *Australian Journal of Botany*, **44**, 297–313.

Youssef, T. & Saenger, P. (1998). Photosynthetic gas exchange and water use in tropical and subtropical populations of the mangrove, *Aegiceras corniculatum*. *Marine and Freshwater Research*, **49**, 329–334.

Ziska, L.H., Drake, B.G. & Chamberlain, S. (1990). Long-term photosynthetic response in single leaves of a $C_3$ and $C_4$ salt marsh species grown at elevated atmospheric $CO_2$ *in situ*. *Oecologia*, **83**, 469–472.

# Chapter 18
# Water use in arid land ecosystems

*J.R. Ehleringer,[1] S. Schwinning[1] and R. Gebauer[1,2]*

## Historical development of physiological ecology within arid lands

The study of physiological ecology within arid land ecosystems has fascinated ecologists and physiologists throughout this century. Haberlandt (1914), Schimper (1898) and Warming (1909) laid the groundwork with descriptions of the morphological attributes that were thought to be of adaptive value. Early field investigations, such as those by Shreve (1923) and Walter (1931) on the metabolism of arid land plants under the natural ranges of drought and temperature, were critical to evaluating initial hypotheses of how plants were adapted to arid lands. More sophisticated instrumentation ultimately followed (Eckardt 1965), that allowed better quantification of metabolic patterns and led to new concepts, such as physiological acclimation (Mooney & West 1964), that were not obvious from previous laboratory studies. In the 1970s, field measurements moved to a new level of sophistication with the introduction of the mobile laboratory (e.g. Mooney *et al.* 1971). Of particular significance was the ability to now measure photosynthesis, transpiration and energy balance with the same precision as was previously possible only under laboratory conditions, so that investigators could better quantify the relationships between environmental shifts and metabolic adjustment (Mooney 1972; Lange *et al.* 1974). Much of the focus of these earlier studies in physiological ecology revolved around measurements of photosynthesis and gas exchange and largely ignored nutrient and water relationships.

From these field observations two decades ago, it became clear that the photosynthetic capacities of plants in low-productivity environments need not be low. In fact, under high soil moisture conditions arid land plants achieved some of the highest photosynthetic rates that had been measured for land plants (Fig. 18.1). At the same time, discussions of the functional significance of $C_3$ *vs.* $C_4$ photosynthesis had gained attention and it was becoming clear that $C_4$ metabolism allowed plants to achieve high rates of carbon gain with less water loss (Osmond *et al.* 1980). What was not clear was why $C_4$ photosynthesis is not more frequent among the taxa of arid lands if it is so advantageous (Ehleringer *et al.* 1997). Field studies in North American and South American arid lands had shown that $C_4$ photosynthesis

[1] *Department of Biology, University of Utah, Salt Lake City, UT 84112-0840, USA. E-mail: ehleringer@biosciences.utah.edu*
[2] *Present address: Department of Biology, Keene State College, Keene, NH 03469, USA.*

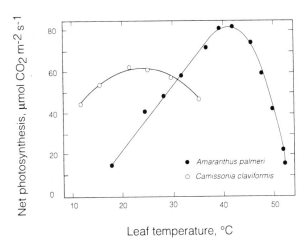

**Figure 18.1** Rates of photosynthetic $CO_2$ uptake as a function of temperature for *Camissonia claviformis* ($C_3$) and *Amaranthus palmeri* ($C_4$), two common annuals found in the arid lands of southwestern North America. Based on data from Mooney *et al.* (1976) and Ehleringer (1983).

was common only among the summer annuals and halophytic shrubs (Mulroy & Rundel 1977; Stowe & Teeri 1978; Winter 1981; Ziegler *et al.* 1981). Equivalent field studies in African and central Asian ecosystems lagged behind, although Winter (1981) had observed a number of $C_4$ taxa in western Asia, and this should have been further explored. It is only now that we are beginning to realize that $C_4$ photosynthesis may be much more common within Old World than New World deserts (P'yankov 1997; Sage & Monson 1999), suggesting that New World deserts may be too young to have reached their potential $C_4$ composition.

Walter (1931) established that there were wide-ranging water potentials in desert plants. Limited ecophysiological advances in arid land plant–water relations followed until the introduction of the pressure bomb (Scholander *et al.* 1965). Within the last two decades, three key observations set into motion a new way of thinking about water relations and the adaptive strategies of desert plants. These observations are hydraulic lift, cavitation and water-source partitioning.

Caldwell and Richards (1989) were among the first to report that water could 'leak' out of roots back into the soil following water potential gradients. Through hydraulic lift, plants are able to move water, from a region of higher water potential at some depth in the soil up into a shallower and drier soil layer. While this phenomenon could have been anticipated theoretically, it was not. The significance of hydraulic lift is still being debated, but it appears to play a role in enhancing gas-exchange capacity under drought and possibly enhancing nutrient cycling (Caldwell *et al.* 1998). Recently, Burgess *et al.* (1998) and Schulze *et al.* (1998) described a novel variation on this theme: inverse hydraulic lift, whereby plants move water from wetter upper soil layers to deeper dry layers. The adaptive significance of this phenomenon remains largely unexplored, but may be associated with the ability of roots to grow through dry soil towards groundwater. If so, it could provide an explanation for the unusually deep root penetration of some desert species through soil regions too deep to be wetted by precipitation (Canadell *et al.* 1996).

The study of plant–water relations under extremes received a boost with the development of methods to measure xylem cavitation on a routine basis (Tyree & Sperry 1989). Decreased water potentials lead to increased xylem cavitation and decreased hydraulic conductivity. How stomata regulate water flow so as to avoid run-away cavitation is an area of very active research. While few data are available for arid land species, it appears that variation in the capacity of xylem to avoid cavitation during drought and winter freeze events may be correlated with the geographical distributions of desert shrubs (Pockman & Sperry 1997).

Lastly, the use of stable isotope analyses of xylem water at natural abundance levels has provided a quantitative means of identifying the specific soil layers from which plants are extracting soil moisture (Fig. 18.2). The approach allows for non-destructive exploration of specific root activity, particularly in response to

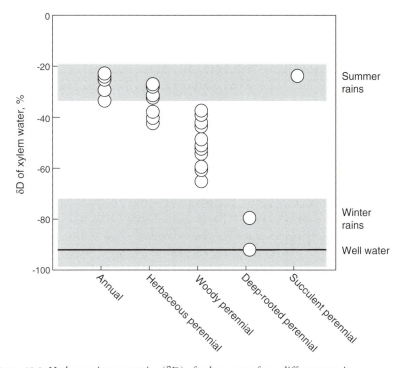

**Figure 18.2** Hydrogen isotope ratios ($\delta$D) of xylem water from different species, representing contrasting life forms, following summer rains in the arid lands of the Colorado Plateau. The data show that different life forms are utilizing different water sources, even following a strong summer precipitation event. Adapted from Ehleringer *et al.* (1991).

dynamic changes in soil moisture availability (Ehleringer & Dawson 1992; Dawson & Ehleringer 1998).

## Emerging linkages between ecophysiology and ecosystem-level processes

While much remains to be learned in the area of basic adaptation in arid land plants, opportunities have arisen over the past decade to extend the principles of physiological ecology from its traditional whole-plant focus to other levels of organization. Among these new research opportunities are studies addressing the molecular basis of ecophysiological functions (Chapter 1). Within arid land ecosystems, exciting progress is being made in such diverse areas as understanding the nature and significance of aquaporins for regulating water movement and in the molecular mechanisms governing photosynthetic pathway switching in response to drought. At the same time, principles of physiological ecology are being extended to higher levels of organization, such as the control of ecosystem fluxes (Chapter 19) and species composition within communities. Which ecophysiological processes are scalable and relevant at higher levels of organization remains an important and interesting challenge to the field (Ehleringer & Field 1993).

The need to address how ecological systems will respond to global change and to understand constraints on ecological systems for land management is widely recognized. However, these increased research opportunities come at a cost to the discipline of physiological ecology: fragmentation resulting from movements into new research opportunities poses a challenge for maintaining communication and integration within the whole-organism science community.

Historically, metabolism and adaptation have been approached through analysis of plant responses to constant resource levels (either high or low). Yet, realistically, arid lands are pulse driven and the focus should be directed at a consideration of how plants exploit pulse-driven resource dynamics. Variability in resource availability occurs on at least three relevant scales: pulses of moisture input within a single season, biseasonal inputs within a single year and interannual variability in precipitation. Analysis of annual precipitation patterns forms the basis for understanding how plant performance and fitness are affected by large-scale climatic phenomena (such as El Niño), which can affect regional climate patterns on a decadal basis.

Arid land ecosystems are among the most sensitive ecosystems to change, whether this change is associated with anthropogenic land-use activities or shifting climate conditions (Schlesinger et al. 1990; IPCC 1996). Globally, arid land ecosystem structure and function have shifted towards increased desertification and biological invasions, largely because of overgrazing (Schlesinger et al. 1990; Kassas 1995). The extent and nature of invasions can alter ecosystem relationships in what appear to be irreversible directions, shifting shrublands to annual grasslands in some regions (e.g. saltbush-bunchgrass to cheatgrass conversions in the northern

Great Basin of North America), while in other locations allowing tree and herbaceous species to establish and dominate (e.g. mesquite, Russian thistle and creosote invasions into the southwestern arid lands of North America). Unfortunately, the factors influencing ecosystem functioning, sensitivity to climate change and resiliency to invasions are not well understood. We believe that advances in the understanding of the dynamics of plant distribution in desert ecosystems will depend critically upon an appreciation of the basis for variability among plants and the susceptibility of these and other ecosystem components to change by outside factors such as invasion and anthropogenic factors altering resource availability on local and regional scales.

## The role of precipitation patterns in arid land ecosystems

Ecosystem primary productivity in arid land ecosystems is a highly correlated linear function of cumulative precipitation pulses, with nitrogen availability modifying the slope of the response (Le Houérou 1984; Noy-Meir, 1985; Gutierrez & Whitford 1987; Gutierrez et al. 1988; Sala et al. 1988; Ludwig et al. 1989; Ehleringer & Phillips 1999). The slope of the relationship between primary productivity and precipitation ('rain-use efficiency', sensu Le Houérou 1984) is steeper for winter-spring precipitation than for summer precipitation, most likely because of temperature-dependent differences on evaporation. Heavily grazed and degraded lands tend to have a lower rain-use efficiency than undisturbed ecosystems (Le Houérou 1984; Varnamkhasti et al. 1995).

Apart from the well-established effect of cumulative precipitation on productivity, there is a less well established, but potentially strong, effect of cumulative precipitation on the composition of desert plant communities. The dominance of many desert vegetation components are known to fluctuate in response to departures from mean climatic conditions (e.g. Goldberg & Turner 1986). Arid land ecosystems also respond strongly to the patterns in which resources are supplied, most importantly, the variability and timing of rainfall events (Noy-Meir 1973; Ehleringer 1999). In addition, different vegetation components may be differentially sensitive to rainfall patterns, as suggested by Ehleringer et al. (1991) and Ehleringer (1999). Lin et al. (1996) and Williams and Ehleringer (1999) tested this hypothesis in field experiments in an arid land system, simulating unusually large summer rain events (25–50 mm). They found that not all dominant perennials took up significant amounts of water from the upper 20–50 cm, although live roots were present in those soil layers.

Extreme-year types, in terms of water relations, can have significant and lasting effects on community composition. An example is the establishment and dominance of the perennial shrubs Encelia farinosa and Prosopis velutina, which replaced Larrea divaricata following unusually strong El Niños in 1941–43 (Turner 1990). More recently, Brown et al. (1997) attributed the expansion of woody perennials in southeastern Arizona to recent, unusually wet winters. There is evidence that in the last 25 years, possibly as a consequence of global change, extreme-year types, such as El Niño, have not only become more common but also more variable in strength (Wang & Ropelewski 1995; Wang 1995). Because of the particular importance of water in the functioning of arid land ecosystems, changes in precipitation patterns are likely to play a much more prominent role in changing arid land ecosystems in the near future than other factors of global change, such as increased atmospheric $CO_2$ or dry deposition.

## Precipitation and climate change

As our understanding of mesoscale climate processes and teleconnections increases, it is becoming increasingly clear that sea surface temperatures (SST) in the Pacific Ocean have a profound influence on both the seasonal intensity and interannual variability of precipitation in ecosystems across western North America (Carleton *et al.* 1990; Cayan & Webb 1992; Cayan 1996; Higgins & Schubert 1996; Zhang & Levitus 1997). Winter storms in western North America are derived from frontal systems out of the Gulf of Alaska in most years, but during El Niño years moisture originates farther south in association with higher SST. Summer monsoon precipitation events in the southwest develop as a result of land–ocean thermal differences, with moisture from the eastern Pacific and the Gulf of California. Stronger monsoons are directly related to elevated SST, which may be further amplified by strong El Niño–Southern Oscillation events (Carleton *et al.* 1990; Hereford & Webb 1992; Higgins *et al.* 1998). Studies of regional precipitation patterns in the southwest indicate trade-offs between the intensity of summer precipitation and winter snowpack from the previous season, suggesting a mechanism to decrease the strength of the monsoonal system (Cayan 1996; Higgins *et al.* 1998). Wetter than average winters are followed by drier than average summers and *vice versa*. One of the most often predicted early climate change events is an increase in the variance of weather and Parker *et al.* (1994) have reported an increased variability in eastern Pacific SST. Modelling studies predict that changes in SST are a crucial variable in climate change scenarios (Rind 1987, 1988; IPCC 1996). Small changes in the climate variability associated with elevated atmospheric $CO_2$ can produce relatively large changes in the frequency of extreme precipitation events (IPCC 1996). Furthermore, the thermal disequilibrium developing between the terrestrial and oceanic surfaces is increasing, leading to predictions of increased spatial and temporal variability in drought events (Rind *et al.* 1990). Thus, climatic changes in intra-seasonal, inter-seasonal and inter-annual precipitation patterns are predicted to occur on time scales that will influence the functioning of arid land ecosystems.

## Pulse patterns, pulse utilization and water-source partitioning

This section explores how both short-term precipitation pulses and longer-term interseasonal and interannual precipitation variability might affect the performance of individual species and thus influence ecosystem structure and function.

The ecosystem that will be focused on is the Colorado Plateau in the inter-mountain region of North America (West 1983). The northern Colorado Plateau represents the dry end of a summer precipitation gradient (Fig. 18.3). While long-term statistics indicate a more or less constant precipitation input through the year, there is an extremely high degree of year-to-year variability (Ehleringer 1994). Geographically, this arises because the Colorado Plateau is located near two cli-matic boundaries: the southern boundary of winter, frontal-system moisture input and the northern boundary of summer monsoon moisture. The high interseasonal and interannual variability in precipitation at this site may be related to modifica-tions of regional weather patterns that result in shifting climatic boundaries, which can enhance precipitation in some years and exclude precipitation in others. The effects of global change are most likely to be seen first here and in other ecosystems located near climatic storm-track boundaries.

Within arid lands, cumulative precipitation during a given season is usually a linear function of the number of storm events and the size–frequency distribution of individual storm events remains constant (Ehleringer 1994). Most storms are small (0–5 mm) and fewer than 8% of the storms are greater than 25 mm. El Niño and La Niña years, however, may deviate from these mean statistical trends. Cayan and Webb (1992) calculated that there were more winter precipitation days during El Niño than La Niña years (24% *vs.* 16%), and that the number of large storms and of flood events were significantly greater in El Niño than La Niña or other years (Cayan & Webb 1992; Hereford & Webb 1992; Higgins *et al.* 1998).

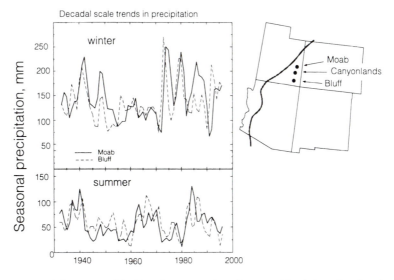

**Figure 18.3** Right: a map of the Four Corners region of western North America, showing Moab, Utah, and the northern boundary of the Arizona summer monsoon. Left: a time series of the total winter precipitation and summer precipitation for Moab and Bluff, Utah, two towns located near the boundary of the Arizona monsoon system.

Winter precipitation is not utilized immediately by most plants of the Colorado Plateau, because cold winter temperatures prevent much plant activity. Instead, winter precipitation infiltrates into deeper soil layers and accumulates there during the course of the winter (Caldwell 1985; Comstock & Ehleringer 1992). By contrast, summer rain generates a brief pulse of elevated soil moisture only in shallow soil layers that persists for hours to several weeks, depending on thermal conditions and the magnitude of the pulse event (Fig. 18.4). Small pulses (<5 mm) are thought to trigger surface processes, such as surface litter mineralization and cryptobiotic crust activity, but may persist for less than a day. Larger pulses (>10–15 mm) are required in order to trigger changes in gas-exchange metabolism of plants, and still larger events (>20 mm) are required to germinate annuals (Noy-Meir 1973, 1985; Beatley 1974).

Stable isotope analyses are a key to unravelling the response patterns to pulses (Chapter 21). Plants do not fractionate against $^2H$ (D) or $^{18}O$ during water uptake, and therefore the isotopic composition of water in roots and suberized stem tissues is an integrated measure of the water-uptake patterns of the roots (Dawson & Ehleringer 1991, 1993; Ehleringer et al. 1993). Moisture pulses derived from winter and summer precipitation events have a different isotopic composition. Therefore analysis of stable isotope ratios of xylem sap provides a quantitative measure of the water sources currently used by plants. This approach was used to study a wide range of species with contrasting life histories in a Colorado Plateau desert scrub ecosystem. The study showed that pulse utilization patterns may be correlated with

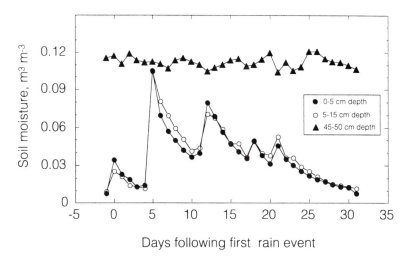

**Figure 18.4** Soil water content near the surface and at an intermediate depth during the summer for Canyonlands National Park, Utah. Note the rapid dynamics and changes in soil moisture in the upper soil layers (J.R. Ehleringer, R. Gebauer & S. Schwinning, unpublished data).

broad life history characteristics (Ehleringer *et al.* 1991 and Ehleringer 1999). For example, several woody perennial shrub species were using little of the moisture which saturated the upper soil layers after a 50-mm storm event. Other woody perennials derived 20–45% of their transpiration water from the upper soil layers, whereas the succulent CAM (crassulacean acid metabolism), annual and herbaceous perennial components fully utilized moisture derived from summer rains. Similar studies have confirmed that herbaceous perennials utilized summer moisture inputs more extensively than did woody perennials, for example in Patagonia shrublands (Sala *et al.* 1989), the Arizona savannah (Weltzin & McPherson 1997), New Mexico rangelands (McDaniel *et al.* 1982) and sagebrush steppe (Donovan & Ehleringer 1994).

Desert species are astonishingly variable in their ability to utilize pulses of water and plant-available nitrogen. Gebauer and Ehleringer (1999) have shown that there is considerable diversity, not just in the amount of water and nitrogen taken up by plants during a pulse event, but also in the season in which plants are best able to utilize pulses (Fig. 18.5). They examined the *in situ* ability of six dominant woody perennials to utilize pulses presented at different times (May, July and September) by simulating a heavy 25-mm pulse event. Stable isotope tracers ($^2$H and $^{15}$N) were used to follow nitrogen and water uptake. In late May all species derived less than 10% of stem water from the simulated rain event, even though the upper 25 cm was saturated by this pulse (data not shown). At other times, a larger fraction of plant water was absorbed by surface roots (up to 65%), and large differences in the amount and timing of summer rain utilization became apparent among perennials (Gebauer & Ehleringer 1999). Interestingly, although the occurrence of water and plant-available nitrogen are highly correlated in time, their uptake is not. For example, in July only *Artemisia* took up appreciable amounts of water and

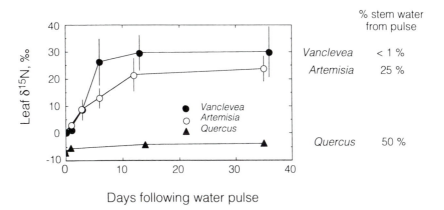

**Figure 18.5** Variations in the ability of desert shrubs to take up either water or nitrogen during July following a pulse label application to the upper soil layers (Gebauer & Ehleringer 1999).

nitrogen. Water, but not nitrogen, uptake occurred in *Quercus*, while *Vanclevea* took up nitrogen but not water. Earlier studies have established that independent uptake patterns for water and nutrients occur naturally (Schulze & Bloom 1984).

The diversity of pulse utilization suggests that niche differentiation may be an important component of ecosystem function and that competition, past or present, is a key factor in the structuring of desert communities. First, the differential use of pulse events may translate into differential sensitivity to interannual and interdecadal variations in moisture input. Ehleringer *et al.* (1991) and Ehleringer (1999) suggest that shifts in precipitation input patterns may move communities away from current species composition. For example, a northerly shift of the monsoon boundary, resulting in increased summer precipitation, should favour primarily those species that utilize summer rain the most and should increase their dominance in the community. Second, and perhaps more importantly, versatility in the exploitation of soil moisture, as exhibited by those species that were equally able to extract soil from the shallow and from the deep soil, may alter the competitive relationships among members of a community. For example, species capable of utilizing both pulse and deeper soil moisture may compete more strongly for water with deep-rooted species during dry summers, when pulses are scarce. Conversely, the same species may compete more strongly with shallow-rooted species in wet summers or following dry winters. Such a reorganization of competitive structure in a community may have complex and much less obvious effects on community composition.

## Theoretical considerations concerning adaptation to pulse-driven resource supply

Functional aspects of water relationships within the soil–plant–atmosphere continuum have been well described by models, such as that of Sperry *et al.* (1999). To gain a better understanding of the trade-offs associated with the observed diverse patterns of pulse utilization by desert plants, we developed a generalized water transport model for plants. This model describes shoot and root function as a function of soil depth in sufficient detail to simulate variation in both absolute and relative pulse utilization by plants. It is assumed that plants have a limited capacity for long-term water storage, and thus do not utilize the adaptive solution of cacti and other succulents (Nobel 1997). Three variable resistances are considered in the transport pathway for water between the soil and the atmosphere: roots in upper soil layers, roots in deeper soil layers and leaves. Leaf resistance to water vapour loss in conjunction with the atmospheric water vapour deficit and leaf area determine the total water loss rate, while the root resistances in conjunction with soil water potential and soil hydraulic characteristics determine how transpiration flux is divided between shallow and deeper soil water sources. The most critical aspect of this model is the identification of the physiological or morphological plant characteristics that determine the three resistance values.

A few general principles have emerged:

1 The ability to utilize pulses is associated with a number of fundamental trade-offs in plant physiology and architecture.

2 Some character combinations work better than others, suggesting that pulse utilization is not determined by isolated characters such as, say, rooting depth, but by entire suites of characters integrating root and shoot function.

3 Plant types with contrasting degrees of pulse utilization have differential sensitivities to year type.

4 Although less explored at this point, diversity in pulse utilization may lead to complex competitive relationships within communities.

Some of the interdependencies between the character trade-offs important to pulse utilization are summarized in Fig. 18.6. To maximize pulse utilization, a plant must, quite obviously, commit a large fraction of its root system to shallow soil. Less obviously, such a plant must also maintain functional roots, despite potentially low soil water potential, and to be able to take up water as soon as it becomes available. Furthermore, a pulse user must be able to support high rates of gas exchange during a pulse, partly through decreased root:shoot ratios and partly through high maximal stomatal conductance. Lastly, pulse users must also be able to avoid desiccation at the end of a pulse event and thus should exhibit a high degree of stomatal sensitivity to declining water status. Modelling results predict that the converse holds for plants that specialize in exploiting water in deeper soil

| functional type | strategy | root/shoot ratio | leaf conductance | increase in carbon gain from pulse |
|---|---|---|---|---|
|  | water uptake from deeper soil | high | low | low |
|  | ability to switch water source | high | moderate | moderate |
|  | water uptake from pulse | low | high | high |

**Figure 18.6** General patterns emerging from a model of plant–water relations and dynamics in response to moisture pulses (S. Schwinning & J.R. Ehleringer, unpublished).

layers. Interestingly, versatile water users, with significant root biomass both in shallow and deeper soil layers (dimorphic roots), have some characteristics in common with deeper-rooted plants, for example higher root:shoot ratios, but are predicted to be more like shallow-rooted plants in other characteristics such as stomatal sensitivity.

The costs and benefits of specific allocation strategies vary with the intensity and the duration of a pulse, with pulse frequency, and with temperature, all of which should vary within and between seasons, and between years. As no plant form can perform equally well under all possible conditions, the evolution of distinct functional types specializing in the utilization of specific components of resource variability in time and space is favoured.

Looking at water as a resource, year types are largely characterized by the timing of rainfall events, the time between events and the amount of rain deposited. For example, a dry summer may be characterized by few rain events and thus long intervals between events and/or small event size. A first step in exploring the effect of year types on functional types is, therefore, to look at the effect of a single pulse and interpulse period on functional types. An example of this is shown in Fig. 18.7, where we evaluate the effect of a single pulse, solely on the basis of the average carbon gain achieved during the pulse and interpulse period, in this case a period of 20 days at typical summer temperatures (40°C daily maximum). We identified a 'dry summer' with a small pulse size (6 mm) and a 'wet summer' with a larger pulse size (12 mm). A 'wet winter' (i.e. with a high degree of recharge in deeper soil layers) is identified by a high soil water potential below 20 cm soil depth (−0.5 MPa) and a 'dry winter' by a lower soil water potential (−1.5 MPa). The calculation of carbon gain is based either on Farquhar et al. (1980) equations for $C_3$ plants

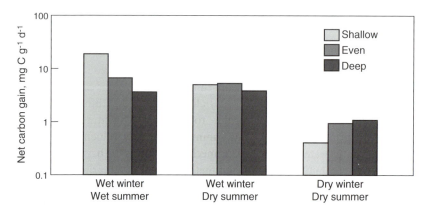

**Figure 18.7** Simulation predictions for differences in carbon gain by the three fundamental life forms in deserts based on root distribution. Carbon gain is averaged over the pulse interval of 20 days and expressed per unit total carbon above and below ground (S. Schwinning & J.R. Ehleringer, unpublished).

(applied to even- and deep-rooted plants) or Collatz *et al.* (1992) simplified equations for $C_4$ plants (applied to shallow-rooted plants).

In these and other simulations we generally found that the wetter the summer, the greater the carbon gain of shallow-rooted plants, as may be expected. These plants, with their low root:shoot ratios, low stem fraction and often $C_4$ photosynthesis, can achieve the highest carbon gains under favourable conditions. However, when summers are dry, they suffer the greatest carbon losses, and may even senesce between pulses. In this case, even-rooted plants with a greater ability to exploit deeper and wetter soil layers have the advantage. Winter and summer drought can favour the deepest-rooted plants (Fig. 18.7). Although deep-rooted plants rely the most on water stored in deeper layers, they are also better able to tolerate low water potentials, because of their high root allocation to that layer. Note that differences in carbon gain, here expressed on the basis of total plant carbon, indicate only one aspect of the advantages and disadvantages of functional types. For example, there may be large differences between functional types in the costs of growth and maintenance: deep-rooted plants allocate a large amount to wood, which is cheap to maintain, whereas shallow-rooted plants may pay a high price to maintain their total biomass. These differences may greatly enhance the year-type trade-offs indicated in Fig. 18.7.

## Competition for pulsed resources

It is generally accepted that arid land plants compete for water, although the contributions of pulse and inter-pulse periods to the competitive relationships among plants remain poorly understood (Casper & Jackson 1997; Goldberg & Novoplansky 1997). However, having gained some conceptual clarification of the processes involved, more specific questions about competition in pulse-driven systems can now be asked. Foremost in the context of this chapter: to what extent do plants compete for summer moisture pulses *vs.* water deposited in the winter and stored in deeper soil layers? Cohen argued as early as 1970 that competition for water should primarily occur in deeper soil layers, because shallower soil layers are depleted not by plant uptake but predominantly by evaporation. This should exclude competition for all but the largest summer pulses. However, Cohen (1970) did not consider that summer pulses may alleviate the competition for deep soil moisture when species with dimorphic root systems constitute a significant portion of the plant community. Thus, summer pulses may influence the strength of competition for water indirectly. This mechanism forms the basis for considering the shifting competitive relationships that were mentioned earlier.

Experimentally, our initial efforts were to examine the role of competition for water in the simplest possible, yet most realistic and important, vegetation type. This was *Coleogyne ramosissima*, which usually forms large nearly monospecific stands on southern portions of the Colorado Plateau, extending west to California (Ehleringer & Phillips 1999). The experiment involved adding 50 mm of winter or summer precipitation each year to target plants that did or did not have their neighbours removed. Distinct ecophysiological differences were observed in response to these treatments, but here only the growth response is shown (Fig. 18.8). Increasing winter moisture and/or removing neighbours significantly increased plant water status, gas exchange and growth, but the impact of a summer pulse was minimal. This was surprising because *C. ramosissima* is unusual among woody perennials in that it derives about half of its transpiration flux from surface layers following a large summer rain event. However, summer pulse utilization did not translate into an immediate growth response. Apparently, only a small fraction of the additional carbon gain derived from the added summer pulse carried over to next year's growth.

The study does suggest that the strength of competition, as reported in the growth disparity between competing and isolated plants, may be strongly affected by single precipitation events. In the case of *C. ramosissima*, the additional water pulse had the greatest effect on the strength of competition when it was administered in winter, suggesting once again that only 'storable' water input can be competed for. However, while plants may not compete for pulse water, they may compete for plant-available nitrogen compounds that are released in shallow soil by a moisture-induced stimulation of microbial metabolism. Unlike water, nitrogen is storable in the shallow soil, so that whatever nitrogen remains unused becomes available to others.

## Towards the integration of ecophysiology with community ecology and ecosystem ecology

It could be said that plant physiological ecologists think of communities as species assemblages selected primarily by climatic constraints and modified by ecological interactions, whereas population ecologists think of communities as organized primarily by ecological interactions and modified by environment. These opposing views are not trivial, because they may lead to very different expectations concerning the sensitivity of ecosystems to change. The study of desert ecosystems provides an opportunity for a more objective appraisal of the relative importance of climate *vs.* population interactions. Moreover, it integrates such diverse research areas as physiological ecology, productivity and fluxes at the ecosystem level, impacts of global circulation patterns on regional climate, and an area of theoretical ecology that is concerned with mechanisms of community stability. Theoreticians have long suggested that environmental variability, in conjunction with niche differentiation and certain forms of competition–environment interactions (Chesson & Huntly 1997), should be crucial for maintaining species diversity and for deciding the ultimate fate of newcomers to a community (Turelli 1981). For unknown reasons, physiological ecologists have largely ignored this hypothesis, which would have obliged them to focus more on the dynamic aspects of resource supply and utilization.

Today, several advances have made it possible to contribute to the conceptual unification of ecology in a substantial way. We have at our disposal long-term

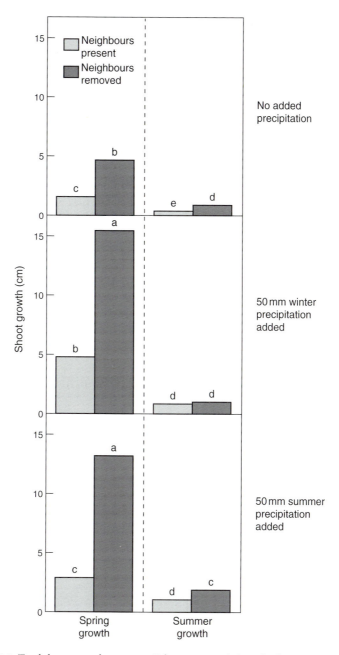

**Figure 18.8** Total shoot growth on target *Coleogyne ramosissima* shrubs under various supplemental precipitation and neighbour removal treatments. Shoot growth is partitioned into spring (March to June) and summer (July to October) periods. Bars not sharing a common letter differ significantly. From Ehleringer and Phillips (1999).

records of climate factors and are increasingly able to reconstruct past, and hypothesize on future, climate. With respect to arid ecosystems, the discipline has made major advances in understanding the impact of dynamic climate patterns on various community members. Lastly, with advances in experimental technology, particularly the use of stable isotopes, estimation of the partitioning of limiting resources among competing plants *in situ* is now possible. Thus, we are in a much better position to quantify the three cornerstones of community dynamics — environmental variability, species functional diversity and competition — and we can begin to test ideas about the origin and the persistence of arid ecosystems and speculate on their future.

## References

Beatley, J.C. (1974). Phenological events and their environmental triggers in Mohave Desert ecosystems. *Ecology*, **55**, 856–863.

Brown, J.H., Valone, T.J. & Curtin, C.G. (1997). Reorganization of an arid ecosystem in response to recent climate change. *Proceedings of the National Academy of Sciences USA*, **94**, 9729–9733.

Burgess, S.S.O., Adams, M.A., Turner, N.C. & Ong, C.K. (1998). The redistribution of soil water by tree root systems. *Oecologia*, **115**, 306–311.

Caldwell, M. (1985). Cold desert. In *Physiological Ecology of North American Plant Communities* (Ed. by H.A. Mooney & B.F. Chabot), pp. 198–212. Chapman & Hall, London.

Caldwell, M.M. & Richards, J.H. (1989). Hydraulic lift: water efflux from upper roots improves effectiveness of water uptake by deep roots. *Oecologia*, **79**, 1–5.

Caldwell, M.M., Dawson, T.E. & Richards, J.H. (1998). Hydraulic lift: consequences of water efflux from the roots of plants. *Oecologia*, **113**, 151–161.

Canadell, J., Jackson, R.B., Ehleringer, J.R., Mooney, H.A., Sala, O.E. & Schulze, E.-D. (1996). A global review of rooting patterns. II. Maximum rooting depth. *Oecologia*, **108**, 583–595.

Carleton, A.M., Carpenter, D.A. & Weser, P.J. (1990). Mechanisms of interannual variability of the southwest United States summer rainfall maximum. *Journal of Climate*, **3**, 999–1015.

Casper, B.B. & Jackson, R.B. (1997). Plant competition underground. *Annual Review of Ecology and Systematics*, **28**, 545–570.

Cayan, D.R. (1996). Interannual climate variability and snowpack in the western United States. *Journal of Climate*, **9**, 928–948.

Cayan, D.R. & Webb, R.H. (1992). El Niño/Southern Oscillation and streamflow in the western United States. In *El Niño Historical and Paleoclimatic Aspects of the Southern Oscillation* (Ed. by H.F Diaz & V. Markgraf), pp. 29–69. Cambridge University Press, Cambridge.

Chesson, P.L. & Huntly, N. (1997). The roles of harsh and fluctuating environments in the dynamics of ecological communities. *American Naturalist*, **150**, 519–553.

Cohen, D. (1970). The expected efficiency of water utilization in plants under different competition and selection regimes. *Israel Journal of Botany*, **19**, 50–54.

Collatz, G.J., Ribas-Carbo, M. & Berry, J.A. (1992). Coupled photosynthesis–stomatal conductance model for leaves of $C_4$ plants. *Australian Journal of Plant Physiology*, **19**, 519–538.

Comstock, J.P. & Ehleringer, J.R. (1992). Plant adaptation in the Great Basin and Colorado Plateau. *Great Basin Naturalist*, **52**, 195–215.

Dawson, T.E. & Ehleringer, J.R. (1991). Streamside trees that do not use stream water. *Nature*, **350**, 335–337.

Dawson, T.E. & Ehleringer, J.R. (1993). Isotopic enrichment of water in the 'woody' tissues of plants: implications for water source, water uptake, and other studies which use the stable isotopic composition of cellulose. *Geochimica Cosmochimica Acta*, **57**, 3487–3492.

Dawson, T.E. & Ehleringer, J.R. (1998). The role of plants in catchment-level hydraulic processes: insights from stable isotope studies. In *Isotope

*Tracers in Catchment Hydrology* (Ed. by J. McDonnell & C. Kendall), pp. 165–202. Elsevier, Amsterdam.

Donovan, L.A. & Ehleringer, J.R. (1994). Water stress and use of summer precipitation in a Great Basin shrub community. *Functional Ecology*, **8**, 289–297.

Eckardt, F.E., Ed. (1965). *Methodology of Plant Ecophysiology. Proceedings of the Montpellier Symposium.* UNESCO, Paris.

Ehleringer, J. (1983). Ecophysiology of *Amaranthus palmeri*, a Sonoran Desert summer annual. *Oecologia*, **57**, 107–112.

Ehleringer, J.R. (1994). Variation in gas exchange characteristics among desert plants. In *Ecophysiology of Photosynthesis* (Ed. by E.D. Schulze & M.M. Caldwell), pp. 361–392. Ecological Studies Series. Springer Verlag, New York.

Ehleringer, J.R. (1999). Deserts In *Primary Productivity in Terrestrial Ecosystems* (Ed. by H.A. Mooney & J. Roy). Academic Press, San Diego.

Ehleringer, J.R. & Dawson, T.E. (1992). Water uptake by plants, perspectives from stable isotope composition. *Plant, Cell and Environment*, **15**, 1073–1082.

Ehleringer, J.R. & Field, C.B. (1993). *Scaling Physiological Processes: Leaf to Globe*. Academic Press, San Diego.

Ehleringer, J.R. & Phillips, S.L. (2000). Impacts of increased precipitation and neighbors on *Coleogyne ramosissima*, a dominant cold-desert shrub. *Journal of Ecology*, in review.

Ehleringer, J.R., Phillips, S.L., Schuster, W.F.S. & Sandquist, D.R. (1991). Differential utilization of summer rains by desert plants, implications for competition and climate change. *Oecologia*, **88**, 430–434.

Ehleringer, J.R., Hall, A.E. & Farquhar, G.D. (1993). *Stable Isotopes and Plant Carbon/Water Relations*. Academic Press, San Diego.

Ehleringer, J.R., Cerling, T.E. & Helliker, B.R. (1997). $C_4$ photosynthesis, atmospheric $CO_2$, and climate. *Oecologia*, **112**, 285–299.

Farquhar, G.D., von Caemmerer, S. & Berry, J.A. (1980). A biochemical model of photosynthetic $CO_2$ assimilation in leaves of $C_3$ species. *Planta*, **149**, 78–90.

Gebauer, R.L.E. & Ehleringer, J.R. (1999). Water and nitrogen uptake patterns following moisture pulses in a cold desert community. *Ecology*, in press.

Goldberg, D. & Novoplansky, A. (1997). On the relative importance of competition in unproductive environments. *Journal of Ecology*, **85**, 409–418.

Goldberg, D. & Turner, R.M. (1986). Vegetation change and plant demography in permanent plots in the Sonoran Desert. *Ecology*, **67**, 695–712.

Gutierrez, J.R. & Whitford, W.G. (1987). Responses of Chihuahuan Desert herbaceous annuals to rainfall augmentation. *Journal of Arid Environments*, **12**, 127–139.

Gutierrez, J.R., DaSilva, O.A., Pagani, M.I., Weems, D. & Whitford, W.G. (1988). Effects of different patterns of supplemental water and nitrogen fertilization on productivity and composition of Chihuahuan Desert annual plants. *American Midland Naturalist*, **119**, 336–343.

Haberlandt, G. (1914). *Plant Physiological Anatomy*, 4th German edn. MacMillan, London.

Hereford, R. & Webb, R.H. (1992). Historical variation in warm-season rainfall on the Colorado Plateau. *Climate Change*, **22**, 239–256.

Higgins, R.W. & Schubert, S.D. (1996). Simulations of persistent North Pacific circulation anomalies and interhemispheric teleconnections. *Journal of Atmospheric Science*, **53**, 188–207.

Higgins, R.W., Mo, K.C. & Yao, Y. (1998). Interannual variability of the United States summer precipitation regime. *Journal of Climate*, **11**, 2582–2606.

IPCC (1996). Climate change 1995. The science of climate change. In *Contribution of Working Group I to the Second Assessment Report of the Intergovernmental Panel on Climate Change*. Cambridge University Press, Cambridge.

Kassas, M. (1995). Desertification: a general review. *Journal of Arid Environments*, **30**, 115–128.

Lange, O.L., Schulze, E.D., Evenari, M., Kappen, L. & Buschbom, U. (1974). The temperature-related photosynthetic capacity of plants under desert conditions I. Seasonal changes of the photosynthetic response to temperature. *Oecologia*, **17**, 97–110.

Le Houérou, H.N. (1984). Rain use efficiency: a unifying concept in arid-land ecology. *Journal of Arid Environments*, **7**, 213–247.

Lin, G., Phillips, S.L. & Ehleringer, J.R. (1996). Monsoonal precipitation responses of shrubs in a cold desert community on the Colorado Plateau. *Oecologia*, **106**, 8–17.

Ludwig, J.A., Whitford, W.G. & Cornelius, J.M. (1989). Effects of water, nitrogen and sulfur amendments on cover, density, and size of Chihuahuan Desert ephemerals. *Journal of Arid Environments*, **16**, 35–42.

McDaniel, K.C., Pieper, R.D. & Donart, G.B. (1982). Grass response following thinning of Broom Snakeweed. *Journal of Range Management*, **35**, 219–222.

Mooney, H.A. (1972). Carbon dioxide exchange of plants in natural environments. *Botanical Review*, **38**, 455–469.

Mooney, H.A. & West, M. (1964). Photosynthetic acclimation of plants of diverse origin. *American Journal of Botany*, **51**, 825–827.

Mooney, H.A., Dunn, E.L., Harrison, A.T. & Morrow, P.A. (1971). A mobile laboratory for gas exchange measurements. *Photosynthetica*, **5**, 128–132.

Mooney, H.A., Ehleringer, J.R. & Berry, J.A. (1976). High photosynthetic capacity of a winter annual in Death Valley. *Science*, **194**, 322–324.

Mulroy, T.W. & Rundel, P.W. (1977). Annual plants: adaptations to desert environments. *BioScience*, **27**, 109–114.

Nobel, P.S. (1997). Root distribution and seasonal production in the northwestern Sonoran Desert for a $C_3$ subshrub, a $C_4$ bunchgrass, and a CAM leaf succulent. *American Journal of Botany*, **84**, 949–955.

Noy-Meir, I. (1973). Desert ecosystems, environment and producers. *Annual Review of Ecology and Systematics*, **4**, 25–41.

Noy-Meir, I. (1985). Desert ecosystem structure and function. In *Ecosystems of the World*, Vol. 12a (Ed. by M. Evenari, I. Noy-Meir & D.W. Goodall), pp. 92–103. Elsevier, Amsterdam.

Osmond, C.B., Björkman, O. & Anderson, D.J. (1980). *Physiological Processes in Plant Ecology. Toward a Synthesis with Atriplex*. Springer-Verlag, Heidelberg.

Parker, D.E., Legg, T.P. & Folland, C.K. (1994). Interdecadal changes of surface temperatures since the late 19th century. *Climate Change*, **31**, 14373–14399.

Pockman, W.T. & Sperry, J.S. (1997). Freezing-induced xylem cavitation and the northern limit of *Larrea tridentata*. *Oecologia*, **109**, 19–27.

P'yankov, V.I. (1997). $C_4$-species of high-mountain deserts of eastern Pamir. *Russian Journal of Ecology*, **24**, 156–160.

Rind, D. (1987). The double $CO_2$ climate: impact of the sea surface temperature gradient. *Journal of Atmospheric Science*, **44**, 3235–3268.

Rind, D. (1988). The doubled $CO_2$ climate and the sensitivity of the modeled hydrologic cycle. *Journal of Geophysical Research*, **93**, 5385–5412.

Rind, D., Goldenberg, R., Hansen, J., Rosenzweig, C. & Ruedy, R. (1990). Potential evapotranspiration and the likelihood of future drought. *Journal of Geophysical Research*, **95**, 9983–10004.

Sage, R. & Monson, R. (1999). *Biology of $C_4$ Plants*. Academic Press, San Diego.

Sala, O.E., Parton, W.J., Joyce, L.A. & Lauenroth, W.K. (1988). Primary production of the central grassland region of the United States. *Ecology*, **69**, 40–45.

Sala, O.E., Golluscio, R.A., Lauenroth, W.K. & Soriano, A. (1989). Resource partitioning between shrubs and grasses in the Patagonian steppe. *Oecologia*, **49**, 101–110.

Schimper, A.F.D. (1898). *Plant Geography Upon a Physiological Basis*. Clarendon Press, Oxford.

Schlesinger, W.H., Reynolds, J.F., Cunningham, G.L., Heunneke, L.F., Jarrell, W.M., Virginia, R.A. *et al.* (1990). Biological feedbacks in global desertification. *Science*, **247**, 1043–1048.

Scholander, P.F., Hammel, H.T., Bradstreet, E.D. & Hemmingsen, E.A. (1965). Sap pressure in vascular plants; negative hydrostatic pressure can be measured in plants. *Science*, **148**, 339–347.

Schulze, E.-D. & Bloom, A.J. (1984). Relationship between mineral nitrogen influx and transpiration in radish and tomato. *Plant Physiology*, **76**, 827–828.

Schulze, E.-D., Caldwell, M.M., Canadell, J., Mooney, H.A., Jackson, R.B., Parson, D. *et al.* (1998). Downward flux of water through roots (i.e. reverse hydraulic lift). *Oecologia*, **115**, 460–462.

Shreve, E.B. (1923). Seasonal changes in the water relations of desert plants. *Ecology*, **4**, 266–292.

Sperry, J.S., Campbell, G.S. & Alder, N. (1999). Hydraulic limitation of flux and pressure in the soil–plant continuum: results from a model. *Plant, Cell and Environment*, in press.

Stowe, L.G. & Teeri, J.A. (1978). The geographic distribution of C$_4$ species of the dicotyledonae in relation to climate. *American Naturalist*, **112**, 609–623.

Turelli, M. (1981). Niche overlap and invasion of competitors in random environments. I. Models without demographic stochasticity. *Theoretical Population Biology*, **20**, 1–56.

Turner, R.M. (1990). Long-term vegetation change at a fully protected Sonoran Desert site. *Ecology*, **71**, 464–477.

Tyree, M.T. & Sperry, J.S. (1989). Vulnerability of xylem to cavitation and embolism. *Annual Review of Plant Physiology and Plant Molecular Biology*, **40**, 19–38.

Varnamkhasti, A.S., Milchunas, D.G., Lauenroth, W.K. & Goetz, H. (1995). Production and rain use efficiency in short-grass steppe: grazing history, defoliation and water resource. *Journal of Vegetation Science*, **6**, 787–796.

Walter, H. (1931). *Die Hydratur der Pflanzen*. Gustav Fischer, Jena.

Wang, B. (1995). Interdecadal changes in El Niño onset in the last four decades. *Journal of Climate*, **8**, 267–285.

Wang, X.L. & Ropelewski, C.F. (1995). An assessment of ENSO-scale secular variability. *Journal of Climate*, **8**, 1584–1599.

Warming, E. (1909). *Oecology of Plants: An Introduction to the Study of Plant Communities*. Clarendon Press, Oxford.

Weltzin, J.F. & McPherson, G.R. (1997). Spatial and temporal soil moisture resource partitioning by trees and grasses in a temperate savanna, Arizona, USA. *Oecologia*, **112**, 156–164.

West, N.E. (1983). Great Basin–Colorado Plateau sagebrush semi-desert. In *Ecosystems of the World. Temperate Deserts and Semi-Deserts*, Vol. 5 (Ed. by N.E. West), pp. 331–349. Elsevier, Amsterdam.

Williams, D.G. & Ehleringer J.R. (2000) Summer precipitation use by three semi-arid tree species along a summer precipitation gradient. *Ecology*, in review.

Winter, K. (1981). C$_4$ plants of high biomass in arid regions of Asia—occurrence of C$_4$ photosynthesis in Chenopodiaceae and Polygonaceae from the Middle East and USSR. *Oecologia*, **48**, 100–106.

Zhang, R.H. & Levitus, S. (1997). Interannual variability of the coupled tropical Pacific Ocean–Atmosphere System associated with the El Niño–Southern Oscillation. *Journal of Climate*, **10**, 1312–1330.

Ziegler, H., Batanouny, K.H., Sankhla, N., Vyas, O.P. & Stichler, W. (1981). The photosynthetic pathway types of some desert plants from India, Saudi Arabia, Egypt, and Iraq. *Oecologia*, **48**, 93–99.

# Chapter 19

# Environmental controls of gas exchange in tropical rain forests

*J. Grace*

## Introduction

Tropical rain forests are characterized by their great stature (exceeding 30 m tall), a wide range of life forms (including many trees with buttresses, thick-stemmed climbers and herbaceous epiphytes), a large number of tree species (30–300 woody species per hectare) and a correspondingly diverse fauna. There is no generally agreed nomenclature or classification (Longman & Jenik 1987; Whitmore 1990; Richards 1996). Lowland evergreen forests occur in equatorial regions where the mean annual temperature is at least 20°C, and the monthly rainfall is usually greater than 100 mm (Whitmore 1990). These forests are rain forests *sensu stricto*. However, many equatorial regions show a marked seasonality in rainfall, with a 'dry season' of several months in which the rainfall is less than 100 mm. Consequent seasonal patterns in phenology and physiological behaviour of the forest are to be expected, even though they may be subtle (Longman & Jenik 1987; Reich 1995; Malhi *et al.* 1998). Where there are more than 4 months with a rainfall below 100 mm, some of the species undergo leaf-shedding in the dry season, and produce shoots and flowers in some relation to the seasonal cycle. Forests in these regions are termed evergreen seasonal forests; and in the more extreme case where there are several months with less than 60 mm rainfall per month, and a substantial part of the canopy sheds its leaves in the dry season, the term semideciduous forest is used. Longman and Jenik's classification additionally includes: tropical montane forests, cloud forests, alluvial forests, swamp forests and peat forests; there are also heath forests, mangrove forests and flooded forests (Longman & Jenik 1987).

Tropical forest occurs in its pristine form as primary forest, and increasingly as secondary forest including logged-over and degraded forest, and regrowth following abandonment of farmland (Brown & Lugo 1990; Anderson & Spencer 1991; Honzak *et al.* 1996; Alves *et al.* 1997). Even remote Amazonian forests have been subjected to cycles of slash and burn agriculture by indians, as well as devastation by floods and fires (Saldarriaga & West 1986). So, much of what is regarded as primary forest is likely to be secondary forest of great age.

*Institute of Ecology & Resource Management, The University of Edinburgh, Kings Building, Mayfield Road, Edinburgh, EH9 3JU, UK. E-mail: jgrace@ed.ac.uk*

Tropical rain forests occupy about $17.5 \times 10^{12}\,\mathrm{m}^2$ of land, corresponding to 12% of the Earth's land surface (Whittaker & Likens 1975; Taylor & Lloyd 1992; Grace *et al.* 1999). They constitute a large store of carbon in the form of biomass, although there is considerable uncertainty in the quantity (Foster Brown *et al.* 1995). A recent compilation of data suggests that they hold, on average, about $150\,\mathrm{t\,C\,ha^{-1}}$ above ground and 35–50 below ground, making a total stock of $185$–$200\,\mathrm{t\,C\,ha^{-1}}$ (Grace *et al.* 1999). Assuming the mid-range figure of $192\,\mathrm{t\,C\,ha^{-1}}$ to apply over the entire $17.5 \times 10^{12}\,\mathrm{m}^2$, this implies a global total carbon stock of $336\,\mathrm{Gt\,C}$, which is more than half of the estimated global biomass stock of $610\,\mathrm{Gt\,C}$ (Schimel 1995).

The current deforestation rate, averaged over the biome, is about 0.7% per year (FAO 1999; Dixon *et al.* 1994). When the land is cleared by the use of fire, not all the carbon in the biomass is oxidized to $CO_2$ or CO on combustion (see Carvalho *et al.* 1995). The standing dead trees decay over a long period. An important fraction of the dead material is elemental carbon, which is not susceptible to microbial decay. Taking account of the incomplete combustion, it is estimated that the carbon flux to the atmosphere through deforestation is globally at least 1.6 Gt each year (Houghton 1996). This is equal to about one-quarter of the current (1990s) emissions from the combustion of fossil fuels. Pastures which replace rain forests have relatively small stores of carbon above ground, although they may have surprisingly deep roots and a high below-ground biomass (Nepstad *et al.* 1994). Most importantly, pastures have different biophysical and physiological properties to the forest they replace. Specifically, forests absorb more solar radiation than pastures, they are aerodynamically rougher, and they have a higher evaporation and transpiration rate (Gash *et al.* 1996; Grace *et al.* 1998). Consequently, it is hypothesized that deforestation may influence regional or even global climates (e.g. Gash *et al.* 1996).

Tropical forests are said to be vertically 'stratified', with a distinct understorey, and an emergent layer (Whitmore 1990; Richards 1996). Textbooks often present idealized profile diagrams which emphasize the stratification. Measured profiles of biomass or leaf area density show that the stratification is diffuse and certainly much less evident than some authors portray (Koike & Syahbuddin 1993). Older literature tended to emphasize the 'deep shade' of rain forests, but measurements show that the light climate in rain forests is not much different from that in temperate forests during the summer months (Chazdon & Fetcher 1984a,b). The leaf area index (LAI) is not very different either: reliable published estimates of LAIs from neotropical forests vary from 4 to 7.5 (Saldarriaga 1985; McWilliam *et al.* 1993; Roberts *et al.* 1996). Lianas may represent as much as 10–30% of the total LAI (Putz 1983; Hegarty 1991).

As well as vertical structure, there is considerable heterogeneity originating partly from the environmental patchiness, caused for example by topography and pedological processes, and partly from the regeneration of forest in the gaps created by the death of trees. These gaps may be quite small, resulting from the shedding of parts of a tree, or large, caused by several trees falling (Brokaw 1987), or very large, caused by blowdown of 100–1000 ha (Nelson *et al.* 1994). It is difficult to estimate the normal longevity of any species, as most tropical trees do not form annual growth rings. Certain fast-growing pioneer species may live for only a few decades. At the other extreme, recent $^{14}C$ determinations from a forest near Manaus, Brazil, show that the age of some emergent trees in undisturbed areas exceed 1000 years (Chambers *et al.* 1998). When old trees fall down, the gaps are themselves heterogeneous, comprising a disturbed 'key hole' shape in which the area occupied by the upturned root is quite different from the area occupied by the fallen canopy, which initially receives an enhanced nutrient supply from the decomposition of the foliage. The gap is filled by young trees, many of which are derived from seed that has been dormant in the soil or from seedlings that have survived the dense shade, often for several years. The trees colonizing the gaps are usually fast-growing pioneers, which are succeeded later by trees that grow tall enough to reach the top of the canopy, as well as shorter and inherently slow-growing understorey species with special adaptations to shade (Bazzaz 1984; Whitmore 1989). Of all the trees with a diameter exceeding 10 cm, approximately 1–2% are lost by mortality each year (Phillips & Gentry 1994).

Tropical forests have many similarities with temperate forests, perhaps more so than many people have imagined. However, they are functionally different in important respects. First, the seasonal fluctuations in the physical climate are relatively small, and do not involve low temperatures. Physiological processes thus proceed at a fast rate all the year around, although most tropical trees do grow intermittently, with distinct 'flushes', even when placed in a constant environment (see Longman & Jenik 1987). Second, soil processes are rapid in the moist and warm conditions of the humid tropics. Most of the soils are completely different from those in the temperate and boreal biomes, usually being highly weathered and strongly acid with a poor supply of phosphorus relative to nitrogen (Greenland & Kowal 1960; Jordan 1982; Proctor 1987; Cuevas & Medina 1988; Huante *et al.* 1995b). In especially poor soils, for example forests on white sand and in tropical mountains, a high degree of sclerophylly is frequent (Grubb 1977; Richards 1996). It is claimed that nutrients other than nitrogen are likely to influence the distribution of species very strongly (Ashton & Hall 1992), notably phosphorus (Medina & Cuevas 1996). When tropical soils are denuded of forest cover they are vulnerable to erosion and loss of fertility as a result of the high intensity of the rain (Bruijnzeel 1991). Finally, there are far more species and life forms in tropical than temperate forests, which is presumably the result of the prolonged history of physical conditions that favour growth throughout the year. This richness of species means that the information content of the ecosystem is immense, biotic interactions are relatively intense and, incidentally, the forest can be used as an especially attractive open-air laboratory in which some of the classical questions in ecology and evolution can be addressed.

## Microclimate: vertical and horizontal gradients

The height of the canopy and density of the foliage is sufficient to cause substantial vertical differentiation in microclimate and in the concentration of $H_2O$ and $CO_2$ (Allen *et al.* 1972; Shuttleworth 1989; Dolman *et al.* 1991; Richards 1996; Buchmann *et al.* 1997). Even in the daytime, when turbulent transport is effective at the top of the undisturbed canopy, there is some degree of decoupling between the emergent layer at the top and the seedling layer at the base of the canopy. By day, the air temperatures at the base of the canopy are lower and more constant than at the top, and the atmosphere at the base of the dense canopy often remains nearly water saturated. By contrast, at the top of the canopy air temperature rises during the day, and the saturated deficit of this may increase to as much as 2 kPa in the late afternoon (Grace *et al.* 1995a), accompanied by considerably negative xylem water potentials in rapidly transpiring trees (Grace *et al.* 1982; Roberts *et al.* 1996). The soil surface is a substantial source of $CO_2$, typically emitting 5–10 mmol $CO_2$ m$^{-2}$ s$^{-1}$ (Meir *et al.* 1996). As a result, the $CO_2$ concentration in the canopy usually builds up at night in the stable meteorological conditions which often prevail, sometimes reaching 550 p.p.m. throughout the vertical profile (Kruijt *et al.* 1996; Buchmann *et al.* 1997). The concentration then falls rapidly in the morning, as turbulent transport flushes the canopy and photosynthetic activity increases (Kruijt *et al.* 1996). In this process, significant reassimilation of respired carbon may occur (Medina & Minchin 1980; Kruijt *et al.* 1996; Lloyd *et al.* 1996).

## The radiation climate

The daily total of incoming solar radiation in the tropics is much less variable than in the temperate zones. For example, in the Brazilian Amazon the monthly mean is 15–18 MJ m$^{-2}$ d$^{-1}$, with the lowest values in the rainy season (Shuttleworth 1989; Gash & Shuttleworth 1991). In units of photosynthetically active radiation (PAR) this corresponds to 34–41 mol photons m$^{-2}$ d$^{-1}$. Tropical rain forests absorb a particularly high fraction of the incoming radiation: the short-wave reflectance is 0.12–0.14, much lower than for pasture (0.17–0.19), and comparable to coniferous forests (Gash & Shuttleworth 1991; Culf *et al.* 1996).

Inside the canopy the radiation is scattered and attenuated, and the pattern of irradiance at the floor of the forest is complex as a result of the heterogeneity of the canopy. Several authors have been able to survey the spatial and temporal variation over the forest floor using quantum sensors, which are ideal for the purpose as they respond rapidly enough to detect even the second-by-second fluctuations, and are relatively inexpensive. However, large numbers of sensors are required to capture the spatial heterogeneity, and so many studies have simply compared contrasting microsites. Some examples are given by Chazdon and Fetcher (1984a,b), showing the contrast between clearings, gaps and understorey. In a clearing in a Costa Rican forest, 75% of the readings were in the PAR range 500–2000 mmol photons m$^{-2}$ s$^{-1}$. In the understorey, about 75% of the data were below 10 mmol photons m$^{-2}$ s$^{-1}$ in Costa Rica (Chazdon & Fetcher 1984a,b) and, elsewhere, the mean PAR in the understorey of a dipterocarp forest was $4.7 \, \text{mmol photons} \, m^{-2} \, s^{-1}$ (Barker *et al.* 1997). These values are close to the typical light compensation point of some shade-tolerant species (Riddoch *et al.* 1991a).

There are also spectral changes in the understorey light climate, such as those reported for temperate forests. The most important are those which stimulate germination of seeds and influence photomorphogenesis, especially the ratio of red:far-red radiation (Vazquez-Yánes & Segovia 1984). Chazdon and Fetcher (1984a,b) report values of the red:far-red ratio that are similar to those in broadleaved temperate forests: 1.2–1.3 in the open, 0.6–1.1 in gaps, 0.4–1.2 in sunflecks and 0.2–0.7 in shadelight.

## Gas exchange of leaves

With a very small number of exceptions, the tree species utilize the $C_3$ photosynthetic pathway. $C_4$ grasses and sedges are absent from the forest itself except in the riparian areas, swampy areas, flooded forest and, occasionally, in large gaps. CAM photosynthesis is found in certain epiphytes, notably the bromeliads.

Leaves at the top of the canopy tend to be thicker than those in the understorey, with a higher maximum stomatal conductance and rate of photosynthesis. Specific leaf areas in an Amazonian forest near Manaus varied from $5 \, m^2 \, kg^{-1}$ for sunlit leaves at the top of the canopy to over $20 \, m^2 \, kg^{-1}$ in the shade near the ground (Roberts *et al.* 1990, 1996). Sunlit leaves in an Amazonian forest had a nitrogen content of $100–200 \, \text{mmol} \, m^{-2}$ and a phosphorus content of $1–2 \, \text{mmol} \, m^{-2}$, falling to half this concentration in the lower canopy (Lloyd *et al.* 1995).

Maximal stomatal conductances are from 100 to $500 \, \text{mmol} \, m^{-2} \, s^{-1}$, with some exceptional values of 1000 or more in large-leaved species (Grace *et al.* 1982; Roberts *et al.* 1990, 1996). Körner (1994) found that the maximum leaf conductance was similar in all forests types irrespective of the biome. The means and standard deviations from his paper are as follows: coniferous forests, $234 \pm 99 \, \text{mmol} \, m^{-2} \, s^{-1}$; temperate deciduous forests, $190 \pm 71 \, \text{mmol} \, m^{-2} \, s^{-1}$; humid tropical forests, $249 \pm 133 \, \text{mmol} \, m^{-2} \, s^{-1}$. Diurnal trends in stomatal conductance are also similar for most forest types (Shuttleworth 1989; Lloyd *et al.* 1995), with a tendency for stomata to open rapidly in the early morning and to reach a maximum aperture before noon, before declining as the vapour pressure deficit increases during the second half of the day (Fig. 19.1). Although lianas have a different hydraulic architecture from trees, with very large vessels, their stomatal conductances are not very different from those of trees (Castellanos 1990; Ewers *et al.* 1990; McWilliam *et al.* 1996; Zotz & Winter 1996).

Photosynthetic rates of tropical rain forest species are low in relation to $C_3$ species as a whole, rarely exceeding $10 \, \text{mmol} \, m^{-2} \, s^{-1}$, except in pioneer species (Riddoch *et al.* 1991a; Roberts *et al.* 1996). Laboratory-based studies reveal that many species have the capacity to acclimate to changes in light and shade, and a less pronounced capacity to acclimate to nutrient availability (e.g. Riddoch *et al.* 1991b; Thompson *et al.* 1992; Lehto & Grace 1994).

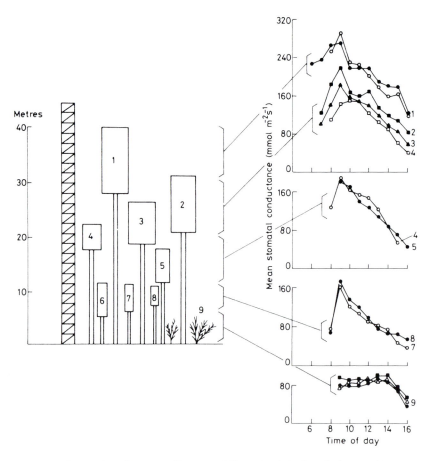

**Figure 19.1** Stomatal conductance of leaves at different levels in the rain forest at Reserva Ducke, Manaus, Brazil (Dolman *et al.* 1991). The symbols in the graphs distinguish different species, 1–9, that occupy different strata of the canopy.

Leaves develop differently according to the light climate of growth. Some of this response is apparently modulated by the differences in red : far red ratio between sunlight and shadelight, through the phytochrome system (Kwesiga & Grace 1986). In leaves of *Nauclea didderichii* developed at high irradiance, an extra layer of palisade mesophyll, a much higher specific leaf area and a higher capacity for photosynthesis was discovered (Riddoch *et al.* 1991b). In the dipterocarp *Bischofia javanica* the palisade layer of cells was over twice as thick in leaves developed in high irradiance compared to those developed in low light, and the maximal photosynthetic rate was correspondingly higher (Kamaluddin & Grace 1992, 1993). The data from *B. javanica* illustrate the considerable capacity for acclimation to their light environment, typical of sun and shade leaves generally (Table 19.1). In low light the plant produces more leaf area per mass of plant (high leaf area ratio) and the leaves are thinner. It is usually found that the quantum efficiency of shade-

**Table 19.1** Capacity for acclimation to light in the dipterocarp species *Bischofia javanica*, grown as seedlings in a controlled environment. Adapted from Kamaluddin and Grace (1993).

| Parameter | High light | Low light |
|---|---|---|
| Relative growth rate (week$^{-1}$) | 0.32 | 0.17 |
| Net assimilation rate (g m$^{-2}$ week$^{-1}$) | 30 | 5.9 |
| Leaf area ratio (m$^2$ kg$^{-1}$) | 8.8 | 30 |
| Specific leaf area (m$^2$ kg$^{-1}$) | 20 | 51 |
| $A_{sat}$ (μmol CO$_2$ m$^{-2}$ s$^{-1}$) | 10.6 | 3.7 |
| $Q_{0.5max}$ (μmol photon m$^{-2}$ s$^{-1}$) | 172 | 81 |
| $R_d$ (μmol CO$_2$ m$^{-2}$ s$^{-1}$) | 0.5 | 0.4 (NS) |
| $\alpha$ (mol CO$_2$ mol photon$^{-1}$) | 0.06 | 0.05 (NS) |
| Stomatal density (mm$^{-2}$) | 266 | 163 |
| Leaf thickness (μm) | 289 | 197 |
| Palisade mesophyll (μm) | 149 | 55 |
| Spongy mesophyll (μm) | 103 | 100 (NS) |
| Chlorophyll $a$ (mg m$^{-2}$) | 340 | 229 |
| Chlorophyll $b$ (mg m$^{-2}$) | 80 | 68 |
| Chlorophyll $a/b$ | 4.3 | 3.4 |

High light = 1000 μmol photon m$^{-2}$ s$^{-1}$, with red : far red of 1.45; Low light = 40 μmol photon m$^{-2}$ s$^{-1}$ with red : far red of 0.1; $A_{sat}$ = is the maximum rate of net photosynthesis; $Q_{0.5max}$ = photon flux requirement to achieve 50% of $A_{sat}$; $R_d$ = respiration of leaves in the dark; $\alpha$ = apparent quantum yield. All comparisons are statistically significant except those marked thus (NS).

grown leaves is no greater than sun leaves, and is sometimes less (Riddoch *et al.* 1991b).

There are large differences between the rates of net photosynthesis between species, which can be related to their ecological distribution, and especially to their status during succession, as a pioneer or a late-stage climax species (Riddoch *et al.* 1991a; Reich *et al.* 1994, 1995; Barker *et al.* 1997). Reich *et al.* (1994) examined the photosynthetic performance of 23 Amazonian tree species. The communities from which they came included tall Terra Firme rain forest on poor soils, periodically flooded tall *caatinga* on sandy soils and short, scrubby *baña* vegetation, also on sandy soils. They also studied species in agricultural or early successional forests. Most of the data form a linear relationship with the foliar nitrogen content, whether expressed on an area or mass basis. The early successional species and those on cultivated sites show the highest rates of photosynthesis, and the greatest sensitivity to foliar nitrogen concentration (Fig. 19.2).

The leaves of seedlings on the forest floor are exposed to large fluctuations in light. Over the course of a day they are likely to be irradiated by a mixture of diffuse and direct light. The full solar beam is about 2000 μmol photons m$^{-2}$ s$^{-1}$. Leaves sometimes experience a 100-fold increase in photon irradiance in less than an hour (as the solar elevation increases and they pass from being shaded to being fully irradiated by direct sunlight). Usually the increase is less than this, as part of the solar disk is obscured by leaves in the main canopy. Not all species may be able to utilize

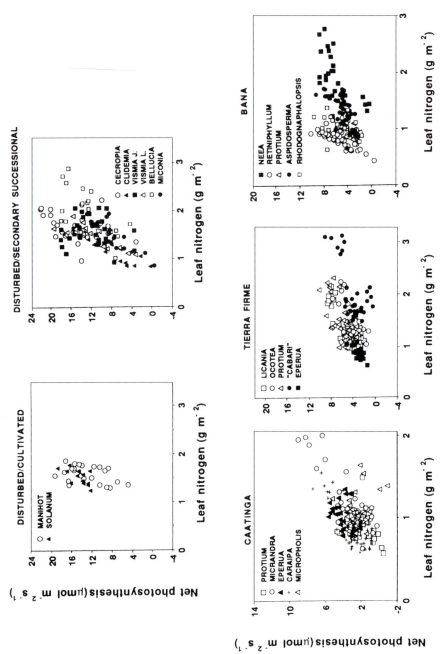

**Figure 19.2** The maximal rates of net photosynthesis plotted against the leaf nitrogen content, at five sites in the upper Rio Negro region of the Amazon basin (Reich *et al.* 1994). The undisturbed communities (*caatinga, tierra firme* and *bana*) show lower rates of maximal photosynthesis, even when compared at the same leaf nitrogen content; and the slope relating net photosynthesis to leaf nitrogen is less steep in the undisturbed sites.

an abrupt increase in light, as maximal photosynthesis is achieved only after an induction period, usually of 5–30 minutes, which may involve light-activation of Rubisco and gradual stomatal opening (Zipperlen & Press 1997). Watling *et al.* (1997) compared the growth of seedlings of a range of species from the Australian rain forest under three light regimes in a growth cabinet: (i) a constant photon irradiance of 4.9 mol m$^{-2}$ day$^{-1}$; (ii) the same regime with added 'sunflecks' made by turning on extra lights at 10 times each day, bringing the total photon irradiance to 7.0 mol m$^{-2}$ day$^{-1}$, and (iii) a constant photon irradiance of 7.0 mol m$^{-2}$ day$^{-1}$. They found contrasting responses among the species. Of the four, two showed a reduction in final biomass when grown in the sunfleck environment rather than in the continuous low irradiance, suggesting that the abrupt increase in light had actually damaged the capacity of the leaves to photosynthesize. One of the species, however, grew faster in the sunfleck regime than in either the continuous low light or the continuous moderate light, suggesting that some species may possess special adaptions to exploit the energy of sunflecks.

The largest increase in photon irradiance occurs when the forest is disturbed by large-scale blowdown of trees or by logging. Foresters sometimes remark on the 'bleaching' or 'scorching' of seedlings, apparently caused by the sudden increase in light. The phenomenon has been studied experimentally. When leaves of *B. javanica* were exposed to a stepwise increase in photon irradiance from 40 to 1000 µmol photons m$^{-2}$ s$^{-1}$, fluorescence characteristics changed within an hour, indicating damage to photosystem II, photosynthesis declined and chlorophyll pigments were bleached over 15 days (Kamaluddin & Grace 1992). Thereafter, however, the extant leaves recovered and acclimated, gaining many of the attributes of leaves grown at high photon irradiance, regreening, becoming thicker and displaying higher rates of photosynthesis. Recently, similar responses in the field have been shown in the species *Shorea johorensis* (Clearwater 1997).

Many of the rain forest species have long-lived leaves, presumably the consequence of a relatively aseasonal climate but perhaps also an adaptive response to nutrient scarcity. For example, the leaf life-span of shade-tolerant species in Panama varies from 1 to 8 years (Kursar & Coley 1993). Long-lived species tend to have lower rates of net photosynthesis and are more tolerant of abrupt increases in irradiance (Reich *et al.* 1994; Lovelock *et al.* 1998).

## Controlling variables and the growth of pioneer and climax species

Several authors have presented views on the physiological differences between the fast-growing species that are the first colonizers of gaps (pioneers) and those that form either the understorey or emergent species in an old forest. In ecological theory these groups correspond to guilds, based on their capacity to utilize resources. It may be useful to distinguish two or three guilds of tree based mainly on their behaviour in response to light (Whitmore 1989; Zipperlen & Press 1996, 1997; Barker *et al.* 1997). Attributes that are believed to be associated with the pioneer guild include: prolonged seed dormancy, photoblastic germination, rapid growth, low wood density, low dependence on mycorrhizas and susceptibility to herbivores (Bazzaz & Pickett 1980; Bazzaz 1984). The climax guild, by contrast, would tend to have opposite characteristics.

In practice, it is difficult to design experiments that are sufficiently comprehensive to test Bazzaz's multiple hypothesis of pioneer behaviour. As far as growth rate is concerned, there have been many studies comparing the growth of pioneers and climax species to a range of environments (Thompson *et al.* 1988; Riddoch *et al.* 1991b; Lehto & Grace 1994), but the comparison is often flawed by the small number of unrelated species employed, and the consequent difficulty in drawing conclusions that relate to pioneers and climax species generally. Recently, however, there have been studies in which the comparison has extended to many species, and in these cases it seems much safer to draw general conclusions when different responses of pioneers and climax species emerge from the 'noise' that is inevitable when the data originate from completely different taxa. Veenendaal *et al.* (1996) compared the growth of seedlings of 15 West African tree species at a range of irradiances. The pioneers had by far the highest growth rates, partly as a result of their enhanced net assimilation rates, and partly because of an elevated leaf area ratio (Fig. 19.3); but even pioneers were apparently unable to take advantage of irradiance exceeding 30% of the 'open' condition of full daylight. The non-pioneer light-demanders behaved in a somewhat similar way, but with lower specific leaf areas and net assimilation rates, and therefore a lower overall relative growth rate. The non-pioneer shade-bearers displayed very low net assimilation rates, and a lack of response of net assimilation rate to irradiances greater than about 20%. They also had notably low specific leaf areas, and relative growth rates which were usually less than one-half that of pioneers. The plants were grown in two types of forest soil, and in a more limited comparison, Veenendaal *et al.* (1996) showed that the growth rate was dependent on the soil type, the more fertile soil usually eliciting a higher net assimilation rate and, consequently, a higher relative growth rate.

Until recently the response of seedlings to nutrient supply was particularly unclear, as there have been indications of a lack of response of dipterocarp seedlings to applied nutrients (Turner *et al.* 1993). In a comparison of 34 species from a tropical deciduous forest in Mexico, Huante *et al.* (1995a) showed that all but one of the species did in fact respond to applied fertilizer, but some were much more responsive than others. Using classical growth analysis she was able to demonstrate a continuum in the responsiveness of net assimilation rate and leaf area ratio, both of which were usually increased by the addition of fertilizer. Moreover, there were correlations between the growth rate and the type of root system. The fast-growing species were usually non-mycorrhizal, or only weakly mycorrhizal, and they displayed dichotomously branching roots, which may be especially effective in nutrient acquisition (Huante *et al.* 1995b).

A third environmental variable to be considered here is the $CO_2$ concentration, which has doubled in the past 20 000 years, and is set to double again over the next few decades. Experiments on the effect of elevated $CO_2$ on woody plants have been

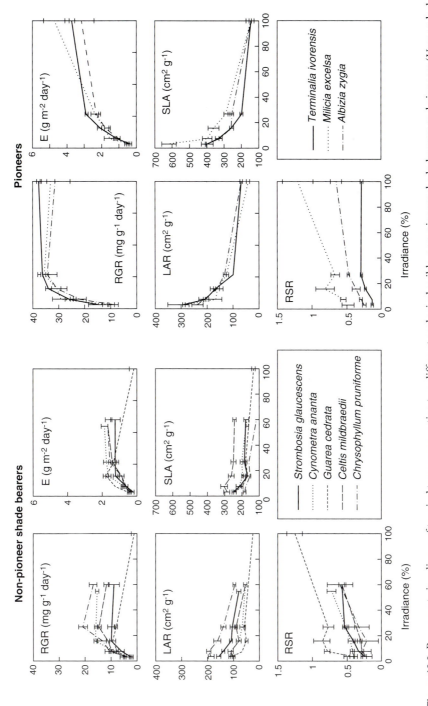

**Figure 19.3** Response to irradiance of tropical trees representing two different ecological guilds: non-pioneer shade-bearers and pioneers (Veenendaal *et al.* 1996). Plants were grown as seedlings in Ghana. Irradiance is expressed as percentage of full daylight. E = net assimilation rate; LAR = leaf area ratio; RGR = relative growth rate; RSR = root/shoot quotient; SLA = specific leaf area.

fraught with difficulty, some of which is attributed to the variability and the artificiality of the experimental conditions that have been used. Growing a collection of tropical plants as a model ecosystem, Körner and Arnone (1992) found very small effects of elevated $CO_2$. Lin *et al.* (1998) report a similar (but larger scale) experiment on the tropical ecosystem that has been established inside 0.2 ha of glasshouse in Arizona, known as Biosphere 2. Although it is impossible to recreate natural conditions exactly, many of the environmental data and the diel patterns of $CO_2$ flux are similar to those found in natural conditions in Brazil (Grace *et al.* 1995a; Malhi *et al.* 1998). Diel patterns of $CO_2$ concentration are not very realistic in Biosphere 2: in the daytime there is a substantial drawdown of the $CO_2$ concentration to 250 p.p.m. Moreover, the experiment is essentially short term as there is only one large glasshouse, and so high and low $CO_2$ concentrations have to be imposed sequentially, for a few days each. The result was a substantial increase in $CO_2$ uptake at elevated (700–800 p.p.m.) $CO_2$, from 3.1 mmol m$^{-2}$ day$^{-1}$ to 5.8 mmol m$^{-2}$ day$^{-1}$. Other work suggests that species from tropical forests may be less responsive to $CO_2$ than other species (Reekie & Bazzaz 1989; Ziska *et al.* 1991; Luxmore *et al.* 1993; Wullschleger *et al.* 1995; Carswell 1998; Würth *et al.* 1998).

Lovelock *et al.* (1996) showed that the shade-tolerant tree *Beilschmeidia pendula* developed an enhanced level of infection by vesicular mycorrhizal fungi at elevated $CO_2$. This may be a general response to $CO_2$, whereby a $CO_2$-induced elevated rate of photosynthesis provides an enhanced flux of carbon to the soil in the form of exudates, which stimulate the rhizosphere and thus increase microbial activity.

## Control of gas exchange at the canopy or ecosystem scale

The introduction of micrometeorological techniques has enabled ecophysiological approaches to be used at the scale of the whole ecosystem (Baldocchi *et al.* 1996; Grace *et al.* 1996; Moncrieff *et al.* 1997; Malhi *et al.* 1998). Early micrometeorological studies using flux-gradient techniques were mainly developed for use over agricultural surfaces, but were not robust enough for use over forests. These techniques have now been superseded by eddy-covariance, which is a simpler measurement concept. Development has been possible through the commercial availability of reliable and fast gas analysers and three-dimensional sonic anemometers (Moncrieff *et al.* 1997). Eddy covariance provides more or less continuous, all-weather measurements of the fluxes of energy, $CO_2$ and water vapour over the land surface, with a spatial average in the order of 1 km$^2$ (100 ha). The data gained in this way are valuable in several respects. First, they reveal the carbon and water balance of the whole ecosystem or a large area of landscape, and demonstrate its sensitivity to climatological and physiological variables, as well as to changes in land usage. Second, they enable testing of models that incorporate the understanding that has been gained from years of ecophysiological research: these include models of forest growth and yield and also regional scale models such as the Simple Biosphere Model (da Rocha *et al.* 1996) which may be used by climatologists. Third, eddy covariance measurements help in the estimation of Gross Primary Productivity and Net Primary Productivity and Net Ecosystem Productivity, and linked with remote sensing, these estimates can be applied to whole biomes or regions, and therefore uncertainties in the global carbon cycle may be reduced.

$CO_2$ fluxes over the rain forest show a maximal net uptake of 15–25 µmol $CO_2$ m$^{-2}$s$^{-1}$ by day (Grace *et al.* 1995a,b; Malhi *et al.* 1998). At night, the wind speed is often very low, mixing is poor, and the fluxes at the top of the canopy are often close to zero, as the $CO_2$ accumulates in the air space of the canopy. There is a 'burst' of $CO_2$ from the canopy in the early morning, as wind speed increases and turbulent transport increases (Grace *et al.* 1995a,b). It is usual to measure in-canopy profiles to estimate the hour-by-hour changes in the $CO_2$ stored within the canopy, and thus correct the fluxes for the 'storage term', to infer the more physiologically meaningful 'biotic fluxes'. It is then possible to plot the biotic uptake by night and by day, corresponding closely to the physiological concept of net photosynthesis and nocturnal respiration (Fig. 19.4). Within the canopy, measurements using chambers attached to leaves, stems and soil show that most of the total respiratory flux of the ecosystem, averaged over night and day, is from the soil (often 5–10 µmol $CO_2$ m$^{-2}$s$^{-1}$).

The diel and seasonal patterns in gas exchange may be related to the climatological variables. The $CO_2$ exchange is found to be a non-saturating function of solar irradiance, and much of the variation can be attributed to variation in humidity (expressed usually as the saturation deficit of the air or, better, the leaf-to-air vapour pressure deficit). Data from a rain forest in Manaus show typical light-response curves (Fig. 19.5). These data imply a quantum requirement of 20 mol photon per mol of $CO_2$. When comparison is made with boreal and temperate forest, it is evident that tropical forests have the highest rate of photosynthesis of all forest types so far examined (Malhi *et al.* 1999).

The longest period of eddy covariance measured is from near Manaus, Amazonia, Brazil. The forest is evergreen, but does experience a 'dry season' with 4 months having rainfall less than 100 mm. The net uptake of $CO_2$ is much lower in the dry season, and there is a very good correlation between the net ecosystem photosynthetic rate and the soil moisture in the rooting zone, measured with a neutron probe (Fig. 19.6).

The water vapour exchange rates may be used to estimate the canopy stomatal conductance by inversion of the Penman–Monteith equation. The stomatal conductance can be described as a function of radiation, vapour pressure deficit and, to a lesser extent, temperature (Lloyd *et al.* 1995). Stomatal models fitted to the data may then be combined with biochemically based models of photosynthesis and some representation of soil respiration, to form a model of carbon and water exchange for that particular forest type (Lloyd *et al.* 1995; da Rocha *et al.* 1996; Williams *et al.* 1998). Such models, with some significant assumptions, are now being used to estimate the possible impact of changes in such climatological variables as temperature and $CO_2$ concentration.

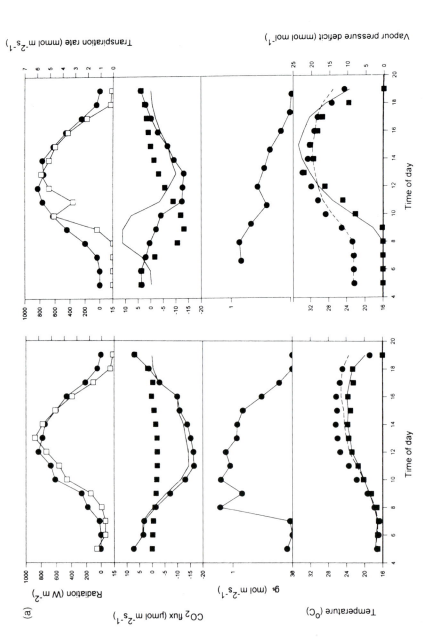

**Figure 19.4** Whole-ecosystem $CO_2$ flux and stomatal conductance, as revealed by eddy covariance measurements at an undisturbed rain forest in Rondonia, in the Brazilian Amazon (Grace et al. 1995a). Note that the convention for plotting micrometeorological fluxes is: fluxes from the atmosphere to the canopy are negative. The left-hand window represents a day of cool windy weather, while the right-hand window represents a more typical day of warm weather with stable nocturnal conditions during which respiratory $CO_2$ accumulates in the canopy at night. The symbols in the top windows are: radiation (□), transpiration (●); in the second windows: $CO_2$ flux as measured above the canopy (line), $CO_2$ flux into or out of the canopy air-space (■), physiological or 'biotic' $CO_2$ flux obtained by correcting the above-canopy flux by the flux into store (●); and in the lower windows: leaf temperature at the top of the canopy (●), air temperature (■), leaf-to-air vapour pressure deficit (- - - -), saturation vapour pressure deficit (——). Note that the storage flux is near zero on the windy day, but that on the calmer day there is a large flux from the canopy to the atmosphere when turbulent transfer begins in the morning.

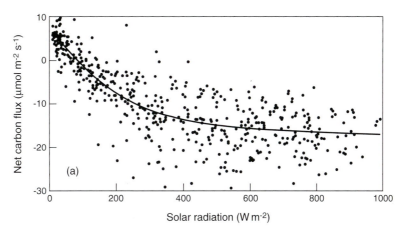

**Figure 19.5** The light–response curve of ecosystem net photosynthesis, as revealed by eddy covariance measurements at an undisturbed rain forest near Manaus, in the Brazilian Amazon (Malhi *et al.* 1998). Note that the convention for plotting micrometeorological fluxes is: fluxes from the atmosphere to the canopy are negative. The large scatter of the data are mainly the result of various degrees of stomatal closure in response to the rising saturation deficit in the afternoon.

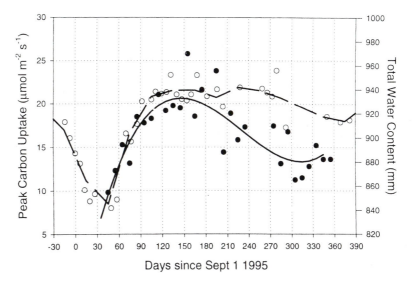

**Figure 19.6** Seasonality in maximal photosynthesis as revealed by eddy covariance measurements at an undisturbed rain forest near Manaus, in the Brazilian Amazon (Malhi *et al.* 1998). Symbols: solid circles and continuous line, maximal net uptake of $CO_2$; open circles and broken line, total water content in the rooting zone measured with a neutron probe.

## Response to global change

There is widespread discussion about the hypothesis that forests worldwide are growing faster in response to increases in $CO_2$ concentration, nitrogen deposition and temperature. If this hypothesis is true, we would expect forests to be making an increasing terrestrial contribution towards the global carbon sink. In temperate and boreal forests, where it has been possible to explore this question using dendrochronological techniques (e.g. Badeau et al. 1996), it is true that many forests are growing faster now than formerly, although it is difficult to disentangle management effects from climatological effects. In the tropics, few trees form annual rings and so dendrochronology is not an option. However, other approaches are possible. The first relies on eddy covariance measurements of $CO_2$ flux, and the second uses the so-called permanent sample plots established over many years by foresters. There is a third possibility, which is to infer the global distribution of sinks from the geographical patterns of $CO_2$ concentration and isotopic composition. Although this third approach is very appealing, as it operates on the regional scale and therefore overcomes the problem of site specificity, the sampling networks that have been operating so far do not adequately cover the equatorial latitudes, and so the results are inconclusive.

Eddy covariance measurements at two undisturbed forests in Brazil (Rondonia and Amazonia) showed the ecosystems to be net sinks of carbon, absorbing $1-5\,t\,C\,ha^{-1}\,yr^{-1}$ (Grace et al. 1995a,b; Malhi et al. 1998). It has previously been considered that mature forest ecosystems in a stable climate are at a steady state with respect to their carbon content per area of land, with gains by photosynthesis being balanced by losses from autotrophic and heterotrophic respiration. If this is correct, then it is reasonable to interpret the sink strength as the result of a shift to more favourable conditions. In fact, all regions of the world show interannual variability in climate, and climatological trends over decades and centuries, so it is difficult to draw firm conclusions about the impact of anthropogenic climate changes.

Permanent sample plots have only recently been used as an indication of long-term change. In a study of forestry inventories across the tropics, Phillips et al. (1998) concluded that there has been a net increase in tree biomass in the neotropics since the 1970s, corresponding to a net carbon sink of $0.45\,Gt\,C\,yr^{-1}$. Their data suggest that there may be a different pattern of change between neotropics and palaeotropics, with most of the biomass increase being localized in South America. This may reflect the impact of differing climatic regimes (e.g. rainfall trends in the various tropical regions have been different over the last two decades), or the differences in human population pressure in these regions.

Not all forests are likely to be behaving in the same way, and in many cases disturbance is likely to have an overwhelming influence. One aspect of environmental change, which has received scant attention so far, is the influence of forest edges that are created during logging and burning. An experimentally fragmented landscape spanning 20 km by 50 km near to Manaus, Brazil was used to create a replicated series of patches of 1, 10 and 100 ha. Recent data from these patches show a loss of up to 36% of the biomass within 100 m of the edge (Laurance et al. 1998).

## Conclusions

Truly experimental studies of the control of gas exchange are usually small scale, in which the unit of study is the leaf or plots with an area less than 0.1 ha. If we are to understand the behaviour of forests over spatial scales of hectares and temporal scales of decades, we have to use models to scale up (Grace *et al.* 1997). Models incorporating known physiological and biochemical mechanisms are nowadays widely used (Lloyd *et al.* 1995; Williams *et al.* 1998).

Such models generally show an increase in the net primary productivity as a result of elevated $CO_2$ and nitrogen deposition (McKane *et al.* 1995; Wang & Polglase 1995; Lloyd & Farquhar 1996). Most of these models are uncalibrated. It is, however, possible to calibrate such models using eddy covariance data, and this has been done in the case of the data from Rondonia (Grace *et al.* 1995a; Lloyd *et al.* 1995). It is useful to plot these model results by showing the photosynthetic and respiratory components separately (Fig. 19.7). They show that the net carbon balance is the small difference between the two large fluxes of photosynthesis and

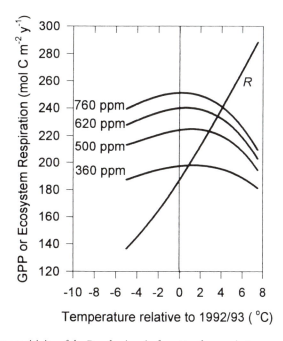

**Figure 19.7** The sensitivity of the Rondonia rain forest to changes in temperature and $CO_2$ concentration, estimated from the model of Lloyd *et al.* (1995) fitted to the data set of Grace *et al.* (1995a). The model has been re-run several times after subtracting or adding a temperature change to the climatological data of 1992 (when the field work took place). The graph provides a rough estimate of the difference between gross primary productivity (labelled according to the $CO_2$ concentration assumed in the model run) and ecosystem respiration (labelled $R$) that may be expected to be observed as the temperature and $CO_2$ concentrations rise, assuming no acclimation or floristic changes. Redrawn from Grace *et al.* (1996).

respiration. In the year of measurement (1992), this difference was positive (i.e. carbon was being taken up). The respiratory flux increases rapidly with temperature, and the photosynthetic flux increases with $CO_2$ concentration, so the balance will depend on the relative rates at which the two variables increase and there may be substantial interannual variability. The likely result is that the tropical forest will continue to be a sink for carbon for several decades before becoming a source, when the effect of temperature on respiration exceeds the effect of the $CO_2$ rise on photosynthesis.

## Acknowledgements

I acknowledge the financial support provided by the Natural Environmental Research Council through its TIGER (Terrestrial Initiative in Global Environmental Research) programme, award number GST/02/605, and for subsequent grants for studies of $CO_2$ fluxes over rain forests. I thank my colleagues, collaborators and students who have contributed variously, especially during the last 5 years: A. McWilliam, A. Miranda, A. Nobre, B. Kruijt, F. Carswell, H. Miranda, J. Gash, J. McIntyre, J. Moncrieff, J. Roberts, P. Meir, P. Jarvis and Y. Malhi.

## References

Allen, L.H., Lemon, E. & Muller, J. (1972). Environment of a Costa Rican forest. *Ecology*, **53**, 102–111.

Alves, D.S., Soares, J.V., Amaral, S., Mello, E.M.K., Almeida, A.S., da Silva, O.F. *et al.* (1997). Biomass of primary and secondary vegetation in Rondonia, Western Brazilian Amazon. *Global Change Biology*, **3**, 451–462.

Anderson, J.M. & Spencer, T. (1991). *Carbon, Nutrient and Water Balances of Tropical Rain Forest Ecosystems Subject to Disturbance*. MAB Digest 7. UNESCO, Paris.

Ashton, P.S. & Hall, P. (1992). Comparison of structure among mixed dipterocarp forests of northwestern Borneo. *Journal of Ecology*, **80**, 459–483.

Badeau, V., Becker, M., Bert, G.D., Dupouey, J.L., Lebourgeois, F. & Picard, J.F. (1996). *Long-Term Growth Trends of Trees: Ten Years of Dendrochronological Studies in France* (Ed. by H. Spiecker, K. Milikaïnen, M. Köhl & J.P. Skovsgaard), pp. 167–181. Springer-Verlag, Berlin.

Baldocchi, D., Valentini, R., Running, S., Oechel, W. & Dahlman, R. (1996). Strategies for measuring and modelling carbon dioxide and water vapour fluxes over terrestrial ecosystems. *Global Change Biology*, **3**, 159–168.

Barker, M.G., Press, M.C. & Brown, N.D. (1997). Photosynthetic characteristics of dipterocarp seedlings in three tropical rain forest light environments: a basis for niche partitioning? *Oecologia*, **112**, 453–463.

Bazzaz, F.A. (1984). Dynamics of wet tropical forests and their species strategies. In *Physiological Ecology of Plants of the Wet Tropics* (Ed. by E. Medina, H.A. Mooney & C. Vázquez-Yánes), pp. 233–243. Junk, The Hague.

Bazzaz, F.A. & Pickett, S.T.A. (1980). Physiological ecology of tropical succession: a comparative review. *Annual Review of Ecology and Systematics*, **11**, 287–310.

Brokaw, N.V.L. (1987). Gap-phase regeneration of three pioneer tree species in a tropical forest. *Journal of Ecology*, **75**, 9–19.

Brown, S. & Lugo, A.E. (1990). Tropical secondary forests. *Journal of Tropical Ecology*, **6**, 1–32.

Bruijnzeel, L.A. (1991). Nutrient input–output budgets of tropical ecosystems: a review. *Journal of Tropical Ecology*, **7**, 1–24.

Buchmann, N., Guehl, J.-M., Barigah, T.S. & Ehleringer, J.R. (1997). Interseasonal comparison of $CO_2$ concentrations, isotopic composition and carbon dynamics in an

Amazonian rainforest (French Guiana). *Oecologia*, **110**, 120–131.

Carswell, F.E. (1998). *The influence of carbon dioxide concentration on carbon assimilation in tropical tree species*. PhD Thesis, University of Edinburgh.

Carvalho, J.A., Santos, J.M., Santos, J.C., Leitão, M.M. & Higuchi, N. (1995). A tropical rainforest clearing experiment by biomass burning in the Manaus region. *Atmospheric Environment*, **29**, 2301–2309.

Castellanos, A.E. (1990). Photosynthesis and gas exchange of vines. In *The Biology of Vines* (Ed. by F.E. Putz & H.A. Mooney), pp. 181–204. Cambridge University Press, Cambridge.

Chambers, J.Q., Higuchi, N. & Schimel, J.P. (1998). Ancient trees in Amazonia. *Nature*, **391**, 135–136.

Chazdon, R.L. & Fetcher, N. (1984a). Light environments of tropical forests. In *Physiological Ecology of Plants of the Wet Tropics* (Ed. by E. Medina, H.A. Mooney & C.Vázquez-Yánes), pp. 27–36. Junk, The Hague.

Chazdon, R.L. & Fetcher, N. (1984b). Photosynthetic light environment in a lowland tropical rain forest in Costa Rica. *Journal of Ecology*, **72**, 553–564.

Clearwater, M.J. (1997). *Growth and acclimation responses of dipterocarp seedlings to logging disturbance.* PhD Thesis, The University of Edinburgh.

Cuevas, E. & Medina, E. (1988). Nutrient dynamics in Amazonian forest ecosystems. II Fine root growth, nutrient availability and leaf litter decomposition. *Oecologia*, **68**, 466–472.

Culf, A.D., Esteves, J.L., Marques Filho, A.,O. & da Rocha, H.R. (1996). Radiation temperature and humidity over forest and pasture in Amazonia. In *Amazonian Deforestation and Climate* (Ed. by J.H.C. Gash, C.A. Nobre, J.M. Roberts & R.L. Victoria), pp. 175–191. Wiley, Chichester.

Dixon, R.K., Brown, S., Houghton, R.A., Solomon, A.M., Trexler, M.C. & Wisniewski, J. (1994). Carbon pools and flux of global ecosystems. *Science*, **263**, 185–190.

Dolman, A.J., Gash, J.H.C., Roberts, J. & Shuttleworth, W.J. (1991). Stomatal and surface conductance of tropical rainforest. *Agricultural and Forest Meteorology*, **54**, 303–318.

Ewers, F.W., Fisher, J.B. & Fichtner, K. (1990). Water flux and xylem structure in vines. In *The Biology of Vines* (Ed. by F.E. Putz & H.A. Mooney), pp. 127–159. Cambridge University Press, Cambridge.

FAO (1999). State of the World's Forests FAO, Rome.

Foster Brown, I., Martinelli, L.A., Thomas, W.W., Moreira, M.Z., Ferreira, C.A.C. & Victoria, R.A. (1995). Uncertainty in the biomass of Amazonian forests: an example from Rondonia, Brazil. *Forest Ecology and Management*, **75**, 175–189.

Gash, C.A., Nobre, J.M., Roberts J.M. & Victoria, R.L. (1996). *Amazonian Deforestation and Climate*. Wiley, Chichester.

Gash, J.H.C. & Shuttleworth, W.J. (1991). Tropical deforestation: albedo and the surface-energy balance. *Climatic Change*, **19**, 123–133.

Grace, J., Okali, D. & Fasehun, E. (1982). Stomatal conductance of two tropical trees during the wet season in Nigeria. *Journal of Applied Ecology*, **19**, 659–670.

Grace, J., Lloyd, J., McIntyre, J., Miranda, A.C., Meir, P., Miranda, H., Moncrieff, J.M. *et al.* (1995a). Fluxes of carbon dioxide and water vapour over an undisturbed tropical rainforest in south-west Amazonia. *Global Change Biology*, **1**, 1–12.

Grace, J., Lloyd, J., McIntyre, J., Miranda, A.C., Meir, P., Miranda, H. *et al.* (1995b). Carbon dioxide uptake by an undisturbed tropical rain forest in South-West Amazonia 1992–93. *Science*, **270**, 778–780.

Grace, J., Malhi, Y., Lloyd, J., McIntyre, J., Miranda, A.C., Meir, P. *et al.* (1996). The use of eddy covariance to infer the net carbon dioxide uptake of Brazilian rain forest. *Global Change Biology*, **2**, 209–218.

Grace, J., van Gardingen, P.R. & Luan, J. (1997). Tackling large-scale problems by scaling-up. In *Scaling Up from Cell to Landscape, Society for Experimental Biology, Seminar Series 63* (Ed. by P.R. van Gardingen, G.M. Foody & P.J. Curran), pp. 7–16. Cambridge University Press, Cambridge.

Grace, J., Lloyd, J., Gash, J.A.H. & Miranda, A.C. (1998). Fluxes of carbon dioxide and water vapour over a $C_4$ pasture in south-west Amazonia (Brazil). *Australian Journal of Plant Physiology*, **25**, 519–530.

Grace, J., Malhi, Y., Higuchi, N. & Meir, P. (1999). Productivity and carbon fluxes of tropical rain forests. In 'Terrestrial Global Productivity' (Ed. by J. Roy, B. Saugier & H.A. Mooney), in press. Springer, Berlin.

Greenland, D.L. & Kowal, J.L. (1960). Nutrient content of the moist tropical forest of Ghana. Plant and Soil, 12, 154–174.

Grubb, P.J. (1977). Control of forest growth and distribution on wet tropical mountains, with special reference to mineral nutrition. Annual Review of Ecology and Systematics, 8, 83–107.

Hegarty, E.E. (1991). Leaf litter production by lianas and trees in a subtropical Australian rain forest. Journal of Tropical Ecology, 7, 201–214.

Honzak, M., Lucas, R.M., do Amaral, I., Curran, P.J., Foody, G.M. & Amaral, S. (1996). Estimation of leaf area index and total biomass of tropical regenerating forests: a comparison of methodologies. In Amazonian Deforestation and Climate (Ed. by J.H.C. Gash, C.A. Nobre, J.M. Roberts & R.L. Victoria), pp. 365–382. Wiley, Chichester.

Houghton, R.A. (1996). Terrestrial sources and sinks of carbon inferred from terrestrial data. Tellus, 48B, 419–432.

Huante, P., Rincon, E. & Acosta, I. (1995a). Nutrient availability and growth rate of 34 woody species from a tropical deciduous forest in Mexico. Functional Ecology, 9, 849–858.

Huante, P., Rincon, E. & Chapin, F.S., III (1995b). Responses to phosphorus of contrasting successional tree-seedling species from the tropical deciduous forest of Mexico. Functional Ecology, 9, 760–766.

Jordan, C.F. (1982). The nutrient balance of an Amazonian tropical rain forest. Ecology, 63, 647–654.

Kamaluddin, M. & Grace, J. (1992). Photoinhibition and light acclimation in seedlings of Bischofia javanica, a tropical forest tree from South-East Asia. Annals of Botany, 69, 47–52.

Kamaluddin, M. & Grace, J. (1993). Growth and photosynthesis of tropical forest tree seedlings (Bischofia javanica Blume) as influenced by a change in light availability. Tree Physiology, 13, 189–201.

Koike, F. & Syahbuddin (1993). Canopy structure of a tropical rain forest and the nature of an unstratified upper layer. Functional Ecology, 7, 230–235.

Körner, Ch. (1994). Leaf diffusive conductances in the major vegetation types of the globe. In Eco-physiology of Photosynthesis (Ed. by E.D. Schulze & M.M. Caldwell), pp. 463–490. Springer, Berlin.

Körner, Ch. & Arnone, J.A. (1992). Responses to elevated carbon dioxide in artificial tropical ecosystems. Science, 257, 1672–1673.

Kruijt, B., Lloyd, J., Grace, J., McIntyre, J.A., Farquhar, G.D., Miranda, A.C. et al. (1996). Sources and sinks of $CO_2$ in Rondonia tropical rain forest. In Amazonian Deforestation and Climate (Ed. by J.H.C. Gash, C.A. Nobre, J.M. Roberts & R.L. Victoria), pp. 331–352. Wiley, Chichester.

Kursar, T.A. & Coley, P.D. (1993). Photosynthetic induction times in shade-tolerant species with long- and short-lived leaves. Oecologia, 93, 165–170.

Kwesiga, F.R. & Grace, J. (1986). The role of the red/far-red ratio on the response of tropical tree seedlings to shade. Annals of Botany, 57, 283–290.

Laurance, W.F., Laurance, S.G., Ferreira, L.V., Rankin-de Merona, J.-M., Gascon, C. & Lovejoy, T.E. (1998). Biomass collapse in Amazonia forest fragments. Science, 278, 1117–1118.

Lehto, T. & Grace, J. (1994). Carbon balance of tropical tree seedlings: a comparison of two species. New Phytologist, 127, 455–463.

Lin, G., Marino, B.D.V., Wei, Y., Adams, J., Tubiello, F. & Berry, J.A. (1998). An experimental and modelling study of responses in ecosystems carbon exchanges to increasing $CO_2$ concentrations using a tropical rainforest mesocosm. Australian Journal of Plant Physiology, 25, 547–556.

Lloyd, J. & Farquhar, G.D. (1996). The $CO_2$ dependence of photosynthesis and plant growth in response to elevated atmospheric $CO_2$ concentration and their interrelationship with soil nutrient status 1. General principles. Functional Ecology, 10, 4–32.

Lloyd, J., Grace, J., Miranda, A.C. et al. (1995). A simple calibrated model of Amazon rainforest productivity based on leaf biochemical properties. Plant, Cell and Environment, 18, 1129–1145.

Lloyd, J., Kruijt, B., Hollinger, D.Y., Grace, J., Francey, R.J., Wong, S.C. et al. (1996). Vegetational effects on the isotopic composition of atmospheric $CO_2$ at local and regional scales — theoretical aspects and a comparison between rain-forest in Amazonia and a boreal forest in

Siberia. *Australian Journal of Plant Physiology*, **23**, 371–399.

Longman, K.A. & Jenik, J. (1987). *Tropical Forest and its Environment*. Longman, London.

Lovelock, C.E., Kylo, D. & Winter, K. (1996). Growth responses to vesicular-arbuscular mycorrhizae and elevated $CO_2$ in seedlings of a tropical tree, *Beilschmiedia pendula*. *Functional Ecology*, **10**, 662–667.

Lovelock, C.E., Kursar, T.A., Skillman, J.B. & Winter, K. (1998). Photoinhibition in tropical forest understorey species with short- and long-lived leaves. *Functional Ecology*, **12**, 553–560.

Luxmore, R.J., Wullschleger, S.D. & Hanson, P.J. (1993). Forest response to $CO_2$ enrichment and climate warming. *Water, Air and Soil Pollution*, **70**, 309–323.

McKane, R.B., Rastetter, E.B., Melillo, J.M., Shaver, G.R., Hopkinson, C.S. & Fernandes, D.N. (1995). Effects of global change on carbon storage in tropical forests of South America. *Global Biogeochemical Cycles*, **9**, 329–350.

McWilliam, A.-L.C., Roberts, J.M., Cabral, O.M.R., Leitao, M.V.B.R., de Costa, A.C.L., Maitelli, G.T. *et al.* (1993). Leaf area index and above-ground biomass of *terra firme* rain forest and adjacent clearings in Amazonia. *Functional Ecology*, **7**, 310–317.

McWilliam, A.-L.C., Cabral, O.M.R., Gomes, B.M., Esteves, J.L. & Roberts, J.M. (1996). Forest and pasture leaf-gas exchange in south-west Amazonia. In *Amazonian Deforestation and Climate* (Ed. by J.H.C. Gash, C.A. Nobre, J.M. Roberts & R.L. Victoria), pp. 265–285. Wiley, Chichester.

Malhi, Y., Nobre, A.D., Grace, J., Kruijt, B., Pereira, M.G.P., Culf, A. *et al.* (1998). Carbon dioxide transfer over a central Amazonian rain forest. *Journal of Geophysical Research*, **103 D24**, 31593–31631.

Malhi, Y., Baldocchi, D.D. & Jarvis, P.G. (1999). The carbon balance of tropical temperate and boreal forests. *Plant, Cell and Environment*, 715–740.

Medina, E. & Cuevas, E. (1996). Biomass production and accumulation in nutrient-limited rain forests: implications for responses to global change. In *Amazonian Deforestation and Climate* (Ed. by J.H.C. Gash, C.A. Nobre, J.M. Roberts & R.L. Victoria), pp. 221–239. Wiley, Chichester.

Medina, E. & Minchin, P. (1980). Stratification of

$\delta^{13}C$ values of leaves in Amazonian rain forests. *Oecologia*, **45**, 377–378.

Meir, P., Grace, J., Miranda, A.C. & Lloyd, J. (1996). Soil respiration measurements in the Brazil forest and cerrado vegetation during the wet season. In *Amazonian Deforestation and Climate* (Ed. by J.H.C. Gash, C.A. Nobre, J.M. Roberts & R.L. Victoria), pp. 319–330. Wiley, Chichester.

Moncrieff, J.B., Massheder, J.M., de Bruin, H., Elbers, J., Friborg, T., Heusinkveld, B. *et al.* (1997). A system to measure surface fluxes of momentum, sensible heat, water vapour and carbon dioxide. *Journal of Hydrology*, **189**, 589–611.

Nelson, B.W., Kapos, V., Adams, J.B., Oliveira, W.J., Braun, P.G. & do Amaral, I.L. (1994). Forest disturbance by large blowdowns in the Brazilian Amazon. *Ecology*, **75**, 853–858.

Nepstad, D.C., de Carvalho, C.R., Davidson, E.A., Jipp, P.H., Lefebvre, P.A., Negreiros, G.H. *et al.* (1994). The role of deep roots in the hydrological and carbon cycles of Amazonian forests and pastures. *Nature*, **372**, 666–669.

Phillips, O.L. & Gentry, A.H. (1994). Increasing turnover through time in tropical forests. *Science*, **263**, 954–958.

Phillips, O.L., Malhi, Y., Higuchi, N., Laurance, W.F., Nuñez, V.P., Vásquez, M.R. *et al.* (1998). Changes in the carbon balance of tropical forests: evidence from long-term plots. *Science*, **282**, 439–442.

Proctor, J. (1987). Nutrient cycling in primary and old secondary rain forests. *Applied Geography*, **7**, 135–152.

Putz, F.E. (1983). Liana biomass and leaf area of a *terra firme* forest in the Rio Negro Basin, Venezuela. *Biotropica*, **15**, 185–189.

Reekie, E.G. & Bazzaz, F.A. (1989). Competition and pasture patterns of resource use among seedlings of five tropical trees grown at ambient and elevated $CO_2$. *Oecologia*, **79**, 212–222.

Reich, P.B. (1995). Phenology of tropical forests: patterns, causes, and consequences. *Canadian Journal of Botany*, **73**, 164–174.

Reich, P.B., Walters, M.B., Ellsworth, D.S. & Uhl, C. (1994). Photosynthesis–nitrogen relations in Amazonian tree species. *Oecologia*, **97**, 62–72.

Reich, P.B., Ellsworth, D.S. & Uhl, C. (1995). Leaf carbon and nutrient assimilation and conservation in species of different successional status in an oligotrophic Amazonian forest. *Functional Ecology*, **9**, 65–76.

Richards, P.W. (1996). *The Tropical Rain Forest*, 2nd edn. Cambridge University Press, Cambridge.

Riddoch, I., Grace, J., Fasehun, F.E., Riddoch, B. & Ladipo, D.O. (1991a). Photosynthesis and successional status of seedlings in a tropical semideciduous rainforest in Nigeria. *Journal of Ecology*, **79**, 491–504.

Riddoch, I., Lehto, T. & Grace, J. (1991b). Photosynthesis of tropical tree seedlings in relation to light and nutrient supply. *New Phytologist*, **119**, 137–147.

Roberts, J., Cabral, O.M.R. & de Aguiar, L.F. (1990). Stomatal and boundary layer conductances in an Amazonian terra firme rain forest. *Journal of Applied Ecology*, **27**, 336–353.

Roberts, J.M., Cabral, O.M.R., da Costa, J.P., McWilliam, A.L.-C. & Sá, T.D.A. (1996). An overview of the leaf area index and physiological measurements during ABRACOS. In *Amazonian Deforestation and Climate* (Ed. by J.H.C. Gash, C.A. Nobre, J.M. Roberts & R.L. Victoria), pp. 287–306. Wiley, Chichester.

da Rocha, H.R., Sellers, P.J., Collatz, G.J., Wright, I.R. & Grace, J. (1996). Estimate of water vapour and carbon exchanges in an Amazonian rain forest by the SiB2 model. In *Amazonian Deforestation and Climate* (Ed. by J.H.C. Gash, C.A. Nobre, J.M. Roberts & R.L. Victoria), pp. 459–472. Wiley, Chichester.

Saldarriaga, J.G. (1985). *Forest succession in the upper Rio Negro of Columbia and Venezuela*. PhD Thesis, University of Tennessee, Knocksville, USA.

Saldarriaga, J.G. & West, D.C. (1986). Holocene fires in the northern Amazon basin. *Quaternary Research*, **26**, 358–366.

Schimel, D.S. (1995). Terrestrial ecosystems and the global carbon cycle. *Global Change Biology*, **1**, 77–91.

Shuttleworth, W.J. (1989). Micrometeorology of temperate and tropical forest. *Philosophical Transactions of the Royal Society of London Series. B*, **324**, 299–334.

Taylor, J.A. & Lloyd, J. (1992). Sources and sinks of atmospheric $CO_2$. *Australian Journal of Botany*, **40**, 407–418.

Thompson, W.A., Huang, L.K. & Kreidemann, P.E. (1992). Photosynthetic response to light and nutrients in sun-tolerant and shade-tolerant rainforest trees II. Leaf gas exchange and component processes of photosynthesis. *Australian Journal of Plant Physiology*, **19**, 19–42.

Thompson, W.A., Stocker, G.C. & Kriedmann, P.E. (1988). Growth and photosynthetic response to light and nutrients of *Flindersia brayleyana* F. Muell a rainforest tree with broad tolerance to sun and shade. *Australian Journal of Plant Physiology*, **15**, 299–315.

Turner, I.M., Brown, N.D. & Newton, A.C. (1993). The effect of fertilizer application on dipterocarp seedling growth and mycorrhizal infection. *Forest Ecology and Management*, **57**, 329–337.

Vazquez-Yánes, C. & Segovia, A.O. (1984). Ecophysiology of seed germination in the tropical humid forests of the world: a review. In *Physiological Ecology of Plants of the Wet Tropic* (Ed. by E. Medina, H.A. Mooney & C. Vázquez-Yánes), pp. 37–50. Junk, The Hague.

Veenendaal, E.M., Swaine, M.D., Lecha, R.T., Walsh, M.F., Abebrese, I.K. & Owusu-Afriyie, K. (1996). Response of West African forest tree seedlings to irradiance and soil fertility. *Functional Ecology*, **10**, 501–511.

Wang, Y.P. & Polglase, P.J. (1995). The carbon balance in the tundra, boreal and humid tropical forests during climate change — scaling up from leaf physiology and soil carbon dynamics. *Plant, Cell and Environment*, **18**, 1226–1244.

Watling, J.R., Ball, M.C. & Woodrow, I.E. (1997). The utilisation of lightflecks for growth in four Australian rain-forest species. *Functional Ecology*, **11**, 231–240.

Whitmore, T.C. (1989). Canopy gaps and the two major groups of forest trees. *Ecology*, **70**, 536–538.

Whitmore, T.C. (1990). *An Introduction to Tropical Rain Forests*. Clarendon Press, Oxford.

Whittaker, R.H. & Likens, G.E. (1975). The biosphere and man. In *Primary Productivity of the Biosphere* (Ed. by R.H. Whittaker & G.E. Likens), pp. 305–328. Springer, Berlin.

Williams, M., Malhi, Y., Nobre, A.D., Rastetter, E.B., Grace, J. & Pereira, M.G.P. (1998). Seasonal variation in net carbon dioxide exchange and evapotranspiration in a Brazilian rain forest: a modelling analysis. *Plant, Cell and Environment*, **21**, 953–968.

Wullschleger, S.D., Post, W.M. & King, A.W. (1995). On the potential for a $CO_2$-fertilisation effect in forests: estimates of the biotic growth factor based on 58 controlled exposure experiments. In *Biotic Feedbacks in the Global Biotic System* (Ed. by G.M. Woodwell & F.T. McKenzie), pp. 85–107. Oxford University Press, Oxford.

Würth, M.K.R., Winter, K. & Körner, Ch. (1998). *In situ* responses to elevated $CO_2$ in tropical forest understory plants. *Functional Ecology*, 865–895.

Zipperlen, S.W. & Press, M.C. (1996). Photosynthesis in relation to growth and seedling ecology of two dipterocarp rain forest tree species. *Journal of Ecology*, **84**, 863–876.

Zipperlen, S.W. & Press, M.C. (1997). Photosynthetic induction and stomatal oscillations in relation to the light environment of two dipterocarp rain forest tree species. *Journal of Ecology*, **85**, 491–503.

Ziska, L.H., Hogan, K.P., Smith, A.P. & Drake, B.G. (1991). Growth and photosynthetic response of nine tropical species with long-term exposure to elevated carbon dioxide. *Oecologia*, **86**, 383–389.

Zotz, G. & Winter, K. (1996). Diel patterns of $CO_2$ exchange in rainforest canopy plants. In *Tropical Forest Plant Ecophysiology* (Ed. by S.S. Mulkey, R.L. Chazdon & A.P. Smith), pp. 89–113. Chapman & Hall, New York.

# Chapter 20
# Comparative plant ecology and the role of phylogenetic information

*D.D. Ackerly*

## Introduction

Comparative ecology has played a prominent and critical role in the development of plant physiological ecology. The correspondence between plant morphology and environmental conditions led to early insights into the functional significance of plant form, and interspecific variation continues to play a central role in examining the association of ecophysiological traits with plant distribution, life histories, climate and environmental conditions. Differences between species reveal patterns of evolutionary divergence, and often provide critical evidence of the adaptive significance of different traits. For example, the impact of the ratio of red : far red light on the elongation response among herbaceous species of open-site and woodland habitats (Morgan & Smith 1979) demonstrated the potential adaptive importance of this trait, and paved the way for extensive study of physiological mechanisms, variation and adaptive significance at the individual and population level (Schmitt & Wulff 1993). Interspecific variation plays a key role in such studies, demonstrating that the functional mechanisms observed at the individual and population level are paralleled by evolutionary divergences under contrasting ecological conditions.

The critical role of phylogenetic information in comparative studies of trait variation and its adaptive significance was recognized from the outset of evolutionary ecology (see Ridley 1992). Studies of convergent evolution focused on distantly related species in order to demonstrate unambiguously that functional similarity was a result of evolutionary convergence and not common ancestry (e.g. Mooney & Dunn 1970). While patterns of phylogenetic relationship were not explicitly addressed, the underlying (and generally correct) premise was that the species were so distantly related that functional similarity did not result from inheritance of a trait from a common ancestor but rather from multiple independent evolutionary events. This conclusion was even more strongly supported by examining similar ecosystems in different parts of the world (in particular the studies of Mediterranean climate vegetation, Cody & Mooney 1978), in which functional adaptations had appeared independently in lineages derived from distinct regional floras.

*Department of Biological Sciences, Stanford University, Stanford, CA 94305, USA. E-mail: dackerly@stanford.edu*

More recently, evolutionary ecology has shifted its attention to the study of eco-logical divergence in closely related species, drawing on two distinct historical tra-ditions in ecology and evolutionary biology. The first line of research, studies of ecological variation in widely distributed species, addressed the relative contribu-tion of ecotypic *vs.* environmentally dependent variation across broad ecological gradients, and their contribution to plant performance under contrasting condi-tions. This work, pioneered by Turesson (1922) in Europe and Clausen and col-leagues in America (Clausen *et al.* 1940), led to the development of reciprocal transplant methodology and the growing realization of the fine scale of ecological specialization and intraspecific local adaptation (e.g. Schmitt & Gamble 1990). These biosystematic studies contributed to the development of research on hybridization, ecological differentiation and evolutionary variation in closely related species (e.g. Stebbins 1974).

The second line of research on closely related species grew out of Darwin's (1859) observation that competition would be most intense between closely related species, because of their ecological similarity. Combined with the competi-tive exclusion theory, and Hutchinson's (1957) conceptualization of the N-dimensional niche, this led to research on the functional basis of niche differences and life history variation in closely related species (e.g. Harper *et al.* 1961; Bazzaz 1987). Closely related species have been particularly valuable for such research because they tend to share many characteristics, due to their origin from a recent common ancestor, isolating the smaller number of divergent traits as the basis of their ecological differentiation. These research traditions, all of which have made very important contributions to the study of adaptive evolution in functional and ecological characters, illustrate two variations on the same theme: the study of ecologically similar but distantly related species, illustrating evolutionary *conver-gence*, and the study of ecologically differentiated but closely related species, illus-trating adaptive *divergence.*

In the past 20 years, evolutionary comparative methods have been developed from both systematic and quantitative perspectives to study these patterns (Harvey & Pagel 1991). On the one hand, the formalization of cladistic systematics provides methods for reconstructing the number and sequence of evolutionary events underlying the distribution of traits within a lineage, with an emphasis on discrete traits (e.g. $C_3/C_4$ photosynthesis; see Monson (1996) for examples from plant physiology and development). From a systematic standpoint, independent origins of a trait (homoplasy) under similar circumstances provide some of the strongest evidence for the action of natural selection. Repeated origins within a lineage have been demonstrated in traits such as $C_4$ photosynthesis in hot and/or high $CO_2$ environments (Ehleringer & Monson 1993), habitat occupancy in the Hawaiian silverswords (Baldwin & Robichaux 1995), carnivory and associated leaf forms (Albert *et al.* 1992), protective prickles in Hawaiian *Cyanea* (Givnish *et al.* 1995) and nitrogen-fixing symbioses within the rosid I clade (Soltis *et al.* 1995). Studies of related species (e.g. within genera) often demonstrate that, even among close relatives, functional and physiological similarities have arisen independently. For example, Monson's (1996) analysis of $C_4$ photosynthesis in *Flaveria*, a genus that exhibits varying degrees of $C_3/C_4$ intermediates, demonstrated two or three independent origins of the $C_4$ pathway among a group of only 20 species. Cladistic methods have led to various suggestions of formal approaches for recognizing traits as adaptations (Wanntorp 1983; Baum & Larson 1991), and for testing adaptationist hypotheses regarding the sequence of evolutionary events (Donoghue 1989).

On the other hand, phylogenetic relationships have been viewed in a statistical context as a framework for analysing interspecific variation. The most powerful of these statistical approaches is the method of independent contrasts (Felsenstein 1985). Several other methods have been introduced that partition interspecific variation into phylogenetic and adaptive components (Gittleman & Kot 1990), or into hierarchical components corresponding to the Linnaean hierarchy (Pagel & Harvey 1988; Bell 1989). These methods are the focus of this review, and are discussed in more detail below.

## Trait correlations: the role of phylogenetic information

Many questions in comparative biology focus on trait correlations as a means to identify adaptive associations between traits or between functional traits and ecological circumstances. Examples include: Are large seeds associated with gap regeneration (Foster & Janson 1985; Kelly 1995)? Does large seed size contribute to greater seedling survival in deep shade or following defoliation (Armstrong & Westoby 1993; Saverimuttu & Westoby 1996)? Does stomatal density vary with leaf thickness (Beerling & Kelly 1996)? Does canopy architecture vary with regeneration strategy and tree stature (King 1990; Kohyama & Hotta 1990; Ackerly & Donoghue 1998)?

Phylogenetic information can play a critical role in such studies to provide greater resolution of correlated evolutionary divergences between closely related species. Comparisons of multiple congeneric species pairs with contrasting ecology were pioneered by Salisbury (1942, 1974), in his study of seed size in relation to regeneration environment (but see Kelly 1996b; Grubb 1998; Thompson & Hodkinson 1998). Each species pair provides an independent instance of evolutionary divergence or functional relationship between the traits, against a constant background of other characters that may be shared within each lineage. Silvertown and Dodd (1996) presented a reanalysis, incorporating phylogenetic information, of two data sets relating reproductive allocation (RA) to life history, and their results illustrate the variable role of phylogeny in comparative analyses. In Fig. 20.1(a), RA was compared between annual and perennial plants, and the differences between the two groups were quite pronounced, with only a few outliers within each group. In this case, an unpaired analysis, which does not incorporate phylogenetic information, results in highly significant differences between the two groups (unpaired $t$-test; $t=3.66$, $P=0.002$). The phylogenetic context of these data can be incorporated using a paired analysis that evaluates only the mean

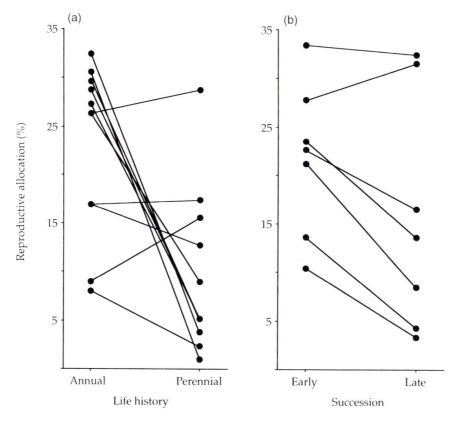

**Figure 20.1** Comparisons of reproductive allocation (RA) in pairs of closely related species. (a) Differences in RA in annual *vs.* perennial species are highly significant using paired or unpaired *t*-tests (*P*=0.0082 *vs.* 0.0016, respectively). (b) Differences in early *vs.* late successional species of Asteraceae are significant in a paired analysis (*P*=0.031) but not in an unpaired analysis (*P*=0.288). Redrawn from Silvertown and Dodd (1996).

differences within each pair, and does not take into account variation between pairs. In this case, however, the paired test actually had a slightly lower significance value (although still highly significant), in part due to the reduced degrees of freedom ($t$=3.29, $P$=0.008). In the second reanalysis, a comparison of RA in early and late successional species of Asteraceae showed no overall difference in an unpaired *t*-test ($t$=1.11, $P$=0.29), but a significant difference when the species pairs were taken into account ($t$=2.81, $P$=0.03; Fig. 20.1b). This means that there is considerable variation among species pairs as well as a significant tendency for RA to be lower in late successional species. The causes of the variation among species pairs, and of the disparity in RA within life history groups, remain unexplained in this sort of analysis and await further study.

In these examples, a continuous variable (RA) was contrasted in two discrete life history groups. In many cases, we are interested in correlations between two continuous characters, such as seed size and plant height (Rees 1996). Ecological studies have routinely presented correlations, regressions, association tests and multivariate analyses of interspecific data in tests of adaptive and functional hypotheses. Such analyses, many of which do not take phylogenetic relationships into account, have been seriously criticized in recent years from two related perspectives. First, if a correlation between two traits is seen as evidence of an adaptive relationship, this suggests that the present-day pattern is the result of correlated changes in the evolutionary history of the two traits. However, it turns out that interspecific correlations provide a very poor estimate of these correlated changes. This is due to what has been termed the 'non-independence' of species, i.e. the fact that phylogenetic history creates a nested hierarchy of relationships and that taxa tend to resemble their close relatives because of shared inheritance of traits from a common ancestor. These shared resemblances will create multiple data points in a correlative analysis, although they may reflect only one set of correlated changes in the ancestral lineages. As a result, it is argued that interspecific correlations may be inflated relative to the correlations in underlying evolutionary changes. A second and related problem arises from the use of trait correlations to address questions of functional linkages between traits. For example, if one asks whether large seeds provide greater resources for survival in deep shade (Saverimuttu & Westoby 1996), the question is not whether these two traits exhibit correlated evolution. Rather, it is a question of present-day functionality and the physiological mechanisms underlying differential survival. In this case, the potential problem arising from non-phylogenetic analyses is that other unmeasured aspects of plant function, rather than seed size, may be the cause of the survival differences. If these unmeasured traits exhibit chance associations with seed size, again because of the similarity of close relatives, then correlations may exist between seed size and survival that do not reflect functional relationships.

The method of independent contrasts (Felsenstein 1985) was introduced to address these concerns. Independent contrasts are constructed from the differences in trait values between pairs of sister taxa, including comparisons of terminal taxa pairs (as in the species pairs approach discussed above) as well as contrasts between sister nodes deeper in the phylogeny. Each contrast represents an independent evolutionary divergence event in the history of a trait, and correlations of independent contrasts provide powerful and accurate measures of correlated evolutionary changes (Pagel 1993). In tests of functional relationships, independent contrasts may help if the hypothetical unmeasured variables are in fact evolutionarily independent of the traits chosen for study, but they will not solve the problem if all the traits coevolve (Price 1997).

The value of independent contrasts may be illustrated from a recent example regarding correlations among leaf functional traits, based on a broad study of over 100 seed plant species from a range of habitats (Reich et al. 1997, 1999). Across all species, there was a negative correlation between leaf life-span and leaf size ($r=-0.42$, Fig. 20.2a), but this was primarily a result of significant differences in both traits between the 22 conifer and 89 angiosperm species in the study. Indepen-

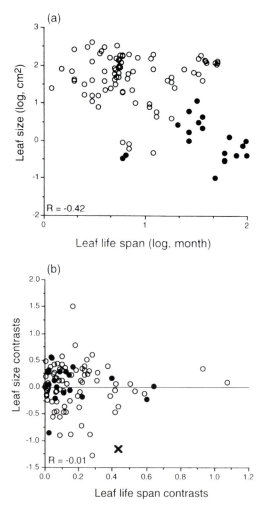

**Figure 20.2** (a) Interspecific correlations between lamina area and leaf life-span in 102 species of seed plants measured in six contrasting environments. Solid symbols are conifers and open symbols are angiosperms. Data from Reich *et al.* (1999) and Ackerly and Reich (1999). (b) Scatterplot of independent contrasts for the relationship between these two traits. The 'X' in the lower right represents the contrast between conifers and angiosperms, and is the primary source of the large negative correlation in (a). There was no significant correlation based on independent contrasts within the conifer or angiosperm groups. Reprinted from Ackerly and Reich (1999).

dent contrasts address this situation by calculating the divergences that have occurred at each node of the phylogeny, so the shift in leaf size and leaf life-span between conifers and angiosperms is represented as only a single data point. Because there was no correlation between these traits within the angiosperms or conifers, the overall correlation of the independent contrasts for these two traits was not significant ($r=-0.01$, Fig. 20.2b; Ackerly & Reich 1999). Do the parallel divergences between leaf size and leaf life-span observed between conifers and angiosperms represent two independent events in the evolution of these groups, or do they reflect a functional association between these two traits? The independent contrasts analysis shows no evidence of correlated evolution between the two characters within conifers or angiosperms, suggesting that there is not a general association between them. However, the small leaf area and long leaf life-span observed in conifers might reflect a shared, ancestral adaptive response to low nutrient and/or cold environments. Under this view, the correlation seen at the species level does reflect a real biological pattern, as these traits have been maintained in a large number of descendants in each group because they continue to occupy ecologically distinct habitats. The statistics alone cannot distinguish these two alternatives, but it is the discrepancy between the two results that leads us to pose this question in the first place, highlighting the value of the phylogenetic information in enhancing our interpretations of trait correlations.

In this example, the correlation between independent contrasts was weaker than the interspecific correlation. There has been a fairly widespread perception that this is generally the consequence of incorporating phylogeny into analyses of trait associations, reinforced by several papers that have re-evaluated ecological data sets and concluded that previously reported correlations did not exist when analysed from a phylogenetic viewpoint (Kelly 1996b and references therein). However, it turns out that correlations calculated with independent contrasts are just as likely to be larger than interspecific correlations as they are to be smaller. This is illustrated in Fig. 20.3. Case 1 shows the situation, as in the example above, of two divergent plant groups, where the overall trait correlation across species is a result of a single divergence between the two clades, with no correlation among species within each group. The independent contrasts calculated from such data will show no correlation, except for a single strong outlier representing the divergence between the groups ('X' in the lower panel). By contrast, the situation in Case 3 illustrates how two clades may diverge from each other, and be displaced with respect to one trait, but still have significant patterns of correlated evolution among the species of each group. The result is that the use of independent contrasts will reveal a significant correlation, because the divergence between the groups is only one datum point in the overall set of changes ('X' in lower panel), and this correlation would not be apparent in the interspecific data when the identity of the two clades was ignored. Case 2 shows the simplest situation, where there has not been any qualitative divergence between the groups, and the correlations of independent contrasts and of interspecific data are quite similar. Several recent theoretical papers have shown that, on average, interspecific and independent contrast correlations should be similar (Case 2), and discrepancies should be balanced as in these hypothetical Cases 1 and 3 (Price 1997; Ackerly unpublished data; cf. Pagel 1993). Recent reviews of the animal literature have indeed shown this to be the case (Ricklefs & Starck 1996; Price 1997; see Ackerly & Donoghue 1998), but similar reviews of applications in plant ecology have not been conducted.

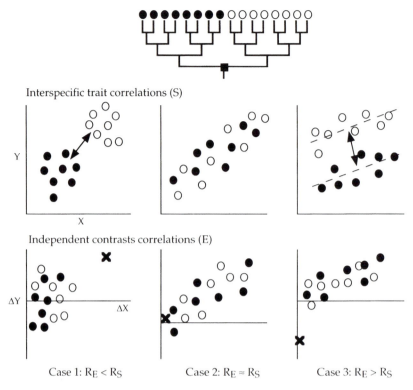

**Figure 20.3** Hypothetical examples of correlated traits to illustrate how non-phylogenetic interspecific correlations may differ from phylogenetically structured comparisons based on independent contrasts. Upper panels: scatterplots of species trait values, with closed and open symbols for the two lineages in the phylogeny above; lower panels: scatterplots of corresponding independent contrasts, showing the contrasts within clades, and the contrast between clades with an 'X'. Case (1): if the correlation across species is primarily a result of a discrete difference between two major groups, then the correlation of independent contrasts will be weaker or non-existent. Case (2): if correlated evolution occurs within both clades, and there is little divergence between them, the correlation of independent contrasts is similar to the interspecific correlations. Case (3): a displacement between two groups, coupled with correlated evolution within groups, results in significant correlations using interspecific contrasts, but no correlation when species data are viewed as a whole.

## Applications of phylogenetic approaches to plant functional ecology: the P-FUNC literature database

In order to evaluate the contribution of phylogenetic and taxonomic perspectives to the study of plant functional ecology, the recent literature in this area has been surveyed. The objectives of this survey are to document the extent and kinds of applications of comparative methods in plant functional ecology, and to review their role in developing and testing hypotheses in this area. Of particular interest is the extent to which the results of studies in comparative ecology have been influenced or altered by phylogenetic perspectives. This review addresses methods of testing trait correlations and methods for quantifying evolutionary convergence, with a focus on continuous characters. The comparative plant functional ecology database (P-FUNC) was compiled from the author's literature database in comparative ecology and plant ecophysiology, together with scanning recent issues of *Journal of Ecology*, *American Journal of Botany*, *Functional Ecology*, *Ecology*, *American Naturalist* and *Evolution*, and extensive cross-referencing through bibliographies of newer papers and Science Citation records for older papers. A total of 108 publications were collected that met the following criteria.

**1** Evolutionary comparative methods incorporating phylogenetic or taxonomic approaches were incorporated into the data analysis.

**2** At least four species were included (thus excluding studies of individual species pairs).

**3** The traits under study included plant functional characters, specifically vegetative morphology, physiology, demography, habitat use, range size and/or seed size; seed size was included because of its importance in seedling regeneration, but studies of other reproductive traits (e.g. floral morphology) and plant–animal interactions were excluded.

Various features of each study were recorded (Table 20.1) in order to generate quantitative summaries of the methodologies and traits included in these studies. The complete publication list and data matrix are available from the author and the World-Wide Web (Ackerly 1999a).

Table 20.1 shows the cumulative number of studies over five time periods and the breakdown in relation to the methods employed by various studies. The results illustrate the rapid proliferation in evolutionary comparative studies over the past 10 years, in parallel with the growth in the scientific literature in recent decades. The largest group represents a mix of miscellaneous approaches that consider the taxonomic groupings of species in some way, such as comparisons of biomechanical scaling across major plant clades (Niklas 1993), comparison of mean ecological traits at the level of families or other higher taxa (Baker 1972; Hodgson & Mackey 1986; Tiffney & Mazer 1995) or comparison of species within genera from different biogeographic ranges (Ricklefs & Latham 1992). Another large group comprises reconstructions of trait evolution on phylogenies, usually with cladistic methodologies (e.g. Monson 1996). Many of the studies included in this group are not very well known in the functional ecology literature, as they appear as part of phylogenetic studies published in the systematic literature (e.g. Baldwin 1993; Soltis *et al.* 1995).

There has also been a rapid rise in the use of phylogenetic and taxonomically structured statistical methods, particularly as many of these methods have been introduced in the past 15 years. In this survey, this is seen in the rapid increase in studies applying taxonomic analysis of variance and independent contrasts, in particular the application of these methods to evaluate and re-evaluate patterns of interspecific trait correlations as discussed above. For these quantitative methods, meta-analyses were conducted to evaluate the impact of phylogenetic approaches

**Table 20.1** Approaches to including taxonomic and phylogenetic information in comparative ecological studies. Each study in the P-FUNC database was scored with respect to the use of different classes of evolutionary comparative methods. Table entries show the number of published studies employing the various methods in each time interval.

| Comparative methods employed by studies | Time interval | | | | | |
|---|---|---|---|---|---|---|
| | <1980 | 1980–84 | 1985–89 | 1990–94 | 1995–98 | Total |
| i) Hierarchical analysis of variance: calculation of variance components at different taxonomic levels | 0 | 0 | 0 | 5 | 10 | 15 |
| ii) Phylogenetic independent contrasts | | | | | | |
| a) Paired approach: contrasts calculated between species pairs | 3 | 0 | 0 | 9 | 16 | 28 |
| b) Full tree approach: contrasts calculated over all nodes of the phylogeny for species in the study | 0 | 0 | 0 | 2 | 17 | 19 |
| iii) Trait reconstructions: parsimony approaches to reconstructing position of trait changes on a phylogeny | 0 | 5 | 4 | 4 | 12 | 25 |
| iv) Taxonomic group methods: comparison of traits among species in different taxonomic groups (genera, families, etc.) | 2 | 3 | 6 | 10 | 11 | 32 |
| v) Other phylogenetic methods: e.g. quantitative convergence index, multivariate methods | 0 | 1 | 3 | 5 | 5 | 14 |
| All studies | 5 | 8 | 13 | 30 | 52 | 108 |

in plant ecological research, and to test for patterns in the distribution of variation among taxa.

### Paired comparisons of related species

In the P-FUNC database, 28 of the 108 studies used paired independent comparisons; for 16 traits reported in 10 studies it was possible to conduct a meta-analysis comparing the results based on paired comparisons of closely related species *vs.* unpaired analyses which do not incorporate species relationships (as in Fig. 20.1). This analysis was conducted by calculating the *t*-statistics for the differences between traits in species of contrasting ecology, using either paired or unpaired *t*-tests (Fig. 20.4; points A and B correspond to the two examples in Figs 20.1a & b, respectively). In the majority of cases, the magnitude of the paired *t*-statistic was greater than the unpaired statistic. In three of the 16 cases the unpaired results were non-significant while the paired analysis provided a significant difference (at $P=0.05$), in eight cases neither result was significant and in five cases both were significant (Fig. 20.4). These results indicate that the study of closely related species

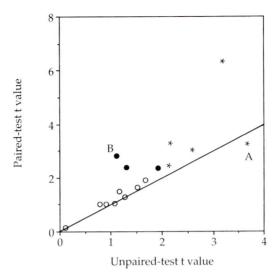

**Figure 20.4** Meta-analysis of studies comparing ecological traits in closely related species pairs (as in Fig. 20.1), plotting *t*-statistics for unpaired *vs.* paired analyses. ○: neither value significant at $P<0.05$; ●: paired value significant, unpaired value not significant; ✻: both values significant. The points labelled A and B correspond to the examples in Figs 20.1(a) and 20.1(b), respectively. Data from Ali (1968), Armstrong and Westoby (1993), Aronson *et al.* (1990), Garnier (1992), Grubb and Metcalfe (1996), Kelly (1996b), Metcalfe and Grubb (1995), Salisbury (1974), Silvertown and Dodd (1996) and Swanborough and Westoby (1996).

pairs, and the incorporation of phylogenetic relationships into the analyses, enhances the power of these studies to demonstrate functional differences between ecological groups.

### Independent contrasts

A smaller number of studies ($n=19$) have considered independent contrasts over the full phylogeny for a particular group of species (e.g. Ackerly & Donoghue 1998), including both the divergences between terminal taxa (as in the species pairs approach above) and between deeper nodes following the methods introduced by Felsenstein (1985; see also Garland *et al.* 1992). This method is particularly powerful for examining correlations between two continuous traits, because contrasts can be calculated at every node, such that $n$ species provide $n-1$ contrasts and there is no loss of degrees of freedom, as is the case with the paired analyses above. For six studies in the P-FUNC database, with a total of 123 pairwise trait correlations, the results were compared for correlation or regression analyses with and without consideration of phylogeny (Fig. 20.5). As is apparent in Fig. 20.5, there is a strong correspondence between the two sets of results, and an approximately equal number of cases in which the interspecific coefficients are greater than the evolutionary coefficients ($n=69$), and *vice versa* ($n=53$). This balance is consistent with the predictions

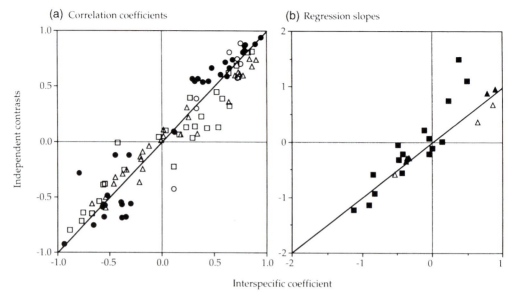

**Figure 20.5** Meta-analysis of correlation coefficients (a) and regression slopes (b), comparing results from non-phylogenetic, interspecific analyses with results of independent contrasts analyses. Note that the results from the two approaches are similar overall, although some individual points depart considerably from the 1 : 1 line. Data from: Ackerly and Donoghue (1998) (●); Franco & Silvertown (1996) (△); Jackson *et al.* (1999) (○); Ackerly and Reich (1999) (□); Beerling and Kelly (1996) (■); Franco and Kelly (1998) (▲).

outlined above, and the causes of discrepancies between interspecific and independent contrast analyses will be discussed in greater detail below. However, this result contrasts with the meta-analysis of paired comparisons (Fig. 20.4), in which the magnitude of the phylogenetically paired comparisons was equal to or greater than the interspecific comparison in most cases. This difference between the two meta-analyses is probably because recently diverged species pairs provide a clearer picture of adaptive and functional relationships, while contrasts estimated at deeper nodes of the phylogeny, which are included in the 'full tree' approach, may provide less reliable estimates of these distant evolutionary divergences. If so, then the comparison of species pairs, which is logistically simpler, may provide a more reliable and powerful approach for ecological tests of adaptive divergence and trait correlations. Conversely, the exclusive study of species pairs excludes these deeper events and draws a rather arbitrary distinction between the divergences of closely related species *vs.* those of higher taxa. Comparison of closely related species pairs may be particularly appropriate when examining functional correlations in a community context, where species will be distantly related and the deeper nodes of the corresponding phylogeny represent very old divergences. The full tree approach to independent contrasts is expected to provide more reliable results when species are sampled intensively from a single clade (e.g. studies of species within a genus), such that the divergences at deeper nodes will be reconstructed with greater accuracy (cf. Ackerly unpublished data). Westoby *et al.* (Westoby *et al.* 1998; Westoby 1998) provide several other guidelines for species sampling in comparative ecology which should prove extremely useful in future research.

## Hierarchical ANOVA

For quantitative traits, several methods have been developed to evaluate the strength of the 'phylogenetic signal' for a particular trait, including the phylogenetic autocorrelation method, hierarchical taxonomic analysis of variance and, most recently, a parsimony based quantitative convergence index (Ackerly & Donoghue 1998). In plant ecology, the most widely used approach for quantitative traits has been the hierarchical taxonomic analysis of variance, in which the Linnaean hierarchy is used to create a nested ANOVA model in order to calculate the proportion of trait variance explained by different levels, i.e. families within orders, genera within families, species within genera, etc. (e.g. Shipley & Dion 1992; Peat & Fitter 1994; Franco & Silvertown 1996). It is argued that if most variation occurs among species within genera, then species may be treated as independent units for the purposes of analysing trait variation because there is little evidence of phylogenetic conservatism. From the P-FUNC literature database, six studies have been examined that collectively have conducted more than 30 analyses using taxonomic ANOVA (several of them for the same trait, e.g. seed size, in different sets of taxa). These are presented in Fig. 20.6 in terms of the proportion of variance reported at three levels: species, genus, and family and above. There is a striking pattern in these results, as some traits exhibit most variation at the species level and others at the family level or above, but only one trait exhibits a predominance of variation among genera within families (the demographic parameter $Hp_x$ in Franco & Silvertown 1996). Traits with most variation at the species level include plant size and related performance measures (height, total seed number, etc.) and species range size. Those with considerable variation at the family level include most of the demographic parameters compiled by Franco and Silvertown (1996) and seed size, with two notable exceptions (Shipley & Dion 1992 and re-analysis of Salisbury 1974 in Kelly 1996b).

The hierarchical ANOVA method has helped to draw attention to phylogenetic patterns in ecological trait variation, and this meta-analysis suggests some consistent differences between different traits which merit further investigation. However, the method is seriously weakened on theoretical grounds by its reliance on the ranking of taxa in the Linnaean hierarchy, i.e. the designation of the rank to which each taxon is assigned (genus, family, order, etc.). Historical analyses reveal that the systematists of the eighteenth and nineteenth century who constructed the early, and still influential, classification systems were in large measure concerned with problems of information storage and retrieval (Stevens 1994). Thus, the number of ranks, the number of groups identified at each rank, and the number of subsidiary taxa within each higher taxon were largely based on non-biological

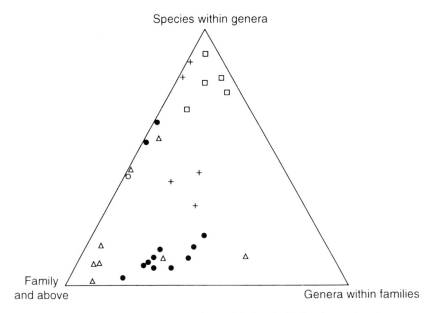

**Figure 20.6** Meta-analysis of studies using hierarchical analysis of variance, showing partitioning of trait variation among species with genera, among genera with families, and among families and higher levels of the taxonomic hierarchy. △: demographic data; ●: seed size; □: range size; plus signs: plant height, weight and fecundity; ○: stomatal distribution. Data from Franco and Silvertown (1996), Kelly (1995, 1996b), Kelly and Woodward (1996), Lord *et al.* (1995), Peat and Fitter (1994) and Shipley and Dion (1992).

criteria. Although many higher taxa are well defined as evolutionary groups, the designation of their ranking remains largely arbitrary. At this stage the question is not whether hierarchical taxonomic systems are conceptually superior, particularly with regard to ranking, but whether despite these problems they turn out to provide biologically meaningful results. It is possible to use traditional taxonomies to construct crude phylogenetic trees for the calculation of independent contrasts (see Kelly & Woodward 1996; Kelly 1996a). However, these analyses are weakened by low phylogenetic resolution because of the small number of ranks as well as errors caused by the many paraphyletic taxa in traditional taxonomies. The need for this approach is being rapidly superseded by improved phylogenies of seed plants and other taxa (e.g. Chase *et al.* 1993; Soltis *et al.* 1997; Nandi *et al.* 1998), and simple methods to conduct sensitivity analyses with respect to phylogenetic uncertainty (Donoghue & Ackerly 1996; Ackerly & Reich 1999).

## The quantitative convergence index

Recently, the quantitative convergence index (QVI) has been introduced for quantifying convergent evolution in continuous characters based on the distribution of trait values across a phylogeny (Ackerly & Donoghue 1998). The QVI is based on

the idea that, for a given set of trait values among taxa, one can construct hypothetical 'best case' and 'worst case' phylogenies in which phenotypically similar
species are either very closely related (minimal convergent evolution) or are distantly related and found on widely separated branches of the phylogenetic tree
(maximum convergent evolution). Using linear parsimony, the lengths of these
minimum and maximum trees can be calculated, based on reconstructions of the
ancestral character states which minimize the total amount of proposed evolutionary change. Figure 20.7 shows hypothetical values of leaf size for six taxa; the tree
showing minimum convergence (Fig. 20.7a) has a length of five units of evolutionary change, while the tree with maximum convergence (Fig. 20.7b) has a length of
nine units. The distribution of trait values on the actual (or proposed) phylogeny
for a group of species can then be examined and the extent of convergent evolution

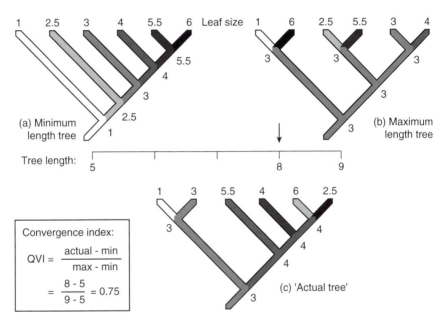

**Figure 20.7** Hypothetical illustration of the quantitative convergence index. Tree lengths
refer to the total amount of evolutionary change in a character, calculated using linear
parsimony methods (total change can be calculated as the sum of the differences between
trait values at each node and its immediate ancestor). (a) The minimum length tree shows
the phylogeny which would correspond to the most conserved distribution of leaf sizes,
with a tree length = 5. (b) The maximum length tree corresponds to the most convergent
pattern, given these trait values, with tree length = 9. (c) The 'actual tree', which might be
derived from a molecular phylogeny or other source, is intermediate between these with
tree length = 8. QVI = 0.75, following the formula in the lower left. QVI values range from 0
to 1, with higher values indicating that related species tend to be most dissimilar (see
Ackerly & Donoghue 1998).

quantified in relation to these two extremes. In this case, the 'actual' tree results in a minimum of eight units of change, resulting in QVI=0.75 (Fig. 20.7c). QVI varies from 0, for a trait in which phenotypically similar species are most closely related, to 1 for a trait in which dissimilar species are closely related, and convergent evolution is maximized (cf. Maddison & Maddison 1992).

The QVI was applied in a study of trait evolution for 32 functional and morphological traits of 17 species of *Acer*, based on an independently derived molecular phylogeny (Ackerly & Donoghue 1998). The resulting values ranged from 0.35 to 1. The lowest values were observed for two aspects of canopy architecture, the bifurcation angle and dominance of the leader shoot in sapling growth (Sakai 1987). The low level of convergent evolution reflects the distinct evolutionary shift in branching architecture among species of the section *Palmata* (including the well-known Japanese maple, *A. palmatum*), in which shoot growth is determinate and sympodial and the apical meristem of each branch either aborts or flowers at the end of each season. This leads to high bifurcation angles (65–80°) between the two laterals that grow out on either side in the subsequent season, and a low dominance index as they tend to be similar in length. This study included six species of section *Palmata*, and *A. carpinifolium*, which exhibits a similar growth pattern, resulting in relatively strong conservation of these traits over the phylogeny (Fig. 20.8a). By contrast, QVI=0.87 was observed in leaf size, indicating a high level of convergent evolution in this trait among these 17 species (Fig. 20.8b).

The QVI analysis has also been conducted on measurements of specific leaf area, leaf life-span, photosynthetic rate, leaf nitrogen concentration and leaf size (in terms of lamina area, using leaflets for compound leaf species) obtained for >100 species distributed across six distinct habitats (Reich *et al.* 1997; Ackerly & Reich 1999). Leaf size and leaf life-span were the most conserved of these traits (QVI= 0.50 and 0.55, respectively), in part because of the marked differences between conifers and angiosperms in this data set, while specific leaf area and nitrogen concentration on an area basis had the highest convergence values (QVI=0.72 and 0.77, respectively). It is interesting to note that leaf size was highly convergent in the maple study above (QVI=0.87), indicating that levels of convergence and conservatism for the same trait can vary at different hierarchical levels.

## Trait conservatism and the importance of independent contrasts

Several authors (e.g. Gittleman & Luh 1992) have suggested that the need for phylogenetically structured analyses, such as independent contrasts, can be evaluated by first asking whether or not a trait exhibits a strong phylogenetic pattern. If a trait is highly divergent in closely related taxa, then it is argued that species may be considered essentially as independent units with respect to that trait, and the results of ahistorical analyses (e.g. simple interspecific correlations) will be less likely to change when phylogenetic relationships are considered (e.g. independent contrasts). The quantitative convergence method introduced above provides a means to evaluate this proposition, by quantifying the degree of convergent evolution and

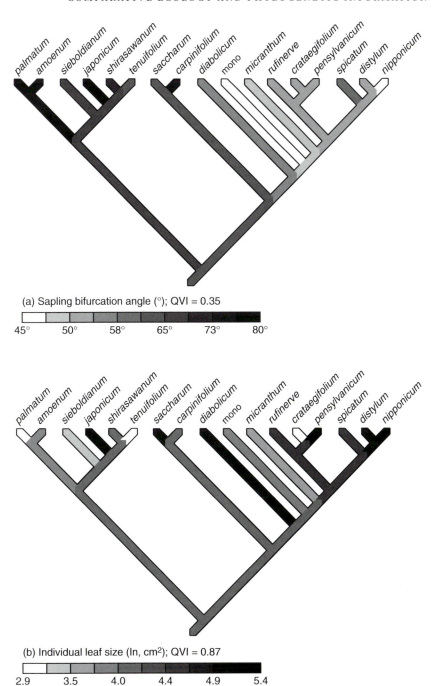

(a) Sapling bifurcation angle (°); QVI = 0.35

45°  50°  58°  65°  73°  80°

(b) Individual leaf size (ln, cm²); QVI = 0.87

2.9  3.5  4.0  4.4  4.9  5.4

**Figure 20.8** Reconstruction of morphological trait evolution in *Acer*. (a) Bifurcation angles of leader branches on saplings, illustrating a trait with a low level of convergence (QVI = 0.35). (b) Leaf size, a trait with a high convergence level (QVI = 0.87). Data from Ackerly and Donoghue (1998).

relating this to the discrepancy between correlation analyses with and without consideration of the phylogeny. Based on the study of 32 functional morphology traits in 17 species of *Acer* (Ackerly & Donoghue 1998) and eight leaf traits in 102 seed plant species (Ackerly & Reich 1999), 524 pairwise correlations were calculated with and without consideration of the phylogeny, the former based on independent contrasts and the latter based on cross-species correlations. The resulting scatterplot (Fig. 20.9a) mirrors the broader literature survey above (Fig. 20.5, cf. Ricklefs & Starck (1996) and Price (1997) for animal studies), illustrating the overall similarity of the two approaches. The absolute value of the difference between the two coefficients was then calculated as a measure of the discrepancy between the ahistorical and phylogenetic correlation analysis. In addition, mean QVI was calculated for the two traits in each correlation to reflect the average level of convergence or conservatism. A regression of the difference between the coefficients *vs.* the mean QVI for the two traits was negative and highly significant ($P<0.001$), although it explained only a moderate portion of the variance ($r^2=0.249$, Fig. 20.9b). Therefore, for traits with more homoplasy, the ahistorical and evolutionary correlations tended to be more similar (falling closer to the 1:1 line in Fig. 20.9a), while the two approaches can differ considerably for traits with low levels of convergence.

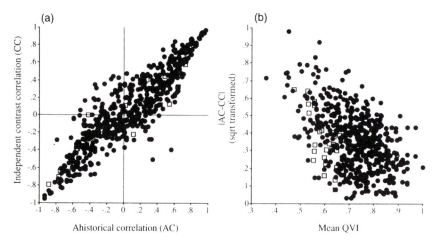

**Figure 20.9** (a) Interspecific correlations *vs.* independent contrast correlations for all 524 pairwise combinations of 32 morphological traits measured in *Acer* species (●, cf. Fig. 20.8) and eight leaf traits measured in a broad survey of seed plants (□, cf. Fig. 20.3). (b) Scatterplot of the absolute difference between the ahistorical and independent contrast correlations (departures from the 1:1 line in (a)) and the mean QVI for the traits involved in each correlation ($r^2=0.249$, $P<0.001$). The negative slope of this relationship illustrates that interspecific correlations are most similar to independent contrast analyses for traits with high levels of convergence. Modified from Ackerly and Donoghue (1998).

## Conclusions

The objective of this chapter was to provide a quantitative overview of the application of taxonomic information in comparative functional ecology of plants, focusing in particular on the analysis of continuous characters. Three points emerge from this analysis. First, the use of species pairs provides a simple and powerful approach to incorporating some phylogenetic structure into comparative analyses, increasing the resolution regarding patterns of divergence in functional traits. However, it is important to recognize that this approach effectively ignores the deeper historical divergences that have occurred between the pairs in a given study. Second, by contrast, the calculation of independent contrasts over the entire phylogeny incorporates all of this variation; interestingly, it turns out that on average trait correlations are similar using cross-species analyses or independent contrasts. However, the conclusions of some analyses may be altered considerably, and it will continue to be extremely important to consider the phylogenetic context of trait variation in comparative functional studies. Third, the quantitative analysis of the extent of evolutionary convergence, taking advantage of parsimony methods and the availability of increasingly resolved phylogenetic trees for seed plants, provides an alternative to the hierarchical analysis of variance as a means of assessing the extent to which related species are ecologically similar. This method has also proven quite useful in supporting the notion that for traits which are highly convergent, such that there is little phylogenetic 'signal' in a particular data set, the use of independent contrasts is less critical for tests of evolutionary trait correlations. Collectively, these phylogenetic approaches to comparative ecology complement and enhance our understanding of interspecific variation, and it is hoped that the antagonism that has marked the supposed debate over 'phylogeny *vs.* ecology' will diminish and be replaced by a more synthetic view (cf. Westoby *et al.* 1995; Mazer 1998; and references therein).

## Acknowledgements

The author is grateful to M.C. Press, K. Preston, J. Silvertown and F.I. Woodward for suggestions that improved the manuscript.

## References

Ackerly, D.D. (1999). P-FUNC: comparative plant functional ecology literature database. www.stanford.edu/~dackerly/PFuncLitDB.html.

Ackerly, D.D. & Donoghue, M.J. (1998). Leaf size, sapling allometry, and Corner's rules: a phylogenetic study of correlated evolution in maples (*Acer*). *American Naturalist*, **152**, 767–791.

Ackerly, D.D. & Reich, P.B. (1999). Convergence and correlations among leaf size and function in seed plants: a comparative test using independent contrasts. *American Journal of Botany*, **86**, in press.

Albert, V.A., Williams, S.E. & Chase, M.W. (1992). Carnivorous plants: phylogeny and structural evolution. *Science*, **257**, 1491–1495.

Ali, S.I. (1968). Correlation between seed weight and breeding system in closely related amphimictic taxa. *Nature*, **218**, 492–493.

Armstrong, D.P. & Westoby, M. (1993). Seedlings

from large seeds tolerate defoliation better: a test using phylogenetically independent contrasts. *Ecology*, **74**, 1092–1100.

Aronson, J.A., Kigel, J. & Shmida, A. (1990). Comparative plant sizes and reproductive strategies in desert and mediterranean populations of ephemeral plants. *Israel Journal of Botany*, **39**, 413–430.

Baker, H.G. (1972). Seed weight in relation to environmental conditions in California. *Ecology*, **53**, 997–1010.

Baldwin, B.G. (1993). Molecular phylogenetics of *Calycadenia* (Compositae) based on its sequences of nuclear ribosomal DNA: chromosomal and morphological evolution reexamined. *American Journal of Botany*, **80**, 223–238.

Baldwin, B.G. & Robichaux, R.H. (1995). Historical biogeography and ecology of the Hawaiian silversword alliance (Asteraceae). In *Hawaiian Biogeography: Evolution on a Hot Spot Archipelago.* (Ed. by W.L. Wagner & V.A. Funk), pp. 259–287. Smithsonian Institution Press, Washington, DC.

Baum, D.A. & Larson, A. (1991). Adaptation reviewed: a phylogenetic methodology for studying character macroevolution. *Systematic Zoology*, **40**, 1–18.

Bazzaz, F.A. (1987). Experimental studies on the evolution of niche in successional plant populations. In *Colonisation, Succession and Stability* (Ed. by A.J. Gray, M.J. Crawley & P.J. Edwards), pp. 245–272. Blackwell Scientific, Oxford.

Beerling, D.J. & Kelly, C.K. (1996). Evolutionary comparative analyses of the relationship between leaf structure and function. *New Phytologist*, **134**, 35–51.

Bell, G. (1989). A comparative method. *American Naturalist*, **133**, 553–571.

Chase, M., Soltis, D.E., Olmstead, R.G. *et al.* (1993). Phylogenetics of seed plants: an analysis of nucleotide sequences from the plastid gene *rbc*L. *Annals of the Missouri Botanical Garden*, **80**, 528–580.

Clausen, J., Keck, D.D. & Hiesey, W.M. (1940). *Experimental Studies on the Nature of Species I. Effect of Varied Environments on Western North American Plants.* Publication no. 520. Carnegie Institute of Washington, Washington, DC.

Cody, M.L. & Mooney, H.A. (1978). Convergence *versus* non-convergence in mediterranean-climate ecosystems. *Annual Review of Ecology and Systematics*, **9**, 265–321.

Darwin, C. (1859). *On the Origin of Species.* Murray, London.

Donoghue, M.J. (1989). Phylogenies and the analysis of evolutionary sequences, with examples from the seed plants. *Evolution*, **43**, 1137–1156.

Donoghue, M.J. & Ackerly, D.D. (1996). Phylogenetic uncertainties and sensitivity analyses in comparative biology. *Philosophical Transactions of the Royal Society of London, Series B*, **351**, 1241–1249.

Ehleringer, J.R. & Monson, R.K. (1993). Evolutionary and ecological aspects of photosynthetic pathway variation. *Annual Review of Ecology and Systematics*, **24**, 411–439.

Felsenstein, J. (1985). Phylogenies and the comparative method. *American Naturalist*, **125**, 1–15.

Foster, S.A. & Janson, C.H. (1985). The relationship between seed size and establishment conditions in tropical woody plants. *Ecology*, **66**, 773–780.

Franco, M. & Kelly, C.K. (1998). The interspecific mass–density relationship and plant geometry. *Proceedings of the National Academy of Sciences USA*, **95**, 7830–7835.

Franco, M. & Silvertown, J. (1996). Life history variation in plants: an exploration of the fast–slow continuum hypothesis. *Philosophical Transactions of the Royal Society of London, Series B*, **351**, 1341–1348.

Garland, T., Jr, Harvey, P.H. & Ives, A.R. (1992). Procedures for the analysis of comparative data using phylogenetically independent contrasts. *Systematic Biology*, **41**, 18–32.

Garnier, E. (1992). Growth analysis of congeneric annual and perennial grass species. *Journal of Ecology*, **80**, 665–675.

Gittleman, J.L. & Kot, M. (1990). Adaptation: statistics and a null model for estimating phylogenetic effects. *Systematic Zoology*, **39**, 227–241.

Gittleman, J.L. & Luh, H.-K. (1992). On comparing comparative methods. *Annual Review of Ecology and Systematics*, **23**, 383–404.

Givnish, T.J., Sytsma, K.J., Smith, J.F. & Hahn, W.J. (1995). Molecular evolution, adaptive radiation, and geographic speciation in *Cyanea* (Campanulaceae, Lobelioideae). In *Hawaiian Biogeography: Evolution on a Hot Spot Archipelago* (Ed. by W.L.

Wagner & V.A. Funk), pp. 288–337. Smithsonian Institution Press, Washington.

Grubb, P.J. (1998). Seed mass and light-demand: the need to control for soil-type and plant stature. *New Phytologist*, **138**, 169–170.

Grubb, P.J. & Metcalfe, D.J. (1996). Adaptation and inertia in the Australian tropical lowland rainforest flora: contradictory trends in intergeneric and intrageneric comparisons of seed size in relation to light demand. *Functional Ecology*, **10**, 512–520.

Harper, J.L., Clatworthy, J.N., McNaughton, I.H. & Sagar, G.R. (1961). The evolution and ecology of closely related species living in the same area. *Evolution*, **15**, 209–227.

Harvey, P.H. & Pagel, M. (1991). *The Comparative Method in Evolutionary Biology*. Oxford University Press, Oxford.

Hodgson, J.G. & Mackey, J.M.L. (1986). The ecological specialisation of dicotyledonous families within a local flora: some factors constraining optimisation of seed size and their possible evolutionary significance. *New Phytologist*, **104**, 497–515.

Hutchinson, G. (1957). Concluding remarks. *Cold Spring Harbor Symposium on Quantitative Biology*, **22**, 415–427.

Jackson, J.B., Abrams, D. & Jackson, U. (1999). Allometry of constitutive defense: a model and a comparative test with tree bark and fire regime. *American Naturalist*, **153**, 614–632.

Kelly, C.K. (1995). Seed size in tropical trees: a comparative study of factors affecting seed size in Peruvian angiosperms. *Oecologia*, **102**, 377–388.

Kelly, C.K. (1996a). Identifying plant functional types using floristic data bases: ecological correlates of plant range size. *Journal of Vegetation Science*, **7**, 417–424.

Kelly, C.K. (1996b). Seed mass, habitat conditions and taxonomic relatedness: a re-analysis of Salisbury (1974). *New Phytologist*, **135**, 169–174.

Kelly, C.K. & Woodward, F.I. (1996). Ecological correlates of plant range size: taxonomies and phylogenies in the study of plant commonness and rarity in Great Britain. *Philosophical Transactions of the Royal Society of London, Series B*, **351**, 1261–1269.

King, D.A. (1990). Allometry of saplings and understorey trees of a panamanian forest. *Functional Ecology*, **4**, 27–32.

Kohyama, T. & Hotta, M. (1990). Significance of allometry in tropical saplings. *Functional Ecology*, **4**, 515–521.

Lord, J., Westoby, M. & Leishman, M. (1995). Seed size and phylogeny in six temperate floras: constraints, niche conservatism, and adaptation. *American Naturalist*, **146**, 349–364.

Maddison, W.P. & Maddison, D.R. (1992). *Macclade: Analysis of Phylogeny and Character Evolution*, Version 3. Sinauer Associates, Sunderland, MA.

Mazer, S. (1998). Alternative approaches to the analysis of comparative data: compare and contrast. *American Journal of Botany*, **85**, 1194–1199.

Metcalfe, D.J. & Grubb, P.J. (1995). Seed mass and light requirements for regeneration in Southeast Asian rain forest. *Canadian Journal of Botany*, **73**, 817–826.

Monson, R.K. (1996). The use of phylogenetic perspective in comparative plant physiology and developmental biology. *Annals of the Missouri Botanical Garden*, **83**, 3–16.

Mooney, J. & Dunn, E. (1970). Convergent evolution of Mediterranean-climate evergreen sclerophyllous shrubs. *Evolution*, **24**, 292–303.

Morgan, D.C. & Smith, H. (1979). A systematic relationship between phytochrome-controlled development and species habitat, for plants grown in simulated natural radiation. *Planta*, **145**, 253–258.

Nandi, O.I., Chase, M.W. & Endress, P.K. (1998). A combined cladistic-analysis of angiosperms using *rbc*L and non-molecular data sets. *Annals of the Missouri Botanical Garden*, **85**, 137–212.

Niklas, K.J. (1993). The scaling of plant height: a comparison among major plant clades and anatomical grades. *Annals of Botany*, **72**, 165–172.

Pagel, M.D. (1993). Seeking the evolutionary regression coefficient: an analysis of what comparative methods measure. *Journal of Theoretical Biology*, **164**, 191–205.

Pagel, M. & Harvey, P. (1988). Recent developments in the analysis of comparative data. *Quarterly Review of Biology*, **63**, 413–440.

Peat, H.J. & Fitter, A.H. (1994). Comparative analyses of ecological characteristics of British angiosperms. *Biological Reviews*, **69**, 95–115.

Price, T. (1997). Correlated evolution and independent contrasts. *Philosophical Transactions of the Royal Society of London, Series B*, **352**, 519–529.

Rees, M. (1996). Evolutionary ecology of seed dormancy and seed size. *Philosophical Transactions of the Royal Society of London, Series B*, **351**, 1299–1308.

Reich, P.B., Walters, M.B. & Ellsworth, D.S. (1997). From tropics to tundra: global convergence in plant functioning. *Proceedings of the National Academy of Sciences USA*, **94**, 13730–13734.

Reich, P.B., Ellsworth, D.S., Walters, M.B. *et al.* (1999). Generality of leaf trait relationships: a test across six biomes. *Ecology*, **80**, 1955–1969.

Ricklefs, R.E. & Latham, R.E. (1992). Intercontinental correlation of geographic ranges suggests stasis in ecological traits of relict genera of temperate perennial herbs. *American Naturalist*, **139**, 1305–1321.

Ricklefs, R.E. & Starck, J.M. (1996). Applications of phylogenetically independent contrasts: a mixed progress report. *Oikos*, **77**, 167–172.

Ridley, M. (1992). Darwin sound on comparative method. *Trends in Ecology and Evolution*, **7**, 37.

Sakai, S. (1987). Patterns of branching and extension growth of vigorous saplings of Japanese *Acer* species in relation to their regeneration strategies. *Canadian Journal of Botany*, **65**, 1578–1585.

Salisbury, E.J. (1942). *The Reproductive Capacity of Plants*. Bell, London.

Salisbury, E.J. (1974). Seed size and mass in relation to environment. *Proceedings of the Royal Society of London, Series B*, **186**, 83–88.

Saverimuttu, T. & Westoby, M. (1996). Seedling longevity under deep shade in relation to seed size. *Journal of Ecology*, **84**, 681–689.

Schmitt, J. & Gamble, S.E. (1990). The effect of distance from the parental site on offspring performance and inbreeding depression in *Impatiens capensis*: a test of the local adaptation hypothesis. *Evolution*, **44**, 2022–2030.

Schmitt, J. & Wulff, R. (1993). Light spectral quality, phytochrome, and plant competition. *Trends in Ecology and Evolution*, **8**, 47–51.

Shipley, B. & Dion, J. (1992). The allometry of seed production in herbaceous angiosperms. *American Naturalist*, **139**, 467–483.

Silvertown, J. & Dodd, M. (1996). Comparing plants and connecting traits. *Philosophical Transactions of the Royal Society of London, Series B*, **351**, 1233–1239.

Soltis, D.E., Soltis, P.S. Nickrent, D.L. *et al.* (1997). Angiosperm phylogeny inferred from 18S ribosomal DNA sequences. *Annals of the Missouri Botanical Garden*, **84**, 1–49.

Soltis, D.E., Xiang, Q.-Y. & Hufford, L. (1995). Relationships and evolution of Hydrangeaceae based on *rbc*L sequence data. *American Journal of Botany*, **82**, 504–514.

Stebbins, G. (1974). *Flowering Plants: Evolution Above the Species Level*. Belknap Press, Cambridge, MA.

Stevens, P.F. (1994). *The Development of Biological Systematics*. Colombia University Press, New York.

Swanborough, P. & Westoby, M. (1996). Seedling relative growth rate and its components in relation to seed size: phylogenetically independent contrasts. *Functional Ecology*, **10**, 176–184.

Thompson, K. & Hodkinson, D.J. (1998). Seed mass, habitat and life history: a re-analysis of Salisbury (1942, 1974). *New Phytologist*, **138**, 163–167.

Tiffney, B.H. & Mazer, S.J. (1995). Angiosperm growth habit, dispersal and diversification reconsidered. *Evolutionary Ecology*, **9**, 93–117.

Turesson, G. (1922). The genotypical response of the plant species to the habitat. *Hereditas*, **3**, 211–350.

Wanntorp, H.-E. (1983). Historical constraints in adaptation theory: traits and non-traits. *Oikos*, **41**, 157–160.

Westoby, M. (1999). Generalisation in functional plant ecology: the species-sampling problem, plant ecology strategy schemes, and phylogeny. In *Handbook of Functional Plant Ecology* (Ed. by F.I. Pugnaire & F. Valladares), pp. 847–872. Marcel Dekker, New York.

Westoby, M., Leishman, M. & Lord, J. (1995). Issues of interpretation after relating comparative datasets to phylogeny. *Journal of Ecology*, **83**, 892–893.

Westoby, M., Cunningham, S.A., Fonseca, C.M., Overton, J.M. & Wright, I.J. (1998).

Phylogeny and variation in light capture area deployed per unit investment in leaves: designs for selecting study species with a view to generalising. In *Inherent Variation in Plant Growth. Physiological Mechanisms and Ecological Consequences* (Ed. by H. Lambers, H. Poorter & M.M.I. Van Vuuren), pp. 539–566. Backhuys Publishers, Leiden, Netherlands.

# Chapter 21

# Stable isotopes reveal exchanges between soil, plants and the atmosphere

*H. Griffiths,*[1] *A. Borland,*[1] *J. Gillon,*[1,2] *K. Harwood,*[1] *K. Maxwell*[1] *and J. Wilson*[1]

## Introduction

Over the past decade there has been an exponential growth in the use of stable isotopes to integrate plant physiological ecology, which has been matched by technical developments in mass spectrometry and theoretical understanding of discrimination processes (Griffiths 1998). Thus, the natural abundance of stable isotopes can provide a quantitative framework for chemical, biological and ecological transformations. Differential reaction rates for light and heavy isotopes leave identifiable signals in soil, water, plant and the atmosphere, representing either discrimination processes or a tracer of original inorganic sources throughout the global biogeochemical cycle. As an example of the former, fractionation against $^{13}C$ and $^{18}O$ during exchanges of $CO_2$ and $H_2O$ between vegetation and atmosphere are regulated by the balance of diffusive exchange and discrimination within the leaf (Farquhar *et al.* 1989, 1993). Thus, exchanges between soils, water, plants and the atmosphere are reflected in the $^{13}C/^{12}C$ and $^{18}O/^{16}O$ composition of organic material and atmospheric $CO_2$. Fluxes associated with photosynthesis and respiration have then been scaled to biospheric exchanges at the global level (Farquhar *et al.* 1993; Lloyd *et al.* 1996; Ciais & Meijer 1998). These approaches have been aided by developments in mass spectrometric systems with high sample throughput, allowing continuous-flow analyses of gases or organic samples following combustion or pyrolysis.

Additionally, isotopes can be used as tracers to reconstruct latitudinal, regional and local gradients in precipitation and groundwater, with the source signal in water ($^2H/^1H$ [D/H] and $^{18}O/^{16}O$) reflecting fractionation processes during evaporation and precipitation (Ehleringer & Dawson 1992; Dawson *et al.* 1998; Yakir 1998; Chapter 18). While the use of nitrogen as a tracer can reveal transformations in soil and plant, this depends in part on relative concentrations and complexities

[1] *Department of Agricultural and Environmental Science, University of Newcastle upon Tyne, NE1 7RU, UK. E-mail: howard.griffiths@ncl.ac.uk*
[2] *Present address: Department of Environmental Research and Energy Research, Weizmann Institute of Science, 76100 Rehovot, Israel*

of inputs (Handley & Scrimgeour 1997; Högberg 1997). Underlying theory and analytical approaches to isotope studies have been extensively reviewed recently (Farquhar *et al.* 1989; Griffiths 1991, 1996, 1998; Ehleringer *et al.* 1993; Deléens *et al.* 1994; Brugnoli & Farquhar 1999). Hence, recent theoretical and methodological advances are illustrated, with examples that relate plant physiological responses under field and laboratory conditions to associated ecological interactions.

Specifically, it will be shown that, at the leaf level, models of discrimination normally fit well with the exchange of water and $CO_2$ under slight water deficits, and that internal leaf mesophyll conductance has a number of morphological and physiological correlates. However, underlying cellular processes such as respiration and photorespiration may be quantitatively more important under extreme environmental conditions. The relationship between the isotopic signal of leaf organic material and the environmental conditions that were prevalent at the time of deposition will be contrasted with more real-time analyses of organic fractions and discrimination during gas exchange. In particular, the interplay between leaf morphology, the regulation of $C_3$ and $C_4$ carboxylation processes and the capacity for CAM will be used to explain habitat preference in a range of tropical epiphytes.

These examples are representative of the integrative power of stable isotopes and illustrate the potential of these techniques to investigate global ecological processes. Thus, the aim is to provide an overview of the uses of stable isotopes for investigating plant physiological ecology, particularly when acting as the interface between soil and water sources and the atmosphere.

## Carbon isotope ratio and discrimination in organic material

Analysis of stable isotopes is traditionally expressed in a differential notation, based on the mass spectrometric comparison of the quotient of heavy to light isotopes for sample and a defined standard. For $\delta^{13}C$, the standard has an absolute $^{13}C:{}^{12}C$ ratio of 0.01118, with atmospheric $CO_2$ $\delta^{13}C$ slightly depleted in $^{13}C$ by $-8‰$ (parts per thousand or per mille). Organic material is more depleted in $^{13}C$ ($-25$ to $-30‰$), with this discrimination, termed $\Delta$, expressed relative to the source $CO_2$ isotope composition (Farquhar *et al.* 1989). Variations in $\delta^{13}C$ allow identification of $C_3$, $C_4$ and CAM pathways in living specimens and herbarium material. The quantitative theoretical framework derived by Farquhar and colleagues (e.g. Farquhar *et al.* 1989; Brugnoli & Farquhar 1999) distinguishes between the inherently high discrimination of Rubisco in relation to limitations imposed by stomata, or interactions with phospho*enol*pyruvate carboxylase (PEPc) and $C_4$ bundle sheath morphology. For plants with contrasting habitat preferences, such as the Bromeliaceae of Trinidad, the bimodal range of discrimination values encompasses constitutive crassulacean acid metabolism (CAM) (i.e less negative $\delta^{13}C$, more $C_4$-like low $\Delta$) through to high $\Delta$ $C_3$ values depleted in $^{13}C$ (Fig. 21.1a). It is important to note that additional analyses of leaf sap titratable acidity are required to identify all $C_3$-CAM intermediates in this type of survey. Such approaches are still yielding valuable

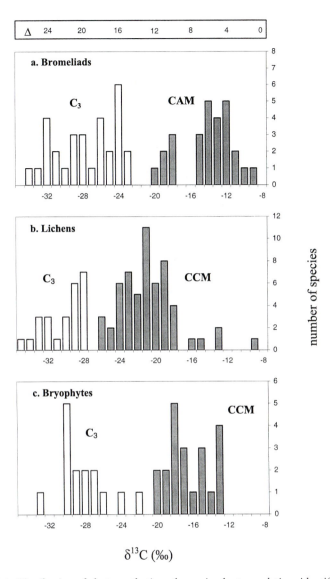

**Figure 21.1** Distribution of photosynthetic pathways in plant populations identified from carbon isotope composition shown as $\delta^{13}C$ and equivalent $\Delta$ values. (a) $C_3$ and CAM members of the Bromeliaceae in Trinidad. (b) $C_3$ and CCM in lichens, distinguished by the occurrence of chloroplast pyrenoid in the phycobiont. (c) $C_3$ and CCM in bryophytes, with CCM in members of the Anthocerotae with pyrenoid. Data from Griffiths and Smith (1983) and Smith and Griffiths (1996).

insights into the way that continental floras have evolved (Akhani *et al.* 1997; Cerling *et al.* 1997; Ehleringer *et al.* 1997) and which may now be threatened by invasion (Chapter 18). Other recent approaches have shown interactions between the residual organic signal in soils, soil respiration and $C_3$–$C_4$ distribution (Follett *et al.* 1997; Rochette & Flanagan 1997; Tieszen *et al.* 1997; Amundsen *et al.* 1998; Buchmann & Ehleringer 1998; Drinkwater *et al.* 1998).

Additional pathways, such as the biophysical carbon-concentrating mechanism (CCM) found in many algae and cyanobacteria, are analogous to the $C_4$ pathway in terms of discrimination characteristics, and result in low $\Delta$ organic material in diverse phylogenetic groups (Griffiths 1996). Differences between cyanobacterial and green algal (phycobiont) lichens have been revealed by thallus organic material $\Delta$, and reflect the activity of a CCM in various assocations (Máguas *et al.* 1993, 1995). The chloroplast pyrenoid has been identified as the site of the CCM in some of the phycobiont associations (Badger *et al.* 1993; Palmqvist 1993; Smith & Griffiths 1996; see Fig. 21.1b). The distribution of carbon isotopes has a similar bimodal distribution to higher plants, albeit shifted towards more negative $\delta^{13}C$ values (i.e a greater discrimination or higher $\Delta$) for all associations, probably associated with refixation of respiratory $CO_2$ derived from the mycobiont.

Most recently, a biophysical CCM has been associated with a family of bryophytes including the genus *Anthoceros* (Smith & Griffiths 1996; Raven *et al.* 1998), with more $C_4$-like organic $\Delta$ values in species containing a pyrenoid (Fig. 21.1c). The carbon isotope composition of liverworts (often appressed to a substrate) or mosses (in hummocks) is usually affected to some extent by additional respiratory inputs (Rice & Giles 1996; Price *et al.* 1997), accounting in part for the slightly higher $\Delta$ values (Fig. 21.1c) than for terrestrial $C_3$ and $C_4$ plants.

## Carbon and oxygen isotope signals in $CO_2$

The interplay between photosynthetic and respiratory processes are described by the inverse relationship between $1/[CO_2]$ and isotope composition for $^{13}C$ and $^{18}O$ in $CO_2$; contributory sources and net ecosystem discrimination can be identified from such a 'Keeling plot' (Buchmann *et al.* 1997a,b; Flanagan *et al.* 1997; Sternberg *et al.* 1997; Bakwin *et al.* 1998; Harwood *et al.* 1999; also see Griffiths 1998). As alluded to above, the isotopic mass balance of the $CO_2$ concentration in the atmosphere reflects the mass balance of photosynthetic, respiratory and anthropogenic exchanges superimposed on the long-term equilibration with inorganic and organic carbon reserves in the lithosphere and hydrosphere. Intra-annual fluctuations in $[CO_2]$ and $\delta^{13}C$ are most evident in the northern hemisphere because of the greater land mass and greater seasonality compared with southern and equatorial regions, and are set against the larger trend of rising $CO_2$ (and more negative $\delta^{13}C$) (Keeling *et al.* 1995; Bakwin *et al.* 1998; Ciais & Meijer 1998). These are similar to those changes seen within a forest canopy on a diel basis (Lloyd *et al.* 1996; Buchmann *et al.* 1997a,b; Flanagan *et al.* 1997; Harwood *et al.* 1999; see Fig. 21.2).

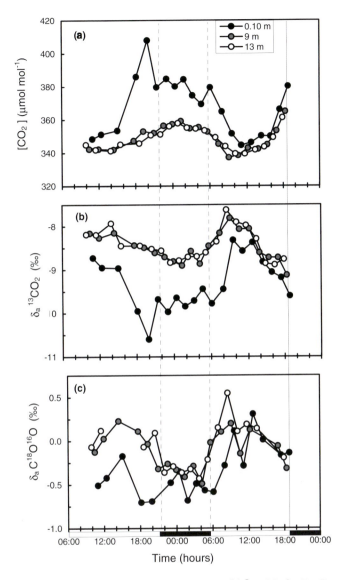

**Figure 21.2** Variation in: (a) ambient $CO_2$ concentration; (b) $\delta_a\,^{13}CO_2$ (*vs.* Pee Dee Belemnite (PDB)); (c) $\delta_a\,C^{18}O^{16}O$ (*vs.* PDB-$CO_2$) within a canopy during the collection period, 12–13 July 1995, at Holystone, Northumberland. Dark horizontal bars represent night whereby dawn and dusk were denoted as the time at which photon flux density (PFD) deviated from zero on the instantaneous light meter in the forest clearing. Data from Harwood *et al.* (1999) reproduced with permission.

At the canopy scale, the concentration of $CO_2$ within a *Quercus petraea* canopy in Northumberland changed in relation to height within the canopy, with a marked increase during the afternoon and overnight, particularly at 10 cm above the ground (Fig. 21.2a). It is evident that the $\delta^{13}C$ of source $CO_2$ varies inversely with the concentration at all heights within the canopy, such that a Keeling plot of all data points has an intercept of –24.3‰, representing the 'net' $^{13}C$ signal from forest to atmosphere. Because the $\delta^{13}C$ of the surrounding leaf material was around –28‰, the canopy signal was enriched by some 4‰, suggesting that additional sources (such as soil) or fractionation processes were contributing to the signal (Harwood *et al.* 1999). Such a shift between the intercept-derived respiratory $\delta^{13}C$ signal and that of leaf material has been observed in tropical and temperate forests (Buchmann *et al.* 1997a,b). In addition, carbon isotope composition has also been used to estimate the proportion of $CO_2$ refixed within forest canopies (Sternberg *et al.* 1997) and to assess the influence on regional carbon budgets (Buchmann *et al.* 1997a,b).

At the global level, the seasonal changes in $[CO_2]$ and $\delta^{13}C$ in the northern hemisphere represent the photosynthetic drawdown in summer with respiratory release predominating in winter (Keeling *et al.* 1995). It has become apparent that $^{13}C$ signals in multiple sources of $CO_2$ are similar: while discrimination at the terrestrial and oceanic level can be detected, it is difficult to distinguish respiratory and fossil fuel sources (Ciais & Meijer 1998). While additional measurements of $^{14}C$ in $CO_2$ or $^{13}C$ in carbon monoxide may help to determine this contribution (Bakwin *et al.* 1998), it is the $\delta^{18}O$ signal in $CO_2$ which has shown most promise to date. Under higher vapour pressure deficits, above-ground $^{18}O$ enrichment of leaf water distinguishes soils and vegetation, allowing contributions from photosynthesis and respiration to be distinguished in the atmosphere.

During gaseous exchange of $CO_2$ within $C_3$ leaves, for every three molecules of $CO_2$ entering the leaf and equilibrating with leaf water, two, on average, diffuse back out to the atmosphere (which also allows discrimination against $^{13}CO_2$ to occur), carrying the leaf water $^{18}O$ signal. The isotopic exchange of $^{18}O$ between leaf water and $CO_2$ is catalysed by the activity of the enzyme carbonic anhydrase (Flanagan *et al.* 1994; Ciais & Meijer 1998), with the signal representing some point in the continuum from chloroplast to evaporative site (Farquhar *et al.* 1993; Flanagan 1998; Yakir 1998). Analogous to $^{13}C$, concurrent measurements of instantaneous gas exchange of $\delta^{18}O$ in $CO_2$ and water vapour can be used to model theoretical and actual discrimination (Farquhar *et al.* 1993; Flanagan *et al.* 1994; Harwood *et al.* 1998, 1999). The $CO_2$ signal is dependent on key assumptions regarding the extent of enrichment at the evaporative site ($\delta_e$), whether the leaf water is in isotopic steady state with source water, and how closely $\delta_e$ reflects the chloroplast water signal and $CO_2$ concentration ($p_c$), and mesophyll conductance ($g_i$) (Flanagan 1998; Yakir 1998). By measuring the isotopic signal of the transpired water vapour ($\delta_t$) directly during gas exchange, $\delta_e$ can be measured under variable conditions in the field (Harwood *et al.* 1998).

At the canopy level, the $\delta^{18}O$ signal of $CO_2$ can be less well coupled to $[CO_2]$, although in general terms it is still inversely related. However, the daytime signal shows larger shifts (relative to $^{13}C$) as transient changes occur in humidity and evaporation rate (Fig. 21.2c). Alternatively, at night, the values at all heights in the canopy tend to converge, suggesting predominance of more stable soil-derived flux (Fig. 21.2c) (Harwood *et al.* 1999). In soils, the $\delta^{18}O$ of $CO_2$ generated or exchanging at the surface is transferred to the atmosphere after exchanging oxygen with soil water, with further modification during diffusion from the soil. Soil respiration is the dominant factor controlling global atmospheric $\delta^{18}O$ of $CO_2$ (Ciais & Meijer 1998), but isotopic exchanges are poorly understood at this level (Amundsen *et al.* 1998).

## Stable isotopes as tracers of water sources: $^{18}O$ and D/H

The contrasting patterns of depletion in evaporating water, or enrichment during condensation, leads to latitudinal and seasonal shifts in $\delta^{18}O$ and $\delta D$ associated with temperature changes, precipitation inland being progressively depleted in the heavier isotope (Ciais & Meijer 1998; Dawson *et al.* 1998). While the $^{18}O$ signal of $CO_2$ only varies by $-1‰$ to $+1‰$ over northern and southern hemispheres, the signal in precipitation is much greater, ranging from around $0‰$ to $-20‰$ between the equator and $60°N/S$. For deuterium, the isotopic effects are greater, with meteoric waters varying by over $400‰$. Because there is (generally) no fractionation during uptake of water by roots, at this point in the hydrological cycle there is a direct link between water sources and xylem constituents (Dawson *et al.* 1998).

The typical range of source water values for $\delta^{18}O$ and $\delta D$ in the soil–plant–atmosphere continuum for mid-latitude, semi-arid conditions is shown in Fig. 21.3. For $\delta^{18}O$, the data are based primarily on values reported by Yakir and colleagues for *Tamarix* (Yakir 1998) and shows that soil surface water may well reflect the immediate precipitation (and surface evaporation processes), with recharge by more enriched 'winter' rains contributing mainly to groundwater storage (Fig. 21.3). Depending on the time of year, there is usually a gradient of $\delta^{18}O$ with depth (Dawson *et al.* 1998), and thus the root water isotope profile depends on whether a particular plant has access to groundwater, surface water or both. Sampling of stem or branch water (by vacuum distillation and equilibration with $CO_2$ for mass spectrometric analysis) is an important adjunct of measurements at the leaf or canopy scale. Thus, for oak leaves in Northumberland, the main signal of transpired water vapour ($\delta_t$) was similar to that of source water, showing in general that leaves were at isotopic steady state (Fig. 21.2; Harwood *et al.* 1999). Similarly at the canopy scale, mixing models of water vapour concentration and $\delta H_2^{18}O$ have revealed that the exchange of water from canopy to atmosphere is also, in general, at isotopic steady state (Moreira *et al.* 1997; Harwood *et al.* 1999).

Analysis of the source water $\delta D$ signal can provide similar information, with the values in Fig. 21.3 representative of semi-arid sites in North America (Ehleringer *et al.* 1991; Dawson 1993). Investigations with deuterium natural abundance have

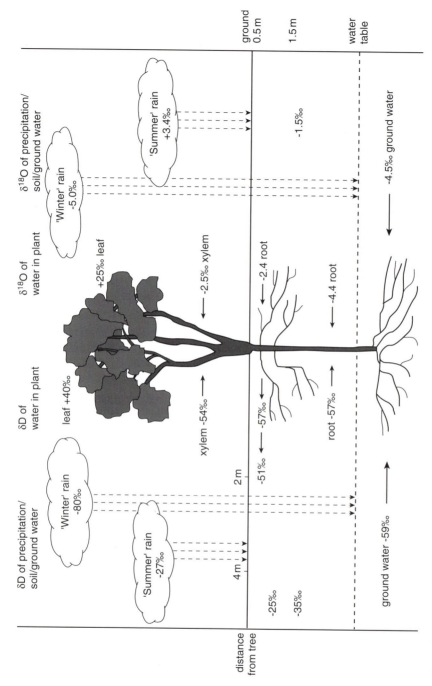

**Figure 21.3** $^{18}O$ and H/D as tracers of water sources. Typical values for $\delta^{18}O$ and $\delta D$ in the soil–plant–atmosphere continuum for mid-latitude, semi-arid conditions. Data derived from Ehleringer *et al.* (1991), Dawson (1993) and Yakir (1998).

given fascinating insights into water sources, whether for trees (Dawson & Ehleringer 1991), or the selection of summer or winter rains by contrasting shrubby lifeforms (Ehleringer *et al.* 1991, 1998). Most elegantly, the extent of hydraulic lift can be quantified from the $\delta D$ signal, which, as illustrated in Fig. 21.3, can lead to the lateral transfer of groundwater *from* surface roots to surrounding soil and herbaceous vegetation (Dawson 1993). Recently, an artificially enriched $\delta D$ water source has been used to indicate the transfer of surface water *to* deeper roots in the Kalahari desert (Schulze *et al.* 1998), a mechanism which perhaps provides the key to phreatophyte establishment. It is important to note that the leaf water enrichment found for both $\delta^{18}O$ and $\delta D$ is only partially transferred to organic constituents, with implications for the timing of carbohydrate deposition and environmental signals (see below).

### Nitrogen sources and incorporation: $\delta^{15}N$

The interrelationship between nitrogen natural abundance ($^{15}N/^{14}N$ signal) in soils and plants has provoked considerable debate, although quantitative relationships only hold under specific conditions. Thus, in a desert cryptobiotic crust where the major input is biological $N_2$ fixation (Evans & Belnap 1998; Lange *et al.* 1998), nitrogen concentration and $\delta^{15}N$ are interrelated, with progressive enrichment of residual nitrogen pools as turnover and transformation lead to losses of the lighter nitrogen from the system. Primary products of biological $N_2$ fixation usually have a low $\delta^{15}N$, similar to source atmospheric nitrogen, which is used as the mass spectrometric standard. A variety of fractionations may occur during nitrogen transformations in soils (Högberg 1997; Hopkins *et al.* 1998) or plants (Evans *et al.* 1996; Handley & Scrimgeour 1997; Robinson *et al.* 1998), which may disguise any simple source–sink relationships and negate the use of natural abundance $^{15}N$ as a tracer. In soils, the pool of organic nitrogen is usually 10–100 times higher than that of inorganic nitrogen as nitrate or ammonium (Fig. 21.4a,c,e).

The low *net* fractionation, typically observed in the $\delta^{15}N$ signal of the soil total-N pool (Fig. 21.4b), reflects the dominance of the large organic nitrogen component, which is predominantly recalcitrant and will tend to mask the effects of fractionating losses from the smaller and more dynamic inorganic nitrogen pools ($NH_4^+$-N and $NO_3^-$-N: Fig. 21.4d,f). The $^{15}N$ natural abundance of ammonium-N and nitrate-N determined by a modified microdiffusion technique (after Sorensen & Jensen 1991), for a Northumberland clay soil under grass, indicated very different $^{15}N$ contents in the $NH_4^+$-N and $NO_3^-$-N pools. Volatilization of $NH_3$ can lead to marked enrichment in the residual $NH_4^+$-N pool following fertilization (Fig. 21.4d). Unravelling the $\delta^{15}N$ signature of the nitrate-N pool is more complex. Nitrogen loss via denitrification can lead to a relative enrichment in $^{15}N$ in the residual nitrate pool; alternatively, the processes of mineralization and nitrification may lead to relative $^{15}N$ depletion.

Despite these difficulties, the $^{15}N$ signal has provided some excellent data on soil

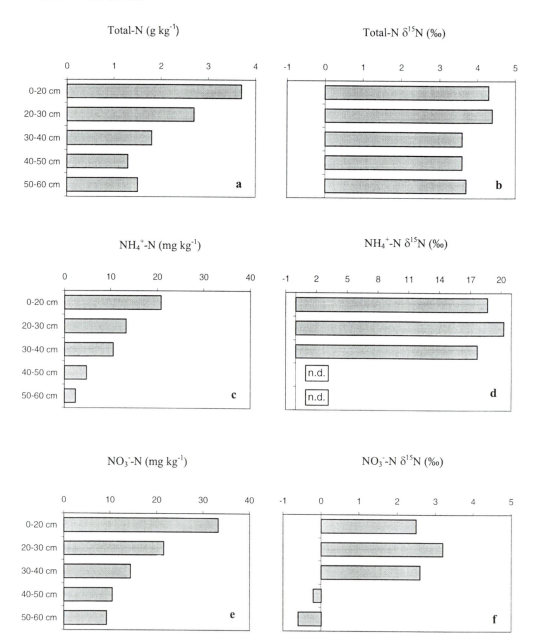

**Figure 21.4** Variations in soil δ15N total-N and inorganic fractions. Vertical profiles of nitrogen concentration and δ15N for a boulder clay soil in Northumberland: (a) total-N profile; (b) δ15N of total-N; (c) ammonium-N profile; (d) δ15N of ammonium-N; (e) nitrate-N profile; (f) δ15N of nitrate-N. Total-N was determined by direct combustion of soil samples, with a modified microdiffusion technique used on NaCl extracts for inorganic fractions. Unpublished data of J.M. Wilson.

and plant nitrogen relations when conducted under carefully defined conditions (Högberg 1997; Evans & Belnap 1998; Chapter 18). The use of isotope pairs in a given compound may prove stronger in separating both source and fractionation effects, particularly the $^{18}O$ signal of nitrate (Kendall *et al.* 1995; Hopkins *et al.* 1998). Ultimately, one aim will be to integrate the plant organic nitrogen signal with that of carbon, whereby the $\delta^{13}C$ integrates environmental conditions and plant growth form. Given the likely cycling between soil carbon and nitrogen pools (Drinkwater *et al.* 1998) and implications globally for tropical, temperate or boreal forests as sources or sinks for carbon (Chapter 19), it is more important than ever that we can evaluate the information available in combined $^{13}C$ and $^{15}N$ signals.

## Correlates of carbon isotope discrimination theory

Having identified the range of $\delta^{13}C$ values associated with different photosynthetic pathways above, we now consider in detail the environmental, morphological and physiological constraints on carbon isotope discrimination ($\Delta$). The quantitative model reported by Farquhar *et al.* (1982) related the extent of discrimination against $^{13}C$ to $p_i/p_a$, the ratio of internal : external $CO_2$ partial pressure, with the indirect link to water-use efficiency via stomatal conductance providing a remarkably robust relationship (Farquhar *et al.* 1989).

Although diagrams have been advanced which link $\Delta$ and WUE via $p_i/p_a$ (O'Leary 1988; Ehleringer 1993), it has become evident that a wide range of external and internal factors influence leaf $\Delta$. At the ecophysiological level internal constraints to diffusion and carboxylation control $\Delta$, in terms of the long-term $^{13}C$ organic signal and that occurring instantaneously during photosynthesis, through either physical or biochemical components of mesophyll conductance (Table 21.1). Thus, the focal point of the scheme is the interplay between $p_i/p_a$, and the drawdown to $p_c$ (the $CO_2$ partial pressure at Rubisco), which is regulated by Rubisco capacity and activity. The general rule is that $\Delta$ is inversely proportional to water-use efficiency (WUE, defined as assimilation/transpiration), mediated via $p_i/p_a$. Measurements of gas exchange can then be used to calculate $\Delta_i$ simply from $\Delta_i = 4.4 + 22.4 \, (p_i/p_a)$ (Farquhar *et al.* 1989). When $\Delta$ is led by conductance, particularly under slight water deficits, and decreasing humidity, discrimination against both $^{13}C$ and $^{18}O$ will be greater, because high rates of transpiration are also coupled to evaporative enrichment of water (Farquhar *et al.* 1997, 1998). Alternatively, when Rubisco capacity is limited, discrimination against $^{13}C$ still reflects the increase in $p_i/p_a$, but $^{18}O$ in organic material is relatively less enriched. When grown under constant atmospheric vapour pressure, a change in the $^{18}O$ cellulose signal reflects evaporative cooling via $g_s$ and hence can be related to yield potential under well-watered conditions (Farquhar *et al.* 1997, 1998). Another approach is to assume that leaf nitrogen content is a surrogate for carboxylation efficiency, and to partition the absolute fluxes of carbon and water to improve the relationship

**Table 21.1** Control of discrimination by external and internal factors.

| Control of Δ | | CO₂ supply and demand | Physiological/ morphological correlates | Environmental conditions |
|---|---|---|---|---|
| Long term | Instantaneous | | | |

Diagram contents (left to right):

Long term column:
- WUE ← $A/g_s$
- organic
- $\delta^{13}C$
- ($\delta^{18}O$, VPD-led)
- $^{13}C$, $^{18}O$ in $CO_2$

Instantaneous column:
- $\Delta_i$
- $p_i$
- $g_i$
- $p_c$
- $\Delta_{observation}$

CO₂ supply and demand:
- Diffusive exchange
- $P_c$ Rubisco capacity[1]
- (Photo)respiration refixation and exchange

Physiological/morphological correlates:
- Lifeform/strategy[2]
- SLM[3]
- Chlorophyll[4]
- Leaf N[5]
- Ash content[6]

Environmental conditions:
- VPD
- PPFD
- Temperature
- Rainfall
- Soil water deficit
- Soil strength/ nutrients

Source references to confirm the physiological and morphological correlates of Δ indicated as: 1, Rubisco capacity; 2, genotype/lifeform; 3, specific leaf mass; 4, chlorophyll content; 5, total leaf nitrogen; 6, ash content. References: Anderson et al. 1996[1,2,3,5]; Araus et al. 1997[3,4,5]; Cordell et al. 1998[2,3]; Damesin et al. 1998[2,3]; Ehleringer 1993; Ebdon et al. 1998[6]; Fleck et al. 1996[1,3,5]; Frank et al. 1997[2,6]; Geber & Dawson 1997[2]; Isla et al. 1998[6]; Kloeppel et al. 1998[1,2,3]; Lauteri et al. 1997[1,2,3,4]; Rao et al. 1995[1,3]; Syvertsen et al. 1997[1,3,5]; Yin & Raven 1998[5].

between yield and water use as prescribed by carbon isotope signals (F.S. Gillion, personal communication).

Thus, we can link the effect of environmental conditions, ranging from the atmospheric (vapour pressure deficit (VPD), temperature, light) to water and nutrient supply, as regulators of leaf ultrastructure and Rubisco activity (Table 21.1). Starting with increased Rubisco capacity ([1] in Table 21.1), an increase in the drawdown to $p_c$ (and hence $p_i$) then leads to an inverse correlation with Δ. Second, lifeform, growth strategy or phenology should all be considered important determinants of Δ, particularly when comparing shrubby perennials and herbaceous forbs with crops ([2] in Table 21.1). Thus, for 'spenders' and 'savers', water or nitrogen limitation under natural conditions may lead to a more conservative strategy for growth, which compromises the survival of spenders (Ehleringer et al. 1991; Ehleringer 1993; Geber & Dawson 1997). By contrast, wild barley provenances from arid regions showed higher Δ (Handley et al. 1994), because the high photosynthetic capacity and water use of a spender allow some seedset before the plant succumbs to salinity or drought under natural conditions. Third, Δ tends to decrease with leaf thickness or specific leaf mass (SLM, $g\,m^{-2}$), which is usually greater under more exposed conditions, or when there are water deficits or temperature limit expansion. Such leaves may have lower stomatal densities (which can affect Δ), while lower stomatal and mesophyll conductances would lead to a greater drawdown of $CO_2$ to $p_i$, $p_c$ and Rubisco ([3] in Table 21.1). Fourth, because light-harvesting capacity in many crop plants responds to nitrogen fertilization and matches Rubisco capacity, many studies find that chlorophyll content is inversely proportional to $\Delta$ ([4] in Table 21.1). Fifth, nitrogen content is usually directly related to Rubisco capacity, and if this results in increased drawdown to $p_c$ and hence a reduction in $p_i/p_a$, then the leaf $\Delta$ and nitrogen content are also inversely related, particularly if conductance remains constant or decreases when water is limiting ([5] in Table 21.1). Conversely, when water is plentiful, if $g_s$ increases to counter photosynthetic capacity, $p_i/p_a$ might increase and hence reverse the relationship between $\Delta$ and leaf nitrogen. Nitrogen source may also have a direct effect on water use, with higher $\Delta$ associated with ammonium assimilation (Yin & Raven 1998). Thus, finally, ash content reflects the rate of transpiration delivering minerals to the shoot, which is often also directly related to canopy temperature (Frank *et al.* 1997; Ebdon *et al.* 1998; [6] in Table 21.1).

At the molecular level recent developments in antisense technology have produced plants with reduced contents of a number of photosynthetic enzymes. Subsequent measurements performed using discrimination against $^{13}C$ and $^{18}O$ have been significant in verifying the physiological functions of Rubisco (Evans *et al.* 1994; Siebke *et al.* 1997) and carbonic anhydrase (Williams *et al.* 1996), among others.

## Mesophyll conductance

Measurement of discrimination against $^{13}CO_2$ during gas exchange ($\Delta_{obs}$) can be compared with $\Delta_i$ predicted from $p_i/p_a$. $\Delta_i - \Delta_{obs}$ represents the lower values of $p_c$ likely to be presented to Rubisco, and is therefore a function of the mesophyll conductance, $g_i$ (Evans *et al.* 1986; Evans & von Caemmerer 1996). The range of $\Delta_{obs}$ values expressed by *Quercus robur*, *Phaseolus vulgaris* and *Triticum aestivum* are shown as a function of the theoretical relationship between $\Delta_i$ and $p_i/p_a$ in Fig. 21.5(a). The gradient of the relationship between $\Delta_i - \Delta_{obs}$ *vs.* $A/p_a$ is inversely proportional to the mesophyll conductance, $g_i$. Generally, fast-growing herbaceous plants have the highest $g_i$, in contrast to woody perennials with more sclerophilous leaves, and this is reflected in the $g_i$ values shown for the three species in Fig. 21.5(b), i.e. 0.57, 0.47 and 0.29 mol m$^{-2}$ s$^{-1}$ bar$^{-1}$ for *T. aestivum*, *P. vulgaris* and *Q. robur*, respectively. Other stable isotope methods can also be used to estimate $g_i$: analysis of the $^{13}C$ composition of carbohydrate pools provides an elegant, daily integrator of mesophyll conductance, which can be used to evaluate responses under field conditions (Lauteri *et al.* 1997; Brugnoli *et al.* 1998; Scartazza *et al.* 1998).

As an illustration of an extreme range of $g_i$ values, we now compare the photosynthetic characteristics of *Nicotiana tabacum* and the succulent CAM plant *Kalanchoë daigremontiana* (Table 21.2). Data for the CAM plant were obtained when net $CO_2$ uptake was mediated entirely by Rubisco during the late afternoon, 'Phase IV' of CAM (see below). The contrasting leaf thickness (SLM: 3.4 *vs.* 0.04 kg m$^{-2}$) was associated with a lower net $CO_2$ assimilation rate ($A$), with low $g_s$ restricting $p_i$, the substomatal $CO_2$ partial pressure. However, the much lower mesophyll

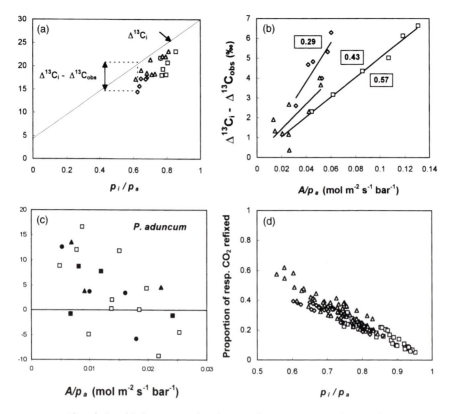

**Figure 21.5** The relationship between carbon isotope discrimination and gas-exchange characteristics. (a) Relationship between $\Delta$ and $p_i/p_a$, showing a theoretical relationship predicted by gas exchange ($\Delta_i = 4.4 + 22.4\, p_i/p_a$, continuous line) and that measured instantaneously, on-line ($\Delta_{obs}$). (b) Use of $\Delta_i - \Delta_{obs}$ offset to derive mesophyll conductance, derived from slope of relationship across a range of photosynthetic rates under 2% $O_2$, from $\Delta_i - \Delta_{obs} = [(b-a)g_i] \cdot (A/p_a)$, with values of Rubisco discrimination, b, set to 29, 30 and 30.5‰ for *Phaseolus vulgaris* (closed triangles), *Triticum aestivum* (open squares) and *Q. robur* (open diamonds), respectively. (c) Inverse relationship between $\Delta_i - \Delta_{obs}$ for *Piper aduncum* in the field in Trinidad, illustrating (photo)respiratory effects. (d) Relationship between rate of respiratory refixation ($\Delta$) and $p_i/p_a$, for data shown in (b) (with additional data for *T. aestivum* at 21% and 40% $O_2$; Gillon & Griffiths 1997).

conductance (0.06 *vs.* 0.37 mol m$^{-2}$ s$^{-1}$ bar$^{-1}$) associated with the low proportion of air spaces in the succulent leaves (8% *vs.* 39%), results in an additional draw-down of $CO_2$ internally to a $p_c$ of 108 µbar in *K. daigremontiana* (Table 21.2). However, there are two important implications for physiological ecologists: one is Rubisco allocation, in that the content on a leaf area basis is similar for both life forms (Table 21.1); second, increased rates of photorespiration must be associated with the restricted $CO_2$ supply, as Rubisco oxygenase activity would maintain electron transport rates when $p_i$ is so low (Maxwell *et al.* 1997, 1998), and indeed,

**Table 21.2** Morphological and photosynthetic correlates with carbon isotope discrimination in *Kalanchoë daigremontiana* and *Nicotiana tabacum*. Data derived from Evans *et al.* (1994) and Maxwell *et al.* (1997).

|  | K. daigremontiana | N. tabacum |
|---|---|---|
| SLM (kg m$^{-2}$) | 3.40 | 0.04 |
| Air space (%) | 8.8 | 39.0 |
| A (μmol $CO_2$ m$^{-2}$ s$^{-1}$) | 3.6 | 19.8 |
| $p_a$ (μbar) | 350 | 350 |
| $p_i$ (μbar) | 168 | 260 |
| $p_c$ (μbar) | 108 | 207 |
| $p_i/p_a$ | 0.48 | 0.74 |
| $g_i$ (mol m$^{-2}$ s$^{-1}$ bar$^{-1}$) | 0.06 | 0.37 |
| Rubisco content (g m$^{-2}$) | 1.2 | 1.4 |
| $C_3 \Delta_i$ (‰) | 16.2 | 22.0 |
| $C_3 \Delta_{obs}$ (‰) | 11.4 | 17.7 |

Note: Steady-state $CO_2$ assimilation (A) was measured at light saturation with values given for the ambient partial pressure of $CO_2$ ($p_a$) and calculated for the substomatal cavity ($p_i$) and at the Rubisco active sites ($p_c$). The mesophyll conductance of $CO_2$ from the stomatal cavity to Rubisco is $g_i$ and SLM is specific leaf mass.

measurements of $\Delta_i - \Delta_{obs}$ should now routinely be performed under 2% $O_2$ (Fig. 21.5b; Evans & von Caemmerer 1996), to reduce the effects of photorespiration.

## Is there any *f* in photorespiration?

Fractionation during respiration and photorespiration (i.e. (photo)respiration), which would alter the residual organic material $^{13}C$ signal, are taken into account in the more detailed equations derived by Farquhar and colleagues (Farquhar *et al.* 1989; Brugnoli & Farquhar 1999; see Griffiths 1998). Thus, two fractionation factors, *e* and *f*, represent the discrimination during respiration and photorespiration, respectively. It has recently been suggested that dark respiration in cultured cells leads to no fractionation (Lin & Ehleringer 1997), although a shift of some 6‰ has now been reported for intact tissues (Duranceau *et al.* 1999), leading to respiratory $CO_2$ being enriched in $^{13}C$ (i.e *e* = −6‰) because of site-specific decarboxylations during carbohydrate breakdown. For photorespiration, the fractionation associated with glycine decarboxylase for intact leaves was 7‰, with $CO_2$ depleted in $^{13}C$ (i.e. *f* = +7‰, Rooney 1988).

In practical terms, the extent of (photo)respiratory processes became apparent when we attempted to measure $\Delta_i - \Delta_{obs}$ under field conditions in Trinidad (Fig. 21.5c; Harwood *et al.* 1998). While investigating the gas-exchange and isotope discrimination characteristics of a tropical forest pioneer *Piper aduncum*, it became apparent that instantaneous carbon isotope discrimination was closely associated with temperature (Gillon *et al.* 1998; Harwood *et al.* 1998) and could not be used to derive $g_i$ as described above. Indeed, $\Delta_{obs}$ declined throughout the day from around

30 to 10‰ and while $p_i$, and hence $\Delta_i$ predicted from gas exchange, remained relatively constant. The resultant positive and negative values of $\Delta_i-\Delta_{obs}$ showed a strong *inverse* correlation with $A/p_a$ (Fig. 21.5c). This unexpected relationship is typical for plants displaying low rates of net $CO_2$ uptake set against the background of potentially high rates of (photo)respiration (Gillon & Griffiths 1997), particularly under stressed conditions. While it is unlikely that the isotopic memory of these events will remain in structural carbon, or even in residual carbohydrate pools, because of the low rates of carbon gain, it may not be possible to derive a value of $g_p$ and hence $p_c$, using isotopic methods under field conditions. Therefore, in order to determine chloroplast water enrichment, and derive the link with the $^{18}O$ signal in atmospheric $CO_2$, the fluorescence method for $g_i$ may be more appropriate under extreme conditions (Maxwell *et al.* 1997; Table 21.2).

In order to account for the effects of photorespiration the on-line discrimination equations may be modified (Gillon & Griffiths 1997; Gillon 1997). Central to this is the partitioning of net gas exchange into its gross components relative to fixation (i.e. $R_d/V_c$ for dark respiration, $\Gamma^*/p_c$ for photorespiration), to which were ascribed the relevant isotopic composition, including fractionation ($e+f$). Because gross assimilation, $V_c$, is a function of ($A+R_d$), then respiratory effects will be greatest when assimilation rates are low (i.e. $R_d/V_c$ is large). By contrast, photorespiration will make a relatively constant contribution (governed by $\Gamma^*/p_c$) as long as temperature is constant. However, photorespiratory effects can be significant under tropical conditions (Fig. 21.5c) and the initial interpretation by Gillon and Griffiths (1997) did not include photorespiratory $CO_2$ which was refixed internally. We now estimate the proportion of respiratory $CO_2$ refixed (as a function of the gross $CO_2$ flux, total leaf conductance ($g$) and photosynthesis via ($V_c/(V_c+g.p_a)$)). Thus, leaf conductance is implied to control refixation of (photo)respiratory $CO_2$ to a large extent (Fig. 21.5d). In the data from Fig. 21.5(b), reported using this formulation as a function of $p_i/p_a$, high rates of assimilation and conductance mean that $\rho$ is low, whereas much higher rates of (photo)respiratory refixation occurred at low conductances (Fig. 21.5d). For the physiological ecologist, when comparing plants with different life-forms and gas-exchange strategies, the relative effects of these processes will depend on $V_c$ and $g_s$ (Fig. 21.5d).

The fluxes from respiration and photorespiration will, however, contribute to the leaf isotope balance, depending on a number of factors. First, if there was no fractionation during these processes, any $CO_2$ produced and leaking from a leaf would simply cancel a proportion of the $^{13}C$ enrichment resulting from net $CO_2$ fixation. Second, if all respiratory $CO_2$ were refixed, effectively in a closed system, then no discrimination would be expressed (by analogy to the $C_4$ bundle sheath) — but why should that apply in a system where the $^{18}O$ composition of $CO_2$ indicates that back-diffusion is facile? Third, in such an open system, should fractionation during respiration or photorespiration be significant (e.g. $e$ and $f$ are large, as shown above), then (photo)respiratory effects would be manifested in both refixation and in any $CO_2$ leaking from the leaf. In the former case, the respiratory $CO_2$

would comprise a second, internal source with a signal distinct from that of ambient air.

Accordingly, the on-line isotope discrimination data shown in Fig. 21.5(a,b) for *Quercus*, *Phaseolus* and *Triticum*, has been corrected to account for refixation using a value of $f=8‰$, close to that of Rooney (1988). This derived value of $f$ yielded convergence of all data and we now suggest that $f=8‰$ for photorespiration in many species (Gillon 1997).

While the complications caused by (photo)respiratory processes may cause dramatic shifts in $\Delta_{obs}$ and need to be included in the interpretation of instantaneous effects, these are transient and do not affect long-term $\delta^{13}C$, which was so effectively modelled by Farquhar *et al.* (1982). As we saw in the field, measurements made later in the growing season probably do not reflect the more optimal conditions for leaf expansion (e.g. Fig. 21.5c), and measurements of instantaneous $\Delta$ may not be entirely representative, whether determined from gas exchange (Harwood *et al.* 1998) or carbohydrate signal (Scartazza *et al.* 1998). Of course, if respiratory and photorespiratory fluxes were similar, it may well be that the reported fractionation during dark respiration ($e=6‰$: Duranceau *et al.* 1999) would cancel out that occurring during photorespiration ($f=+8‰$; Rooney 1988; Gillon 1997). Additionally, it is important to remember that the generalizations implicit in the simplified Farquhar *et al.* (1989) model hold consistently across a range of conditions, particularly under modest water deficits. Hence genotypic differences in plant gas-exchange and water-use characteristics relate well to $\delta^{13}C$ in terms of growth or tree ring characteristics and refinement of palaeoclimatic reconstruction (Griffiths 1998; Brugnoli & Farquhar 1999). However, it is important to note that the isotope signal in organic material may not always reflect prevalent environmental conditions.

## Integrating across time scales: synchronicity between organic and instantaneous signals

As discussed above, many of the global climate models used to evaluate climate change make broad generalizations, e.g. including the assumption that $p_i \approx p_c$ within terrestrial vegetation. Perhaps this is much to the consternation of physiological ecologists, who evaluate either genotypic differences in relation to habitat preference, or phenotypic plasticity during acclimation to extreme environmental conditions. An example of the way that photosynthetic characteristics vary within a single genus in relation to leaf morphology and biochemistry is now provided. The gas-exchange and on-line discrimination characteristics for three hemi-epiphytic stranglers in the genus *Clusia*, which can be found in the seasonal tropical forests of Trinidad and show a range of CAM characteristics (Borland *et al.* 1998), is considered. *Clusia aripoensis* is restricted to moist, upper montane forest formations and shows a typical $C_3$ gas-exchange pattern with $C_3$ values of $\Delta_{obs}$ tracking $\Delta_i$ throughout the day (Fig. 21.6). While *C. aripoensis* may induce limited CAM activity under extreme stress, *C. minor* shows a remarkable capacity for CAM

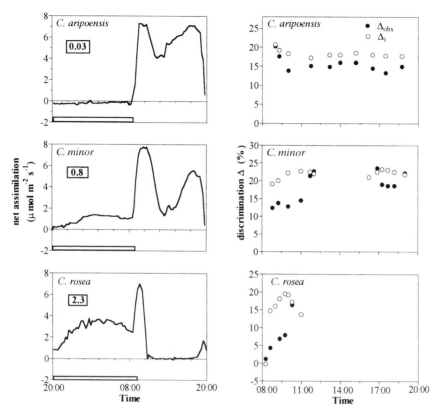

**Figure 21.6** Patterns of gas exchange and on-line discrimination for three species of *Clusia* with differing capacities for CAM: *C. aripoensis*; *C. minor* and *C. rosea*. The boxed numbers indicate the ratio of *in vitro* PEPc : Rubisco activities, with $\Delta_i$ predicted from gas exchange and $\Delta_{obs}$ measured under high assimilation rates. Data from Borland *et al.* (1998).

induction during the dry season. Even under relatively well-watered conditions, CAM gas-exchange characteristics may prevail, with notably an extended early-morning peak of $CO_2$ uptake ('Phase II' of CAM) dominated by PEPc activity (seen in low values of $\Delta_{obs}$) and continued acid accumulation (Fig. 21.6; Borland & Griffiths 1997; Roberts *et al.* 1998). The $C_4$ on-line signal ($\Delta_{obs}$) shifts to being dominated by Rubisco at the end of 'Phase II' (Fig. 21.6). While $\Delta_{obs}$ cannot be measured during the reduced gas-exchange rates at midday when $CO_2$ would also be generated internally from CAM activity, the continued activity of Rubisco can be inferred from the initial $\Delta_{obs}$ values during the late afternoon ('Phase IV'). The final species, *C. rosea*, is a constitutive CAM plant, displaying the traditional phases of CAM, with a strong PEPc $\Delta_{obs}$ signal at dawn, which then shifts to Rubisco (Fig. 21.6).

One means of showing the progressive commitment to CAM can be seen in the increasing PEPc : Rubisco ratios analysed biochemically, which supports the observed morphological shift in leaf succulence and concomitant decrease in mes-

**Table 21.3** Comparison of mesophyll conductance ($g_i$) for species showing a range of leaf succulence and CAM activities. The drawdown of $CO_2$ from stomata to the chloroplast is indicated by gas-exchange measurements of internal $CO_2$ concentration ($p_i$) and $CO_2$ concentration at the chloroplast ($p_c$) as calculated from $g_i$. Data derived from Gillon *et al.* (1998) and A.M. Borland (unpublished results).

| Species | Photosynthetic type | Leaf succulence (kg m$^{-2}$) | $g_i$ (mol m$^{-2}$ s$^{-1}$ bar$^{-1}$) | $p_i$ (μ bar) | $p_c$ (μ bar) |
|---|---|---|---|---|---|
| *Kalanchoë daigremontiana* | Obligate CAM | 3.40 | 0.061 | 224 | 159 |
| *Clusia fluminensis* | Obligate CAM | 1.15 | 0.082 | 217 | 181 |
| *Clusia rosea* | Obligate CAM[1] | 0.85 | 0.077[1] | 267 | 175 |
| *Clusia minor* | C$_3$-CAM intermediate | 0.77 | 0.100 | 262 | 199 |
| *Clusia aripoensis* | C$_3$ | 0.62 | 0.192 | 280 | 242 |

[1]Values determined from late in 'Phase II' may reflect some PEPc activity.

**Table 21.4** The carbon isotope ratios of biochemical fractions isolated from leaves and fruits of *Clusia minor* and *Clusia rosea*. Samples were collected during the dry season in Trinidad (A.M. Borland, unpublished data).

| Organic fraction | Carbon isotope ratio (‰) | |
|---|---|---|
| | *C. minor* | *C. rosea* |
| Leaf soluble sugars | −23.1 | −15.8 |
| Leaf organic acids | −17.0 | −13.9 |
| Leaf starch | −23.5 | −17.4 |
| Leaf structure | −25.3 | −20.0 |
| Fruit soluble sugars | −17.9 | −29.1 |
| Fruit organic acids | −19.3 | −29.3 |
| Fruit starch | −22.3 | −26.9 |
| Fruit structure | −25.6 | −25.8 |

ophyll conductance ($g_i$, Table 21.3). With additional data for two other constitutive CAM species, *K. daigremontiana* and *C. fluminensis* for comparison, Table 21.3 shows that the gradation in leaf succulence is directly associated with $g_i$, with the drawdown of $CO_2$ inside the leaf (to $p_i$ and $p_c$) greatest in the most succulent leaves during the periods of C$_3$ photosynthesis. The progressive reliance on CAM, as leaves become more succulent, shows the advantages of internally elevated $CO_2$ partial pressures, overcoming the problems of being thick (Maxwell *et al.* 1997, 1998; Osmond *et al.* 1999).

Having seen the wide-ranging interaction between leaf morphology and expression of CAM, we now consider whether the degree of CAM is reflected in leaf and fruit organic material (Table 21.4). Here there are two components which reflect photosynthetic processes over two stages of development: one, the leaf $\delta^{13}C$, records environmental constraints during expansion; the other, the fruit, reflects growth conditions later in the life cycle. The two species were sampled under field conditions in Trinidad, where *C. minor* normally induces CAM in the dry season when flowering and seedset occur. Leaf organic material and biochemical components are predominantly $C_3$-like in *C. minor* (Table 21.4), reflecting the environmental conditions during leaf expansion and, perhaps, the $C_3$ signal likely to be associated with the majority of carbon exported from $C_3$ activity in 'Phase IV' (Borland *et al.* 1994). Only the organic acids pool, likely at the time of sampling to be participating in the CAM cycle, bear a strong, $C_4$-like $\delta^{13}C$ signal at $-17‰$ (Table 21.4). This CAM signal was reflected in fruit carbohydrate and organic acids, but starch and structural material were more $C_3$-like in *C. minor*. In some respects the situation was reversed in *C. rosea*: leaf carbon pools showed a much stronger CAM, $C_4$-like signal, but fruit components were clearly $C_3$-like, ranging from $-25.8$ to $-29.3‰$ (Table 21.4). Individual fruits of *C. rosea* were five to 10 times the size of *C.minor*, and so the shift was not consistent with diffusion-limited $C_3$ autotrophy of the fruit, but could reflect a high degree of respiratory refixation (see above). While stable isotopes provide an insight into contrasting ecological strategies, it is apparent that more detailed carbohydrate budgets and compound-specific isotope composition are needed to provide additional detail. This is particularly relevant for analysing the extent of autotrophy and interaction with refixation of respiratory $CO_2$, which can make a significant contribution to the carbon balance of developing cereal grains (Bort *et al.* 1996; Araus *et al.* 1997).

As we have seen, the carbon isotope signal is subject to a number of external and internal constraints, while the $^{18}O$ composition of organic material, and particularly leaf cellulose, traditionally thought to provide a more faithful indicator of local humidity (see above). Having examined the gas-exchange processes within an oak canopy, data are now presented on the gradation in $\delta^{18}O$ of leaf cellulose and leaf water. Expectations that the evaporative enrichment would be greater in the upper canopy were confounded by the gradation in $\delta^{18}O$ of cellulose, which, over 3 years, consistently *decreased* with height (Fig. 21.7a). A more detailed study in 1998 has now shown that early in the growing season, during May, both leaf water enrichment and relative humidity (Fig. 21.7b,c) *are* correlated with $\delta^{18}O$ cellulose. Later in the growing season (June) the expected gradient of humidity and leaf water enrichment was established in the canopy (i.e. higher VPD and enrichment in the upper canopy): however, by now, leaf $\delta^{18}O$ has already been laid down, reflecting those environmental conditions earlier in the year. Thus, future attempts at palaeoclimatic reconstruction using leaf water and cellulose should bear in mind the likely timing of organic material deposition.

## And finally: are you now or were you ever . . .?

Growing applications of stable isotope analyses continue to augment the armoury of techniques available to the plant physiological ecologist. Refinements in our understanding of theoretical discrimination processes will allow broader applica-

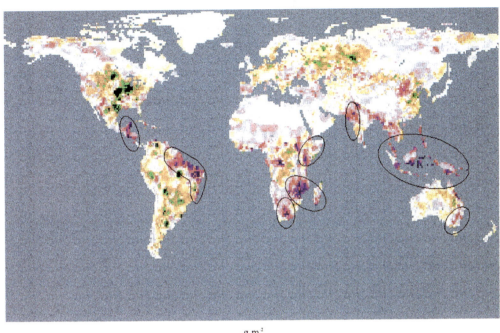

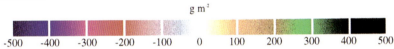

**Plate 22.1** Terrestrial net ecosystem productivity (NEP) for potential vegetation for 1987 (g m$^{-2}$). Positive values indicate $CO_2$ sinks. Areas surrounded by ellipses indicate areas of drought (from Glantz (1996)).

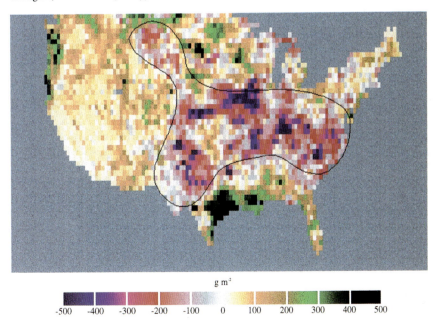

**Plate 22.2** Terrestrial NEP for the potential vegetation of the conterminous USA in 1983.

[Facing p. 434]

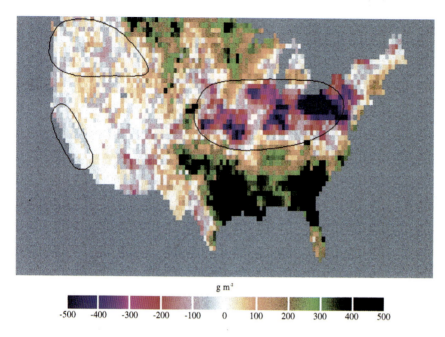

g m²

-500  -400  -300  -200  -100  0  100  200  300  400  500

**Plate 22.3**  Terrestrial NEP for the potential vegetation of the conterminus USA in 1991.

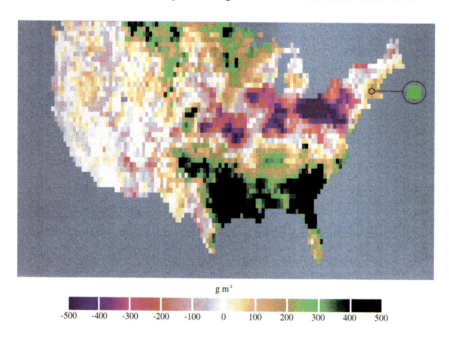

g m²

-500  -400  -300  -200  -100  0  100  200  300  400  500

**Plate 22.4**  Simulation and observation (Goulden *et al.* 1996) of NEP for the Harvard Forest in northeast USA, 1991.

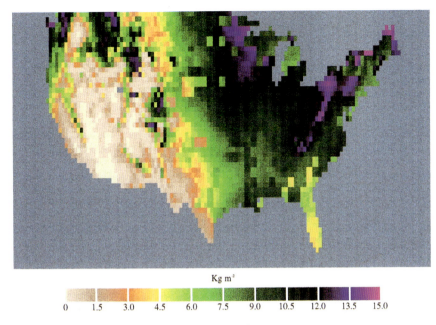

Kg m²

| 0 | 1.5 | 3.0 | 4.5 | 6.0 | 7.5 | 9.0 | 10.5 | 12.0 | 13.5 | 15.0 |

**Plate 22.5** Simulated soil carbon content (kg m⁻²) for 1993.

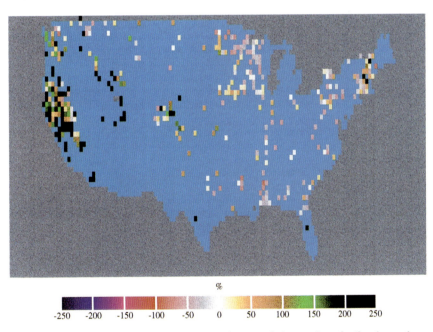

%

| -250 | -200 | -150 | -100 | -50 | 0 | 50 | 100 | 150 | 200 | 250 |

**Plate 22.6** Error term for a comparison of simulation and observation of soil carbon, where the error term is calculated as (observed-simulated)/simulated, as a percent. The light blue colour indicates the land mass of the USA but for which there are no soil measurements.

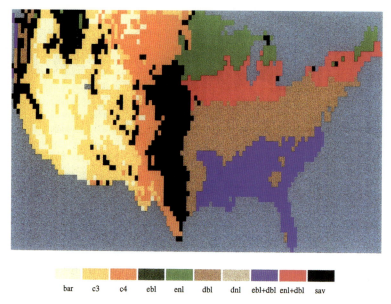

**Plate 22.7** Predicted dominant functional types of vegetation. Bar = bare ground; c3 = herbs and shrubs with $C_3$ photosynthesis; c4 = herbs and shrubs with $C_4$ photosynthesis; dbl = deciduous broad-leaf forest; dnl = deciduous needle-leaf forest; ebl = evergreen broad-leaf forest; ebl + dbl = mixed forest of evergreen broad-leaf and deciduous broad-leaf trees; enl = evergreen needle-leaf forest; enl + dbl = mixed forest of evergreen needle-leaf forest and deciduous broad-leaf trees; sav = savanna.

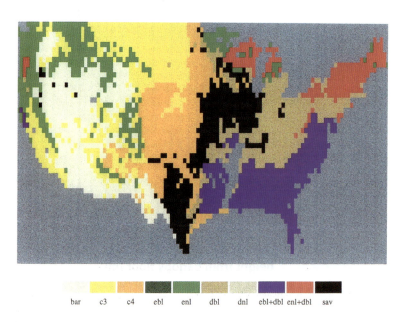

**Plate 22.8** Potential vegetation map of USA, as derived from the VEMAP project (symbols as for Plate 22.7). The area of light blue is an area classed as inland wetland and is only defined for the VEMAP classification.

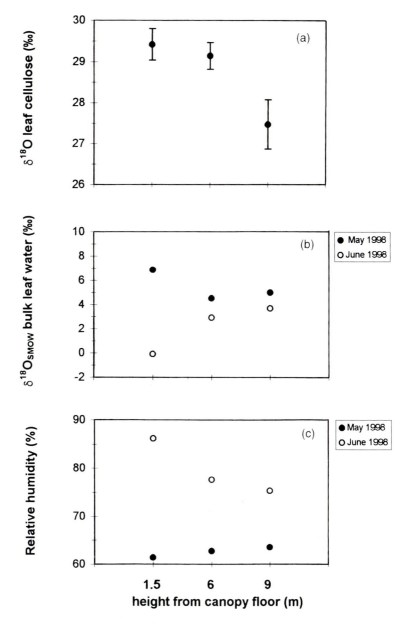

**Figure 21.7** Relationship between $\delta^{18}O$ of leaf cellulose and evaporative enrichment of leaf water within a *Q. petraea* canopy: (a) variation in mean leaf cellulose $\delta^{18}O$ with height for years 1994, 1995 and 1997; (b) change in $\delta^{18}O$ bulk leaf water with height in May and June 1998; (c) contrasting gradients of humidity in May and June 1998. Unpublished data of K.G. Harwood.

tions under more extreme conditions, but we should not forget how consistently well the simplest theoretical predictions of gas exchange and isotope discrimination relate particularly to plants subjected to slight water deficits. Improvements in mass spectrometer technology now allow analysis of small gaseous samples and more detailed compound-specific analyses. Such measurements are required to provide the additional details from cellular to community level, so as to support the generalizations needed to be made in scaling responses for global climate models. Ultimately, it is hoped that this chapter will help to refine some approaches and accommodate discrepancies which are particularly pertinent for the analysis of relatively slow growing vegetation under field conditions, with low assimilation capacity and high rates of (photo)respiration. In conclusion, the most important message in a similar review some time ago cautioned that, in the best of journalistic traditions, one should be sure of one's sources (*viz.* inorganic signal in $CO_2$, $H_2O$ or N; Griffiths 1991). Perhaps we would advise that in addition, investigators should now question whether any organic signal is contemporaneous with environmental conditions now, or at some distant time in the development of the plant!

## Acknowledgements

We are extremely grateful for support from the Natural Environmental Research Council (NERC) and the Royal Society, and we particularly wish to thank I. Blackburn for her highly efficient editing and interpretation of hieroglyphs, figures and tables.

## References

Akhani, H., Trimborn, P. & Ziegler, H. (1997). Photosynthetic pathways in Chenopodiaceae from Africa, Asia and Europe with their ecological, phytogeographical and taxonomical importance. *Plant Systematics and Evolution*, **206**, 187–221.

Amundsen, R., Stern, L., Baisden, T. & Wang, Y. (1998). The isotopic composition of soil and soil-respired $CO_2$. *Geoderma*, **82**, 83–115.

Anderson, J.E., Williams, J., Kriedemann, P.E., Austin, M.P. & Farquhar, G.D. (1996). Correlations between carbon isotope discrimination and climate of native habitats for diverse eucalypt taxa growing in a common garden. *Australian Journal of Plant Physiology*, **23**, 311–320.

Araus, J.L., Amaro, T., Zuhair, Y. & Nachit, M.M. (1997). Effect of leaf structure and water status on carbon isotope discrimination in field-grown durum wheat. *Plant, Cell and Environment*, **20**, 1484–1494.

Badger, M.R., Pfanz, H., Budel, B., Heber, U. &

Lange, O. (1993). Evidence for the functioning of a photosynthetic $CO_2$ concentrating mechanism in lichens containing green algal and cyanobacterial photobionts. *Planta*, **191**, 59–72.

Bakwin, P.S., Tans, P.P., White, J.W.C. & Andres, R.J. (1998). Determination of the isotopic ($^{13}C/^{12}C$) discrimination by terrestrial biology from a global network of observations. *Global Biogeochemical Cycles*, **12**, 555–562.

Borland, A.M. & Griffiths, H. (1997). A comparative study on the regulation of $C_3$ and $C_4$ carboxylation processes in the constitutive crassulacean acid metabolism (CAM) plant *Kalanchoë daigremontiana* and the $C_3$-CAM intermediate *Clusia minor*. *Planta*, **201**, 368–378.

Borland, A.M., Griffiths, H., Broadmeadow, M.S.J., Fordham, M.C. & Maxwell, C. (1994). Carbon-isotope composition of biochemical fractions and the regulation of carbon balance in leaves of the $C_3$-CAM intermediate *Clusia minor* L.

growing in Trinidad. *Plant Physiology*, **106**, 493–501.

Borland, A.M., Tecsi, L.I., Leegood, R.C. & Walker, R.P. (1998). Inducibility of crassulacean acid metabolism (CAM) in *Clusia* species: physiological/biochemical characterisation and intercellular localisation of carboxylation and decarboxylation processes in three species which exhibit different degrees of CAM. *Planta*, **205**, 342–351.

Bort, J., Brown, R.H. & Araus, J.L. (1996). Refixation of respiratory $CO_2$ in the ears of $C_3$ cereals. *Journal of Experimental Botany*, **47**, 1567–1575.

Brugnoli, E. & Farquhar, G.D. (1999). Carbon isotope fractionation. In *Photosynthesis: Physiology and Metabolism* (Ed. by R.C. Leegood, T.D. Sharkey & S. von Caemmerer), Kluwer Academic Publishers, the Netherlands, in press.

Brugnoli, E., Scartazza, A., Lauteri, M., Monteverdi, M.C. & Máguas, C. (1998). Carbon isotope discrimination in structural and non-structural carbohydrates in relation to productivity and adaptation to unfavourable conditions. In *Stable Isotopes: Integration of Biological, Ecological and Geochemical Processes* (Ed. by H. Griffiths), pp. 133–147. BIOS Scientific, Oxford.

Buchmann, N. & Ehleringer, J.R. (1998). $CO_2$ concentration profiles, and carbon and oxygen isotopes in C-3 and C-4 crop canopies. *Agricultural and Forest Meteorology*, **89**, 45–58.

Buchmann, N., Guehl, J.M., Barigah, T.S. & Ehleringer, J.R. (1997a). Interseasonal comparison of $CO_2$ concentrations, isotopic composition, and carbon dynamics in an Amazonian rainforest (French Guiana). *Oecologia*, **110**, 120–131.

Buchmann, N., Kao, W.Y. & Ehleringer, J.R. (1997b). Influence of stand structure on carbon-13 of vegetation, soils, and canopy air within deciduous and evergreen forests in Utah (USA). *Oecologia*, **110**, 109–119.

Cerling, T.E., Harris, J.M., MacFadden, B.J., Leakey, M.G., Quade, J., Elsenmann, V. *et al.* (1997). Global vegetation change through the Miocene/Pliocene boundary. *Nature*, **389**, 153–158.

Ciais, P. & Meijer, H.A.J. (1998). The $^{18}O/^{16}O$ isotope ratio of atmospheric $CO_2$ and its role in global carbon cycle research. In *Stable Isotopes:*

*Integration of Biological, Ecological and Geochemical Processes* (Ed. by H. Griffiths), pp. 409–438. Bios Scientific, Oxford.

Cordell, S., Goldstein, G., MuellerDombois, D., Webb, D. & Vitousek, P.M. (1998). Physiological and morphological variation in Metrosideros polymorpha, a dominant Hawaiian tree species, along an altitudinal gradient: the role of phenotypic plasticity. *Oecologia*, **113**, 188–196.

Damesin, C., Rambal, S. & Joffre, R. (1998). Co-occurrence of trees with different leaf habit: a functional approach on Mediterranean oaks. *Acta Oecologica-International Journal of Ecology*, **19**, 195–204.

Dawson, T.E. (1993). Hydraulic lift and water use by plants: implications for performance, water balance and plant–plant interactions. *Oecologia*, **95**, 565–574.

Dawson, T.E. & Ehleringer, J.R. (1991). Streamside trees that do not use stream water. *Nature*, **350**, 335–337.

Dawson, T.E., Pausch, R.C. & Parker, H.M. (1998). The role of hydrogen and oxygen stable isotopes in understanding water movement along the soil–plant–atmosphere continuum. In *Stable Isotopes: Integration of Biological, Ecological and Geochemical Processes* (Ed. by H. Griffiths), pp. 169–183. Bios Scientific, Oxford.

Deléens, E., Cliquet, J.-B. & Prioul, J.L. (1994). Use of $^{13}C$ and $^{15}N$ plant label near natural abundance for monitoring carbon and nitrogen partitioning. *Australian Journal of Plant Physiology*, **21**, 133–146.

Drinkwater, L.E., Wagoner, P. & Sarrantonio, M. (1998). Legume-based cropping systems have reduced carbon and nitrogen losses. *Nature*, **396**, 262–265.

Duranceau, M., Ghashghaie, J., Badeck, F., Deléens, E. & Cornic, G. (1999). $\delta^{13}C$ of $CO_2$ respired in the dark in relation to $\delta^{13}C$ of leaf carbohydrates in *Phaseolus vulgaris* L. under progressive drought. *Plant, Cell and Environment*, **22**, 515–523.

Ebdon, J.S., Petrovic, A.M. & Dawson, T.E. (1998). Relationship between carbon isotope discrimination, water use efficiency, and evapotranspiration in Kentucky bluegrass. *Crop Science*, **38**, 157–162.

Ehleringer, J.R. (1993). Gas exchange implications of isotopic variation in arid-land plants. In *Water*

*Deficits Plant Responses from Cell to Community* (Ed. by J.A.C. Smith & H. Griffiths), pp. 265–284. Bios Scientific, Oxford.

Ehleringer, J.R. & Dawson, T.E. (1992). Water uptake by plants: perspective from stable isotopes. *Plant Cell and Environment*, **15**, 1073–1082.

Ehleringer, J.R., Philips, S.L., Schuster, W.F.S. & Sandquist, D.R. (1991). Differential utilization of summer rains by desert plants, implications for competition and climate change. *Oecologia*, **88**, 430–434.

Ehleringer, J.R., Hall, A.E. & Farquhar, G.D., Eds (1993). *Stable Isotopes and Plant Carbon–Water Relations.* Academic, London.

Ehleringer, J.R., Cerling, T.E. & Helliker, B.R. (1997). $C_4$ photosynthesis, atmospheric $CO_2$ and climate. *Oecologia*, **112**, 285–299.

Ehleringer, J.R., Evans, R.D. & Williams, D. (1998). Assessing sensitivity to change in desert ecosystems—a stable isotope approach. In *Stable Isotopes: Integration of Biological, Ecological and Geochemical Processes* (Ed. by H. Griffiths), pp. 223–237. Bios Scientific, Oxford.

Evans, J.R. & von Caemmerer, S. (1996). Carbon dioxide diffusion within leaves. *Plant Physiology*, **110**, 339–346.

Evans, J.R., Sharkey, T.D., Berry, J.A. & Farquhar, G.D. (1986). Isotope discrimination measured concurrently with gas exchange to investigate $CO_2$ diffusion of leaves in higher plants. *Australian Journal of Plant Physiology*, **9**, 121–137.

Evans, J.R., von Caemmerer, S., Setchell, B.A. & Hudson, G.S. (1994). The relationship between $CO_2$ transfer conductance and leaf anatomy in transgenic tobacco with a reduced Rubisco content. *Australian Journal of Plant Physiology*, **21**, 475–495.

Evans, R.D. & Belnap, J. (1998). Long-term consequences of disturbance on nitrogen dynamics of an arid ecosystem. *Functional Ecology*, **12**, 195–202.

Evans, R.D., Bloom, A.J., Sukrapanna, S.S. & Ehleringer, J.R. (1996). Nitrogen isotope composition of tomato (*Lycopersicon esculentum* Mill. cv. T-5) grown under ammonium of nitrate nutrition. *Plant Cell and Environment*, **19**, 1317–1323.

Farquhar, G.D., O'Leary, M.H. & Berry, J.A. (1982). On the relationship between carbon isotope discrimination and the intercellular carbon dioxide concentration in leaves. *Australian Journal of Plant Physiology*, **9**, 121–137.

Farquhar, G.D., Ehleringer, J.R. & Hubick, K.T. (1989). Carbon isotope discrimination and photosynthesis. *Annual Review of Plant Physiology and Molecular Biology*, **40**, 503–537.

Farquhar, G.D., Lloyd, J., Taylor, J.A., Flanagan, L.B., Syversten, J.P., Hubick, K.T. *et al.* (1993). Vegetation effects on the isotope composition of oxygen in atmospheric $CO_2$. *Nature*, **363**, 439–442.

Farquhar, G.D., Henry, B.K. & Styles, J.M. (1997). A rapid on-line technique for determination of oxygen isotope composition of nitrogen-containing organic matter and water. *Rapid Communications in Mass Spectrometry*, **11**, 1554–1560.

Farquhar, G.D., Barbour, M.M. & Henry, B.K. (1998). Interpretation of oxygen isotope composition of leaf material. In *Stable Isotopes: Integration of Biological, Ecological and Geochemical Processes* (Ed. by H. Griffiths), pp. 27–62. Bios Scientific, Oxford.

Flanagan, L.B. (1998). Oxygen isotope effects during $CO_2$ exchange: from leaf to ecosystem processes. In *Stable Isotopes: Integration of Biological, Ecological and Geochemical Processes* (Ed. by H. Griffiths), pp. 185–201. Bios Scientific, Oxford.

Flanagan, L.B., Phillips, S.L., Ehleringer, J.R., Lloyd, J. & Farquhar, G.D. (1994). Effect of changes in leaf water oxygen isotopic composition on discrimination against $C^{18}O^{16}O$ during photosynthetic gas exchange. *Australian Journal of Plant Physiology*, **21**, 221–234.

Flanagan, L.B., Brooks, R.J., Varney, G.T. & Ehleringer, J.R. (1997). Discrimination against $C^{18}O^{16}O$ during photosynthesis and the oxygen isotope ratio of respired $CO_2$ in boreal forest ecosystems. *Global Biogeochemical Cycles*, **11**, 83–98.

Fleck, I., Grau, D., Sanjose, M. & Vidal, D. (1996). Carbon isotope discrimination in Quercus ilex, resprouts after fire and tree-fell. *Oecologia*, **105**, 286–292.

Follett, R.F., Paul, E.A., Leavitt, S.W., Halvorson,

A.D., Lyon, D. & Peterson, G.A. (1997). Carbon isotope ratios of great plains soils and in wheat-fallow systems. *Soil Science Society of America Journal*, **61**, 1068–1077.

Frank, A.B., Ray, I.M., Berdahl, J.D. & Karn, J.F. (1997). Carbon isotope discrimination, ash, and canopy temperature in three wheatgrass species. *Crop Science*, **37**, 1573–1576.

Geber, M.A. & Dawson, T.E. (1997). Genetic variation in stomatal and biochemical limitations to photosynthesis in the annual plant, *Polygonum arenastrum*. *Oecologia*, **109**, 535–546.

Gillon, J. (1997). *Carbon isotope discrimination interactions between respiration, leaf conductance and photosynthetic capacity*. PhD Thesis, University of Newcastle upon Tyne, UK.

Gillon, J.S. & Griffiths, H. (1997). The influence of (photo)respiration on carbon isotope discrimination in plants. *Plant, Cell and Environment*, **20**, 1217–1230.

Gillon, J.S., Borland, A.M., Harwood, K.G., Roberts, A., Broadmeadow, M.S.J. & Griffiths, H. (1998). Carbon isotope discrimination in terrestrial plants: carboxylations and decarboxylations. In *Stable Isotopes: Integration of Biological, Ecological and Geochemical Processes* (Ed. by H. Griffiths), pp. 111–131. Bios Scientific, Oxford.

Griffiths, H. (1991). Application of stable isotope technology in physiological ecology. *Functional Ecology*, **5**, 254–269.

Griffiths, H. (1996). Evaluation and integration of environmental stress using stable isotopes. In *Environmental Stress and Photosynthesis* (Ed. by N.R. Baker), pp. 451–468. Kluwer Academic, Dordrecht.

Griffiths, H. (1998). *Stable Isotopes: Integration of Biological, Ecological and Geochemical Processes*. Bios Scientific, Oxford.

Griffiths, H. & Smith, J.A.C. (1983). Photosynthetic pathways in the Bromeliaceae of Trinidad: relations between life-forms, habitat preference and occurrence of CAM. *Oecologia*, **60**, 176–184.

Handley, L.L. & Scrimgeour, C.M. (1997). Terrestrial plant ecology and [15]N natural abundance: the present limits to interpretation for uncultivated systems with original data from a Scottish Old Field. *Advances in Ecological Research*, **27**, 133–212.

Handley, L.L., Nevo, E., Raven, J.A., Martinez-

Carrasco, R., Scrimgeour, C.M., Pakniyat, H. *et al.* (1994). Chromosome 4 controls potential water use efficiency ([13]C) in barley. *Journal of Experimental Botany*, **45**, 1661–1663.

Harwood, K.G., Gillon, J.S., Griffiths, H. & Broadmeadow, M.S.J. (1998). Diurnal variation of $\Delta^{13}CO_2$, $\Delta C^{18}O^{16}O$ and evaporative site enrichment of $\delta H_2^{18}O$ in *Piper aduncum* under field conditions in Trinidad. *Plant, Cell and Environment*, **21**, 269–283.

Harwood, K.G., Gillon, J.S., Roberts, A. & Griffiths, H. (1999). Stable isotopes ([13]C and [18]O) in ambient $CO_2$ and water vapour reflect the relationships between sources and sinks for gas exchange within a temperate *Quercus petraea* canopy. *Oecologia*, **119**, 109–119.

Högberg, P. (1997). [15]N natural abundance in soil–plant systems. *New Phytologist*, **137**, 179–203.

Hopkins, D.W., Wheatley, R.E. & Robinson, D. (1998). Stable isotope studies of soil nitrogen. In *Stable Isotopes: Integration of Biological, Ecological and Geochemical Processes* (Ed. by H. Griffiths), pp. 75–88. Bios Scientific, Oxford.

Isla, R., Aragues, R. & Royo, A. (1998). Validity of various physiological traits as screening criteria for salt tolerance in barley. *Field Crops Research*, **58**, 97–107.

Keeling, C.D., Whorf, T.P., Wahlen, M. & van der Pilcht, J. (1995). Interannual extremes in the rate of rise of atmospheric carbon dioxide since 1980. *Nature*, **375**, 666–670.

Kendall, C., Campbell, D.H., Burns, D.A., Shanley, J.B., Silva, S.R. & Chang, C.C.Y. (1995). Tracing sources of nitrate in snowmelt runoff using the oxygen and nitrogen isotopic compositions of nitrate. In *Biogeochemistry of Seasonally Snow-Covered Catchments* (Proceedings of a Boulder Symposium, July 1995), pp. 339–347. IAHS Publishers no. 228, Wallingford.

Kloeppel, B.D., Gower, S.T., Treichel, I.W. & Kharuk, S. (1998). Foliar carbon isotope discrimination in Larix species and sympatric evergreen conifers: a global comparison. *Oecologia*, **114**, 153–159.

Lange, O.L., Belnap, J. & Reichenberger, H. (1998). Photosynthesis of the cyanobacterial soil-crust lichen *Collema tenax* from arid lands in southern Utah, USA: role of water content on light and temperature responses of $CO_2$ exchange. *Functional Ecology*, **12**, 195–202.

Lauteri, M., Scartazza, A., Guido, M.C. & Brugnoli, E. (1997). Genetic variation in photosynthetic capacity, carbon isotope discrimination and mesophyll conductance in provenances of *Castenea sativa* adapted to different environments. *Functional Ecology*, **11**, 675–683.

Lin, G.H. & Ehleringer, J.R. (1997). Carbon isotopic fractionation does not occur during dark respiration in C-3 and C-4. *Plant Physiology*, **114**, 391–394.

Lloyd, J., Kruijt, B., Hollinger, D.Y., Grace, J., Francey, R.J., Wong, S.-C. *et al.* (1996). Vegetation effects on the isotopic composition of atmospheric $CO_2$ at local and regional scales: theoretical aspects and a comparison between rain forest in Amazonia and a Boreal Forest in Siberia. *Australian Journal of Plant Physiology*, **23**, 371–399.

Máguas, C., Griffiths, H., Ehleringer, J.R. & Serodio, J. (1993). Characterisation of photobiont associates in lichens using carbon isotope discrimination techniques. In *Stable Isotopes and Plant Carbon–Water Relations* (Ed. by J. Ehleringer, A.E. Hall & G.D. Farquhar), pp. 201–212. Academic, Oxford.

Máguas, C., Griffiths, H. & Broadmeadow, M.S.J. (1995). Gas exchange and carbon isotope discrimination in lichens: evidence for interactions between $CO_2$-concentrating mechanisms and diffusion limitation. *Planta*, **196**, 95–102.

Maxwell, K., von Caemmerer, S. & Evans, J.R. (1997). Is a low internal conductance to $CO_2$ diffusion a consequence of succulence in plants with Crassulacean Acid Metabolism? *Australian Journal of Plant Physiology*, **24**, 777–786.

Maxwell, K., Badger, M.R. & Osmond, C.B. (1998). A comparison of $CO_2$ and $O_2$ exchange patterns and the relationship with chlorophyll fluorescence during photosynthesis in $C_3$ and CAM plants. *Australian Journal of Plant Physiology*, **25**, 45–52.

Moreira, M.Z., Sternberg, L.S.L., Martinelli, L.A., Victoria, R.L., Barbosa, E.M., Bonates, L.C.M. *et al.* (1997). Contribution of transpiration to forest ambient vapour based on isotopic measurements. *Global Change Biology*, **3**, 439–450.

O'Leary, M.H. (1988). Carbon isotopes in photosynthesis. *BioScience*, **38**, 325–336.

Osmond, B., Maxwell, K., Popp, M. & Robinson, S. (1999). On being thick: fathoming apparently futile pathways of photosynthesis and carbohydrate metabolism in succulent CAM plants. In *Carbohydrate Metabolism in Plants* (Ed. by M. Burrel, J. Bryant & N. Kruger), pp. 183–200. Bios Scientific, Oxford.

Palmqvist, K. (1993). Photosynthetic $CO_2$ use efficiency in lichens and their isolated photobionts: the possible role of a $CO_2$-concentrating mechanism in cyanobacterial lichens. *Planta*, **191**, 48–56.

Price, G.D., McKenzie, J.E., Pilcher, J.R. & Hoper, S.T. (1997). Carbon-isotope variation in *Sphagnum* from hummock-hollow complexes: implications for Holocene climate reconstruction. *Holocene*, **7**, 229–233.

Rao, R.C.N., Udaykumar, M., Farquhar, G.D., Talwar, H.S. & Prasad, T.G. (1995). Variation in carbon-isotope discrimination and its relationship to specific leaf-area and ribulose-1, 5-bisphosphate carboxylase content in groundnut genotypes. *Australian Journal of Plant Physiology*, **22**, 545–551.

Raven, J.A., Griffiths, H., Smith, E.C. & Vaughn, K.C. (1998). New perspectives in the biophysics and physiology of bryophytes. In *Proceedings of the Centenary Symposium of the British Bryological Society* (Ed. by J.W. Bates, N.W. Ashton & J.G. Duckett), pp. 261–275. Many Publishing and the British Bryological Society, Leeds.

Rice, S.K. & Giles, L. (1996). The influence of water content and leaf anatomy on carbon isotope discrimination and photosynthesis in Sphagnum. *Plant Cell and Environment*, **19**, 118–124.

Roberts, A., Borland, A.M., Maxwell, K. & Griffiths, H. (1998). Ecophysiology of the $C_3$-CAM intermediate *Clusia minor* L. in Trinidad: seasonal and short-term photosynthetic characteristics of sun and shade leaves. *Journal of Experimental Botany*, **326**, 1563–1573.

Robinson, D., Handley, L.L. & Scrimgeour, C.M. (1998). A theory for [15]N/[14]N fractionation in nitrate-grown vascular plants. *Planta*, **205**, 397–406.

Rochette, P. & Flanagan, L.B. (1997). Quantifying rhizosphere respiration in a corn crop under field conditions. *Soil Science Society of America Journal*, **61**, 466–474.

Rooney, M.A. (1988). *Short-term carbon isotope fractionation by plants.* PhD Thesis, University of Wisconsin, USA.

Scartazza, A., Lauteri, M., Guido, M.C. & Brugnoli, E. (1998). Carbon isotope discrimination in leaf and stem sugars, water-use efficiency and mesophyll conductance during different developmental stages in rice subjected to drought. *Australian Journal of Plant Physiology*, **25**, 489–498.

Schulze, E.-D., Caldwell, M.M., Canadell, J., Mooney, H.A., Jackson, R.B., Parson, D. *et al.* (1998). Downward flux of water through roots (i.e. inverse hydraulic lift) in dry Kalahari sands. *Oecologia*, **115**, 460–462.

Siebke, K., von Caemmerer, S., Badger, M. & Furbank, R.T. (1997). Expressing an RbcS antisense gene in transgenic *Flaveria bidentis* leads to an increased quantum requirement for $CO_2$ fixed in photosystems I and II. *Plant Physiology*, **115**, 1163–1174.

Smith, E.C. & Griffiths, H. (1996). The occurrence of the chloroplast pyrenoid is correlated with the activity of a $CO_2$-concentrating mechanism and carbon isotope discrimination in lichens and bryophytes. *Planta*, **198**, 6–16.

Sorensen, P. & Jensen, E.S. (1991). Sequential diffusion of ammonium and nitrate from soil extracts to a polytetrafluoroethylene trap for [15]N determination. *Analytica Chimica Acta*, **252**, 201–203.

Sternberg, L.D.S., Moreira, M.Z., Martinelli, L.A.,

Victoria, R.L.S., Barbosa, E.M., Bonates, L.C.M. *et al.* (1997). Carbon dioxide recycling in two Amazonian tropical forests. *Agricultural and Forest Meteorology*, **88**, 259–268.

Syvertsen, J.P., Smith, M.L., Lloyd, J. & Farquhar, G.D. (1997). Net carbon dioxide assimilation, carbon isotope discrimination, growth and water-use efficiency of citrus trees in response to nitrogen status. *Journal of the American Society for Horticultural Science*, **122**, 226–232.

Tieszen, L.L., Reed, B.C., Bliss, N.B., Wylie, B.K. & DeJong, D.D. (1997). C-3 and C-4 production, and distribution in great plains grassland land cover classes. *Ecological Applications*, **7**, 59–78.

Williams, T.G., Flanagan, L.B. & Coleman, J.R. (1996). Photosynthetic gas exchange and discrimination against $^{13}CO_2$ and $C^{18}O^{16}O$ in tobacco plants modified by an antisense construct to have low chloroplastic carbonic anhydrase. *Plant Physiology*, **112**, 319–326.

Yakir, D. (1998) Oxygen-18 of leaf water: a crossroad for plant-associated isotopic signals. In *Stable Isotopes: Integration of Biological, Ecological and Geochemical Processes* (Ed. by H. Griffiths), pp. 147–168. Bios Scientific, Oxford.

Yin, Z.H. & Raven, J.A. (1998). Influences of different nitrogen sources on nitrogen- and water-use efficiency, and carbon isotope discrimination, in C-3 *Triticum aestivum* L. and C-4 *Zea mays* L. plants. *Planta*, **205**, 574–580.

# Chapter 22

# Issues when scaling from plants to globe

*F.I. Woodward*

## Introduction

Krebs (1972) defined ecology as the scientific study of the interactions that determine the distribution and abundance of organisms. The use of the word *interactions* implies interactions with the abiotic and biotic environments. Begon *et al.* (1996) expanded the definition around the term *organisms* to consider, explicitly, individual organisms, populations and communities, an expansion which is wholly in keeping with the title of their book. The accent on interactions given by Krebs is more in sympathy with the drive by Tansley (1935) to initiate the use and consideration of the ecosystem concept, with greater emphasis on process rather than species identity. Present-day ecosystem ecology (e.g. Aber & Melillo 1991) provides even greater emphasis on process than organism identity.

Global-scale ecology (e.g. Woodward 1987) has developed markedly over the last decade, in response to the need for ecologists to make predictions about the future effects of global change on terrestrial vegetation (Walker & Steffen 1996). One of the major limitations to progress in this area is the small and limited data set defining the responses of plant species to the wide range of conditions implied by the term global change. Given the paucity of data about the global suite of (plant) species, the immediate response has been to simplify and not consider species but groups of similarly responsive species or functional types (Reich *et al.* 1997; Smith *et al.* 1997), an approach with a long pedigree (Raunkiaer 1934). The definition of functional types tends to depend on the question being addressed and so different types will emerge in different programmes of research. However, the key term is *functional*, which indicates process. Chlorophyll and the $CO_2$-fixing enzyme ribulose-1,5-*bis*phosphate carboxylase/oxygenase (Rubisco) are universal in photosynthetic plants and account for global $CO_2$ fixation (Tolbert & Preiss 1994). As the process of $CO_2$ fixation is universal, then a quantitative description of this process (e.g. Farquhar *et al.* 1980) could be used at the global scale (e.g. Woodward *et al.* 1995; Foley *et al.* 1996). In addition, it should provide accurate predictions of the past, present and future capacity of the terrestrial biosphere to sequester anthropogenic releases of $CO_2$, an important aspect of present-day attempts to mitigate against $CO_2$ releases from human activities (Schimel 1995). It follows, at least as a first step, that it should also be possible to model the processes of $CO_2$

*Department of Animal and Plant Sciences, University of Sheffield, Sheffield, S10 2TN, UK. E-mail: f.i.woodward@sheffield.ac.uk* uptake and release by anonymous terrestrial vegetation and provide an accurate answer, without the need to consider species and functional types. Therefore, predictive and testable developments in modelling can be made without the inevitable and almost compulsory need to first define plant or vegetation functional type, with attendant defined and constrained process characteristics (e.g. Melillo *et al.* 1993). If this is ecology then organisms do not figure overtly. At the global scale this appears a useful simplification as we will never characterize the full range of responses by the Earth's species to varying environments.

Krebs' definition of ecology can now be recast as the scientific study of the interactions between the environment and biological processes. Biological processes cover many features, from genetic adaptation, as studied in well-recognized organisms, to ecophysiological explanations of species distributions. Although it is the organism that we generally see and recognize as a discrete package, in general it seems to be the various processes within the package that we investigate and these range from fluxes of $CO_2$ in physiological ecology, through resource capture and mortality in population biology, to genetic change in communities. Thus, this chapter describes aspects of scaling up, by modelling, from a chloroplast-level process, photosynthesis, to global-scale net terrestrial production (the global equivalent of net ecosystem productivity, net primary production less heterotrophic respiration). The method is compared with measurements, in order to test its validity. Finally, to show that process and package are not independent worlds, the approach is used to predict dominant functional types, as outcomes of the modelling of biological processes but with no initially defined constraints, such as seen in biogeochemical models (e.g. Melillo *et al.* 1993). This chapter does not aim to describe the model used for this scaling up beyond necessary detail, rather the aim is to view typical problems which arise when scaling up. Model detail can be found elsewhere (e.g. Woodward *et al.* 1995, 1998, 1999).

## Irradiance responses of leaf photosynthesis

The Farquhar *et al.* (1980) model of $C_3$ photosynthesis has stood the test of time in its capacity to make accurate predictions and provide understanding of photosynthetic responses to the environment (e.g. Harley *et al.* 1992; Wullschleger 1993; Woodward *et al.* 1995). The model predicts the well-known observation that $C_3$ photosynthesis responds asymptotically to irradiance (Fig. 22.1). This response is fundamental to predicting plant productivity but the non-linearity of the response imposes a need for frequent simulations of 1 hour's duration or less. Models aim to minimize the simulation frequency and one simulation per day would be desirable when making simulations at the global scale. However, simply estimating the mean irradiance of the day and then the photosynthetic rate at that irradiance significantly overestimates the actual mean daily photosynthetic rate (by 15% in Fig. 22.1).

The Farquhar *et al.* model of photosynthesis predicts the rate of photosynthesis as the minimum of the rate of carboxylation and the rate of photosynthetic elec-

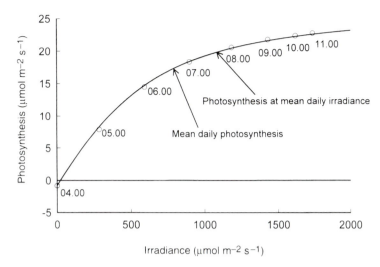

**Figure 22.1** Simulation of $C_3$ photosynthesis (Woodward *et al.* 1995) during the morning, showing predictions on the hour from 04.00 to 11.00. The simulations assume a perfect sine wave of irradiance for a 16-hour day, with no effects of changing temperature or water vapour pressure deficit on photosynthetic response. The arrow between 06.00 and 07.00 indicates the model estimate of the mean daily rate of photosynthesis. The arrow between 07.00 and 08.00 indicates the rate of photosynthesis calculated from the mean daily irradiance.

tron transport, $J_m$. These rates are also measures of the investment by the plant in photosynthetic material and this investment can change, such as through responses to changes in nutrient availability (Harley *et al.* 1992). No matter how this investment changes, measured, for example, as variations in $J_m$, the relationship between integrated daily irradiance and mean daily photosynthetic rate, just like the hourly response (Fig. 22.1), is also asymptotic (Fig. 22.2) at all levels of the maximum rate of photosynthetic electron transport ($J_m$). Therefore, scaling to longer time intervals does not remove non-linearity.

However, these simulations raise a more fundamental issue which is that, with decreasing irradiance, leaves with high rates of $J_m$, and high costs of synthesis and maintenance, will be making suboptimal use of the photosynthetic system; a high construction cost for a low photosynthetic benefit. If the leaf maximizes the photosynthetic return for a given investment in photosynthetic machinery, then both the daily photosynthetic response and the actual value of $J_m$ should decline linearly with irradiance (Fig. 22.2), assuming that the construction costs are a constant fraction of $J_m$. Such an optimization response will maximize the availability of nutrients such as nitrogen, which often limit plant growth, to other parts of the plant.

Observations of the efficiency of the photosynthetic machinery at different growth irradiances (Fig. 22.3) clearly indicate that, in terms of irradiance, the leaf

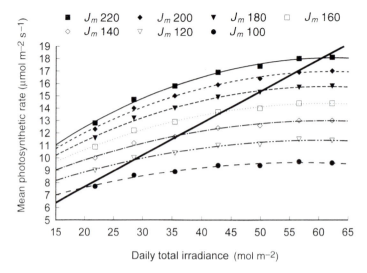

**Figure 22.2** Simulation of the mean daily photosynthetic rate for different daily irradiances, derived by changing daylength. No effects of changing temperature or water vapour pressure deficit are included. $J_m$ is the maximum rate of photosynthetic electron transport ($\mu mol\,m^{-2}\,s^{-1}$) and is closely correlated with the maximum rate of carboxylation. A line of equal efficiency (photosynthetic rate divided by $J_m$) is also shown.

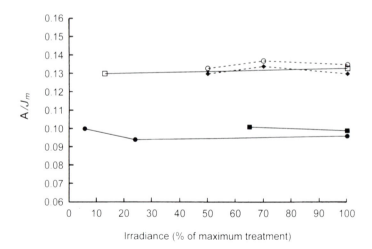

**Figure 22.3** The influence of growth irradiance on the ratio of mean daily photosynthetic rate (A) to the maximum rate of electron transport ($J_m$). (——) = species grown under differing degrees of shade (Evans *et al.* 1988); (-----) = seasonal observations on *Senecio jacobaea*; ● = mycorrhizal; ◆ = non-mycorrhizal.

operates close to an optimal response to light, with decreasing photosynthetic rate and $J_m$ in step with irradiance, as proposed in Fig. 22.2. Species differences in the ratio of photosynthetic rate to $J_m$ are seen, but it is not the aim of this paper to investigate these aspects. Other approaches to the environmental responses of photosynthesis, including responses to changes in nitrogen availability, temperature and irradiance, all support the proposal that leaves tend to maximize their photosynthetic returns in terms of necessary photosynthetic machinery (Terashima & Hikosaka 1995; Haxeltine & Prentice 1996).

An important outcome of the consideration of photosynthetic efficiency is that, although the daily asymptotic nature of photosynthesis (Fig. 22.1) still holds, this is not the case in the longer term of days and weeks, when the system adapts optimally (Fig. 22.3) to new environmental conditions (Reich et al. 1991). This response to irradiance over the longer time scale is an important physiological consideration, in addition to the relaxation of the frequency of photosynthetic model simulations from hourly, or less to daily, or even weekly. This response is integrated into the vegetation model by interactions between the capacity to take up nitrogen by the plant (Woodward et al. 1995) and the capacity of the soil to provide nitrogen through the coupled Century model (Parton et al. 1993). If irradiance decreases over daily to weekly intervals, the photosynthetic rate and requirement for nitrogen decreases, as do the rates of carboxylation and electron transport ($J_m$). As a consequence, the relationship between photosynthetic rate and $J_m$ will follow an optimum line, such as shown in Fig. 22.2.

## Defining photosynthetic capacity in a global context

Given environmental conditions of irradiance, temperature and water vapour pressure deficit, it is possible to predict leaf and canopy net primary productivity, based upon the Farquhar et al. model, with reasonable accuracy (Woodward et al. 1995). Only the influences of irradiance have been considered here, although similar approaches exist that consider maximization for addressing responses to temperature (Haxeltine & Prentice 1996).

When scaling from the plant to the globe it is necessary to predict the coefficients of the Farquhar et al. model, particularly maximum rates of carboxylation by Rubisco and electron transport, and this entails new scaling considerations. As Rubisco requires nitrogen for its construction, and this may often be a limiting nutrient, then there should be a relationship, or set of relationships, between leaf nitrogen and photosynthesis (e.g. Field & Mooney 1986; Terashima & Hikosaka 1995). The previous section indicated that the photosynthetic responses to irradiance are mediated by variations in the uptake of nitrogen and so it follows that the coefficients of the Farquhar et al. model should be derived from the kinetics of nitrogen uptake.

Predicting rates of nitrogen mineralization and the rate of organic nitrogen supply are therefore critical in defining photosynthetic rate, an approach which has previously been described (Woodward & Smith 1994; Woodward et al. 1995) and based around the importance of mycorrhizal associations for nitrogen uptake (Schimel & Chapin 1996). This importance is seen as a close correspondence between leaf nitrogen concentration, maximum photosynthetic rate and mycorrhizal association (Fig. 22.4). The lowest photosynthetic rates and leaf nitrogen concentrations are associated with ericoid mycorrhizas and the highest with plants associated with arbuscular mycorrhizas (Woodward & Smith 1994).

The lowest rates of photosynthesis in the ericoid plants are correlated with the presence of ericoid mycorrhizas, but the limiting process relates to low rates of nitrogen uptake from the heavily organic soils associated with ericoid communities (Read 1991; Woodward & Smith 1994; Chapter 7). There is a tendency for an increase in the proportion of nitrogen taken up in organic form in the series from arbuscular, through ecto- to ericoid mycorrhizal systems (Read 1991). This appears to exert a significant impact on leaf nitrogen concentration which decreases as the proportion of organic nitrogen in the xylem sap increases (Stewart et al. 1992). However Näsholm et al. (1998) clearly demonstrated that on a highly organic soil, plants with all three types of mycorrhizal association have the capacity to take up organic nitrogen from the soil, avoiding the well-characterized process of nitrogen mineralization (Vitousek & Howarth 1991). Therefore, it does not appear that it is the mycorrhizal association *per se* which determines leaf nitrogen concentration and consequent photosynthetic activity but the capacity of the soil environment to provide inorganic and organic nitrogen.

Therefore, at the global scale, photosynthetic capacity appears strongly limited by the rate and nature of the nitrogen supply, although phosphorus supply may also prove to be important (Schimel 1998). The relationship between photosynthetic capacity and nitrogen is a critical feature of scaling from the plant to the

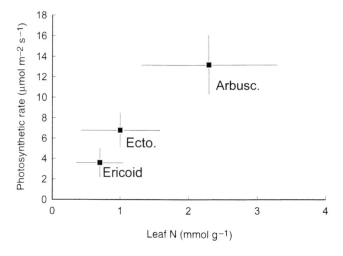

**Figure 22.4** Relationship between maximum photosynthetic rate, leaf nitrogen concentration and mycorrhizal association. Mean values are shown with 95% confidence limits. Data from Woodward and Smith (1994).

globe. It provides a relatively easy method of predicting the coefficients of the Farquhar *et al.* model globally (Woodward & Smith 1994), from the distribution of mycorrhizal types (e.g. Read 1991) and from models of soil carbon dynamics (e.g. Parton *et al.* 1993).

## Global-scale net ecosystem productivity

The purpose of the global-scale modelling presented here is not to address specifically future potential changes but rather to investigate and probe past interannual changes in the productivity of terrestrial vegetation. The emphasis on the past allows model testing as well as the potential for an improved understanding of the nature and scale of global-scale changes. The model simulations in this chapter will address the carbon cycle, as this is influenced strongly by vegetation processes and has a direct impact on past and future climates (e.g. Mitchell *et al.* 1995; Randerson *et al.* 1997; Cao & Woodward 1998; Woodward *et al.* 1998). The magnitude and sign of the carbon cycle fluxes depend on vegetation net primary productivity (NPP) and on heterotrophic respiration by the soil. Whereas the emphasis on many vegetation models is, quite rightly, on NPP (e.g. Woodward *et al.* 1995), the integrated emphasis for the carbon cycle must be on net ecosystem productivity (NEP), which is NPP minus heterotrophic respiration. Therefore NEP will be the emphasis of this discussion, because it lends itself to testing against field measurements of $CO_2$ fluxes by the eddy covariance technique (Baldocchi *et al.* 1996) and also against analyses of trends in atmospheric $CO_2$ (Keeling *et al.* 1995) and $O_2$ concentration (Keeling *et al.* 1996). NEP is viewed here as the pinnacle when scaling from photosynthesis at the level of the leaf, and on a time scale of hours, to NEP to the global scale and on an annual time scale.

The estimates of terrestrial NEP by Keeling *et al.* (1995), using time series of atmospheric $CO_2$ and $\delta^{13}C$, have been selected for the purposes of testing the model projections of global terrestrial NEP. These calculations compare well with estimates of terrestrial NEP from atmospheric $O_2$ concentrations (Keeling *et al.* 1996), although data from the newer and latter technique are only available from 1989. Fluctuations of the global terrestrial NEP are marked (Fig. 22.5), with the terrestrial biosphere a significant source of $CO_2$ in El Niño years (1980, 1983, 1987) but not in 1992, which was a weak El Niño event. The model simulations of global NEP were carried out with the Sheffield Dynamic Global Vegetation Model (SDGVM), a variant of a previously published model (Woodward *et al.* 1995), with updated detail including vegetation dynamics (Woodward *et al.* 1999). The global-scale database of climate and soil characteristics (only soil texture) used for the simulations was the ISLSCP (International Satellite Land Surface Climatology Project) data set (Meeson *et al.* 1995). The data are interpolated, by various modelling procedures (Meeson *et al.* 1995), to a common grid of 1° by 1° resolution. Therefore, the vegetation model simulations will always include errors due to the processes of spatial extrapolation between data sources (e.g. Woodward *et al.* 1995).

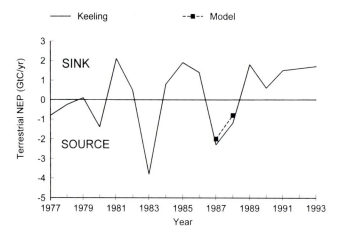

**Figure 22.5** Global terrestrial net ecosystem productivity (NEP) from Keeling *et al.* (1995) (———) and determined by modelling for 1987 and 1988 (■ - - - - - ■). Climate data for modelling were taken from Meeson *et al.* (1995). Positive values of NEP indicate $CO_2$ sinks.

Only 2 years of simulation are possible because the ISLSCP data source is limited to 1987 and 1988. 1987 was an El Niño year and it is clear that both the Keeling *et al.* deconvolutions of the $CO_2$ time series and the SDGVM model simulations are close in the agreement of the source capacity of the terrestrial biosphere in this year. The subsequent year of 1988 also shows close agreement as the El Niño event subsides.

### Spatial variation in net ecosystem productivity

The spatial variation of NEP (Plate 22.1, facing page 434) indicates that the impacts of El Niño years can be both positive and negative, in addition to other natural climatic variability unconnected with El Niño. However the El Niño phenomenon is a useful tool for investigating the effects of changes in climate on vegetation with a response time in the order of 1 year. Testing the model's spatial predictions has not proved simple or ideal, however, Glantz (1996) has published maps, using varying sources of information, indicating areas of drought impacts during El Niño years. These areas (Plate 22.1) coincide closely with model simulations of terrestrial $CO_2$ source activity, even though the SDGVM models the NEP of potential vegetation (no agriculture), while the drought data are usually obtained for crops. In addition, the areas in the central USA also coincide with significant flooding, presumably with the potential for subsequent stimulation of NEP after flood subsidence.

### Simulations and tests for the USA

The data for testing global simulations and for providing the necessary driving climatic variables are currently limiting further progress in NEP simulations at the global scale. However data are available for testing the SDGVM at the scale of the coterminous USA. El Niño events vary in timing, location and extremity of impact, effects which are clear at the global scale (Fig. 22.5). The variations have also led to poor predictions of crop yields; for example, Glantz (1996) notes that crop yields for the northern USA were predicted to be very high for the El Niño year of 1983, even after knowledge of the start of the El Niño in 1982. Reality was rather different, with maize yields down by about 50% from average. The 1983 El Niño year was the most extreme on record, causing severe droughts on all continents (Glantz 1996). Unfortunately, global climate data for 1982 and 1983 are not yet fully available, however, gridded data over this period are available for the USA, through the VEMAP project (e.g. VEMAP 1995).

Simulated NEP for 1983 (Plate 22.2) is negative in the eastern-central states, including the maize belts and in an area where marked effects of drought were recorded. As for the global simulations (Plate 22.1) it should be noted that the SDGVM simulations are for potential vegetation, while knowledge of the extent of drought in the USA is predominantly agricultural in origin.

The weak El Niño event of 1991 and 1992 exerted a rather different impact on global NEP (Fig. 22.5) than in 1983. In the USA the drought of 1991 was less extensive (Plate 22.3), with notable impacts in the west of the USA, a region of generally low productivity. An additional reason for simulating NEP for 1991 is that this was the first year in which the eddy covariance technique for measuring NEP was operated for a full calendar year over a deciduous forest (Wofsy et al. 1993). It is also interesting to note that the first estimation of the annual NEP was high at $370\,g\,m^{-2}$, but further analysis reduced this value to $280\,g\,m^{-2}$ (Goulden et al. 1996). It is instructive to note the discussions in Goulden et al. (1996) about the various sources of error in the estimation of NEP and to realize that data for testing models are not without significant errors. For the year 1994, with an annual NEP of $210\,g\,m^{-2}$, the 90% confidence interval is asymmetrical at $-30$ and $+80\,g\,m^{-2}$, with the larger error term resulting from the problems of estimating respiration during calm nocturnal periods. In 1991, the SDGVM simulates a value of $235\,g\,m^{-2}$ for NEP (Plate 22.4), which is close to the estimate of Goulden et al. (1996) of $280\,g\,m^{-2}$, and likely to be within the error limits of measurement. It should soon be possible to test the SDGVM simulations against a broader geographical spread of NEP measurements (Baldocchi et al. 1996), once the climate data for 1993 and 1994 become available.

### Further model testing

The tests of the model so far employed have been useful but, apart from the tests of NEP, qualitative rather than quantitative. A further test is a comparison between model predictions of soil carbon accumulation and actual measurements. This test measures the time integration capacity of the model, in particular the production and decay of vegetation litter. Measurements at the global scale are limited, however there is a large data set of observations for the USA (Zinke et al. 1984), which is also available on the World Wide Web (HTTP://cdiac.esd.ornl.gov/ftp/ndp018/ndp018.dat). The data include more than 3500 soil profiles, of which just over 2000 measurements have been made in the coterminous USA. In many cases these measurements are repeats for the same latitude and longitude, providing estimates of variation. Investigations of 25 areas with replicates, recorded at the same latitude and longitude, indicated large variation. Overall, the ratio of the largest to the smallest measurement of soil carbon, at the same latitude and longitude, was 5.5, with a 95% confidence limit of $\pm 2.3\,kg\,m^{-2}$. Therefore, the typical and maximum range of soil carbon for small pixels of 0.1° resolution is about 550%, or $\pm 275\%$, assuming a symmetrical distribution of variation. This suggests that the ideal of providing data for quantitative testing of the model cannot be realized, because the observed data are too variable.

A qualitative test has been arranged by simulating soil carbon accumulation (Plate 22.5) and then comparing these simulations with observations, in terms of percentage agreement (Plate 22.6). The agreement between observation and simulation is generally within $\pm 50\%$ east of the Rockies, but large errors in terms of a simulated underestimation by the model can be seen for the west coast. There are many likely causes of these large differences, but clearly samples will have been taken beneath different vegetation types and under agriculture, all of which exert significant impacts on soil carbon (Post *et al.* 1982). However, the systematic underestimation by simulation for the west coast suggests a need for further investigation.

## Functional types

This chapter started by suggesting that ecology should be defined as the study of the interactions between the environment and biological processes, with no overt consideration of organisms, or packages of processes. This final section takes the simulations of processes, along with established functional-type specific responses to the environment (Woodward 1987; Woodward *et al.* 1998, 1999), to define dominant functional types, or mixtures of functional types, for the coterminous USA. In brief, the dominant life form, in terms of either tree or non-tree (shrub, grass, herb), is determined by the competitive growth of these two life forms in a pixel, fractions of which are also disturbed (cleared to bare soil level) by fire or severe drought. The balance between $C_3$ and $C_4$ (non-tree) vegetation is based on the annual NPP. The physiognomic classes of functional types, i.e. evergreen and deciduous, and broad-leaved and needle-leaved vegetation, are based on survival responses to absolute minimum temperature. The map of simulated functional types (Plate 22.7) can then be compared with a map of the potential functional types of the USA (Plate 22.8), as defined for the VEMAP project (VEMAP 1995).

When the processes become packaged, similarities and differences with ground observations become clear. There are agreements between the simulated geographical distributions and observations for the eastern USA, for southern Texas, for the arid areas of the southwest USA and for the forests of the northern Rockies. There are areas of notable differences, in particular the simulation of savanna for the northern prairies and for the simulation of a broad-leaf deciduous forest for the Pacific Northwest, instead of an evergreen needle-leaf forest. This last and inaccurate simulation accounts for the large underestimate of soil carbon (Plate 22.6) and indicates that the identity of the functional-type package can have critical impacts on processes, such as soil carbon sequestration. It is also interesting to note that large underestimates of soil carbon in the central Rockies coincide with arid areas, with expected low accumulations of soil carbon (Woodward & Smith 1994) on both the simulated and VEMAP vegetation maps. Differences between observed and simulated soil carbon estimates would clearly occur, even with what is identified as the appropriate and dominant vegetation type, a problem caused at least in part by the low spatial resolution of the simulations, compared with the very high resolution of the observations on the ground.

## Conclusions

This chapter addresses some of the issues that emerge when scaling plant processes to the global scale. For short-term processes, such as NEP, it also appears that observations and simulations agree quite closely and that the identity of the vegetation is not greatly critical. Over the longer term, with the capacity for increased sensitivity through time integrals, it has proved difficult to test the model as field observations are particularly variable. However, even through this noise it is clear that functional types are important, but only in extreme cases, such as the huge-stature Pacific Northwest forests, with significantly greater capacities to sequester soil carbon than broad-leaf deciduous forests. Even here it is uncertain whether the incorrect functional-type simulation is actually the result of poor climate simulations, for a climate where the vegetation may accumulate significant water from fog (Waring & Franklin 1979), which is not predicted by climatic interpolations. Appropriate simulation of these processes may still allow accurate vegetation simulations, without recourse to considering functional-type packages.

It is important to note that the simulations discussed here include no disturbances through changes in human land use. These changes may exert very significant impacts on patterns of NEP, both global and regional. The emphasis on NEP also integrates the responses of both vegetation biomass and soil carbon to changes in climate. In dry years, such as 1983 (Plate 22.2) and 1991 (Plate 22.3), the marked source regions of carbon generally show positive gains in vegetation biomass, but these are more than offset by very significant losses of soil carbon. Such a response will also be modified by human land use, particularly if soil carbon reserves are diminished during forest harvesting (Houghton 1995).

## Acknowledgements

I am grateful to D. Beerling and C. Osborne for their comments on the manuscript.

## References

Aber, J.D. & Melillo, J.M. (1991). *Terrestrial Ecosystems*. Saunders College Publishing, Philadelphia.

Baldocchi, D., Valentini, R., Running, S., Oechel, W. & Dahlman, R. (1996). Strategies for measuring and modelling carbon dioxide and water vapour fluxes over terrestrial ecosystems. *Global Change Biology*, **2**, 159–168.

Begon, M., Harper, J.L. & Townsend, C.R. (1996). *Ecology: Individuals, Populations and Communities*, 3rd edn. Blackwell Science, Oxford.

Cao, M. & Woodward, F.I. (1998). Dynamic responses of terrestrial ecosystem carbon cycling to global climate change. *Nature*, **393**, 249–252.

Evans, J.R., Von Caemmerer, S. & Adams, W.W. III, Eds (1988). *Ecology of Photosynthesis in Sun and Shade*. CSIRO, Melbourne, Australia.

Farquhar, G.D., Von Caemmerer, S. & Berry, J.A. (1980). A biochemical model of photosynthetic $CO_2$ assimilation in leaves of $C_3$ species. *Planta*, **149**, 78–90.

Field, C. & Mooney, H.A. (1986). The photosynthesis–nitrogen relationship in wild plants. In *On the Economy of Form and Function* (Ed. by T.J. Givnish), pp. 25–55. Cambridge University Press, Cambridge.

Foley, J.A., Prentice, I.C., Ramankutty, N., Levis, S., Pollard, D., Sitch, S. *et al.* (1996). An integrated biosphere model of land surface processes, terrestrial carbon balance, and vegetation dynamics. *Global Biogeochemical Cycles*, **10**, 603–628.

Glantz, M.H. (1996). *Currents of Change: El Niño's Impact on Climate and Society*. Cambridge University Press, Cambridge.

Goulden, M.L., Munger, J.W., Fan, S.-M., Daube, B.C. & Wofsy, S.C. (1996). Measurements of carbon sequestration by long-term eddy covariance: methods and a critical evaluation of accuracy. *Global Change Biology*, **2**, 169–182.

Harley, P.C., Thomas, R.B., Reynolds, J.F. & Strain, B.R. (1992). Modelling photosynthesis of cotton grown in elevated $CO_2$. *Plant, Cell and Environment*, **15**, 271–282.

Haxeltine, A. & Prentice, I.C. (1996). A general model for the light-use efficiency of primary production. *Functional Ecology*, **10**, 551–561.

Houghton, R.A. (1995). Land-use and the carbon cycle. *Global Change Biology*, **1**, 275–287.

Keeling, C.D., Whorf, T.P., Wahlen, M. & van der Plicht, J. (1995). Interannual extremes in the rate of rise of atmospheric carbon dioxide since 1980. *Nature*, **375**, 666–670.

Keeling, R.F., Piper, S.C. & Heimann, M. (1996). Global and hemispheric $CO_2$ sinks deduced from changes in atmospheric $O_2$ concentration. *Nature*, **381**, 218–221.

Krebs, C.J. (1972). *Ecology*. Harper & Row, New York.

Meeson, B.W., Corprew, F.E., McManus, J.M.P., Myers, D.M., Closs, J.W., Sun, K.-J. *et al.* (1995). *ISLSCP Initiative I — Global Data Sets for Land-Atmosphere Models*. 1987–88, s 1–5, published on CD-ROM. NASA, Greenbelt, USA.

Melillo, J.M., McGuire, A.D., Kicklighter, D.W., Moore, B., Vorosmarty, C.J. & Schloss, A.L. (1993). Global climate change and terrestrial net primary production. *Nature*, **363**, 234–240.

Mitchell, J.F.B., Johns, T.C., Gregory, J.M. & Tett, S.F.B. (1995). Climate response to increasing levels of greenhouse gases and sulphate aerosols. *Nature*, **376**, 501–504.

Näsholm, T., Ekblad, A., Nordin, A., Giesler, R., Högberg, M. & Högberg, P. (1998). Boreal forest plants take up organic nitrogen. *Nature*, **392**, 914–916.

Parton, W.J., Scurlock, J.M.O., Ojima, D.S., Gilmanov, T.G., Scholes, R.J., Schimel. D.S. *et al.* (1993). Observations and modeling of biomass and soil organic matter dynamics for the grassland biome. *Global Biogeochemical Cycles*, **7**, 785–809.

Post, W.M., Emanuel, W.R., Zinke, P.J. & Stangenberger, A.G. (1982). Soil carbon pools and world life zones. *Nature*, **298**, 156–159.

Randerson, J.T., Thompson, M.V., Conway, T.J., Fung, I.Y. & Field, C.B. (1997). The contribution of terrestrial sources and sinks to trends in the seasonal cycle of atmospheric carbon dioxide. *Global Biogeochemical Cycles*, **11**, 535–560.

Raunkiaer, C. (1934). *The Life-Forms of Plants and Statistical Plant Geography*. Oxford University Press, Oxford.

Read, D.J. (1991). Mycorrhizas in ecosystems. *Experientia*, **47**, 376–391.

Reich, P.B., Walters, M.B. & Ellsworth, D.S. (1991). Leaf age and season influence the relationships between leaf nitrogen, leaf mass per area and photosynthesis in maple and oak trees. *Plant, Cell and Environment*, **14**, 251–259.

Reich, P.B., Walters, M.B. & Ellsworth, D.S. (1997). From tropics to tundra: global convergence in plant functioning. *Proceedings of the National Academy of Sciences USA*, **94**, 13730–13734.

Schimel, D.S. (1995). Terrestrial ecosystems and the carbon cycle. *Global Change Biology*, **1**, 77–91.

Schimel, D.S. (1998). The carbon equation. *Nature*, **393**, 208–209.

Schimel, J.P. & Chapin, F.S., III (1996). Tundra plant uptake of amino acid and $NH_4^+$ nitrogen *in situ*: plants compete well for amino acid N. *Ecology*, **77**, 2142–2147.

Smith, T.M., Shugart, H.H. & Woodward, F.I., Eds (1997). *Plant Functional Types. Their Relevance to Ecosystem Properties and Global Change*. IGBP Book Series. Cambridge University Press, Cambridge.

Stewart, G.R., Joly, C.A. & Smirnoff, N. (1992). Partitioning of inorganic nitrogen assimilation between roots and shoots of cerrado and forest trees of contrasting plant communities of south east Brasil. *Oecologia*, **91**, 511–517.

Tansley, A.G. (1935). The use and abuse of vegetational concepts and terms. *Ecology*, **16**, 284–307.

Terashima, I. & Hikosaka, K. (1995). Comparative ecophysiology of leaf and canopy photosynthesis. *Plant, Cell and Environment*, **18**, 1111–1128.

Tolbert, N.E. & Preiss, J., Eds (1994). *Regulation of Atmospheric $CO_2$ and $O_2$ by Photosynthetic Carbon Metabolism*. Oxford University Press, New York.

VEMAP Members. (1995). Vegetation/ecosystem modeling and analysis project: comparing bio-geography and biogeochemistry models in a continental-scale study of terrestrial ecosystem responses to climate change and $CO_2$ doubling. *Global Biogeochemical Cycles*, **9**, 407–437.

Vitousek, P.M. & Howarth, R.W. (1991). Nitrogen limitation on land and in sea. How can it occur? *Biogeochemistry*, **13**, 87–115.

Walker, B.H. & Steffen, W.L., Eds (1996). *Global Change and Terrestrial Ecosystems*. IGBP Book Series. Cambridge University Press, Cambridge.

Waring, R.H. & Franklin, J.F. (1979). Evergreen coniferous forests of the Pacific northwest. *Science*, **204**, 1380–1386.

Wofsy, S.C., Goulden, M.L., Munger, J.W., Fan, S.-M., Bakwin, P.S., Daube, B.C. *et al.* (1993). Net exchange of $CO_2$ in a mid-latitude forest. *Science*, **260**, 1314–1317.

Woodward, F.I. (1987). *Climate and Plant Distribution*. Cambridge University Press, Cambridge.

Woodward, F.I. & Smith, T.M. (1994). Global photosynthesis and stomatal conductance: modelling the controls by soil and climate. *Advances in Botanical Research*, **20**, 1–41.

Woodward, F.I., Smith, T.M. & Emanuel, W.R. (1995). A global land primary productivity and phytogeography model. *Global Biogeochemical Cycles*, **9**, 471–490.

Woodward, F.I., Lomas, M.R. & Betts, R.A. (1998). Vegetation-climate feedbacks in a greenhouse world. *Philosophical Transactions of the Royal Society*, **353**, 29–39.

Woodward, F.I., Lomas, M.R. & Lee, S.E. (1999). Predicting the future production and distribution of global terrestrial vegetation. In *Terrestrial Global Productivity* (Ed. by J. Roy, B. Saugier & H. Mooney), in press. Academic, London.

Wullschleger, S.D. (1993). Biochemical limitations to carbon assimilation in $C_3$ plants—a retrospective analysis of the A/$C_i$ curves from 109 species. *Journal of Experimental Botany*, **44**, 907–920.

Zinke, P.J., Strangenberger, A.G., Post, W.M., Emanuel, W.R. & Olson, J.S. (1984). *Worldwide Organic Soil Carbon and Nitrogen Data*. ORNL/TM-8857. Oak Ridge National Laboratory, Oak Ridge, TN.

# Index